BRADY

Forensic Fire Scene Reconstruction

BRADY

Forensic Fire Scene Reconstruction

Second Edition

David J. Icove
The University of Tennessee

John D. DeHaan
Fire-Ex Forensics, Inc.

Upper Saddle River, New Jersey 07458

Library of Congress Control Number: 2007939051

Publisher: Julie Levin Alexander
Publisher's Assistant: Regina Bruno
Executive Editor: Marlene McHugh Pratt
Senior Acquisitions Editor: Stephen Smith
Associate Editor: Monica Moosang
Editorial Assistant: Patricia Linard
Director of Marketing: Karen Allman
Executive Marketing Manager: Katrin Beacom
Marketing Specialist: Michael Sirinides
Marketing Assistant: Lauren Castellano
Managing Production Editor: Patrick Walsh
Production Liaison: Julie Li
Production Editor: Karen Fortgang, bookworks publishing services
Manufacturing Manager: Ilene Sanford
Manufacturing Buyer: Pat Brown
Senior Design Coordinator: Christopher Weigand
Cover Designer: Bruce Kenselaar
Cover Photo: Courtesy of ATF Fire Research Laboratory
Composition: Aptara, Inc.
Printing and Binding: Edwards Brothers
Cover Printer: Phoenix Color Corporation

Credits and acknowledgments borrowed from other sources and reproduced, with permission, in this textbook appear on appropriate pages within text.

Note: A portion of this book was prepared by David J. Icove, a former employee of the Tennessee Valley Authority (TVA). The views expressed in this book are the views of Dr. Icove and his coauthor. They are not necessarily the views of TVA or the United States Government. Reference in this book to any product, process, or service by trade name, trademark, manufacturer, or otherwise does not necessarily constitute its endorsement or recommendation by TVA or the United States Government.

Pearson Education LTD.
Pearson Education Singapore Pte. Ltd
Pearson Education, Canada, Ltd
Pearson Education–Japan

Pearson Education Australia PTY, Limited
Pearson Education North Asia Ltd
Pearson Educación de Mexico, S.A. de C.V.
Pearson Education Malaysia, Pte. Ltd

10 9 8 7 6 5 4 3 2 1

ISBN 13: 978-0-13-222857-2
ISBN 10:   0-13-222857-2

To Dougal Drysdale, for carrying it forward

Contents

Preface to the Second Edition

For centuries the investigation of fires was limited to inspecting the ruins, asking questions of witnesses, and applying common sense to conclude how it all began. Fire investigation is a particularly challenging endeavor because of fire's transient power and destructiveness. It is tempting to apply "common knowledge," because fire has been part of the human experience for thousands of years, and everyone "knows" about fires. But it turns out that some of that "knowledge" was not well based and sometimes led to erroneous conclusions. Unlike the situation in many other disciplines, there was usually no way to do controlled tests of the conclusions to determine whether they were reliable. There were too many variables and no way to duplicate all the necessary conditions. Over the years more systematic knowledge of science and engineering was applied as chemists and fire scientists became involved. Forensic scientists like Paul Kirk and fire scientists like Howard Emmons and Philip Thomas contributed greatly to improving the accuracy of investigations by demonstrating the usefulness of scientific analysis, but concerned professional fire investigators realized that even more science was needed.

The confluence of events that dramatically influenced fire investigation began in the 1980s with the publication of several keynote textbooks: the National Bureau of Standards' *Fire Investigation Handbook* in 1980, the first edition of *Combating Arson-for-Profit* in 1980, the first revision of Kirk's classic text as *Kirk's Fire Investigation* in 1983, Cooke and Ide's *Techniques of Fire Investigation* in 1984, and Drysdale's epic *Introduction to Fire Dynamics* in 1985.

A marked change in the investigation of fires occurred when Harold E. "Bud" Nelson, of the National Institute of Standards and Technology (NIST), assisted federal agencies in the investigation of the tragic fire at the Dupont Plaza Hotel in Puerto Rico in 1986. Bud offered to study various scenarios to help explain the growth of the fire using computer modeling in exchange for data and information from the scene. That tentative collaboration was very successful and led to the confirmation of not only the area of origin but also the manner of fire and smoke spread that led to the terrible loss of life.

From that time on, the interaction among fire engineers, fire scientists, and investigators grew rapidly. By 1992, the National Fire Protection Association (NFPA) had published its first collaboration among those disparate groups with *NFPA 921*, and the (then) Bureau of Alcohol, Tobacco and Firearms (ATF) had made fire engineering an essential part of its training for Certified Fire Investigators (CFIs) with the development of Quintiere's *Principles of Fire Behavior* in 1998. Each successive edition of *NFPA 921* and *Kirk's Fire Investigation* has strengthened the bonds, and it is hoped that this book will extend the capacity of investigators to apply engineering and criminalistics to the accurate investigation of fire.

Fire scene reconstruction had been part of good fire investigation for many years, but the concept was generally limited to the physical scene. This meant identifying fuel packages by their postfire remains and replacing them in their positions during the fire. This reconstruction could be confirmed by interviews with residents, tenants, visitors, or customers and by reviewing prefire photos or videos.

Forensic fire scene reconstruction goes well beyond establishing what furnishings were present and where they were placed. Reconstruction involves identification and documentation of all relevant features of the fire scene—materials, dimensions, location, and physical evidence—that help identify fuels and establish human activities and contacts. This information is then placed in context with principles of fire engineering and human behavior and is used to evaluate various scenarios of the origin, cause, and development of the fire and the interaction of people with it.

Identification of the cause of an incendiary fire is often much more straightforward than the elimination of possible accidental ignition sources, but this is a necessary element in all arson cases. Systematically employed in accidental fires, forensic reconstructions using the scientific method can eliminate all but one conflicting hypothesis. Such a systematic approach can be used to test for expected consequences of each possible hypothesis. For instance, if a transmission failure in a car being driven caused it to catch fire, one would expect there to be physical evidence of such a failure—burned transmission fluid, melted plastic, metal particles, and so forth. If examination reveals no such defects, another cause or mechanism must be sought out. It is the intent of this text to explore and explain the process of applying the full spectrum of forensic reconstruction to fire events. Today, fire scene reconstruction is based primarily on physical evidence of burn patterns on the remnants of structures and vehicles. What is not always taken into consideration when looking at the pattern of fire damage is the physical evidence of pre- and postfire human activities and the behaviors that controlled these activities.

This textbook is intended for use by fire and law enforcement investigators, attorneys, forensic scientists, and fire protection professionals. A thorough understanding of fire dynamics is valuable in applying these forensic engineering techniques. This book describes and illustrates a new systematic approach for reconstructing fire scenes. The approach applies the principles of fire protection engineering along with forensic science and behavioral science.

Using historical fire cases, the authors provide new lessons and insight into the ignition, growth, development, and outcome of those fires. All documentation in the case examples follows or exceeds the methodology set forth by the NFPA in *NFPA 921— Guide for Fire and Explosion Investigations*. The best features of this text include the use of real-world case examples to illustrate the concepts discussed herein. These examples shed new light on the forensic science, fire engineering, and human factor issues. Each example is illustrated using the guidelines from *NFPA 921*. In cases where fire engineering analysis or fire modeling is applicable, these techniques are explored.

The authors acknowledge the numerous essential references from the Society of Fire Protection Engineers (SFPE) that form the basis for much of the material included in this textbook. These essential references, ranging from core principles of fire science to human behavior, include the *SFPE Handbook of Fire Protection Engineering* and many of SFPE's engineering practice guides.

Chapters 1–9 also conclude with a set of short self-study problems. The references in Appendix A include the latest public-domain publications on major fire investigations, U.S. government major fire reports, fire modeling software, and other documents and sources that may be of interest to the reader.

ABOUT THIS BOOK

The first edition of this book started as a series of lectures for members of fire and police agencies and the insurance industry, fire protection engineering professionals, and academicians. These lectures were used over time in college classrooms, at law enforcement training academies, at in-service fire investigation seminars, and in forensic laboratories.

The scientific approach that the authors presented during these lectures is consistent with present-day expert witness guidelines in federal and state courts. Based on inquiries from both our colleagues and our students, we decided that this textbook would take the form of a stand-alone volume capable of furthering the expertise of any individual or group actively involved in the pursuit of fire and arson investigation. This target audience includes

- public safety officials charged with the responsibility of investigating fires;
- prosecutors of arson and fire-related crimes who seek to add to their capabilities of evaluating evidence and presenting technical details to a nontechnical judge and jury;
- judicial officials seeking to comprehend better the technical details of cases over which they preside;
- private-sector investigators, adjusters, and attorneys representing the insurance industry who have the responsibility of processing claims or otherwise have a vested interest in determining responsibility for the start or cause of a fire;
- citizens and civic community service organizations committed to conducting public awareness programs designed to reduce the threat of fire and its devastating effect on the economy; and
- scientists, engineers, academicians, and students engaged in the education process pertaining to forensic fire scene reconstruction.

In the few years since the first edition of *Forensic Fire Scene Reconstruction* was released, a great deal has happened in the fire investigation community. Extensive new research has been published on fire behavior, ignition mechanisms, and fire patterns from the United States, Canada, the United Kingdom, and elsewhere. Major fire cases (in some of which the authors have been directly involved) have attracted national attention to fire dangers and investigations. There have been many new court cases, challenges, and decisions that affect how analyses and investigations should be conducted. The authors have endeavored to monitor all these new events, as reflected in the extensive changes to the text and greatly expanded reference citations. The major published texts *NFPA 921* and *Kirk's Fire Investigation* have both been released in new, expanded versions, and references to both have been revised.

One of the biggest changes to this text is the greatly expanded peer review it has undergone. Starting with informal comments generated by the first edition (which was the first text of its kind to be completely peer-reviewed), the authors circulated drafts of this text to three sets of reviewers. These reviewers, both anonymous and those acknowledged here, contributed all manner of valuable suggestions, text, and photos. This text reflects many improvements as a result of their efforts. There are also many interesting new case studies and worked examples to reinforce critical points. The website lists and math refresher have also been updated to improve their usefulness to our valued readers.

SCOPE OF THE BOOK

Forensic Fire Scene Reconstruction is divided into the following chapters.

Chapter 1, **"Principles of Reconstruction,"** describes a systematic approach to reconstructing fire scenes in which investigators rely on the combined principles of fire protection engineering along with forensic and behavioral science. Using this approach, the investigator can more accurately document a structural fire's origin, intensity, growth, direction of travel, and duration as well as the behavior of the occupants.

Chapter 2, **"Basic Fire Dynamics,"** provides the investigator with a firm understanding of the phenomenon of fire, heat release rates of common materials, heat transfer, growth and development, fire plumes, and enclosure fires.

Chapter 3, **"Fire Pattern Analysis,"** describes the underpinnings of how fire patterns are used by investigators in assessing fire damage and determining a fire's origin. Fire patterns are often the only remaining visible evidence after a fire is extinguished. The ability to document and interpret fire pattern damage accurately is a skill of paramount importance to investigators when they are reconstructing fire scenes.

Chapter 4, **"Fire Scene Documentation,"** details a systematic approach needed to support forensic analysis and reports. The purpose of forensic fire scene documentation includes recording visual observations, emphasizing fire development characteristics, and authenticating and protecting physical evidence. The underlying theme is that thorough documentation produces sound investigations and courtroom presentations.

Chapter 5, **"Arson Crime Scene Analysis,"** reviews the techniques used in the analysis of arsonists' motives and intents. It presents nationally accepted motive-based classification guidelines along with case examples of the crimes of vandalism, excitement, revenge, crime concealment, and arson-for-profit. The geography of serial arson is also examined, along with techniques for profiling the targets selected by arsonists.

Chapter 6, **"Fire Modeling,"** discusses the use of various mathematical and computer-assisted techniques for modeling fires, explosions, and the movement of people. Numerous models are explored, along with their strengths and weaknesses. Several case examples are also presented.

Chapter 7, **"Fire Death and Injuries,"** provides an in-depth examination of the impact and tenability of fires on humans. The chapter examines what kills people in fires, particularly their exposure to by-products of combustion, toxic gases, and heat. It also examines the predictable fire burn pattern damage inflicted on human bodies and summarizes postmortem tests and forensic examinations desirable in comprehensive death investigations.

Chapter 8, **"Fire Testing,"** reviews the applicable standard fire testing methods that are important to forensic fire scene analysis and reconstruction. These can range from benchtop "lab" tests to full-scale fire reconstructions of varying complexity.

Chapter 9, **"Case Studies,"** profiles many of the concepts used in forensic fire scene investigations using sanitized real-world cases.

Chapter 10, **"Future Tools for the Fire Investigator,"** describes various concepts recently introduced in fire investigation and presents proactive strategies being used to increase investigative effectiveness and strengthen successful case documentation in the future.

Appendix A contains the references and a list of websites that are cited in the text. Appendix B is a glossary covering common terminology used in the fire investigation field. Appendix C is a mathematics refresher.

PEER REVIEWERS

Peer review is important for ensuring that a textbook is well balanced, useful, authoritative, and accurate. The following agencies, institutions, companies, and individuals provided invaluable support during the peer-review process in the first and second editions.

Dr. Vytenis (Vyto) Babrauskas, Fire Science and Technology Inc., Issaquah, Washington

Dr. John L. Bryan, Professor Emeritus, University of Maryland, Department of Fire Protection Engineering. College Park, Maryland

Guy E. "Sandy" Burnette, Jr., Attorney, Tallahassee, Florida

Community College of South Nevada, Nevada

Coosa Valley Technical College, Georgia

Robert F. Duval, National Fire Protection Association, Quincy, Massachusetts

Fishers Fire Department, New York

Greg Gorbett, John Kennedy & Associates, Sarasota, Florida

Patrick M. Kennedy, National Association of Fire Investigators, Sarasota, Florida

Daniel Madrzykowski, National Institute of Standards and Technology, Gaithersburg, Maryland

John E. "Jack" Malooly, Bureau of Alcohol, Tobacco, Firearms, and Explosives, Chicago, Illinois (retired)

Michael Marquardt, Bureau of Alcohol, Tobacco, Firearms, and Explosives, Grand Rapids, Michigan

J. Ron McCardle, Bureau of Fire and Arson Investigations, Florida Division of State Fire Marshal

Lamont "Monty" McGill, McGill Investigations, Gardnerville, Nevada

Robert R. Rielage, former State Fire Marshal, Ohio Division of State Fire Marshal, Reynoldsburg, Ohio

Michael Schlatman, Fire Consulting International Inc., Shawnee Mission, Kansas

Robert K. Toth, Iris Fire, LLC, Denver, Colorado

Luis Velazco, Brunswick, Georgia

ACKNOWLEDGMENTS

We have many people to thank for both their help on and their inspiration for both editions of this book, including the many present and past employees at the following institutions and agencies.

Jim Allen

Roger Berrett

Bureau of Alcohol, Tobacco, Firearms, and Explosives (ATF): Steve Carman, Steve Avato, Dennis C. Kennamer, Dr. David Sheppard, Jack Malooly (retired), Wayne Miller (retired), Michael Marquardt, Luis Velaszco (retired), John Mirocha (retired), Ken Steckler, Brian Grove, and John Wheeler

J. H. Burgoyne & Partners (U.K.): Robin Holleyhead, and Roy Cooke

California State Fire Marshal's Office: Joe Konefal (retired)

Eastern Kentucky University: Ronald L. Hopkins (retired)

Federal Bureau of Investigation (FBI): Richard L. Ault (retired), Steve Band, S. Annette Bartlett, John Henry Campbell (retired), R. Joe Clark (retired), Roger L. Depue (retired), Joseph A. Harpold (retired), Timothy G. Huff (retired), Sharon A. Kelly (retired), John L. Larsen (retired), James A. O'Connor (retired), John E. Otto (retired), William L. Tafoya (retired), and Arthur E. Westveer

Fire Safety Institute: Dr. John M. "Jack" Watts, Jr.

Fire Science and Technology: Dr. Vyto Babrauskas

Forensic Fire Analysis: Gerald A. Haynes, Lester Rich

Gardiner Associates: Mick Gardiner, Jim Munday, Jack Deans

Rodger H. Ide

Iris Fire, LLC, Denver, Colorado: Robert K. Toth

Kent Archaeological Field School, UK: Dr. Paul Wilkinson

Knox County (Tennessee) Sheriff's Office: Det. Michael W. Dalton

Leica Geosystems: Rick Bukowski, Tony Grissim

McGill Consulting: Monty McGill

McKinney (Texas) Fire Department: Chief Mark Wallace

National Institute of Standards and Technology: Richard W. Bukowski, Dr. William Grosshandler, Dan Madrzykowski, and Dr. Kevin McGrattan

New South Wales Fire Brigades: Ross Brogan

Novato Fire Protection District: Assistant Chief Forrest Craig

Ohio State Fire Marshal's Office, Reynoldsburg: Eugene Jewell (deceased), Charles G. McGrath (retired), Mohamed M. Gohar (retired), J. David Schroeder (retired), Jack Pyle (deceased), Harry Barber, Lee Bethune, Joseph Boban, Kenneth Crawford, Dennis Cummings, Dennis Cupp, Robert Davis, Robert Dunn, Donald Eifler, Ralph Ford, James Harting (retired), Robert Lawless, Keith Loreno, Mike McCarroll, Matthew J. Hartnett, Brian Peterman, Mike Simmons, Rick Smith, Stephen W. Southard, and David Whitaker (retired)

Panoscan: Ted Chavalas

Richland (Washington) Fire Department: Glenn Johnson and Grant Baynes

Sacramento County (California) Fire Department: Jeff Campbell (retired)

Saint Paul (Minnesota) Fire Department: Jamie Novak

Santa Ana (California) Fire Department: Jim Albers (retired) and Bob Eggleston (retired)

Seneca College School of Fire Protection Engineering Technology: David McGill

Tennessee State Fire Marshal's Office: Richard L. Garner (retired), Robert Pollard, Eugene Hartsook (deceased), and Jesse L. Hodge (retired)

Tennessee Valley Authority (TVA): Carolyn M. Blocher, James E. Carver, R. Douglas Norman (retired), Larry W. Ridinger (retired), Sidney G. Whitehurst (retired), and Norman Zigrossi (retired)

University of Arkansas, Department of Anthropology: Dr. Elayne J. Pope

University of Edinburgh, Department of Civil Engineering: Professor Emeritus Dr. Dougal Drysdale

University of Maryland, Department of Fire Protection Engineering: Dr. John L. Bryan (Emeritus), Dr. James A. Milke, Dr. Frederick W. Mowrer, Dr. James G. Quintiere, Dr. Marino di Marzo, and Dr. Steven M. Spivak (Emeritus)

University of Tennessee, College of Engineering: Dr. A. J. Baker, J. Douglas Birdwell, Samir M. El-Ghazaly, Dr. Rafael C. Gonzalez, Dr. M. Osama Soliman, Dr. Jerry Stoneking (deceased), Dr. Tse-Wei Wang, and the many students in the ME 495 and ECE 599 courses, Enclosure Fire Dynamics and Computer Fire Modeling

University of Tennessee, Medical Group, Memphis: Dr. O'Brien C. Smith

U.S. Consumer Products Safety Commission: Gerard Naylis (retired) and Carol Cave

U.S. Fire Administration, Federal Emergency Management Agency: Edward J. Kaplan, Kenneth J. Kuntz, and Denis Onieal.

Our sincere, heartfelt thanks go to Carolyn M. Blocher, Dr. Angi M. Christensen, Vivian Woodall, Edith De Lay, Wendy Druck, and Larry Harding, who reviewed the early manuscripts, addressed technical questions, and made many beneficial suggestions as to the format and content of this textbook.

A very special mention goes to our copy editor Barbara Liguori; our development editor, Andrea Edwards; and our ever-patient associate editor at Pearson/Brady, Monica Moosang.

Also, special thanks go to our families and close friends over the years for their patient support.

SPECIAL ACKNOWLEDGMENT

In addition to our ongoing admiration of Bud Nelson, to whom the first edition was dedicated, the authors specifically acknowledge in this second edition the contributions and encouragement of Professor Emeritus Dougal Drysdale, one of the most influential fire protection engineering educators of the twentieth century. Through all his innovative work at the University of Edinburgh, United Kingdom, and with the International Association for Fire Safety Science, he has advanced the science of fire scene analysis through his development and application of fire dynamics to fire investigation.

AUTHOR BACKGROUNDS

This textbook is coauthored by two of the United States' most experienced fire scientists. Their combined talents total more than 70 years of experience in the fields of behavioral science, fire protection engineering, fire behavior, investigation, criminalistics, and crime scene reconstruction.

David J. Icove, PhD, PE, CFEI

An internationally recognized forensic fire engineering expert with more than 38 years of experience, Dr. Icove is coauthor of *Combating Arson-for-Profit,* the leading textbook on the crime of economic arson. Since 1992 he has served as a principal member of the *NFPA 921 Technical Committee on Fire Investigations.* As a retired career federal law enforcement agent, Dr. Icove served as a criminal investigator on the federal, state, and local levels. He is a Certified Fire and Explosion Investigator (CFEI).

He retired in 2005 as an Inspector in the Criminal Investigations Division of the U.S. Tennessee Valley Authority (TVA) Police, Knoxville, Tennessee, where he was assigned for the last 2 years to the Federal Bureau of Investigation (FBI) Joint Terrorism Task Force (JTTF). In addition to conducting major case investigations, Dr. Icove oversaw the development of advanced fire investigation training and technology programs in cooperation with various agencies, including the Federal Emergency Management Agency's (FEMA's) U.S. Fire Administration.

Before transferring to the U.S. TVA Police in 1993, he served 9 years as a program manager in the elite Behavioral Science and Criminal Profiling Units at the FBI, Quantico, Virginia. At the FBI, he implemented and became the first supervisor of the Arson and Bombing Investigative Support (ABIS) Program, staffed by FBI and ATF criminal profilers. Prior to his work at the FBI, Dr. Icove served as a criminal investigator at arson bureaus of the Knoxville Police Department, the Ohio State Fire Marshal's Office, and the Tennessee State Fire Marshal's Office.

His expertise in forensic fire scene reconstruction is based on a blend of on-scene experience, conduction of fire tests and experiments, and participation in prison interviews of convicted arsonists and bombers. He has testified as an expert witness in civil and criminal trials, as well as before U.S. congressional committees seeking guidance on key arson investigation and legislative initiatives.

Dr. Icove holds BS and MS degrees in Electrical Engineering and a PhD in Engineering Science and Mechanics from the University of Tennessee. He also holds a BS degree in Fire Protection Engineering from the University of Maryland–College Park. He is presently an Adjunct Assistant Professor in Department of Electrical Engineering and Computer Science at the University of Tennessee, Knoxville; serves on the faculty of the University of Maryland's Professional Master of Engineering in Fire Protection; and is a Registered Professional Engineer in Tennessee and Virginia.

John D. DeHaan, PhD, FABC, CFI–IAAI, CFEI, FFSS, FSSDip

An internationally recognized forensic science expert, Dr. DeHaan is the author of *Kirk's Fire Investigation,* the leading textbook in the field of fire and arson investigation. He is also a former principal member of the *NFPA 921 Technical Committee on Fire Investigations.*

Dr. DeHaan has been a criminalist for more than 39 years and has gained considerable expertise in fire and explosion evidence as well as human hair, shoe print, and instrumental analysis and crime scene reconstruction. He has been employed as a criminalist by the Alameda County Sheriff's Office, the U.S. Treasury Department, and the California Department of Justice.

His research into forensic fire scene reconstruction is based on first-hand fire experiments on fire behavior involving more than 500 observed full-scale structure, and 100 vehicle fires under controlled conditions, as well as laboratory-scale studies. Dr. DeHaan has testified as an expert witness in civil and criminal trials across the United States and overseas. He is currently the president of Fire-Ex Forensics Inc. and consults on civil and criminal fire and explosion cases across the United States and Canada and overseas.

Dr. DeHaan graduated from the University of Illinois–Chicago Circle in 1969 with a BS degree in Physics and a minor in Criminalistics. He was awarded a PhD in Pure and Applied Chemistry (Forensic Science) by Strathclyde University in Glasgow, Scotland, in 1995. Dr. DeHaan is a Fellow of the American Board of Criminalists (Fire Debris), a Fellow of the Forensic Science Society (UK), and holds Diplomas in Fire Investigation from the Forensic Science Society and the Institution of Fire Engineers and a Certified Fire Investigator certification from the International Association of Arson Investigators. He is also a Certified Fire and Explosion Investigator from the National Association of Fire Investigators.

Principles of Reconstruction

It is a capital mistake to theorize before one has data. Insensibly one begins to twist facts to suit theories, instead of theories to suit facts.

—Sir Arthur Conan Doyle,
"A Scandal in Bohemia"

The goal of this chapter is to introduce a new look on the principles and concepts of science-based fire scene reconstruction. The use of the scientific method in forming an expert opinion as to the fire's origin and cause, development, and impact on human lives is discussed from several aspects.

The process of fire scene reconstruction is used to determine the most likely development of a fire using a scientifically based methodology. Reconstruction follows the fire from ignition to extinguishment; and it explains aspects of the fire and smoke development, the role of fuels, effects of ventilation, the impact of manual and automatic extinguishment, the performance of the building, life safety features, and manner of injuries or death.

Underlying principles of forensic fire scene reconstruction rely firmly on a comprehensive review of the fire pattern damage, sound fire protection engineering principles, human factors, physical evidence, and an appropriate application of the scientific method. These factors often form the basis for an expert opinion as to the most probable origin and cause of the fire or explosion. The expert opinion may be part of a written report or the basis for oral testimony in depositions or courtrooms.

To be effective, these expert opinions must be able to pass the eventual scrutiny of cross-examination during peer review, sworn depositions, Daubert hearings, and courtroom testimony. Recent court decisions place more weight on expert forensic testimony based on scientific, rather than merely experience-based, knowledge.

When determining the origin and cause of a fire, a comprehensive reconstruction often involves a fire engineering analysis that tests various scenarios. This analysis may use fire modeling to compare actual events with predicted outcomes by varying causes and growth scenarios. This engineering analysis adds value, understanding, and clarity to complex fire scene investigation.

◆ 1.1 NEED FOR SCIENCE IN FIRE SCENE RECONSTRUCTION

An international conference in November 1997 assessed the current state of the art and identified technical gaps in fire investigation (Nelson and Tontarski 1998). The International Conference on Fire Research for Fire Investigation concluded that many scientific gaps existed in the methodology and principles used to reconstruct fire scenes.

This conference cited research, development, training, and education needs including the following:

- **Fire incident reconstruction**—laboratory methods for testing ignition source hypotheses
- **Burn pattern analysis**—methods for validation and training for evaluating patterns on walls and ceilings and patterns resulting from liquids on floors and comparing them to patterns generated by radiant heat
- **Burning rates**—determination of burning rates for different items, development of a burning rate database
- **Electrical ignition**—validation of means to identify electrical faults as an ignition source
- **Flashover**—impacts of flashover on fire patterns and other indicators
- **Ventilation**—effects of ventilation on fire growth and origin determination
- **Fire models**—methods for validation of, training in, and education on the use of fire models
- **Health and safety**—methods to evaluate fire investigator occupational health and safety
- **Certification**—development of training and certification programs for investigators and laboratory personnel
- **Train-the-trainers**—methods to pass along training objectives and expertise to groups of qualified trainers

The primary goal of the conference was to encourage the development and use of scientific principles and methodologies by fire investigators and determine what facilities could improve investigation by public-sector investigators. The work resulted in the issuance of a white paper under the auspices of the Fire Protection Research Foundation (FPRF 2002). That study concluded that there was a basic lack of a scientific foundation for many methods now in use to identify the area of origin and the cause of fires. Complex types and forms of burning materials, building geometry, ventilation, and firefighting actions were listed as contributing effects that complicate these determinations.

◆ 1.2 THE SCIENTIFIC METHOD

Although he began his professional career as a barrister, Francis Bacon (1561–1626) is best known for establishing and popularizing the method of **inductive logic** for conducting critical scientific inquiries. His method of inductive logic extracted knowledge through observation, experimentation, and testing of hypotheses.

The underpinning of forensic fire scene reconstruction is the use and application of relevant scientific principles and research in conjunction with a systematic examination of the scene. This is particularly true in cases that later require expert witness opinions.

Insight as to whether a fire investigator can be considered an expert was addressed by the American Society for Testing and Materials (ASTM). In a now-withdrawn yet factually accurate standard, ASTM defined a "technical expert" as "a person educated, skilled, or experienced in the mechanical arts, applied sciences, or related crafts" (ASTM 1989). A properly trained and qualified fire investigator, who applies and practices the application of fire science, could clearly be considered a "technical expert" under the ASTM's definition.

The basic concepts of the **scientific method** are simply to **observe, hypothesize, test**, and **conclude.** The scientific method, which embraces sound scientific and fire protection engineering principles combined with peer-reviewed research and testing, is considered the best approach for conducting fire scene analysis and reconstruction. Furthermore, present job performance requirements under *NFPA 1033—Standard for Professional Qualifications for Investigator,* specify that fire investigators "shall employ all elements of the scientific method as the operating analytical process throughout the investigation and for drawing of conclusions" (NFPA 2003a).

NFPA 921—Guide for Fire and Explosion Investigations authoritatively defines the application of the scientific method as

> the systematic pursuit of knowledge involving the recognition and formulation of a problem, the collection of data through observation and experiment, and the formulation and testing of a hypothesis. (NFPA 2004a, pt. 3.3.129)

The process of the scientific method continuously refines and explores a working hypothesis until arriving at a final expert conclusion or opinion, as illustrated in Figure 1.1. An important concept in the application of the scientific method to fire scene investigations is that all fires should be approached by the investigator without a presumption of the cause. Until sufficient data have been collected by the examination of the scene (assuming that it is still available), no specific hypothesis should be formed (NFPA 2004a, pt. 4.3.6.7).

Listed next is an annotated seven-step systematic process, based on *NFPA 921* (NFPA 2004a, pt. 4.3) for applying the scientific method to fire investigation and reconstruction. Properly applied, this systematic process addresses how to correctly formulate, test, and validate a hypothesis.

1. **Recognize the Need.**   A paramount responsibility of first responders is to protect a scene until a full investigation can be started. After initial notification, the investigator should proceed to the scene as soon as possible and determine the resources needed to conduct the investigation thoroughly. It is necessary not only to determine the origin and cause of this event but to determine whether future fires, explosions, or loss of life can be prevented through new designs, codes, or enforcement strategies. In public-safety sectors, determination of the origin and cause of every fire is often a statutory requirement. The identification of unsafe products related to the cause of the fire can also result in recalls and prevention of future incidents.
2. **Define the Problem.**   Devise a tentative investigative plan to preserve and protect the scene, determine the cause and nature of the loss, conduct a needs assessment, formulate and implement a strategic plan, and prepare a report. This step also includes determining who has the primary responsibility and authority to interview witnesses, protect evidence, and review preliminary findings and documents describing the loss.
3. **Collect Data.**   Data are "facts or information to be used as a basis for a discussion or decision" (ASTM 1989). Collect facts and information about the incident through direct observations, measurements, photography, evidence collection, testing, experimentation, historical case histories, and witness interviews. All collected data should be subject to

FIGURE 1.1 ◆ A flowchart outlining the scientific method as it applies to fire scene investigation and reconstruction. *Derived from NFPA 921 [NFPA 2004, pt. 4.3].*

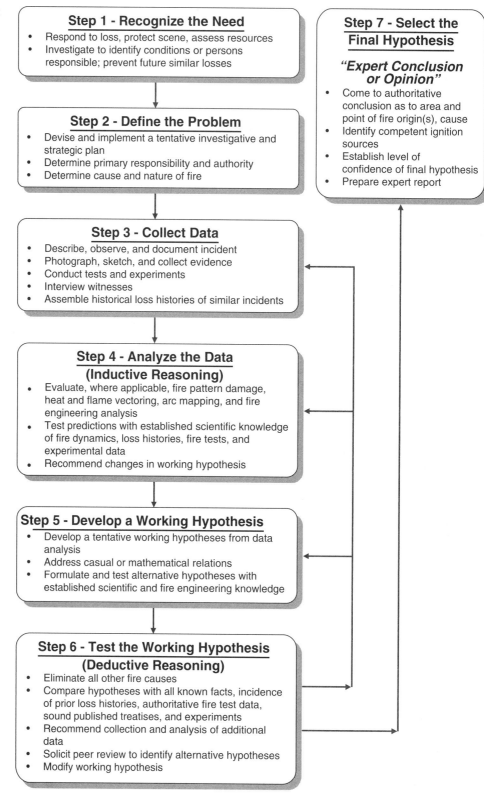

Step 1 - Recognize the Need
- Respond to loss, protect scene, assess resources
- Investigate to identify conditions or persons responsible; prevent future similar losses

Step 2 - Define the Problem
- Devise and implement a tentative investigative and strategic plan
- Determine primary responsibility and authority
- Determine cause and nature of fire

Step 3 - Collect Data
- Describe, observe, and document incident
- Photograph, sketch, and collect evidence
- Conduct tests and experiments
- Interview witnesses
- Assemble historical loss histories of similar incidents

Step 4 - Analyze the Data (Inductive Reasoning)
- Evaluate, where applicable, fire pattern damage, heat and flame vectoring, arc mapping, and fire engineering analysis
- Test predictions with established scientific knowledge of fire dynamics, loss histories, fire tests, and experimental data
- Recommend changes in working hypothesis

Step 5 - Develop a Working Hypothesis
- Develop a tentative working hypotheses from data analysis
- Address casual or mathematical relations
- Formulate and test alternative hypotheses with established scientific and fire engineering knowledge

Step 6 - Test the Working Hypothesis (Deductive Reasoning)
- Eliminate all other fire causes
- Compare hypotheses with all known facts, incidence of prior loss histories, authoritative fire test data, sound published treatises, and experiments
- Recommend collection and analysis of additional data
- Solicit peer review to identify alternative hypotheses
- Modify working hypothesis

Step 7 - Select the Final Hypothesis

"Expert Conclusion or Opinion"
- Come to authoritative conclusion as to area and point of fire origin(s), cause
- Identify competent ignition sources
- Establish level of confidence of final hypothesis
- Prepare expert report

verification of how it was legally obtained, its chain of custody, and notation as to its reliability and authoritative nature. Data collection should include the comprehensive documentation of the building, vehicle, or wildlands involved; construction and occupancy; fuel loading; the processing and layering of the debris and evidence as found; examination of the fire (heat and smoke) patterns; char depths, calcination, and the application of electrical arc mapping surveys where relevant.

4. **Analyze the Data (Inductive Reasoning).**   Using inductive reasoning, which involves formulating a conclusion based on a number of observations, analyze all the data collected. The investigator relies on his or her knowledge, training, and experience in evaluating the totality of the data. This subjective approach to the analysis may include knowledge of similar loss histories (observed or obtained from references), training and understanding of fire dynamics, fire testing experience, and experimental data. Data evaluated can include the pattern of fire damage, heat and flame vectoring, arc mapping, and fire engineering and modeling analysis.

5. **Develop a Working Hypothesis.**   A hypothesis is defined as "a supposition or conjecture put forward to account for certain facts, and used as a basis for further investigation by which it may be proved or disproved" (ASTM 1989). Based on the data analysis, develop a tentative working hypothesis to explain the fire's origin, cause, and development that is consistent with on-scene observations, physical evidence, and testimony from witnesses. The hypothesis may address a causal mechanism or a mathematical relation (e.g., plume flame height, impact of differing fuel loads, locations of competent ignition sources, room dimensions, impact of open or closed doors and windows).

6. **Test the Working Hypothesis (Deductive Reasoning).**   Using deductive reasoning, which involves a conclusion based on previously known facts, compare the working hypothesis with all other known facts, the incidence of prior loss histories, relevant fire test data, authoritative published treatises, and experiments. Use the working hypothesis to eliminate all other reasonable origins and causes for the fire or explosion, recommend the collection and analysis of additional data, seek new information from witnesses, and develop or modify the working hypothesis. This may involve reviewing the analysis with other investigators with relevant experience and training (peer review). Interactively repeat steps 4, 5, and 6 until there are no discrepancies with the working hypothesis. A critical feature of hypothesis testing is to create alternative hypotheses that also can be tested. If the alternatives are in opposition to the working hypothesis, their evaluation may reveal issues that need to be addressed. By testing all hypotheses rigorously against the data, those that cannot be conclusively eliminated should still be considered viable.

7. **Select Final Hypothesis (Conclusion or Opinion).**   An **opinion** is defined as "a belief or judgment based upon facts and logic, but without absolute proof of its truth" (ASTM 1989). When the working hypothesis is thoroughly consistent with evidence and research, it becomes a final hypothesis and can be authoritatively presented as a conclusion or opinion of the investigation.

NFPA 921 currently suggests two levels of confidence with respect to opinions (NFPA 2004a, pt. 18.6). A **probable opinion** ranks at a level of certainty to being more likely true, or the likelihood of the hypothesis being true greater than 50 percent. A **possible opinion** ranks at the level of certainty to be feasible, but not probable, particularly if two or more hypotheses are equally likely. If a final hypothesis cannot be determined, the cause should be reported as "undetermined."

Levels of confidence may also vary with the profession of the person preparing the report. For example, engineers commonly state in their written reports that their **expert opinion** is "to a reasonable degree of engineering certainty." Forensic opinions should be offered only at a very high level of certainty; that is, no other logical solutions offer the same level of agreement with the available data. Curiously, *NFPA 921* currently does not include such opinion in its hierarchy. In criminal cases such opinions are scrutinized under the same "beyond a reasonable doubt" standard as are the ultimate decisions of the triers of fact. In civil cases, opinions may rest on an evaluation as to whether the event was more likely than not. The examiner must be aware of all the potential uses to which the opinion may be put.

When applying the scientific method in stating an expert opinion on the origin and cause of a fire or explosion, the investigator sets a level for the degree of confidence in forming the opinion at a prescribed threshold. Specifically, when this degree of confidence in the data or hypothesis is at the "possible" or "suspected" level, "undetermined" should be cited as the cause of the fire or explosion. An opinion should be expected to withstand the challenges of a reasonable examination by peer review or a thorough cross-examination in the courtroom (NFPA 2004a, pt. 18.7).

THE WORKING HYPOTHESIS

The concept of the **working hypothesis** is central within the framework of the scientific method. In the case of fire scene reconstruction, the working hypothesis is based on how an investigator describes or explains the fire's origin, cause, and subsequent development. It is subject to continuous review and modification and is not the final conclusion.

As shown in Figure 1.2, the investigator molds a working hypothesis and then modifies and refines it by drawing on and examining many sources of information, including past investigative experiences. The working hypothesis may include peer reviews by other qualified investigators, who themselves bring institutional knowledge and experiences. Alternative hypotheses may be created and tested.

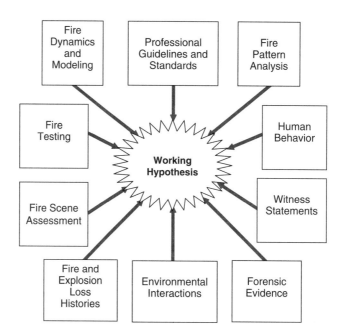

FIGURE 1.2 ◆ Sources of information that may contribute to a working hypothesis. *Courtesy of D. J. Icove.*

New hypotheses are often built on their predecessors, particularly when the knowledge base of a discipline broadens. When an accepted working hypothesis cannot explain new data, we strive to construct a new hypothesis and in some cases generate completely new insight into the problem. A historical example of the application of the hypothesis dealt with proof of the geocentric versus heliocentric concept of the solar system, or simply, Does the earth revolve around the sun—or vice versa? The early hypothesis that the sun revolved around the earth was replaced with a new model proposing that the earth revolved around the sun that better fitted new data.

One of the underpinnings of the scientific method is the concept of reproducibility, whereby a hypothesis can be independently tested and verified by others. Even when a working hypothesis agrees with all known experimental evidence, it may, nevertheless, be disproved through subsequent experimentation or an emergent discovery. One area of recent experimental evidence is in the interpretation of fire patterns (FEMA 1997a). Certain classes of fire pattern damage once thought to be produced solely by accelerants at fire scenes have been disproved by new principles, practice, and testing (DeHaan 2007; NFPA 2004a). These include, for example, spalling of concrete, collapsed furniture springs, and crazed glass (Lentini 1992; Tobin and Monson 1989).

The knowledge of the application of **fire dynamics and modeling** is a significant advantage that can clearly set novices apart from expert fire investigators. Such application is not necessarily new or novel, since the basic tenets of fire spread are based on the established principles of fire dynamics. Our understanding of fire dynamics is improving steadily and is often improved by modeling fire situations. Since investigators are applying fire science to their profession, the use of consistent terminology, descriptions of the scene, and interpretations are important in developing and communicating the working hypothesis and in conducting peer review.

Professional guidelines and standards are often relied on in developing the hypotheses, particularly those that are authoritative. Engineering handbooks in wide use include the NFPA (2003a) *Fire Protection Handbook;* the *SFPE Handbook of Fire Protection Engineering* (SFPE 2002b); and the *Ignition Handbook: Principles and Applications to Fire Safety Engineering, Fire Investigation, Risk Management and Forensic Science* (Babrauskas 2003). Guides that are widely recognized include the NFPA's (2004a) *NFPA 921—Guide for Fire and Explosion Investigations;* the National Institute of Justice's (NIJ 2000) *Fire and Arson Scene Evidence: A Guide for Public Safety Personnel; Kirk's Fire Investigation,* 6th ed. (DeHaan 2007); and *Combating Arson-for-Profit: Advanced Techniques for Investigators,* 2nd ed. (Icove, Wherry, and Schroeder 1998). Other NIJ national guidelines cover such topics as general crime scenes, death investigations, eyewitness identification, bombings, digital images, and electronic evidence. ASTM has several standard practices applicable to fire investigations, which are discussed later in this chapter.

The Society of Fire Protection Engineers (SFPE) publishes a series of engineering guides, many of which are directly applicable to better understanding fire and human behavior and to making realistic assessments. These publications include the *SFPE Engineering Guide for Assessing Flame Radiation to External Targets from Pool Fires, Predicting 1st and 2nd Degree Skin Burns, Piloted Ignition of Solid Materials under Radiant Exposure, Performance-Based Fire Protection Analysis and Design of Buildings, Human Behavior in Fires, Fire Exposures to Structural Elements*, and *Evaluation of the Computer Fire Model DETACT-QS.*

Fire pattern analysis is an important aspect of interpreting the impact of the fire plume on the structure and contents of the building, vehicle, forest land, victim, or other property. Because liquid accelerants are frequently used by arsonists, authoritative fire testing continues to extend the body of knowledge into fire pattern analysis, particularly those comparing flammable and combustible liquid to those from ordinary combustible burn patterns (DeHaan 1995; Kennedy et al. 1997; Putorti 2001).

An understanding of **human behavior** is paramount when evaluating the movement and tenability of persons involved in fires. Not all people respond to perceptions of fire danger in the same way. The investigator must be able to understand and describe how occupants may respond to smoke and fire conditions that change in time and location, resulting in either their successful escape or their death from exposure to various by-products of combustion. The evaluation should include not only behavior patterns of individuals but also interactions with their primary group (family) or with strangers. This understanding should include gender-based differences of persons in their response to the interpretation of the fire cues, decision making, decisions, and their resulting coping actions (SFPE 2003).

Witness statements giving the accounts of actions taken prior to, during, and after the fire are important to include when developing and testing the working hypothesis. Often, witnesses may need to be reinterviewed to bring out additional details left out of the initial statements given to authorities. This information may include observations seen by a witness prior to the fire and as the fire was discovered and progressed.

Forensic science can add additional information to a working hypothesis. Examples include the use of traditional forensic examinations of fire debris, the analysis of impression and trace evidence, and, lately, the increased use of DNA in cases of death investigations. Knowledge of **environmental interactions** of the initial temperature, humidity, wind speed, and direction on the developing fire may play an important role in developing the working hypothesis. For example, in a high-rise building fire, the environmental factors of wind and temperature along with information about the location of the fire, occurred whether above or below the neutral plane, may affect the initial discovery of odors, smoke, and fire development (SFPE 2002a, 4–275). Lightning strikes are an environmental interaction and potential cause for ignition that should be considered. U.S. meteorological studies shown the Louisiana and Florida as the leading states for the highest density of lightning strikes, followed by the adjacent areas of Tennessee, Mississippi, and Kentucky (Orvill and Huffines 1999).

Fire and explosion loss histories are significant to the development of any working hypothesis. The knowledge of similar historical case histories may shed new knowledge in developing the hypotheses that should be tested before selecting the final hypothesis. Note that such statistical data should not be relied on as the means to prove the cause of a particular fire under investigation; rather, they serve a better purpose in assisting in formulating, examining, or refining the working hypotheses.

Many public and private organizations collect fire data for statistically profiling occupancies to develop risk assessment and fire prevention efforts. For example, Table 1.1 is a profile of average annual fire losses occurring in eating and drinking establishments by cause for the years 1994–1998 (NFPA 2003c). NFPA also focuses on fires in residences and publishes statistics on the forms of material first ignited, data that are essential when developing hypotheses (Rohr 2001, 2005). Note that other industry sources of fire loss data are also available, such as

TABLE 1.1 ◆ Profile by Cause for Average Annual Fire Losses in Eating and Drinking Establishments, 1994–1998

Fire Cause	No. of Losses	Percentage
Cooking equipment (stove, fryer, grill)	4700	42.5
Electrical distribution (lights, lamps, wiring)	1500	13.1
Arson	1200	10.7
Other equipment	1100	9.7
Heating equipment (central, water heater)	700	6.5
Appliance (tool, air conditioning)	600	5.1
Smoking materials	500	4.6
Open flame, ember, torch	400	3.3
Exposure to other hostile fire	200	1.8
Natural causes (spontaneous heating, etc.)	200	1.5
Other heat sources	200	1.4
Total losses	11,100	100.0

Source: National Fire Protection Association, U.S. Fire Problem Overview Report, June 2003.

the periodically issued Factory Mutual Global "Property Loss Prevention Data Books" (Factory Mutual 1996).

These loss histories may also be included in the investigator's vast experience of **fire scene assessments** while he or she is working on, reviewing, or reading about similar cases. The investigator's heuristic knowledge (rules of thumb), training, and experience, combined with professional investigative guidelines and standards, are also drawn on during the development of a working hypothesis. Note that such statistical histories are useful for developing alternative hypotheses to be tested and not for eliminating a cause for a specific fire because it does not appear on the list.

The combination of generally accepted scientific principles and research, when supported by empirical **fire testing,** can yield valuable insight into fire scene reconstruction. Scientifically conducted test results can normally be reproduced and their error rates validated. When this is true, the results are considered to be **objective**—neither prejudiced nor biased by the researcher. Accepted fire testing procedures are developed and maintained by the Committee on Fire Standards (E05) of the American Society for Testing and Materials (ASTM 1999). Fire testing will be discussed later in greater detail in another chapter.

APPLYING THE WORKING HYPOTHESIS

It is vital for every experienced investigator to become proficient in the development of the working hypothesis. This can be accomplished by consistently documenting and evaluating evidence from the fire scene, recording observations and behaviors of witnesses to the fire, comparing similar case histories, reading the fire literature, actively participating in authoritative fire testing, and writing and publishing peer-reviewed research.

The scientific method as it applies to forensic fire scene reconstruction provides fundamental principles for developing a modifiable (working) hypothesis about a fire, testing the hypothesis, and deriving a sound theory as to its origin, development, damage, and cause. Through an iterative process, the working hypothesis is revisited, revised, and reformulated until it is reconciled with all the available data. At this point, the hypothesis becomes a final theory and can be presented as a conclusive opinion of the investigation.

Struggling with the concept of the working hypothesis is not new in fire investigations. Example 1.1 relates the many hypotheses that arose during the 1942 Cocoanut Grove Night Club fire. Some of the theories appeared far-fetched, but investigators examined each one during the case. Clearly, the authorities pursued all theories that surfaced in this case. This is an example of how the scientific method can be used to discount those theories that cannot be supported by analytical data, fire protection principles, and witness observations.

EXAMPLE 1.1 Hypothesis Testing

The Cocoanut Grove Night Club fire of November 28, 1942, in Boston cost the lives of 492 people, with many others seriously injured, including 131 treated at nearby hospitals. The fire was first discovered in the basement area called the Melody Lounge. It spread quickly through the lounge and up the stairs to the entrance lobby at street level. See Figure 1.3.

Investigating authorities probing the fire were flooded with publicly expressed opinions explaining how the fire started and progressed. Even though the exact source of ignition is still disputed today, many of the publicly expressed opinions had to be addressed in the report. The major issues were the use of combustible decorations, the rapid rate of fire spread, the lack of adequate exits, and the cause of so many postfire deaths. What is known is that witnesses first reported that the fire originated in the lounge, ignited adjacent combustible decorations and cloth-covered ceilings and walls, and progressed up the stairs—thereby blocking the only visible means of exiting the building.

Many of the theories on the fire's origin, cause, and rapid spread were reported in the official NFPA report issued January 11, 1943 (NFPA 1943). Even though the published report does not go into sufficient technical detail on these theories, it still serves as an example of a working hypothesis in fire scene reconstruction and analysis. The initial NFPA report listed the following theories and eliminated some based on their improbability.

Ordinary Fire Theory.   The ordinary fire theory stated that the cloth and paper decorations on the ceiling and walls, and other available combustible materials, generated an abundance of smoke and carbon monoxide and a means for rapid fire spread. The report noted the possibility of another unexplained element that could have accelerated the growth of the fire. The report did not further discuss the unexplained element. In retrospect, the high number of deaths of "survivors" hours after the fire led to suspicion of unusual factors in the toxicity of the smoke.

Alcohol Theory.   The alcohol theory suggested that vapors from the mouths of heavy drinkers contributed to the rapid fire spread. This included the proposal that warm alcohol evaporating from drinks on tables could also have contributed to the fire. This theory was discounted as a possibility, as the concentration of such vapors would be far too low to enhance flame spread.

Pyroxylin Theory.   The pyroxylin (nitrocellulose) theory suggested that the extensive use of artificial leather used on the walls of the building caused the rapid fire spread and production of nitrous by-products of combustion, accounting for the numerous fatalities. A scientific assessment led to the elimination of this theory. These materials contained only a small percentage of pyroxylin, which was insufficient to contribute to the fire spread or to the number of fatalities.

Motion Picture Film Theory.   The motion picture film theory was raised when it was determined that the building had previously been used as a motion picture film exchange. The theory suggested that a quantity of decomposing scrap nitrocellulose film might have remained in a hidden storage area and was ignited by a source of ignition. The observations of witnesses did not support this theory.

Refrigerant Gas Theory.   The air conditioning supplying the Melody Lounge was located in a false corner wall. Historically, air-conditioning systems have used ammonia or sulfur dioxide as refrigerant. Such refrigerant gases are combustible and toxic. A theory suggesting that these refrigerant gases ignited was discounted. Ammonia is a flammable gas but at concentrations far above the tolerable level for human exposure, and no survivors reported ammonia odors. Furthermore, the refrigerant gas ran in copper tubing, which would have needed to be thermally or mechanically breached to release the gas during the initial stages of the fire.

Flameproofing Theory.   The flameproofing chemicals theory was based on the suggestion that additives to the combustible decorations may have given off ammonia and other gases when heated. The investigation failed to determine whether any flame retardants were ever applied to the decorations, and the surviving victims did not report any symptoms associated with this type of exposure. Fire inspectors at the time routinely exposed the edge of a portion of the decorative fabric to a match flame as a field test. Decorations tested prior to the fire reportedly passed such a limited test.

Fire Extinguisher Theory.   The fire extinguisher theory came about when it was learned that fire extinguishers were used on an incipient fire in an artificial palm tree and that toxic gases may have been emitted by the extinguishing liquid. The investigation found no supporting evidence for this theory.

Insecticide Theory.   The insecticide theory emerged when it was learned that insecticides were used in the basement kitchen. It was suggested that if flammable vapors had collected in concealed wall spaces, they might have accounted for the reported initial flash fire. There was no information as to whether the insecticide used contained an ignitable liquid, and there were no indications of excess quantities of any insecticide.

Gasoline Theory.   The gasoline theory was suggested when it was determined that the building had previously been used as a garage and that gasoline tanks still existed under the basement floor. The theory that gasoline vapor emerged from these tanks was discounted, since such fumes are heavier than air and would have remained at floor level. Furthermore, the physical scene examination proved that there was no source to create an adequate concentration of vapors to support the flame spread seen in this fire. Vapors would have been detected by occupants at levels far below ignitable concentrations.

Electric Wiring Theory.   The electric wiring theory came about when it was learned that an unlicensed electrician had installed part of the wiring in the building. The theory suggested that the heated insulation on overloaded wiring was responsible for forming flammable and toxic vapors. No evidence was found that could support the theory of overloading or that an electrical failure was responsible for ignition.

Smoldering Fire Theory.   The smoldering fire theory was suggested when it was learned that witnesses reported that several walls were hot to the touch and that a smoldering fire went undetected. Investigators physically examining the fire scene found no physical evidence to support this theory.

(a)

(b)

FIGURE 1.3 ◆ (a) Exterior of Cocoanut Grove fire scene (November 1942), with 492 dead and hundreds more injured. *Copyright © 1942, National Fire Protection Association, Quincy, MA 02169. All rights reserved.* (b) Interior of Cocoanut Grove fire scene. Note the limited fire damage to furniture and decorations. *Courtesy of AP Images.*

As a postscript to these theories, over half a century later the cause of the fire is still listed as "undetermined." Some witnesses reported that the initial fire was in an artificial tree, where it could have been started by an errant flame from a match, lighter, or candle. Hypotheses are still being tested as to the cause of the Cocoanut Grove fire. Recent reviews of the investigation by the NFPA beginning in 1996 suggest that fire modeling combined with the use of the scientific method may shed new insights into this case (Beller and Sapochetti 2000). For further reading, several recent texts are available (Schorow 2005; Esposito 2005).

BENEFITS OF USING THE SCIENTIFIC METHOD

The International Conference on Fire Research for Fire Investigation (Baltimore, MD 1997) concluded that many scientific gaps existed in the methodologies and principles used to reconstruct fire scenes (Nelson and Tontarski 1998). Major conference goals included an assessment of the use of scientific principles and methodologies in fire investigation and identification of fire investigation needs and education. There are numerous benefits to using the scientific method to examine fire and explosion cases, three of which follow:

- ◆ Acceptance of the methodology in the scientific community
- ◆ Use of a uniform, peer-reviewed protocol of practice, such as *NFPA 921* (NFPA 2004a)
- ◆ Improved reliability of testimony from opinions formed using the scientific method

First, the use of the scientific method is well received in both the technical and the research communities. An investigation conducted using this approach is more likely to be embraced by those who would tend to doubt a less thoroughly conducted probe.

Second, the scientific method is a generally accepted protocol of practice in the literature. Those persons ignoring or deviating from recommended practices would bring closer scrutiny to their reports and opinions. Modern fire investigation texts endorse the application (DeHaan 2007).

A guide that repeatedly cites the scientific method is the NIJ's (2000) *Fire and Arson Scene Evidence: A Guide for Public Safety Personnel*. The NIJ guide notes that actions taken at the scene of a fire or arson investigation can play a pivotal role in the resolution of a case. A thorough investigation is key to ensuring that potential physical evidence is neither tainted nor destroyed, nor potential witnesses overlooked.

Third, and most important, expert testimony in fire and explosion cases has come to rely more heavily on opinions formed using the scientific method. Recent U.S. Supreme Court decisions underscore these principles, with many state courts following the trend. The following section discusses at length many issues surrounding expert testimony.

ALTERNATIVES TO THE SCIENTIFIC METHOD

There are alternative approaches to the scientific method that use other forms of logic apart from inductive and deductive reasoning. For example, **abductive reasoning** involves the process of reasoning to the best explanations for a phenomenon. An example of an abductive reasoning of a known rule to explain an observation might be, "If it rains, the grass is wet." This abductive rule explains why the grass is wet. But note that an alternative explanation might answer the question, Can the grass be wet if it has not rained? Yes, the lawn sprinklers were used, or the dew was heavy. In this approach, the reasoning starts with a set of given facts and derives their most likely explanations or hypotheses. Such reasoning leads to testable alternatives.

Modern-day applications of abductive reasoning include artificial intelligence, fault tree diagnosis, and automated planning. For a thorough discussion on the comparison of inductive, deductive, and abductive reasoning, see the Joint Military Intelligence College paper entitled "Critical Thinking and Intelligence Analysis" (Moore 2006).

◆ 1.3 FOUNDATIONS OF EXPERT TESTIMONY

FEDERAL RULES OF EVIDENCE

For the purposes of this textbook, emphasis is placed on testimony given primarily in the federal court system. The *Federal Rules of Evidence* (FRE) clearly state who may offer testimony in federal court (FRCP 2000). Many state courts model their guidelines on the federal rules.

Recent amendments to the FRE affected the admissibility of evidence and opinion testimony under *Rules 701, 702, 703, 704, 705*, and *706*. These changes state the following.

Rule 701. Opinion Testimony by Lay Witnesses

If the witness is not testifying as an expert, the witness' testimony in the form of opinions or inferences is limited to those opinions or inferences which are (a) rationally based on the perception of the witness, (b) helpful to a clear understanding of the witness' testimony or the determination of a fact in issue, and (c) not based on scientific, technical, or other specialized knowledge within the scope of Rule 702.

Rule 702. Testimony by Experts

If scientific, technical, or other specialized knowledge will assist the trier of fact to understand the evidence or to determine a fact in issue, a witness qualified as an expert by knowledge, skill, experience, training, or education may testify thereto in the form of an opinion or otherwise, if (1) the testimony is sufficiently based upon reliable facts or data, (2) the testimony is the product of reliable principles and methods, and (3) the witness has applied the principles and methods reliably to the facts of the case.

Rule 703. Bases of Opinion Testimony by Experts

The facts or data in the particular case upon which an expert bases an opinion or inference may be those perceived by or made known to the expert at or before the hearing. If of a type reasonably relied upon by experts in the particular field in forming opinions or inferences upon the subject, the facts or data need not be admissible in evidence in order for the opinion or inference to be admitted. Facts or data that are otherwise inadmissible shall not be disclosed to the jury by the proponent of the opinion or inference unless the court determines that their probative value in assisting the jury to evaluate the expert's opinion substantially outweighs their prejudicial effect.

Rule 704. Opinion on Ultimate Issue

(a) Except as provided in subdivision (b), testimony in the form of an opinion or inference otherwise admissible is not objectionable because it embraces an ultimate issue to be decided by the trier of fact.
(b) No expert witness testifying with respect to the mental state or condition of a defendant in a criminal case may state an opinion or inference as to whether the defendant did or did not have the mental state or condition constituting an element of the crime charged or of a defense thereto. Such ultimate issues are matters for the trier of fact alone.

Rule 705. Disclosure of Facts or Data Underlying Expert Opinion

The expert may testify in terms of opinion or inference and give reasons therefor without first testifying to the underlying facts or data, unless the court requires otherwise. The expert may in any event be required to disclose the underlying facts or data on cross-examination.

Rule 706. Court Appointed Experts

(a) Appointment.
    The court may on its own motion or on the motion of any party enter an order to show cause why expert witnesses should not be appointed, and may request the parties to submit nominations. The court may appoint any expert witnesses agreed upon by the parties, and may appoint expert witnesses of its own selection. An expert witness shall not be appointed by the court unless the witness consents to act. A witness so appointed shall be informed of the witness' duties by the court in writing, a copy of which shall be filed with the clerk, or at a conference in which the parties shall have opportunity to participate. A witness so appointed shall advise the parties of the witness' findings, if any; the witness' deposition may be taken by any party; and the witness may be called to testify by the court or any party. The witness shall be subject to cross-examination by each party, including a party calling the witness.

(b) Compensation.
    Expert witnesses so appointed are entitled to reasonable compensation in whatever sum the court may allow. The compensation thus fixed is payable from funds which may be provided by law in criminal cases and civil actions and proceedings involving just compensation under the Fifth Amendment. In other civil actions and proceedings the compensation shall be paid by the parties in such proportion and at such time as the court directs, and thereafter charged in like manner as other costs.

(c) Disclosure of appointment.
    In the exercise of its discretion, the court may authorize disclosure to the jury of the fact that the court appointed the expert witness.

(d) Parties' experts of own selection.
    Nothing in this rule limits the parties in calling expert witnesses of their own selection.

Note that these federal guidelines do not apply to all state courts or international situations, and the reader is encouraged to seek guidance from an appropriate legal advisor.

SOURCES OF INFORMATION FOR EXPERT TESTIMONY

In general, experts will normally use three major sources of information on which to base their opinions (Kolczynski 2000):

- Firsthand observations
- Facts presented to experts prior to trial
- Facts supplied in court

Examples of **firsthand observations** include those observed at a fire scene, in a laboratory examination or fire test, or from evaluation of evidence under *FRE 703*. It is important that experts try to actually visit the fire scene and not merely rely on sketches and photographic evidence taken by another party. Although valuable, it is not always required or possible to actually visit the scene, particularly if the scene has been considerably altered or destroyed. If photographic documentation is sufficiently comprehensive and can be documented or validated by other means, the expert can rely on it (see *United States of America v. Ruby Gardner* 2000).

Experts usually critically review the case and gain insight and knowledge of **facts prior to trial.** This knowledge may be based on facts gleaned from scientific manuals, learned treatises, or results of historical testing. Examples include information learned through expert treatises such as *NFPA 921, Kirk's Fire Investigation* and the *SFPE Handbook*.

Rule 701 allows for lay witnesses to offer nonexpert opinion evidence testimony. These opinions might include the state of intoxication, a vehicle's speed, and the memory of an identifiable odor of gasoline. For example, firefighters who are not experts in fire investigation may be allowed to testify that they detected a strong odor of gasoline when entering a certain room of a burning structure.

Rule 702 requires that the expert's testimony must assist the trier of fact, and its admissibility depends in part on the connection between the scientific research or test to be presented and the particular disputed factual issues in this case (*In Re: Paoli Railroad Yard PCB Litigation* (1994) citing *U.S. v. Downing* 1985). Furthermore, each step in the expert's analysis must reliably connect the work of the expert to that particular case.

A lay witness might also offer expert testimony under *Rule 702* when the opinion is based on an accepted evaluation method, as long as the witness is qualified to make this evaluation. Examples of accepted evaluation methods include the guidelines in *NFPA 921* and various ASTM standards.

Rule 703 historically (since its enactment in 1975) allowed for experts to form an opinion based on facts, whether or not in evidence, as long as these facts were necessary to form professional judgments from a nonlitigation point of view. The recent change clarifies that just because an expert witness uses information to form an opinion, its use does not automatically make that information admissible in court. Under *FRE 703*, experts may also rely on hearsay through statements of other witnesses. For example, the expert may use information from the persons who witnessed the fire, testimony of other witnesses, documents and reports, videotapes of the fire in progress, and related written reports submitted in the case to help form an opinion. (See also *United States of America v. Ruby Gardner* 2000.)

Facts supplied in court may also be a basis for expert witness testimony. These facts may include information from testimony of other witnesses and evidence presented. The expert witness may also be asked a hypothetical question. For example, an expert may be asked, "Assume that three separate gasoline-filled plastic jugs with partially burned wicks were found in separate rooms of the house. What would be your conclusions based on these facts and your 25 years of experience as a fire investigator?"

DISCLOSURE OF EXPERT TESTIMONY

The following is an excerpt regarding guidelines on the requirement to disclose expert testimony in civil cases from *Rule 26(a)(2)(B)* of the *Federal Rules of Civil Procedure* (FRCP 2000) effective December 1, 2000, as amended.

(2) Disclosure of Expert Testimony

A. In addition to the disclosures required by paragraph (1), a party shall disclose to other parties the identity of any person who may be used at trial to present evidence under *Rules 702, 703,* or *705* of the *Federal Rules of Evidence.*

B. Except as otherwise stipulated or directed by the court, this disclosure shall, with respect to a witness who is retained or specially employed to provide expert testimony in the case or whose duties as an employee of the party regularly involve giving expert testimony, be accompanied by a written report prepared and signed by the witness. The report shall contain a complete statement of all opinions to be expressed and the basis and reasons therefor; the data or other information considered by the witness in forming the opinions; any exhibits to be used as a summary of or support for the opinions; the qualifications of the witness, including a list of all publications authored by the witness within the preceding ten years; the compensation to be paid for the study and testimony; and a listing of any other cases in which the witness has testified as an expert at trial or by deposition within the preceding four years.

C. These disclosures shall be made at the times and in the sequence directed by the court. In the absence of other directions from the court or stipulation by the parties, the disclosures shall be made at least 90 days before the trial date or the date the case is to be ready for trial or, if the evidence is intended solely to contradict or rebut evidence on the same subject matter identified by another party under paragraph (2)(b), within 30 days after the disclosure made by the other party. The parties shall supplement these disclosures when required under subdivision (e)(1).

FRCP *Rule 705* addresses the disclosure of facts or underlying data on which an expert forms his/her expert opinion. The expert may testify as to an opinion without first introducing the underlying facts or data used to develop that opinion, unless otherwise directed by the court. However, the expert may be required to later disclose the underlying facts or data they used during a cross-examination.

Under FRCP *Rule 26(a)(2)(B)*, experts are required to provide a written expert witness disclosure report. This report, which must be prepared and signed by the expert witness, must contain the following.

- ◆ Complete statement of all opinions to be expressed along with the basis for these opinions
- ◆ Data or information considered by the expert used in forming the opinions
- ◆ Exhibits planned to be used in summarizing or supporting the opinions

- ◆ Qualifications of the expert witness, including a list of all publications authored within the preceding 10 years
- ◆ Compensation to be paid for the report and testimony
- ◆ Listing of all cases in which the expert has testified in trial or deposition within the preceding 4 years

The outlined format of this written expert disclosure report varies. Table 1.2 is an example of an outline for a summary disclosure letter sent by an expert to an attorney. Table 1.3 lists areas to cover when qualifying as a fire expert (Beering 1996).

TABLE 1.2 ◆ Sample Expert Disclosure

I. Qualifications

I,_____, P.E., am a consulting engineer and have been retained as an expert witness by the Defendant. A copy of my résumé is attached as Exhibit A. In the course of my professional practice, I have over XX years of experience as a specialist in fire scene investigation and reconstruction. In the past XX years, I have testified as an expert witness in XX cases. Information about those cases is listed in Exhibit B. In the past XX years, I have coauthored XX textbooks, XX chapters in textbooks, and XX articles; and the titles are listed in my vitae in Exhibit C.

II. Scope of Engagement

I am retained by the Defendant in this case to evaluate the initial growth and development of a fire on XXXXXX XX, XXXX, at a XXXX located at XXXX Street, XXXXX, XXXXX County, XXXXXX; the impact of smoke detectors to notify the victims of the incipient fire; and the impact of conditions and human behavior during the fire which claimed the life of XXX X. XXX. The analysis performed for this case conforms to the generally accepted fire protection engineering methods of fire scene investigation and reconstruction.

III. Information Reviewed

To prepare my opinion, I reviewed the following documents: (LIST)

IV. Methodology Used

The methodology used to form my opinion is based upon the use of the scientific method along with other accepted fire engineering principles and practices.

V. Summary of Opinion

The analysis and opinions expressed in this report are based on my knowledge of the facts and information made available to date. If additional information becomes available which has a bearing on these opinions as expressed below, I will amend or supplement these opinions appropriately. Based on my understanding of the issues in the complaint and on the scope of this engagement, it is my opinion that: (LIST)

VI. Compensation

I have billed $X,XXX to the attorneys representing the Defendant for my services through the date of this report. Billings for future services, including testifying at depositions and trial, will be at $XXX.XX per hour.

TABLE 1.3 ◆ Areas to Cover When Qualifying as a Fire Expert

Category	Example
Identification—Who is the expert and how long has he or she worked in the field?	Name Title, department Employment—current and previous
Education and training—What formal specialized training has the expert received?	Formal education Fire training academies National Fire Academy FBI Academy Federal Law Enforcement Training Center Annual international and state seminars Specialized training
Certifications—Is the expert licensed, certified, or otherwise credentialed in his or her field?	Fire investigator Police officer Fire protection specialist Professional engineer Instructor
Experience—How has the expert acquired firsthand knowledge and experience about fire?	Fire suppression Fire investigations Fire testing Laboratory
Prior testimony—Has the expert given prior testimony?	Criminal, civil, and administrative courts Lay and expert witness Jurisdictions (state, federal, international)
Professional associations—Of which professional associations is the expert a member, and what standing (e.g., fellow, life member)?	Professional fire investigation Fire and forensic engineering
Teaching experience—Has the expert delivered certified and credible training to others in the field?	Local, state, federal, international schools and organizations National academies Colleges and universities
Professional publications—Has the expert written and published peer-reviewed quality articles?	Peer-reviewed articles, technical papers Books National standards
Awards and honors—What significant honors and awards has the expert received?	Local, state, federal, and international professional organizations and societies

Source: Summarized and updated from Beering 1996.

◆ 1.4 RECOGNITION OF FIRE INVESTIGATION AS A SCIENCE

DAUBERT CRITERIA

The recent trend in U.S. Supreme Court decisions is that the Court is continuing to define and refine the admissibility of expert scientific and technical opinions, particularly as they relate to fire scene investigations. These decisions affect how expert testimony is accepted and interpreted.

TABLE 1.4 ◆ Criteria and Pertinent Issues Examined by *Daubert*

Criteria	Pertinent Issues Examined
Testing	Has the methodology, theory, or technique been tested?
Peer review and publication	Has the methodology, theory, or technique been subjected to peer review and publication?
Error rates and professional standards	What are the known or potential error rates of this methodology, theory, or technique?
	Does the methodology, theory, or technique comply with controlling standards, and how are those standards maintained?
General acceptance	Is the methodology generally accepted in the scientific community?

The bottom line is that although much controversy exists over these Court decisions, fire and explosion investigation is being considered more as a science and less than an art. This is particularly true when the scientific method is combined with relevant engineering principles and research in providing expert witness testimony (Chesbro 1994; Ogle 2000).

A judge has the discretion to exclude testimony that is speculative or based on unreliable information. In the case *Daubert v. Merrell Dow Pharmaceuticals* (*Daubert* 1993), the Court placed the responsibility on a trial judge to ensure that expert testimony was not only relevant but also reliable. The judge's role is to serve as a "gatekeeper" to determine the reliability of a particular scientific theory or technology. The Court defined four criteria to be used by the gatekeeper to determine whether the expert's theory or underlying technology should be admitted. *Daubert* allows the judge to gauge whether the expert testimony aligns with the facts of the case as presented. The central issue is often whether the expert has followed an identified, accepted, and peer-reviewed method as the basis for his or her conclusion.

In a more recent decision, *Kumho Tire Co. Ltd. v. Carmichael* (1999), the Court applied the four-guideline criteria to expert testimony to determine whether it was based on science or experience. The four-guideline *Daubert* criteria consist of testing, peer review and publication, error rates and professional standards, and general acceptability, as shown in Table 1.4, which lists the pertinent issues examined by each criterion. In short, expert testimony must rely on a balance of valid peer-reviewed literature, testing, and acceptable practices and a demonstration that those practices were followed if it is to be considered credible by the courts.

A recent federal case citing *Daubert* and *Rules 702* and *703* allowed the expert testimony of Dr. John DeHaan pertaining to the origin and cause of an arson fire based on his review of case materials relied on by other experts (*United States of America v. Ruby Gardner* 2000). Dr. DeHaan relied on a review of reports, photographs, and third-party observations, some of which may not have been directly admissible as evidence.

The court determined that this procedure was reliable and appropriate, confirming the conviction. Furthermore, the court drew a parallel to other testimony, including hearsay and third-party observations used by an arson investigation expert in forming an opinion or the ability of a psychiatrist to testify as an expert by relying only on staff reports, interviews with other doctors, and background information.

In *U.S. v. Downing* (1985), a Third Circuit case, both *Daubert* and *Downing* factors were used in determining whether an expert's methodology was reliable. *Downing* factors consider (1) the relationship of the technique to methods that have been established to be reliable; (2) the qualifications of the expert witness testifying based on methodology; and (3) the nonjudicial issues to which the method has been put.

FRYE STANDARD

At present some 16 states and the District of Columbia (Gianelli 2007, 1–15) rely on the *Frye* test (or a local variant of it such as the *Kelly-Frye* test in California, *Reed-Frye* in Maryland, and *Davis-Frye* in Michigan) as the standard for admissibility of expert testimony instead of the *Daubert* standard. The original *Frye* decision was reached in 1923 over the issue of admissibility of the (then) new scientific discipline of polygraph examinations, for which a limited number of adherents maintained its reliability in the face of general skepticism. The U.S. Court of Appeals for the District of Columbia stated:

> Just when a scientific principle or discovery crossed the line between experimental and demonstrable stages is difficult to define. Somewhere in this twilight zone, the evidential force of the principle must be recognized and, while courts will go a long way in admitting expert testimony deducted from a well-recognized scientific principle or discovery, the thing from which the deduction is made must be sufficiently established to have gained general acceptance in the particular field in which it belongs." (*Frye v. United States* 1923)

The *Frye* decision established the judge to be the gatekeeper to exclude junk science. It has been suggested that the reasoning behind the "general acceptance" standard is to assist the trier of fact in six ways:

1. Promote a degree of uniformity of decision across the judiciary
2. Avoid the interjection of a time-consuming and often misleading determination of the reliability of a scientific technique during litigation
3. Assure that the scientific evidence introduced will be reliable and thus relevant
4. Ensure that a pool of experts exists who can critically examine the validity of a scientific determination
5. Provide a preliminary screening to protect against the natural inclination of the jury to assign significant weight to scientific techniques presented as evidence where the trier of fact is in a poor position to evaluate the reliability of that evidence accurately
6. Impose a threshold standard of reliability for scientific expert testimony in new areas when cross-examination by opposing counsel may be unable or unlikely to bring inaccuracies to light (Graham 1994)

This standard is a conservative one, when strictly applied, requiring new technologies or techniques to have been in use long enough for them to have been tested by a number of practitioners. It is the offering party's burden to prove the concept of general acceptance. There are no methods suggested in *Frye* by which a judge should determine general acceptance. The testimony of a single expert (particularly one who might have a bias or vested interest) is rarely acceptable. Presentation of multiple peer-reviewed, published papers, testimony of an independent expert retained by the court, or testimony of multiple experts have all been used.

"The trial judge determines whether the *Frye* test has been satisfied under a rule comparable to *Federal Rules of Evidence 104(a)*. The 'preponderance' standard applies" (Gianelli 2007, 1–5). Some courts have applied *Daubert*-type considerations to establishing "hallmarks" such as independent peer review, thorough testing, establishing error rates, or promulgation of objective standards applicable to the technique. One general principle is that general acceptance is met if the use of the technique is supported by a clear majority of the scientific community.

Another major issue with *Frye* is the difficulty in establishing what is the "relevant" scientific field or field in which the technique falls. Fire investigation and reconstruction both rely on multiple fields—fire dynamics, fire engineering, chemistry, and, in some cases, human behavior; they do not fall within a single academic discipline or professional discipline. The judge may turn to the published fire investigation literature or to resources in fire engineering, fire safety, or chemistry (including local college instructors who may or may not have the necessary familiarity with the concepts). Prior judicial decisions have also been used as a basis for such decisions (even when they do not necessarily apply to the technique at hand).

Some courts have extended the *Frye* standard beyond the reliability of the technique to considerations of the underlying theory or theories on which the technique is based. That could cause real problems for fire investigators if called on to prove the validity of underlying theories for events such as V patterns, spalling, failure of glass, flashover, or halo or ring patterns whose underlying theories, if known, are couched in the fire dynamics and materials science disciplines.

Other courts have avoided the *Frye* issue by deciding that some evidence does not fall within the "scientific" realm but rather relies on demonstrable evidence (physical models, photographs, X-rays, and the like). In those cases the jury does not have to place blind faith in the assurances of a testifying expert but can verify the results by their own observations. Such reliance raises the specter of the multitude of courtroom tricks of past decades—in which substitutions, mischaracterization, and even sleight of hand had been used by unscrupulous counsel to sway a jury.

IMPACT OF *NFPA 921* ON SCIENCE-BASED EXPERT TESTIMONY

The importance of *NFPA 921* has also been cited along with *Daubert* as an interlinking element of expert testimony, since it establishes guidelines for the reliable and systematic investigation or analysis of fire and explosion incidents (Campagnolo 1999; Icove and DeHaan 2004). Several clusters of recent federal court opinions and rules fall into the following areas.

- Use of investigative protocols, guidelines, and peer-reviewed citations
- Methodological explanations for burn patterns
- Qualifications to testify

Professional education is paramount in any profession. It is incumbent on all professional fire investigators continuously to read and keep abreast of all relevant fire, engineering, and legal publications and critically evaluate their conclusions with this ever-changing knowledge. As with science, today's knowledge may change with new developments and impact on the reviews of an established hypothesis.

References to *NFPA 921*.   In 1997, a federal judge upheld a motion by a defendant insurance company in a civil case barring the plaintiff's expert fire investigator from mentioning *NFPA 921* during his testimony. *NFPA 921* was first published in 1992, after the November 16, 1988, fire in question. The judge requested that the plaintiffs provide him a copy of *NFPA 921*. The judge found that under *Rule 703* of the *Federal Rules of Evidence,* any reference to *NFPA 921* would have little probative value and would confuse the jury (*LaSalle National Bank et al. v. Massachusetts Bay Insurance Company et al.* 1997). This ultraconservative finding has not been widely accepted.

In a November 18, 1999, motor vehicle accident, an individual was trapped for approximately 45 minutes in the front passenger seat after the car collided with a utility pole. Eyewitnesses reported a bluish flickering and fire that engulfed the vehicle's interior and severely burned the trapped passenger. A product liability lawsuit against the vehicle's manufacturer alleged a defective design, since a safety device would have disconnected the battery after the collision (*John Witten Tunnell v. Ford Motor Company* 2004). The plaintiffs alleged that the energized wiring harness suffered a high-resistance fault, which could have caused a fire without blowing a fuse. The judge reviewed and overruled motions to exclude the testimony of fire origin experts and repeatedly cited the applicability of the *NFPA 921* methodology to this case.

Investigative Protocols.   In a 1999 federal criminal case, the defendant moved that the court exclude a state fire marshal's testimony on the incendiary nature of a fire, arguing that the testimony did not meet the standards for admissibility under *Daubert* (*U.S. v. Lawrence R. Black Wolf, Jr.* 1999). The defendant argued that the testimony should be excluded because the investigator did not arrive at the scene until 10 days after the fire occurred and did not take samples for laboratory analysis, the theory of the fire's origin and cause had not been subject to accurate and reliable peer review, and no scientific methods or procedures were used to reach the conclusions. The defendant's Exhibit A was the 1998 edition of *NFPA 921*.

In this case, the U.S. magistrate determined that the investigator's proffered testimony was reliable, since he indeed used an investigative protocol consistent with the basic methodologies and procedures recommended by *NFPA 921,* and he passed the requisite scrutiny demanded of fire origin and cause experts (Figure 1.4). The magistrate

FIGURE 1.4 ◆ An investigator's proffered testimony may be deemed reliable if it is in accordance with an investigative protocol, such as the one published by the U.S. Department of Justice, that cites methodologies and procedures recommended by *NFPA 921. Courtesy of U.S. Department of Justice;* www.usdoj.gov.

ruled that the proffered testimony was reliable according to the first part of the *Rule 702 Daubert* test.

Regarding the relevancy test, the magistrate determined that the investigator's testimony would help a jury understand the circumstances surrounding the fire's origin and cause. The magistrate further found that *Rule 703* had been met, since the investigator's proffered testimony relied on facts and data normally collected by experts when forming opinions as to the fire's origin and cause. The magistrate concluded that the defendant would have ample opportunity to address the expert's opinion during cross-examination, through presentation of contrary evidence, and in instructions to the jury.

Use of Guidelines and Peer-Reviewed Citations.   In a case involving a November 16, 1994, residence fire, a number of independent investigators attempted to determine the exact origin and cause of the fire. One investigator, an electrical engineer, offered the opinion that a television set located in the basement family room caused the fire. The plaintiffs sued the television's manufacturer, claiming product liability, negligence, and breach of warranties (*Andronic Pappas et al. v. Sony Electronics, Inc. et al.* 2000).

A federal judge hearing this case held a 2-day *Daubert* hearing and concluded that the one investigator's causation testimony was inadmissible. The judge cited that the issue in this case was the second *Rule 702* factor, that an expert's opinion should be based on a reliable methodology. The judge noted that the investigator did not use a fixed set of guidelines in determining the cause of the fire. Notably, the investigator did confirm that even though he was aware of the existence of established guidelines in *NFPA 921* and *Kirk's Fire Investigation,* he relied on his own experience and knowledge. During the hearing, the plaintiffs did not submit any books, articles, or witnesses on proper fire causation techniques other than the testimony of one investigator. The judge specifically commented on the fact that the one investigator referred to his reliance on a burn pattern inside the television set. No citations were made to peer-reviewed sources to support this opinion.

In a products liability case arising from a fire on July 16, 1998, which destroyed a business, the plaintiff's only expert placed the origin of the fire to be a product manufactured by the defendant (*Chester Valley Coach Works et al. v. Fisher-Price Inc.* 2001). Based on a *Daubert* hearing held June 14, 2001, the court excluded the expert's testimony as to his opinion regarding the fire's origin and cause. At the hearing, the expert indicated his conclusions were based primarily on his experience and education and not based on any testing, experimentation, or generally accepted texts or treatises. Several of the steps he undertook in his investigation were contrary to the *NFPA 921* guidelines.

In a federal court decision on April 16, 2003, a plaintiff's *Daubert* motion to exclude the testimony of a defendant's expert witness was overruled (*James B. McCoy et al. v. Whirlpool Corp. et al.* 2003). The plaintiffs experienced a fire at their residence on February 16, 2000, alleging that it was caused by a dishwasher. The plaintiff pointed out that the expert's report on the origin and cause of the fire did not explicitly cite the methodology set forth in *NFPA 921.* In its decision, the court noted that the expert did not need to cite *NFPA 921*, since he had adequately provided a complete statement of all opinions and a substantive rationale for the basis of these opinions.

Peer Review of Theory.   In a case involving a residence fire that occurred around December 22, 1996, an insurance company hired an investigator to determine the cause of the fire, which all parties agreed had originated in an area to the right of a fireplace firebox. In dispute by the defendants was the cause of the fire, and they filed a motion to exclude the testimony of the plaintiff's expert (*Allstate Insurance Company v. Hugh Cole et al.* 2001).

Even though the plaintiff filed his complaint on December 21, 1998 (prior to the December 1, 2000, effective date of the amended *Rule 702*), the judge applied the amended *Rule 702* to the admissibility of the testimony (FRCP 2000). The plaintiff asserted that in his trial testimony the investigator would rely on data from *NFPA 921* to support his opinion.

The opinion of the plaintiff's investigator was that heat from a metal pipe ignited adjacent combustibles. The judge noted that the theory, based on *NFPA 921,* had already been subject to peer review, was generally accepted in the scientific community, and met the sufficient reliability guidelines to satisfy the admissibility requirements of *Rule 702.* Peer review can also include consultation with other qualified fire experts, both scientific and investigative.

Methodology Needed.   On September 13, 1996, a fire in a residence started in the corner of a kitchen containing a dishwasher, toaster oven, and microwave oven. The municipal fire marshal concluded that the fire was caused by the microwave oven, whereas in a federal civil case, the plaintiff's expert asserted that a defective toaster oven caused the fire (*Jacob J. Booth and Kathleen Booth v. Black and Decker, Inc.* 2001).

A senior federal judge held a *Daubert* hearing in which evidentiary and testimonial records were reviewed and ruled that the opinion of the plaintiff's investigator was not admissible under *Rule 702.* The judge concluded that the plaintiff's investigator did not provide sufficient reliable evidence to support the methodology for investigating the cause of the fire. The judge also noted that the comprehensive nature of *NFPA 921* contained a methodology that could have supported the opinion and tested the hypothesis for the fire's cause.

On December 8, 1999, a circuit court judge in the State of Michigan granted a plaintiff's motion *In Limine* to exclude the testimony of the defendant's expert witness (*Ronald Taepke v. Lake States Insurance Company* 1999). In testimony and written reports, the defendant's expert admitted the authoritative nature of *NFPA 921,* yet admitted that he did not follow its accepted methodology. Furthermore, the methodology used to conclude that the fire was arson is not recognized in any reliable source of literature in the fire investigation field. In other areas, the expert admitted that he speculated that high-temperature accelerants were used without a reliable evidence basis for the opinion.

In a federal civil case opinion on March 28, 2002, involving automobile ignition switch fires, a U.S. District Court judge examined the proposed use of a statistical fire loss database of vehicle fires in litigation. Also examined in detail was the importance for reliance on *NFPA 921* in fire investigations when forming a hypothesis using the scientific method (*Snodgrass et al. v. Ford Motor Company and United Technologies Automotive, Inc.* 2002). Note that since this court case was within the Third Circuit, a determination of whether an expert's methodology was reliable used the combined *Daubert* and *Downing* factors. In *United States of America v. Downing* (1985), the court considered the admissibility of expert testimony concerning the reliability of eyewitness identification.

In a federal civil case opinion on July 10, 2002, an insurance company sued to recover subrogation paid by the company for fire damage to the property of its subrogors. The judge denied the defendant's motion to exclude the trial testimony of the plaintiff's origin and cause expert witness. The judge noted that the expert's testimony was the product of reliable principles based on the scientific method as outlined in *NFPA 921* and that the expert applied these principles and methods in a reliable and relevant manner to satisfy the admissibility requirements of *Daubert* and *Rule 702* (*Royal v. Daniel Construction* 2002).

In a state court of appeals opinion decided June 16, 2005, the judge ruled that the testimony of the appellee's independent fire investigator was correctly permitted under *Daubert* as to the origin and cause of the fire (*Abon Ltd. et al. v. Transcontinental Insurance Company* 2005). The court noted that the independent investigator followed the *NFPA 921* methods and principles to reach his opinion as to the incendiary nature of the fire's origin and cause.

In an apartment fire on October 23, 2000, the occupant died and her personal representatives and relatives brought an action against the manufacturer of an electric blanket, alleging it had a defective safety circuit that caused the deadly fire (*David Bryte v. American Household Inc.* 2005). The fire marshal who determined the origin and cause of the fire was not certified as a fire investigator, yet he had attended training in the fire and explosion investigation field. He traced the fire's path to the deceased's left side, took photographs, made a fire scene sketch, and took oral statements from witnesses. He did not inspect the electrical wiring or take any evidence, noted an electrical cord draped across the victim's body, and concluded the cause of the fire to be the improper use of an electric blanket. During the trial on December 8, 2003, under *Daubert* the fire marshal and another expert were not allowed to provide expert testimony because they did not have a sufficiently reliable basis for their testimony.

In a federal court of appeals ruling in a product liability case alleging responsibility in a fire-related death, both the magistrate judge and district court concluded by the *Daubert* and *Federal Rules of Evidence* standards that the methodologies employed by the plaintiff's expert witnesses were too unreliable to serve as the basis for admissible testimony (*Bethie Pride v. BIC Corporation, Société BIC, S.A.* 2000). The case alleged that the death of a 60-year-old man who sustained 95 percent burns was caused by an explosion of a disposable lighter he was carrying while inspecting a pipe behind his house. The court noted that the plaintiff's experts failed to conduct timely, replicable laboratory experiments demonstrating how the explosion was consistent with a manufacturing defect in the lighter. One of the plaintiff's experts testified he began a laboratory experiment to test his hypothesis but "chickened out and shut the experiment down."

In a federal case in which an insurance company attempted to recover subrogation for fire damage to the property of one of its insureds, the defendant construction company filed a motion to preclude the trial testimony of the plaintiff's fire origin and cause expert (*Royal Insurance Company of America as Subrogee of Patrick and Linda Magee v. Joseph Daniel Construction, Inc.* 2002). The plaintiff had hired a construction company to work on a garage on their property on December 14, 1998, which involved the use of an acetylene torch to install new beams above the second floor. During the operation of the torch, two fires occurred and were extinguished. Approximately 14 hours after the construction company left, a fire was discovered in the garage, which sustained damage.

Approximately 1 year after the fire, the plaintiffs hired an expert to investigate the origin and cause of the fire. Based on his investigation, the expert concluded that the fire was caused by the careless use by the construction company employees of the welding and cutting equipment. The court noted that the expert's investigation was conducted in accordance with methodology set forth by *NFPA 921*. To develop his hypothesis that the garage fire was caused by smoldering ignition by molten slag dropped by careless construction workers, the investigation used data consisting of information gleaned from copies of photographs, insurance investigative and engineering reports, reviews of depositions and his own witness interviews, and eliminated other common causes for the fire. The court ruled, citing the *Daubert* standard, that the expert's testimony was relevant in establishing a direct relationship between the proffered theory that molten slag resulted from the carelessness of the defendant caused the fire.

An unpublished Third Circuit Court of Appeals case concluded that a district court did not err in granting and excluding the testimony of a plaintiff's expert witness (*State Farm and Casualty Company as subrogee of Rocky Mountain and Suzanne Mountain v. Holmes Products; J.C. Penney Company, Inc.* 2006). On April 5, 1999, a fire destroyed a residence. The insurance company's investigator placed the fire's area of origin where a halogen lamp (which lacked a safety guard to prevent contact with the high-temperature portions from the lamp) ignited nearby curtains, clothing, or other flammable materials. The investigator, after eliminating all other potential causes, concluded that the fire was caused by nearby draperies coming into contact with the defective energized lamp. The investigator further hypothesized that the dog belonging to the residence's owners might have accidentally pulled the window draperies over the energized lamp or knocked the lamp as to come into contact with the draperies. The judge concluded that the fire investigator's testimony did not satisfy *Daubert* because the conclusion on causation was based on assumptions and was not supported by any methodology.

In a subrogation lawsuit, an insurance company brought an action against a clothes dryer manufacturer to recover damages and relief in connection with 23 fires allegedly caused by a design defect (*Travelers Property and Casualty Corporation v. General Electric Company* 2001). A *Daubert* hearing was necessary when the defendant manufacturer submitted a motion *In Limine* to prevent the testimony of the plaintiff's expert, who had written and issued a three-page report alleging that accumulated lint in an undetectable location could be ignited by the dryer's heating elements. The judge overruled the motion finding that the expert's proffered opinion testimony met the requirements of both *Rule 702* and *Daubert,* that the opinion was consistent with the principles of *NFPA 921* and the scientific method, and his qualifications were sufficient. However, the judge found that the plaintiff's three-page expert disclosure report amounted to bad faith, and he both sanctioned and ordered the plaintiff to reimburse the defendant for one-third of its costs and expenses for taking the first 12 days of the expert's deposition and permitted up to another 2 more days of depositions.

Methodological Explanations for Burn Patterns.   In the case of an April 23, 1997, building fire, a federal U.S. magistrate denied the plaintiff's motion to bar opinion testimony as to the origin and cause of the fire at the time of trial (*Eid Abu-Hashish and Sheam Abu-Hashish v. Scottsdale Insurance Company* 2000). In this case, both a municipal fire department investigator and a private insurance investigator concluded

FIGURE 1.5 ◆ Investigators should be able to provide adequate and methodical explanations of the fire pattern analysis methods they use to reach conclusions as to the origin of a fire. *Courtesy of D. J. Icove.*

from their examinations of the scene that the fire was incendiary in origin. No physical evidence was taken for laboratory examination.

The plaintiff argued that these investigators were not reliable and their testimonies were inadmissible under *Rule 702* because they did not use the scientific method as outlined in *NFPA 921* and relied on only physical evidence observed at the scene. The case also contained parallels with the *Benfield* case (*Michigan Miller's Mutual Insurance Company v. Benfield* 1998). In denying the plaintiff's motion, the federal magistrate noted in his decision that the investigators were able to provide an adequate methodological explanation for the analysis of burn patterns that led to how they reached their conclusion as to the fire's incendiary origin (Figure 1.5).

A products liability case arising from a fire on October 16, 2000, which destroyed a video rental store, two of the plaintiff's fire causation experts determined the origin and cause of the fire to be a copier machine (*Fireman's Fund Insurance Company v. Canon U.S.A., Inc.* 2005). The two experts relied on burn patterns inside the copier and testing of an exemplar heater assembly for the copier. The federal appeals court upheld the district court's opinion that documentation of the experimentation by the plaintiff's experts did not meet the standards of *NFPA 921*, as well as being unreliable and potentially confusing to a jury. Tests must be carefully designed to replicate actual conditions, conducted properly, and documented thoroughly.

A state civil court opinion and order entered February 1, 2006, precluded *In Limine* the testimony of the defendant's expert alleging the incendiary nature of a fire he investigated (*Marilyn McCarver v. Farm Bureau General Insurance Company* 2006). On August 16, 2003, a fire damaged the plaintiff's residence. A report authored August 23, 2003, by the defendant insurance company's expert concluded that the fire was incendiary in origin after visiting and viewing the remains of the residence, taking photographs, taking samples for laboratory examination, and performing other techniques. The state court noted that although the defendant's expert stated his investigation complied with *NFPA 921*, several procedural areas were not followed. The expert failed to document any depth of char physical measurements; failed to document his vector analysis, which formed the basis for flame patterns; and failed to retrieve samples submitted for chemical testing, which resulted in their destruction. After a *Daubert* hearing, the court concluded that a fatal flaw in the expert failure to collect and memorialize depth of char data undermined his methodology for determining the origin and cause of the fire.

Methodology and Qualifications. In the case of a July 1, 1998, residential fire, which resulted in a federal lawsuit for product liability, the judge granted the defendant's motion to exclude the testimony of the electrical engineer citing *Daubert*, *Rule 702*, and *Rule 704* (*American Family Insurance Group v. JVC America Corp.* 2001). In this case, an insurance company investigator called on an electrical engineer to remove and examine the charred remains of a bathroom exhaust fan, clock, lamp, timer, compact disc player, computer with printer and monitor, ceiling fan, power receptacle, and power strip. The engineer came to an opinion that a defect in the compact disc player caused the fire. His observations were based on burn patterns in the room, on appliance remains, and on his experience, education, and training. In his ruling, the judge noted that the training and experience of the engineer did not qualify him to offer an analysis of burn patterns and theory of fire origin. Furthermore, the judge noted that the engineer did not use the scientific methodology recommended by *NFPA 921* to form a hypothesis from the analysis of the data, nor did he satisfy the requirements for expert testimony under *Daubert*.

In a case of a March 1, 2001, fire that originated in a building's janitorial closet was alleged to have been caused by a short circuit in a contaminated electrical bus duct (*103 Investors I, L.P. v. Square D Company* 2005). The plaintiff retained a licensed mechanical engineer who was also a certified fire investigator to investigate the origin and cause of the fire. The engineer testified that he investigated the fire in accordance with the scientific method protocol set forth in *NFPA 921*. In an opinion decided May 10, 2005, the federal judge ruled that the engineer's testimony could include that the short circuit resulted from the presence of contaminants. However, since the engineer failed to follow established methods of inquiry for hypothesis testing of how the contamination actually occurred, he could not testify that the fire specifically resulted from short circuiting in the bus duct.

In a case before a U.S. magistrate court a third-party defendant's motion to exclude the testimony of the plaintiff's expert was denied (*TNT Road Company, et al. v. Sterling Truck Corporation, Sterling Truck Corporation, Third-Party Plaintiff v. Lear Corporation, Third-party Defendant* 2004). The third-party defendant alleged that the witness was not qualified to render an expert opinion, lacked a college degree, was not a certified fire investigator, was not a licensed private investigator, did not interview every conceivable witness, quickly came to his conclusion, relied heavily on theory and circumstantial evidence, had testified only once as an expert witness, and that his investigation was too faulty to be admissible under *Federal Rule of Evidence 702*. After significant review, the U.S. magistrate ruled that the plaintiff's witness was indeed qualified to present expert testimony on the origin and cause of the vehicle fire.

The judge noted that the witness had demonstrated sufficient knowledge, skill, training, and education to qualify as an expert and that his methodology appeared reliable. The judge decided that the witness's investigation substantially complied with *NFPA 921* in meeting with the vehicle's owners, systematically examining and photographing the scene along with all the components that may have been potential causes of the fire, obtained and reviewed the maintenance records, systematically removed and photographed the fire debris, formed no immediate opinions as to the cause of the fire, and narrowed his focus onto a suspected ignition switch only after several hours. Once he suspected the switch may have been the cause of the fire, the expert suspended his investigation to preserve the evidence until other interested parties could be available to also inspect the vehicle.

On February 15, 2001, a fire broke out in the kitchen of a residence in the vicinity of a coffeemaker, causing extensive fire, smoke, and water damage. The insurance company paid the claim and pursued subrogation against the manufacturer of the coffeemaker (*Vigilant Insurance v. Sunbeam Corporation* 2005). The testimonies of several witnesses were limited based on a *Daubert* hearing on March 1, 2004. The testimony of the investigator was found inadmissible under *Rule 702* because his opinion was not based on data that he might have gleaned from testing the coffeemaker and toaster cords, safety devices, or the fluorescent light. He was able to testify under *Rule 701* as a lay witness as to his personal observations that the coffeemaker was at the base of the V burn pattern. During the hearing, the court found that another witness's burn test was simply a reenactment of the kitchen fire and was not substantially similar owing to differences in the heat applied to the coffeemaker, the amount of water in the coffeepot, and the thickness of the countertop. A jury trial held from March 2, 2005, to March 17, 2005, resulted in a verdict in favor the of the plaintiff insurance company.

In a federal civil lawsuit, the plaintiffs sought to hold the defendant, the manufacturer of a freezer, liable for damages from a fire that allegedly originated in the freezer (*Mark and Dian Workman v. AB Electrolux Corporation et al.* 2005). An investigator for the plaintiff's insurance company failed to determine what caused the fire, using *NFPA 921* to guide his investigative process. The investigator then arranged for a mechanical engineer, who conducted a further destructive examination and determined that the fire originated inside the freezer as the result of a short circuit in the evaporator compartment. The court's review of the engineer's testimony based on the criteria set forth in *Daubert* and *Rule 702* found the engineer's testimony reliable. However, the engineer did not mention or describe a defect in the freezer.

In a federal civil lawsuit, a June 1, 2001, fire damaged the home, garage, and property of the plaintiffs (*Theresa M. Zeigler, individually; and Theresa M. Zeigler, as mother and next friend of Madisen Zeigler v. Fisher-Price, Inc.* 2003). The fire allegedly originated in a toy vehicle parked in the garage and plugged into a charger. Two expert witnesses looked into the causation of the fire. The court determined that the insurance company investigator followed *NFPA 921*, an appropriate and generally accepted methodology, and found that his opinions were reliable. The second investigator was an engineer who did not perform an origin and cause analysis but had an opinion that the fire's ignition source was located on a connector to the toy vehicle. The court ruled that the engineer could not give an opinion as to the manufacturer's record keeping, or the origin or cause of the fire.

AUTHORITATIVE SCIENTIFIC TESTING

The *Daubert* criteria list testing as one of the primary considerations in evaluating and demonstrating the reliability of a scientific theory or technique. In fire investigations, the scientific theory relies heavily on established and proven aspects of nature, such as those demonstrated in the science of fire dynamics. The combination of generally accepted scientific principles and research, when supported by empirical fire testing, can yield valuable insight into fire scene reconstruction. Much of the accumulated knowledge of fire scene reconstruction is impossible for individual investigators to gain from personal experience alone, regardless of the skill and diligence with which their analyses are conducted.

There are case citations that reliance based on tests conducted prior to the event is less likely to be biased. Some experts develop opinions based solely on the results of tests conducted specifically to support expert testimony. The *Daubert II* (1995) court placed greater weight on testimony based on preexisting research that used the scientific method, as it is considered more reliable (Clifford 2000).

Scientifically conducted test results can normally be reproduced and their error rates validated or estimated. When this is true, the results are considered to be *objective*, neither prejudiced nor biased by the researcher. An example of peer-reviewed published guidelines is incorporated in the accepted fire testing procedures developed and maintained by the Committees on Fire Standards (E05) and Forensic Sciences (E30) of the ASTM (1999). Table 1.5 lists many of the ASTM testing standards applicable to fire investigations. Thorough documentation of all phases of testing will help the courts evaluate the reliability and admissibility of the test results.

TABLE 1.5 ◆ Testing Standards Applicable to Fire Investigations

Technique	Instruction	Standard
Reporting	Reporting opinions of scientific or technical experts	ASTM E 620
	Standard practice for the evaluation of technical data	ASTM E 678
	Examining and testing items that are or may become involved in litigation	ASTM E 860
	Standard practice for reporting incidents	ASTM E 1020
	Terminology of technical aspects of products liability litigation	ASTM E 1138
	Collection and preservation of information and physical items by a technical investigator	ASTM E 1188
	Physical evidence labeling and documentation	ASTM E 1459
	Receiving, documenting, storing, and retrieving forensic evidence	ASTM E 1492
Laboratory testing	Debris Samples by Steam Distillation	ASTM E 1385
	Debris Samples by Solvent Extraction	ASTM E 1389
	Debris Samples by Gas Chromatography	ASTM E 1387
	Debris Samples by Headspace Vapors	ASTM E 1388
	Debris Samples by Passive Headspace	ASTM E 1412
	Debris Samples by GC-MS	ASTM E 1618

(Continued)

TABLE 1.5 ◆ Testing Standards Applicable to Fire Investigations (*continued*)

Technique	Instruction	Standard
Flash and fire point	Tag Closed Tester	ASTM D 56
	Cleveland Open Cup	ASTM D 92
	Pensky-Martens Closed Tester	ASTM D 93
	Tag Open Cup Apparatus	ASTM D 1310
	Setaflash Closed Tester	ASTM D 3278
	Ignition Temperature of Plastic	ASTM D 1929
Autoignition temperature	Liquid Chemicals	ASTM E 659
Heat of combustion	Hydrocarbon Fuels by Bomb Calorimeter	ASTM D 2382
Flammability	Apparel Textiles	ASTM D 1230
	Apparel Fabrics by Semi-restraint Method	ASTM D 3659
	Finished Textile Floor Covering	ASTM D 2859
	Aerosol Products	ASTM D 3065
	Chemical Concentration Limits	ASTM E 681
	Standard Methods of Fire Tests for Flame Propagation of Textiles and Films	NFPA 701
	Field Flame Test for Textiles and Flames	NFPA 705
Cigarette ignition	Mockup Upholstered Furniture Assemblies	ASTM E 1352
	Upholstered Furniture	ASTM E 1353
Surface burning	Building Materials	ASTM E 84
Fire tests and experiments	Roof Coverings	ASTM E 108
	Floor/Ceiling, Floor/Roof; Walls, Columns	ASTM E 119
	Room Fire Experiments	ASTM E 603
	Measurement of Gases Present or Generated	ASTM E 800
	Windows	ASTM 2010
	Doors	ASTM 2074
Critical radiant flux	Floor Covering Systems	ASTM E 648
Release rates	Heat and Visible Smoke	ASTM E 906
	Heat and Visible Smoke Using an Oxygen Consumption Calorimeter	ASTM E 1354
Rate of pressure rise	Combustible Dusts	ASTM E 1226
Electrical	Dielectric Withstand Voltage	ASTM D 495
	Insulation Resistance	MILSTD202F
	Investigation of electrical incidents	ASTM E 2345
Self-heating	Spontaneous Heating Values of Liquids and Solids (Differential Mackey Test)	ASTM D 3523 UN N.1, N.4

ASTM E 620 outlines the content that should be included in a typical report of a technical expert: names and addresses of authors, descriptions of items examined, date and location of exam, scope of activities performed, pertinent facts relied on and sources for them, all opinions and conclusions rendered, logic and reasoning of the expert by which each of the conclusions was reached, and signature of the expert with affiliations. Note that these requirements follow Federal Rule 26a.

ASTM E 678 is essentially a description of the scientific method. It recommends including definition of the problem or issue addressed, identification and explanation of hypotheses addressed (including alternatives), a description of the data collected and analyzed (and sources for that data), estimation of reliability of data used, and opinion reached. Conclusions expressed must be consistent with known facts and accepted scientific principles.

ASTM E 860 is the standard practice for examining and testing items that are or may become involved in product liability litigation. The standard addresses actions that need to be taken if testing will alter or destroy the evidence in any way. Such destructive testing may limit how additional testing may occur. It recommends documenting the condition of the evidence prior to and after any examination or testing and establishing the chain of custody. If proposed tests are expected to alter the item, procedures are outlined for notifying the client and other interested parties, giving them the opportunity to respond or attend the proposed tests.

ASTM E 1020 is a guide to the information that should be included in reports of accidents or events that may become the subject of an investigation or litigation. Its provisions are very similar to the content of fire incident report forms such as those reprinted in Appendix H of *Kirk's Fire Investigation* (DeHaan 2007).

ASTM E 1188 offers an outline of evidence documentation and preservation that is reflected in good crime scene practices that establish a defensible chain of custody for physical evidence recovered.

PEER REVIEW AND PUBLICATIONS

A credible, reliable theory must take into account the body of research that has been compiled, verified, and published by experts in the field. The necessity for credibility underscores advantages of using the scientific method. Recently, there has been a tendency for courts to hold experts to the same standards that scientists use in evaluating each other's work, sometimes referred to as peer review (Ayala and Black 1993). Peer review is common among respected journals that publish the results of theories, testing, and methodology. The common approach to peer in review technical journals is to use reviewers who are quite familiar with the published scientific data and methodologies to determine whether the author's views are of the quality of these publications and whether the data presented support the conclusions of the author. Peer-reviewed journals useful in the field of forensic fire investigation include *Fire Technology, Journal of Fire Protection Engineering, Fire and Materials, Fire Safety Journal, Combustion and Flame,* and *Journal of Forensic Sciences*. Other publications include the *Fire and Arson Investigator*, a journal produced by the International Association of Arson Investigators.

The Society of Fire Protection Engineers published in 2002 their *Guidelines for Peer Review in the Fire Protection Design Process* (SFPE 2002e). These guidelines cover the scope, standard of care, confidentiality, and reporting of the results of a peer review. Although these guidelines concentrate on design, they share significant parallels with the fire investigation field.

Investigators can also participate in peer review when their cases are submitted to supervisors for review. In law enforcement, this is the primary function of supervisory review: to assure that all questions, logical investigative leads, laboratory examinations, and plausible theories are addressed. Private investigation companies working in civil cases also often have supervisors peer-review fire reports. Some corporations also use qualified experts to peer-review fire investigative reports when their companies are too small or do not have an established peer-review process. Peer review is often also carried out by other equally qualified employees in the same agency or company.

In these instances, it must be stressed that the initial investigator still has the primary responsibility to collect and report relevant information following the scientific method. The investigator, not the peer reviewer, will still be the primary witness in any litigation.

Peer review also applies to establishing the credibility of published information (Icove and Haynes 2007), There are several peer-reviewed textbooks and references that are recommended for inclusion in both conducting and assessing complex fire investigations. These texts are written by prominent credentialed fire scientists and engineers, are widely cited in scholarly publications, are used and distributed in the fire investigation community, are included in association publication lists, are cited in court cases, and have often been reviewed by forensic journals.

Accordingly, these texts have become or have the potential of becoming expert or learned treatises in the field of fire investigation. Learned treatises are those texts that rise to the level of authoritative acceptance, so that they can be admitted as evidence in court to support or rebut the testimony given by an expert witness. Furthermore, authoritative references are those which the fire investigator frequently uses as a reference or guide because they have been found to be reliable. Some fire investigators when asked in court what they consider to be an authoritative text will list only those that have legal or administrative impact on their work. Although not totally inclusive, Table 1.6 lists several peer-reviewed reference documents that are considered expert treatises in the fire investigation field:

NEGATIVE CORPUS

The Eleventh Circuit Court of Appeals applied *Daubert* and excluded the testimony of a fire investigator in the *Benfield* case (*Michigan Miller's Mutual Insurance Company v. Benfield* 1998). The court held that the investigation of fires is science based and that the *Daubert* criterion applies.

The use of the **negative corpus** or **arson by default** approach (the process of ruling out all accidental causes for a fire without sufficient scientific and factual basis to determine what did cause the fire) is rarely an acceptable methodology for determining that a fire was intentionally set. In the *Benfield* case, the investigator was not able to articulate the methodology of how he eliminated possible ignition sources and had no scientific basis for his opinion.

The term *negative corpus* has no recognized definition. It *implies* that the corpus delicti of the event (not necessarily of a crime) has not been proven. The logic of negative corpus cause determination was recently examined in great detail by Smith (2006). He pointed out that if it has any application at all, it is only where the origin is "clearly defined" (*NFPA 921*) or is identified as the "exact point of origin" (*Kirk's Fire Investigation*).

It is argued that if the fire damage is too severe, the surrounding materials will be too degraded and the clearly defined point of origin cannot be determined. This may

TABLE 1.6 ◆ Expert Peer-Reviewed Treatises in the Fire Investigation Field

- Babrauskas, V., and S. J. Grayson, eds. 1992. *Heat Release in Fires*. Basingstoke, UK: Taylor and Francis. ISBN 0-4191-6100-7.
- Babrauskas, V. 2003. *Ignition Handbook*. Issaquah, WA: Fire Science, and Bethesda, MD: Society of Fire Protection Engineers. ISBN 0-9728111-3-3.
- Beveridge, A. D., ed. 1998. *Forensic Investigation of Explosions*. Basingstoke, UK: Taylor & Francis. ISBN 0-7484-0565-8.
- Brannigan, F. L. 1992. *Building Construction for the Fire Service,* 3rd ed. Quincy, MA: National Fire Protection Association. ISBN 0-87765-381.
- Cole, L. S. 2001. *Investigation of Motor Vehicle Fires,* 4th ed. San Anselmo, CA: Lee Books. ISBN 0-939818-29-9.
- Cooke, R. A., and R. H. Ide. 1985. *Principles of Fire Investigation*. Gloucestershire, UK: Institution of Fire Engineers.
- DeHaan, J. D. 2007. *Kirk's Fire Investigation*, 6th ed. Upper Saddle River, NJ: Prentice Hall. ISBN 0-13-171922-X.
- Drysdale, D. 1999. *An Introduction to Fire Dynamics*, 2nd ed. New York: Wiley. ISBN 0-471-97290-8.
- Icove, D. J., and J. D. DeHaan. 2004. *Forensic Fire Scene Reconstruction*, 1st ed. Upper Saddle River, NJ: Pearson-Prentice Hall.
- Icove, D. J., V. B. Wherry, and J. D. Schroeder, 1998. *Combating Arson-for-Profit: Advanced Techniques for Investigators*, 2nd ed. Columbus, OH: Battelle Press. ISBN 1-5747-7023-3.
- Iqbal, N., and M. H. Salley. 2004. *Fire Dynamics Tools (FDTs): Quantitative Fire Hazard Analysis Methods for the U.S. Nuclear Regulatory Commission Fire Protection Inspection Program*. Washington, DC.
- Janssens, M. 2000. *Introduction to Mathematical Fire Modeling*, 2nd ed. Lancaster, PA: Technomic. ISBN 1-5667-6920-5.
- Karlsson, B., and J. G. Quintiere. 1999. *Enclosure Fire Dynamics*. Boca Raton, FL.: CRC Press. ISBN 0-8493-1300-7.
- Quintiere, J. G. 1997. *Principles of Fire Behavior*. Albany, N.Y.: Delmar. ISBN 0-8273-7732-0.
- Quintiere, J. G. 2006. *Fundamentals of Fire Phenomena*. West Sussex, England: Wiley. ISBN 0-4700-9113-4.

be true when postfire burn patterns are the entirety of the evidence of origin location. When a reliable witness (or several witnesses) observes an established fire only in one area and can view other nearby areas and reliably conclude there was no observable fire there, then that should be taken as strong evidence that an ignition occurred in that location. Sometimes color and density of smoke or height of flames will offer a clue as to the nature of the first fuel package involved, further narrowing the area of interest. Video surveillance cameras have recorded the area of ignition (if not the point) in many fires that have gone on to cause great destruction. Once a narrow area of origin can be established, then possible sources of heat can be sought out. It should be remembered that such an origin determination is a hypothesis that requires testing, like any other inductive conclusion. This testing involves careful physical examination of the area for remains of either fuel packages or processes known to have been in that vicinity. Before blindly pursuing possible ignition sources only within the defined area, the scientific method demands testing by posing and testing alternatives. These could take the form of asking, "This looks like my origin, but I have a heat source over *there.* Could the damage I'm seeing (or the events a witness saw) have been the result of something (wall covering, drapery, cabinet) igniting from *that* and falling into *this* area?" Smith points out that the better the knowledge base of the investigator, the more likely he or she will be to see (and test) these alternatives. The poorly prepared investigator is far more likely to seize on an apparent point of origin based on burn pattern and then look for heat sources only in that area, never considering that his or her hypothesis about the area of origin is wrong.

It should be remembered that *all* aspects of origin and cause determination should be subjected to careful (thorough) analysis by hypothesis testing, including the formation and testing of alternatives, especially contrary ones, by all means available. The cause of a fire is a determination of first fuel ignited (not necessarily an identification), the source of heat, and the circumstances that brought them together. The determination of first fuel ignited can rely on laboratory tests (ignitable liquid or chemical incendiary residues), physical remains, or witness observations. In a criminal case, where the standard of proof is necessarily very high, circumstantial evidence of first fuel or even heat source may not be adequate. In a civil case where the standard of proof may be "preponderance of evidence" or "more likely than not" (NFPA 2004a, pt. 11.5.5) circumstantial exclusion of all but one competent ignition source and one susceptible fuel may be adequate. The tests for "competent" and "susceptible" are, of course, part of the fire engineering analysis conducted as part of the scientific method of fire investigation.

Presently a reference to "negative corpus" does not explicitly appear in *NFPA 921;* rather, the concept is handled based on the existence rather than the absence of evidence in determining the cause of a fire (NFPA 2004, pt. 16.2.5). The concept of elimination of all other potential causes uses the scientific method of testing and rejecting alternative hypotheses. *NFPA 921* specifically cautions that "the elimination of all accidental causes to reach a conclusion that a fire was incendiary is a finding that can rarely be justified scientifically." If pursued vigorously and within bounded known experimental limits, the scientific method can be used to demonstrate successfully that the only mechanism for ignition had to be deliberate by demonstrating that all relevant accidental mechanisms were specifically evaluated, tested, and eliminated, and that deliberate ignition (and its expected aftermath) fits all the available data. If new data are presented, the conclusion must be reevaluated, possibly resulting in a different conclusion.

ERROR RATES, PROFESSIONAL STANDARDS, AND ACCEPTABILITY

Under *Daubert,* the court considers the known or potential rate of error and the existence of acceptable professional standards on the techniques used by the expert. Error rates from repeated tests are available for many equations, relationships, and models used to describe fire and explosion dynamics. These error rates are evaluated during fire test development, such as those listed in Table 1.5.

A particular practice that has a broad impact on investigators is *ASTM E 860— Standard Practice for Examining and Testing Items That Are or May Become Involved in Products Liability Litigation* (ASTM 1999). This practice covers evidence (actual items or systems) that may have future potential for testing or disassembly and are involved in litigation.

This practice sets forth the following guidance:

- Documentation of evidence prior to removal and/or disassembly, testing, or alteration
- Notification of all parties involved
- Proper preservation of evidence after testing

This practice also stresses the importance of safety concerns associated with testing and disassembly of evidence. This is particularly important when dealing with energized equipment or evidence containing potentially hazardous chemicals.

Other relevant ASTM professional standards include *E 620—Practice for Reporting Opinions of Technical Experts, E 678—Practice for Evaluation of Technical Data, E 1020—Practice for Reporting Incidents,* and *E 1188—Practice for Collection and Preservation of Information and Physical Items by a Technical Investigator.* These practices are a part of the *ASTM E 30* committee's responsibility and are presently being revised to make them suitable in all litigation, both civil and criminal.

◆ 1.5 FORENSIC FIRE SCENE RECONSTRUCTION

Forensic fire scene reconstruction extends beyond the systematic evaluation of the scene by using the scientific method to conduct a comprehensive review of the fire pattern damage, witness accounts, and evidence supporting human activities. According to *NFPA 921,* **fire scene reconstruction** is defined as

> the process of recreating the physical scene during fire scene analysis through the removal of debris and the replacement of contents of structural elements in their pre-fire positions. (NFPA 2004a, pt. 3.3.60)

Specifically, the *forensic* fire scene reconstruction approach is much more comprehensive than that described in *NFPA 921,* as it goes well beyond the concept of merely physically placing the items back into the scene. The scientific method mandates the testing of hypotheses by applying appropriate tools for their evaluation. This includes not only the testing of the physical processes that create visible fire patterns but also the many methods of applying fire engineering and fire dynamics principles.

These principles are well documented in the peer-reviewed literature, often much more so than are the mechanisms that produce the postfire physical effects on which we regularly rely. If sufficient, accurate, reliable detail is available to create reasonable hypotheses to test, there is no reason analyses to predict layer temperatures, plume heights, flame spread, toxic gas concentrations, smoke movement, and other similar factors should not be carried out. In fact, the scientific method demands that such conceptual tests be carried out where appropriate. Fire investigators have long applied some of the "rules" expressed by the engineering relationships to origin and cause determinations but have lacked the specialist knowledge to fully apply them accurately in the light of important variables and constraints. These steps may be taken in different sequences when conditions require; some may be conducted in parallel with others. Yet others form an iterative process in which the results of one step may require repetition of earlier ones.

The applicability of using modern fire dynamics relationships to fire scene investigations has been advanced over the years. The impact is documented in key textbooks and articles by Babrauskas (1997), Drysdale (1999), Grosselin (1998), Lilley (1995), and Nelson (1987).

The steps in Table 1.7 form the basis for the remaining portion of this textbook and are illustrated by example and references. In cases where suitable fire patterns or damage exists, investigators can often accurately determine the area and point of origin of the fire after extinguishment. In this process, the investigator traces the impact of fire plumes on the fire's area, growth, development, and direction of travel taking into account the identifiable fire patterns left behind of smoke deposits and damage

TABLE 1.7 ◆ Steps in Forensic Fire Scene Reconstruction

- ◆ Step 1—Document the fire growth, burn pattern indicators, and processing of the scene.
- ◆ Step 2—Evaluate the heat transfer damage by fire plumes and gases.
- ◆ Step 3—Establish the starting conditions.
- ◆ Step 4—Correlate the human observations and factors.
- ◆ Step 5—Conduct a fire engineering analysis.
- ◆ Step 6—Formulate and evaluate conclusions.

due to heat transfer. Other areas of concern taken equally into consideration are human factors, forensic physical evidence, effects of fire suppression, and a broad knowledge of case histories of similar scenarios.

Reconstruction may point to obvious areas to collect fire debris evidence, particularly when flammable or combustible liquids are suspected to have been used to start the fire (DeHaan 2007; Stickevers 1986). In evidence collection, reconstruction should be combined with a suitable knowledge of fire dynamics. Armed with this knowledge, an investigator who suspects that an ignitable liquid was used to start a fire will look for an area at floor level close to the edge of a fire plume's fuel source that may have remained cooler than the surrounding area during the fire. These cooler areas are where residues of liquids sometimes survive, especially when absorbed and protected by a floor covering such as carpet. These areas are usually evidenced by a demarcation patterns on the material. The analysis of fire patterns left by fire plumes can also provide valuable information, especially in complicated fires where multiple points of origin may exist, as is often the case in arson fires.

Forensic fire scene reconstruction involves establishing what the scene looked like at the start of the fire, not only physically (what was present and where was it) but also considering environmental conditions (temperatures, weather, etc.). A very useful concept for testing accidental scenarios is for the investigator to consider, What was different the day of (or just before) the fire? A change in environmental conditions, equipment operation, time of use, or source of material can sometimes be the factor that shifts a marginally "stable" system into a finally unstable state, triggering a fire by accidental means. Reconstruction may also indicate specific evidence of accidental or natural fires.

STEP 1: DOCUMENT THE FIRE PATTERNS OF THE FIRE SCENE AND ITS PROCESSING

A combination of photographs, sketches, and fire scene analysis is necessary to record accurately the information obtained during the fire scene reconstruction. This step is explored in depth in Chapter 4.

NFPA 921 and *Kirk's Fire Investigation* provide numerous examples of how to graphically record and document photographic, witness, and interpretive fire scene analysis. This documentation should also identify the operation of fire protection equipment, such as heat and smoke detectors, sprinklers, and audible alarms.

The systematic documentation of a fire scene from its initial stages can play an essential role in a professional effort to record the events for all parties involved. This approach captures all available information that is needed for later use in criminal, civil, or administrative matters. The bottom line is that systematic documentation, when it is correctly conducted, will be better suited to pass the *Daubert* challenges for courtroom admissibility and will ensure that an independent investigator will arrive at the same opinions given the total facts and circumstances.

FIGURE 1.6 ◆ A single photograph may not be sufficient to document the fire scene processing. *Courtesy of D. J. Icove.*

Fire scenes often contain complex information that must be documented, a task of paramount importance to the investigator. A single photograph and scene diagram are not often sufficient to capture information on fire dynamics, building construction, evidence collection, and avenues of escape for its occupants, as illustrated in Figure 1.6.

This task is properly accomplished through a comprehensive effort of forensic photography, sketches, drawings, and analysis. In summary, the purposes of forensic fire scene documentation include documenting visual observations, emphasizing development characteristics, and authenticating physical evidence found at the scene. On fire scene sketches, it is important to record the investigator's observations as to the significant fire pattern damage, location of the fire plumes, and the area and point of fire origin. The evidence and explanations contained in this documentation should be sufficiently clear and precise so that an independent and qualified investigator who is unfamiliar with the fire scene can independently arrive at the same conclusions as to its origin and cause.

STEP 2: EVALUATE THE HEAT TRANSFER DAMAGE BY FIRE PLUMES

Heat is transferred by three fundamental methods: **conduction, convection,** and **radiation.** The relative importance of these three methods that dominate a fire depends on not only the intensity and size of the fire but also the physical environment. The heat transferred increases the surface temperature of the target, which can cause observable and measurable effects. Heat transfer is covered in more detail in Chapter 2.

Rapid detection and suppression normally account for fires being contained within their rooms or compartments of origin. In these simple fires of limited extension,

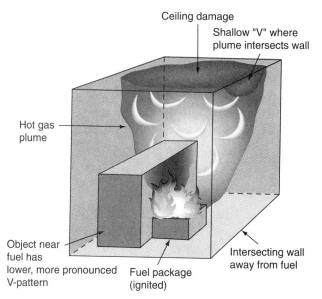

Ceiling damage

Shallow "V" where plume intersects wall

Hot gas plume

Object near fuel has lower, more pronounced V-pattern

Fuel package (ignited)

Intersecting wall away from fuel

FIGURE 1.7 ◆ Typical fire plume damage may produce a circular burn pattern on the ceiling and V or inverted-V patterns on intersecting vertical surfaces. The area of most intense damage on the ceiling is centered over the highest-temperature region of the plume's flame. *From J. D. DeHaan,* Kirk's Fire Investigation, *5th ed., 2002. Reprinted by permission of Pearson Education, Inc., Upper Saddle River, NJ.*

the heat transfer relationships of conduction, convection, and radiation can be used to explain further the fire plume burn pattern damage (Quintiere 1994). Understanding plume damage is a key to reconstruction, and a simplified diagram showing a typical plume-damaged room is shown in Figure 1.7. These processes are examined in more detail in Chapter 3.

STEP 3: ESTABLISH STARTING CONDITIONS

Nearly all rooms contain some sort of fuel. The location, distribution, and type of fuel are critical in evaluating the ignition and spread of a fire. These fuels may be in the form of furniture (live load) but also in the form of carpet and pad, draperies, and wall and ceiling coverings, as well as the building materials themselves (e.g., wood floors, walls and ceilings, fiberboard walls). It is vital to know what these materials were, where they were located in the room, and what their ignition and combustion properties were.

It is also important to know the initial conditions before the fire. For example, the investigation may need to document the prefire conditions of environment (temperature, wind, humidity), ventilation (doors, windows, furnace, HVAC), and operability of potential ignition sources. The first fuel ignited is a vital piece of information in most reconstructions. This fuel should be assessed to determine what the fuel was, how it could be ignited, how long ignition took, and how large a fire would be produced.

STEP 4: CORRELATE THE HUMAN OBSERVATIONS AND FACTORS

It is important to document and correlate the observation of witnesses and victims to the fire scene reconstruction. Investigators' fire analysis should include eyewitness testimony, news videos, and firefighters' descriptions as to the fire's growth during their presence. This documentation can be used to develop and test hypotheses correlating the actions of persons prior to, during, and after the fire, as well as those related to origin and cause.

Early research work evaluated the decision-making behaviors of the occupants and the various factors that affected their decisions in large fires involving high-rise buildings (Paulsen 1994). An investigator who is involved with forensic fire scene reconstructions should become knowledgeable about the existing research into the wide range of human behavior in fire situations. Unlike decisions made during violent crimes (fight, hide, or run), decisions made by people in fire environments can have numerous possibilities—ignore, investigate, approach and observe, run away and call for help, run away and return for property, fight the fire, and so on.

These human factors may include acts or omissions with respect to the ignition of fires either deliberately or by acts of carelessness, improper extinguishment of cigarettes, impaired judgment while under the influence of drugs or alcohol, or reactions to the environment. Other factors include the reactions and decisions, either proper or improper, to trigger alarms and escape to areas of refuge or to remain to fight the fire.

Some times in criminal cases the confession of an arsonist comes into question, particularly when investigators attempt to determine the veracity of the statement. For example, the type, quantity, and distribution of an accelerant used to set the fire or its identification can arise from the admissions of someone involved in the ignition of the fire and comparison with the physical evidence.

The location and intensity of the fire origin can sometimes be established by combining witness observations with an understanding of fire dynamics and physical evidence. This strategy is particularly important when the fire has caused extensive damage to the building or compartment or when the damage extends beyond the compartment of origin, making the determination of the location of the initial fire plume difficult.

Thermal burns on a fire victim's body or carboxyhemoglobin (COHb) levels in the blood may be indicative of their location at the time of the incipient fire, direction of travel, activities, and escape route. For example, an arsonist running away from a premature and rapidly developing flash fire (often characteristic of a flammable liquid) may suffer radiant heat burns on exposed skin of the neck and arms or to the backsides of legs and clothing.

STEP 5: CONDUCT A FIRE ENGINEERING ANALYSIS

The behavior of many fires can be analyzed by applying fundamental fire protection engineering calculations to estimate a fire's size, growth, and damage effects from heat and smoke. Even though these calculations are never 100 percent accurate, they can assist in the interpretation and explanation of fire behavior (Quintiere 1997). Common factors used in fire engineering calculations useful to investigators include heat release rates, flame height of a fire plume, location of the virtual source, heat flux, and flashover.

Fire modeling can also be an important tool in fire engineering analysis. A fire model is generally a computer program used to predict the environment in single- and multicompartment structures subjected to a fire. Models usually calculate temperatures associated with the distribution of smoke and fire gases throughout a building. Fire calculations and models are discussed in Chapter 6.

STEP 6: FORMULATE AND EVALUATE THE CONCLUSION

This final step is the comprehensive conclusion. This final step should include the most complete documentation of how each of the hypotheses, tests, witness observations, and so forth points to the final conclusion for the case. The information that helped exclude other hypotheses should also be described.

◆ 1.6 BENEFITS OF FIRE ENGINEERING ANALYSIS

The primary goal of this text is to extend and explain scientific approaches to fire analysis for use by the fire investigator. However, investigators need a reasonable degree of training in the principles of fire protection engineering and fire dynamics to construct and interpret the results in their cases correctly. Analysis by the application of these principles greatly increases the hypothesis testing ability of a capable investigator, as it may provide answers to queries about a fire's ignition, spread, or effects that would otherwise go unresolved.

FIRE ENGINEERING ANALYSIS

Fire engineering analysis, which can range from a basic back-of-the-envelope first-level calculation to a sophisticated fire model, is an important step in fire scene reconstruction that provides numerous benefits. Historical investigations using fire engineering analysis have accurately assessed fire development, measured the performance of fire protection features and systems, and predicted the survivability and behavior of people during the incident.

Several significant fire engineering analysis studies have been undertaken by the NIST over the years. These detailed reports consist of the Dupont Plaza fire (Nelson 1987), First Interstate Bank Building fire (Nelson 1989), Pulaski Building fire (Nelson 1990), Hillhaven Nursing Home fire (Nelson 1991), Happyland Social Club fire (Bukowski 1992), 62 Watts Street fire (Bukowski 1996), Cherry Road fire (Madrzykowski and Vettori 2000), Cook County Administration Building Fire (Madrzykowski and Walton 2004), and The Station Nightclub Fire (Grosshandler et al. 2005). Some of these cases will be described in Chapter 8.

Some form of fire engineering analysis is now considered a prudent step in comprehensive fire scene analysis and reconstruction (Schroeder 2004). Fire engineering analysis combined with modeling, which is explored in detail later in this text, offers the following benefits.

- Establishes the basis for the collection of data needed to construct event timelines, ignition sequences, and failure modes and effects analysis (FMEA)
- Invites application of the scientific method when testing hypotheses and validating fire scene reconstruction
- Provides a viable alternative to full-scale fire testing and can extend full-scale test results to differing ranges of conditions (sensitivity analysis)
- Provides answers to many important questions raised about human factors, ignition sequences, equipment failure, and fire protection systems (detectors, alarms, and sprinklers)
- Identifies important future research areas in fire investigation

Historically, the use of engineering analysis and modeling in fire scene reconstruction was conducted on a case-by-case basis, owing mainly to the complexity of the process. A vast amount of knowledge and time were required to collect and analyze

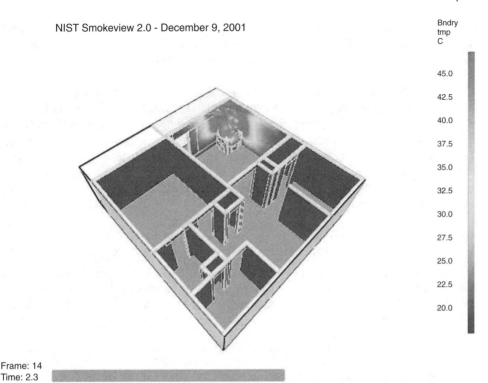

NIST Smokeview 2.0 - December 9, 2001

Bndry
tmp
C

45.0

42.5

40.0

37.5

35.0

32.5

30.0

27.5

25.0

22.5

20.0

Frame: 14
Time: 2.3

FIGURE 1.8 ◆ A fire engineering analysis often uses fire modeling to compare actual events with predicted outcomes. Heat transfer contact areas from a fire initiated in a chair in the upper-right corner of the living room are shown at 2.3 min of the computer simulation. This pattern can be compared with witness accounts and postfire patterns. *Courtesy of D. J. Icove.*

the information on the development of the fire. Note that engineering analysis does not necessarily involve mathematical modeling, but modeling whether by computer or hand calculation offers important support to most such analyses. All such analyses are based on the fundamental physics of mass and energy transfer and convection.

Engineering analysis and modeling are becoming more commonplace. Despite the complexity of the processes, fire engineering analysis and modeling should be applied to cases involving multiple deaths, cases in which code deficiencies contributed to the fire, or cases that will probably involve extensive civil or criminal litigation.

FIRE MODELING

A fire engineering analysis often uses fire modeling to compare actual events with predicted outcomes using varying fire causes and growth scenarios, as illustrated in Figure 1.8, involving a chair fire in an apartment. Surface temperatures on the walls are calculated by the fire model and displayed using a color-coded gradient scale. Results from the surface temperatures can be compared with burn patterns to confirm the area of origin and sequence of the fire. This analysis often adds value, understanding, and clarity to complex fire scene investigation. Fire modeling, a less expensive and more environmentally sensitive approach, can also serve as an alternative to full-scale fire testing to explain burn patterns and the fire dynamics.

Results can also be extended to predict what effects different conditions will have on the fire, a process called **sensitivity analysis**. For example, a model based on heat release rates, plume height, fuel configuration, and placement of the origin within the room (middle, wall, or corner) could potentially resolve whether the fire was the result of an accidental trash can fire or a fire caused by a pool of intentionally spread flammable liquid.

The model can also predict the impact of an open door or window on the fire. In addition, the comparison of a timeline is a beneficial product of the fire model. Example 1.2 shows the successful application of fire modeling as a tool in litigation. However, it should be understood that modeling is less realistic than full-scale fire testing. The exact fuel arrangement that was present can be accurately tested in full scale; but fire modeling is often dependent on fuel data such as heat release rates, which are available in the published literature and which may be quite different from that of the supposedly similar item.

EXAMPLE 1.2   Fire Model Helps Defend U.S. Government against Lawsuit

At 6:45 A.M. in a three-bedroom town house on a military reservation, just before the winter holidays, a 5-year-old was involved in a storage closet fire. The smoke alarm notified the parents, and the father tried unsuccessfully to extinguish the fire. Because their telephone was out of order, the husband told his wife to awaken their children while he went next door to call the fire department.

The husband's wife and two youngest children perished in the fire. Three older children survived. The responding fire department was slightly delayed, by 2.5 min, when they had to yield to troops on the road. This delay formed the basis for a $36 million wrongful death lawsuit against the U.S. Government. The U.S. Department of Justice asked the National Institute of Standards and Technology (NIST) to model the fire to determine what, if any, role the reported delay played in the deaths. A NIST fire protection engineer used HAZARD I (which at the time contained the earlier version 2.0 of CFAST) to model the fire. He relied on construction plans, fire incident reports, radio logs, and witness statements to input the variables into the model.

The fire model accurately predicted areas of fire and smoke damage, the safe egress routes of the survivors, and the impact that attempts by neighbors to gain access by opening a bedroom window had on the resulting carbon monoxide and heat intensity levels. These predictions agreed with the autopsy findings, blood toxicity tests, and thermal injuries to the victims.

A further timeline analysis showed that the delayed response played no significant role in the fatalities. After depositions were taken from the NIST engineer, the lawsuit was settled out of court for less than $200,000 (Bukowski 1991).

Fire engineering analysis and modeling may also provide answers to important questions raised during fire investigations. These questions might include the following:

- What was the most probable cause of the fire (i.e., can several possible causes be eliminated)?
- How long did it take to activate fire or smoke detectors and water sprinkler heads?
- What were the smoke and carbon monoxide levels in each room after 10 min?
- Why didn't the occupants of the building survive the fire?
- Was an accelerant used in the fire and, if so, what type?
- How much time elapsed from when the owner of the structure left the building until the fire reached flashover?
- Did a negligent building design or failed fire detection and suppression system contribute to the growth of the fire?
- What changes to policy, building, or fire codes are necessary to ensure that similar fires will not occur in the future?

Discussion and cooperative research between fire investigators and fire protection engineers underscore the need to include fire modeling as an integral and required step in fire scene reconstruction. Fire investigators and engineering professionals need to collaborate further to expand their knowledge and validate the methods shared in common by both groups.

◆ 1.7 CASE STUDY 1: *DAUBERT* AND THE FIRE INVESTIGATOR

The debate over the application of *Daubert* to fire scene investigations has intensified at the point of deciding whether origin and cause determination is to be considered scientific evidence or nonscientific technical evidence. The advocates of the strict scientific approach bristle at the suggestion that fire scene investigation is in any way nonscientific, pointing to the many misconceptions previously used by fire investigators (spalling, floor penetrations, etc.), which were exposed by fire scientists only in recent years. They advocate the use of *Daubert* in fire scene analysis as the only means of preventing a return to the improper fire scene methodologies employed by unqualified investigators lacking proper scientific training. In contrast, the "technicians" argue that fire scene investigation has never been a pure science like chemistry or physics, as it employs elements of both disciplines. The term *nonscientific* in the context of *Daubert* is a legal distinction, rather than a scientific one. It is not to say that fire investigation is *un*scientific or devoid of any application of scientific principles. Instead, it is a recognition of the objective and subjective components that form a part of every fire scene investigation, more precisely, the human component in examining, analyzing, and, ultimately, interpreting fire scene evidence to reach a conclusion about the fire's origin and cause.

Early on in *Daubert*'s history, the Tenth Circuit directly addressed the testimony of a fire investigator in an arson case. In *United States v. Markum* (1993), the court found the admission of a fire chief's testimony that a fire was the result of arson to be proper based primarily on his extensive experience in fire investigations. There the court said,

Experience alone can qualify a witness to give expert testimony. See *Farner v. Paccar, Inc.*, 562 F. 2d 518, 528–29 (8th Cir. 1977); *Cunningham v. Gans*, 501 F. 2d 496, 500 (2nd Cir. 1974).

Chief Pearson worked as a firefighter and Fire Chief for 29 years. In addition to observing and extinguishing fires throughout that period, he attended arson schools and received arson investigation training. The trial court found that Chief Pearson possessed the experience and training necessary to testify as an expert on the issue whether the second fire was a natural rekindling of the first fire or was deliberately set. That finding was not clearly erroneous. (*Markum* at 896)

Another case that directly addressed the application of *Daubert* to fire investigation was *Polizzi Meats, Inc.* (PMI) *v. Aetna Life and Casualty* (1996). In that case, the Federal District Court of New Jersey said,

PMI's counsel argues that because of a lack of "scientific proof" of the fire's causation, none of Aetna's witnesses may testify at trial. This astounding contention is based on a seriously flawed reading of the United States Supreme Court's decision in *Daubert v. Merrell Dow Pharmaceuticals, Inc. Daubert* addresses the standards to be applied by a trial judge when faced with a proffer of expert scientific testimony based upon a novel theory or methodology. Nothing in *Daubert* suggests that trial judges should exclude otherwise relevant testimony of police and fire investigators on the issues of the origins and causes of fires. *Polizzi* at 336–37. (citations omitted)

These two decisions were the only reported cases considering *Daubert* in the specific context of fire investigation until the Eleventh Circuit announced its decision in a case that took an entirely different view on the process of fire scene investigation—the Joiner case. The *Markum* case analysis was provided courtesy of Guy E. Burnette, Jr., Tallahassee, Florida (Burnette 2003).

♦ 1.8 CASE STUDY 2: THE *JOINER* CASE; A CLARIFICATION OF *DAUBERT*

As courts from various jurisdictions are still trying to shed light on the full meaning of *Daubert,* the U.S. Supreme Court recently took up the issue again and provided some guidance and insights. In *General Electric Company v. Joiner* (1997) the Court reviewed a case in which the trial judge had entered summary judgment in favor of the defendant in a lawsuit alleging that the plaintiff had contracted cancer as the result of exposure to PCB chemicals. The scientific evidence in support of the plaintiff's claim was derived from laboratory studies of mice that had been injected with massive doses of PCB chemicals and certain limited epidemiological studies suggesting a causal connection between PCB chemicals and cancer in humans. The trial judge ruled that the evidence offered by the plaintiff failed to satisfy the requirements of *Daubert,* describing the evidence offered by the plaintiff's experts as "subjective belief or unsupported speculation." It was noted that *Joiner* failed to present any credible scientific evidence of a direct causal connection between exposure to PCB chemicals and cancer.

On appeal, the ruling was reversed by the Eleventh Circuit, which held that the evidence should have been presented to the jury for a decision. The appellate court observed that the *Federal Rules of Evidence* favor the admissibility of expert testimony as a general rule. Further, the appellate court applied a more stringent standard of review of the trial court's ruling, since the ruling was "outcome determinative" (i.e., resolved the entire case).

The U.S. Supreme Court overturned the decision of the Eleventh Circuit and reinstated the ruling of the trial court. In doing so, the Court reiterated and clarified some of the points made in the *Daubert* decision. First, the role of the trial judge as "gatekeeper" was reaffirmed. In particular, the trial judge not only was allowed to draw his own conclusions about the weight of evidence offered by an expert witness, but was expected to do so. It was noted that this had been a function of the trial judge long before the *Daubert* decision itself. Since this was a proper role of the trial judge, the decision to accept or reject expert testimony would not be subjected to a more stringent standard of review on appeal. The decision of the trial judge would be given deference on appeal and it would require showing an "abuse of discretion" for the decision of the trial judge to be overturned.

The Supreme Court held that the application of *Daubert* to expert testimony is not merely a review and approval of the methodology employed, but includes scrutiny of the ultimate conclusions reached by the expert witness based on the methodologies and data employed to reach those conclusions. Notably, the Court did not clarify the controversy over scientific evidence versus technical evidence. The Court did not address the issue in *Joiner* because it was clearly a "scientific evidence" case. That remains a major part of the controversy in construing Daubert and the admissibility of expert testimony. The *Benfield* decision did directly address this issue and demonstrated a new perspective on this critical aspect of fire investigation.

This case history was provided courtesy of Guy E. Burnette, Jr., Tallahassee, Florida (Burnette 2003).

◆ 1.9 CASE STUDY 3: THE *BENFIELD* CASE

In *Michigan Miller's Mutual Insurance Company v. Janelle R. Benfield* (1998), the U.S. Court of Appeals for the Eleventh Circuit applied the *Daubert* analysis to a fire scene investigation. This case has attracted great attention within the fire investigation community and has become a focal point of the *Daubert* controversy.

In January 1996 the *Benfield* case was tried in federal district court in Tampa, Florida. The case involved a house fire in which the insurance company, Michigan Miller's Mutual, refused to pay on the policy based, in part, on the fire's being incendiary and on the apparent involvement of an insured party in setting the fire. As a part of the insurer's case, a fire investigator with over 30 years of experience in fire investigations was called as an expert witness to present his opinion of the origin and cause of the fire. He testified that the fire was started on top of the dining room table where some clothing, papers, and ordinary combustibles had been piled together. He examined the fire scene primarily by visual observation and concluded that the fire was incendiary based on the absence of any evidence of an accidental cause, along with other evidence and factors noted at the scene. After cross-examining the investigator, the plaintiff moved to exclude the testimony under *Daubert,* and the trial court agreed. In the trial court's ruling striking the expert's testimony, the judge specifically found that the witness

> cites no scientific theory, applies no scientific method. He relies on his experience. He makes no scientific tests or analyses. He does not list the possible causes, including arson, and then using scientific methods exclude all except arson. He says no source or origin can be found on his personal visual examination and, therefore, the source and origin must be arson. There is no question but that the conclusion is one to which *Daubert* applies, a conclusion based on the absence of accepted scientific method. . . .

> And finally, it must be noted that [his] conclusion was not based on a scientific examination of the remains, but only on his failure to be able to determine a cause and origin from his unscientific examination. This testimony is woefully inadequate under *Daubert* principles and pre-*Daubert* principles, and his testimony will be stricken and the jury instructed to disregard the same. (*Daubert* motion hearing transcript at 124–26)

Interestingly, the court in *Benfield* initially found the expert to be qualified to render opinions in the area of origin and cause of fires and allowed him to testify, based on his qualifications and credentials as a fire investigator. However, the judge struck the expert's testimony after it was presented based on his methodologies in conducting the particular fire scene investigation in that case. Having stricken the expert's testimony, the judge then found that, as a matter of law, arson had not been proved by Michigan Miller's and directed a verdict against them on the arson issue.

The evidence was undisputed that the area of origin was on top of the dining room table. Therefore, the only issue was the cause of the fire. The expert testified that while he was conducting his investigation, he spoke with Ms. Benfield, who told him that when she was last in the house before the fire there was a hurricane lamp and a half-full bottle of lamp oil on the top of the table. He further testified that he examined photographs taken by the fire department before the scene was disturbed and observed an empty, undamaged bottle of lamp oil lying on the floor with the cap removed (also undamaged), indicating that it had been opened and moved from the table prior to the setting of the fire. He also explained his observations at the fire scene that enabled him to rule out all possible accidental causes. He concluded that the fire was incendiary, using the "elimination method" long recognized as a valid method of determining fire origin and cause. He could not, however, determine the source of ignition for the fire. More importantly, he did not "scientifically document" his findings on various points and relied primarily on his 30 years' experience as a fire investigator, even as he held himself out as an expert in fire science adhering to the scientific method in conducting his investigation.

On cross-examination by Ms. Benfield's attorney, the expert was asked to define the scientific method and was asked the "scientific basis" for the taking of certain photographs apparently unrelated to the fire itself. The cross-examination continued by attacking each piece of evidence used by the expert that could not be said to be scientifically objective and scientifically verified. The investigator's determination of the smoldering nature of the fire and the time he estimated it burned before being discovered were discredited as not being based on scientific calculations of heat release rate and fire spread, but merely the investigator's observations of the smoke damage and other physical evidence. The court noted those points from the cross-examination in finding that the expert's methodology was not in conformity with the scientific method, relying instead almost exclusively on the expert's own training and experience, which it held to be inadmissible under *Daubert*.

On May 4, 1998, the Eleventh Circuit issued its ruling in the *Benfield* case. Contrary to the Tenth Circuit decision in *Markum* and the federal district case in *Polizzi Meats*, the court found the investigator's fire scene analysis to be subject to the *Daubert* test of reliability. In reaching this conclusion, the court noted that the investigator in *Benfield* held himself out as an expert in the area of "fire science" and claimed that he had complied with the "scientific method" under *NFPA 921*. Thus, by his own admission he was engaged in a "scientific process," which the court held to be subject to *Daubert*.

Under the *Daubert* test of reliability, the appeals court upheld the decision of the trial judge to strike his testimony. Noting that it is a matter of the trial court's discretion to admit expert testimony, such a decision will be affirmed on appeal absent a showing of an "abuse of discretion" or that the decision was "manifestly erroneous." Under such a daunting standard, the trial judge is effectively given the "final word" on whether both the qualifications and findings of an expert witness will be considered reliable enough to be presented to the jury. It is not simply a matter of having the power to decide if a witness is qualified to testify as an expert; the substance of the expert's testimony and his or her professional conclusions will first have to meet the approval of the trial judge before they can be presented to the jury. The trial judge acting as gatekeeper can summarily reject the findings and conclusions of an expert witness, preempting the jury from making that decision. In the *Benfield* decision, various scientifically unsupported and scientifically undocumented conclusions of the investigator were cited as grounds for the determination that his observations and findings failed the *Daubert* reliability test. A chandelier hanging over the dining room table where the fire started showed no signs of having caused the fire, but the investigator had not conducted any tests or examinations to eliminate it scientifically as a potential cause of the fire. His observations alone were held inadequate. Similarly, his opinion that the fire had likely been accelerated with the lamp oil contained in the bottle was rejected under *Daubert,* since he could not scientifically prove that there had been oil in the bottle before the fire and he had not taken any samples from the fire debris to prove scientifically its presence or absence at the time the fire was ignited. These and other observations of the investigator were held to demonstrate that there was no *scientific* basis for his conclusions, only his personal opinion from experience in investigating other fires.

It was not that the investigator was found to be "wrong" in the *Benfield* case. Indeed, there was never any evidence of an accidental cause of the fire. Ironically, although the appeals court upheld the decision of the trial judge to strike the testimony of the insurance investigator, it granted a new trial for Michigan Miller's on the arson defense. The appeals court felt that a prima facie case of arson had been shown at trial through the fire department investigator, who initially classified the fire only as "suspicious" (with virtually no challenge to the scientific documentation of his opinion), and the many incriminating circumstances surrounding the fire itself. Those circumstances included the fact Ms. Benfield claimed that she had not locked the deadbolts when she left the house, yet the deadbolts were locked when she returned to discover the fire. Ms. Benfield and her daughter (who had been out of town) had the only keys to those locks. Her assertion that the fire was extinguished by her boyfriend with a garden hose was refuted by the observations of the responding firefighters. Her insurance claim appeared to be significantly inflated. She had tried to sell the house and could not do so. She was trying to convince her estranged husband to transfer the house to her but could not do so. Ms. Benfield had given conflicting and contradictory accounts of her activities immediately before the fire. In listing all these reasons, the appeals court found that there was compelling evidence of arson even as it discredited the findings of the insurance investigator that the fire was incendiary. The critical point in the case appears to be the distinction between the insurance investigator's testifying as an expert in fire *science* and the fire department representative's testifying as an expert in fire *investigation*.

◆ 1.10 CASE STUDY 4: THE *MAGNETEK* CASE: LEGAL AND SCIENTIFIC ANALYSES

MAGNETEK—A LEGAL CASE ANALYSIS

A recent decision of the U.S. Court of Appeals for the Tenth Circuit has underscored the importance of properly evaluating and presenting expert testimony in fire litigation cases. *Truck Insurance Exchange v. MagneTek, Inc.* (2004) not only demonstrated the application of the *Daubert* standard (*Daubert v. Merrell Dow Pharmaceuticals, Inc.* (1993) for the admissibility of expert testimony but effectively contradicted a long-standing principle of fire science that had been used in a significant number of cases to prove the cause of a fire.

The *MagneTek* case involved a subrogation action filed by Truck Insurance Exchange against the manufacturer of a fluorescent light ballast that was alleged to have caused a fire which destroyed a restaurant in Lakewood, Colorado. Responding to the alarm, the fire department first found heavy smoke in the restaurant but no open flames. The fire subsequently broke through the kitchen floor in the restaurant from the ceiling of a storage room in the basement. Before the fire could be controlled and extinguished, it destroyed the restaurant and caused damages in excess of $1.5 million.

Investigations by both the local fire protection district and a private fire investigation firm hired by the insurer determined that the fire had started in a void space between the basement storage room ceiling and the kitchen floor. In the basement, the investigators found the remains of a fluorescent light fixture that had been mounted to the ceiling of the storage room. They determined the light fixture had been located in the area of origin of the fire and concluded that the fire had been caused by an apparent failure of the ballast in the light fixture.

The remains of the fluorescent light fixture were examined by the investigators and a physicist. They determined the ballast had been manufactured by MagneTek. They observed oxidation patterns on the light fixture indicating an internal failure, along with discoloration of the heating coils of the ballast suggesting it had shorted to cause overheating, which resulted in the fire. The ballast contained a thermal protector designed to shut off power to the fixture when the internal temperature exceeded 111°C (232°F). The thermal protector in the ballast appeared to function properly, even after the fire. However, the investigators remained convinced that the ballast had somehow failed, overheated, and started the fire.

Tests were conducted with similar ballasts manufactured by MagneTek, which showed that at least one of the exemplar ballasts when shorted would not cut off power to the fixture until the internal temperatures had reached at least 171°C (340°F) and would continue to provide power to the fixture even when the ballast maintained constant temperatures of 148°C (300°F) or more.

The investigators theorized that the heat from the ballast had caused "pyrolysis" to occur in the adjacent wood structure of the ceiling, causing "pyrophoric carbon" to form over a prolonged time, which would be capable of ignition at temperatures substantially below the normal ignition range of 204°C (400°F) or more for the fresh whole wood. The phenomenon of pyrolysis in the formation of "pyrophoric carbon" has been the subject of a number of studies, reviews, and articles by fire investigators and fire scientists. It has been cited as the cause of a number of fires having no other apparent explanation, often linked to heated pipes in structures within walls, ceilings, and floor areas. As the MagneTek court case would note, however, the validity of this phenomenon has been discussed and debated by fire investigators and fire scientists for a number of years.

The investigators in this case acknowledged that electrical wiring ran through the ceiling area of the storage room near the fluorescent light fixture but discounted the possibility of a failure in the electrical wiring. They reported finding no evidence of arcing or shorting in the electrical wiring, although the fire at the restaurant resulted in the destruction of much of the electrical wiring and other evidence in the area. Because the investigations concluded the fire had originated in the immediate area of the light fixture, they concluded the only source of ignition for the fire was the light fixture and its ballast.

Pyrolysis and the formation of "pyrophoric carbon" was the foundation of the plaintiff's case against MagneTek. The ballast in the light fixture showed no signs of failure in the thermal protector, which would have limited the heat generated by the ballast to about 111°C (232°F). Even with the exemplar ballast whose thermal protector failed to perform as it had been designed, the temperatures generated did not exceed 171°C (340°F). The investigators admitted the ignition temperature of *fresh whole wood* is typically at least 204°C (400°F) and the temperatures from the ballast alone would not have been sufficient to cause ignition. Their theory that the ballast had caused the fire depended on the concept of pyrolysis to allow ignition to occur at a lower temperature within the range of the temperatures generated by the ballast.

Following discovery in the case, MagneTek filed a *Daubert* motion to exclude the testimony of the experts that the ballast had caused the fire. MagneTek asserted that the theory of "pyrolysis" was not sufficiently reliable and scientifically verifiable to be offered by the experts in support of their conclusion for the cause of the fire. It was a challenge to the "reliability" of the experts' theory, which required a consideration of whether the reasoning or methodology underlying the testimony was scientifically valid, as mandated by *Daubert* and *Rule 702* of the *Federal Rules of Evidence.*

The Supreme Court in *Daubert* had outlined a number of factors that, although not an exclusive list of considerations for a trial court, should be examined in making the determination of reliability (see Table 1.4).

In proving the scientific validity of an expert's reasoning or methodology, the Court noted that "the plaintiff need not prove that the expert is undisputably correct or that the expert's theory is 'generally accepted' in the scientific community. Instead, the plaintiff must show that the method employed by the expert in reaching the conclusion is scientifically sound and that the opinion is based on facts which sufficiently satisfy *Rule 702*'s reliability requirements."

The Tenth Circuit applied the standard of appellate review for a trial court's ruling on a *Daubert* issue: the showing of an "abuse of discretion" demonstrating that the appellate court has "a definite and firm conviction that the lower court made a clear error of judgment or exceeded the bounds of permissible choice in the circumstances" *United States v. Ortiz* (1986). The court then looked to the ruling of the trial judge finding that the testimony of the experts failed to satisfy the reliability standard under *Rule 702* and the *Daubert* decision. The physicist testifying on behalf of Truck Insurance Exchange was a highly credentialed expert with an advanced degree in physics from Oxford University and over 20 years of experience in the study of fire and explosion incidents. Both the trial court and the appellate court observed that this expert was unquestionably qualified to testify as an expert witness under *Rule 702*. However, his hypothesis of pyrolysis and pyrophoric carbon as the cause of ignition of the wood in the area surrounding the light fixture could not be considered a reliable basis for the admission of his expert testimony on the cause of the fire.

The reliability standard under *Daubert* applies to the reliability of both the theory itself and its application to the facts of the case. The court focused on the first component of this reliability criteria. In support of the theory of its experts, the insurer had introduced into evidence three publications on the theory of pyrolysis and pyrophoric carbon. Those articles were written by some of the most respected fire scientists in the world, but those articles and case studies acknowledged that the process of pyrolysis occurred over an undefined period of time described as "a period of years" or "a very long time" with no specific parameters for the timing and sequence of events involved in pyrolysis. One of those articles acknowledged that there are "a number of things not known about the process" and that "it may be many decades before it will be solved by sufficiently improving theory." The article concluded by stating that "the phenomenon of long-term, low-temperature ignition of wood has neither been proven nor successfully disproven at this time."

The plaintiff's engineer had stated in his deposition that the process of pyrolysis "depends on a lot of factors, as yet quantitatively unidentified." He went on to testify that "you would have to have a good theory of pyrophoric carbon and formation and the chemical kinetics of that; and there isn't one. . . ." The other experts testifying for the plaintiff in this case based their theory of pyrolysis and "pyrophoric carbon" on their experience in the investigation of fires without any reference to a specific scientific basis for that theory.

The court noted that under *NFPA 921* an investigator offering the hypothesis of an appliance fire must first determine the ignition temperature of the available fuel in the area and then must determine the ability of the appliance or device to generate temperatures at or above the ignition temperature of the fuel. In that regard, the experts failed on both counts. The ignition temperature of the fuel (wood) exposed to the fixture in this case could not be scientifically proved to be below the 204°C (400°F) threshold for the ignition of most types of wood, and the tests of the ballasts had shown that even with a failed thermal protector, a ballast could not generate temperatures anywhere near that range. Their hypothesis would have to be based on either an unreliable theory (pyrolysis) or unsubstantiated assumptions and speculation about the temperature of the ballast that contradicted their own test results. As such, their testimony could not be admitted.

The appellate court affirmed the ruling of the trial court that the testimony of all the experts had been shown to be not sufficiently reliable to be admitted under the standards of the *Daubert* decision. Without the testimony of the experts, Truck Insurance Exchange could not make a prima facie case for establishing the cause of the fire. Accordingly, the trial court entered a summary judgment in favor of MagneTek, and the appellate court affirmed this decision, as well.

This decision has significant implications for the litigation of fire cases everywhere. It demonstrates the importance of developing sound and scientifically verifiable theories for proving the cause of a fire as required by the standards of the *Daubert* decision. Moreover, it provides a compelling example of how a theory that has not been validated and generally recognized by others in the scientific community may not withstand a *Daubert* challenge in court.

The lessons from this decision are many. First and foremost, experts must be prepared to prove the reliability of their investigative methodologies and theories to the satisfaction of the trial court. Experience alone is not sufficient. Even a theory that appears on the surface to be a reasonable and logical theory for the cause of a fire must be shown to be scientifically verifiable. Without a foundation in science, even the most experienced investigator will never be allowed to testify in court. Parties hiring those investigators in their cases must be aware of the requirements for proving a case under investigative methodologies and theories that will meet the reliability standards of the *Daubert* decision, to guide them in both the selection of the expert used to investigate the fire and the decision to litigate the case. Attorneys handling those cases must be aware of those issues to successfully litigate that case at trial. The investigator, the client, and the attorney all have a responsibility to ensure that cases are properly investigated and properly litigated. The MagneTek case is a striking example of the consequences of not doing so.

This case history was provided courtesy of Guy E. Burnette, Jr., Tallahassee, Florida (Burnette 2005).

MAGNETEK—A SCIENTIFIC CASE ANALYSIS

A scientific analysis and assessment of the MagneTek case decision indicates that the court may not have understood several important aspects of fire science, including pyrolysis, which is not newly discovered nor scientifically disputed. In the MagneTek case, the Court rejected under *Daubert* the "pyrolysis theory" of long-term, low-temperature ignition as being unreliable. Fire investigators who wish to explain this phenomenon in court must be prepared to make clear and convincing arguments using authoritative treatises and clearly defined definitions.

In the *Ignition Handbook* (Babrauskas 2003), pyrolysis is clearly defined as "the chemical degradation of a substance by the action of heat." This definition includes both oxidative and nonoxidative pyrolysis, taking into account the cases where oxygen plays a role. Pyrolysis is a process that has been studied for many decades. Without pyrolysis to break down their molecular structures, most solid fuels will not burn. Unfortunately, the published court decision declares "pyrolysis" to be inadequately studied and documented (presumably intending to refer to "pyrophoric process"). It is unclear what effect this misstatement of the court will have on future deliberations.

Under the scientific method, describing the conditions under which ignition can take place may use (1) a scientific theory, test, or calculation or (2) refer to authoritative data on the substance. For example, in the second case, the peer-reviewed literature (Babrauskas 2005) can be used to show that the ignition temperatures of wood exposed to long-term, low-temperatures are documented as low as 77°C (170°F). Excluding other competent sources of energy, it could be substantiated in the MagneTek case that a heat-producing product in excess of 77°C (170°F) in contact with wood for an extended period of time created a potential risk. Some substances have spontaneous combustion models. Unfortunately for the fire investigator there does not exist at this time a model that predicts ignition of wood by pyrolysis with time/temperature curves that would allow the accurate "prediction" of necessary conditions, primarily owing to the complex chemical nature of wood as a fuel and the wide variability in the physical nature of the types and forms of wood.

Even more problematically, the court effectively denied the existence of any ignition due to self-heating, because it declared that ignitable substances have a "handbook" ignition temperature, and a hot object below that temperature will not be a competent source of ignition. In actual fires, self-heating substances do not have a unique ignition temperature. Furthermore, if a substance (e.g., wood) can ignite owing to external, short-term heating or to a prolonged process involving self-heating, the minimum temperature for the latter has no relation to the published ignition temperatures for the former. In this respect, the citation from *NFPA 921* regarding ignition temperatures is in error.

Finally, in the future, fire investigators who testify as to "pyrophoric" ignition (i.e., low-temperature) ignition of wood should be prepared to discuss the phenomenon by referring to this collected set of observations on the topic and the hard scientific data. In addition to the *Ignition Handbook*, there exist ample peer-reviewed publications and case studies on the ignition process of wood. Also, additional information is available on the charring rate and ignition of wood (Babrauskas 2005; Babrauskas, Gray, and Janssens 2007).

This scientific case analysis is primarily based on information courtesy of Dr. Vytenis Babrauskas, Fire Science and Technology, Inc., Issaquah, Washington. For additional information, Zicherman and Lynch (2006) and Babrauskas (2004) also address the technical merits of this case.

◆ 1.11 SUMMARY AND CONCLUSIONS

This chapter introduced the application of the scientific method formed by validated principles and research (fire testing, dynamics, suppression, modeling, pattern analysis, and historical case data) to fire investigation and reconstruction. This approach relies on combined principles of fire protection engineering, forensic science, and behavioral science. In using the scientific method, the investigator can more accurately estimate a fire's origin, intensity, growth, direction of travel, and duration, as well as understand and interpret the behavior of the occupants.

Fire scene reconstruction can also bring in additional fire testing and similar cases to support the final theory. In addition to *NFPA 921,* the fire investigator should become familiar with other authoritative references and sources of data and information. It should be remembered that *NFPA 921* is a summary of practices, extracted and developed from thousands of research efforts, some extending back over a century. There is a universe of reliable information available to fire investigators that is not included in *921.*

The strengths of this approach make a thorough examination of many hypotheses more effective. The reliability of the result of a specific method of investigation is a function of the number of alternative hypotheses that have been tested and eliminated. Fire engineering analysis also offers many advantages. In particular, it allows the investigator to evaluate the potential of several scenarios without repeated full-scale fire testing. A sensitivity analysis can also extend the analysis to different ranges of conditions.

Problems

1.1. Obtain an adjudicated case from your local fire marshal's office involving the conviction of an arsonist who pleaded guilty to setting a fire. Analyze the fire investigator's opinion as to the origin and cause of the fire. Demonstrate how it did or did not meet the guidelines for the scientific method.

1.2. Obtain a photograph of a fire plume against a wall. Trace the lines of demarcation for char, scorching, and smoke deposition onto a plastic overlay. What can you learn from this exercise?

1.3. Search the Internet or current published periodical fire investigation literature and list three examples of peer-reviewed fire testing research.

1.4. Search the Internet or current published periodical fire investigation literature and list three federal and state court cases mentioning *NFPA 921* and *Daubert.*

1.5. Identify scientific treatises that describe pyrolysis. What is the earliest date?

1.6. Review recent fire investigation publications and find five references to the scientific method. What is the described impact on the investigations?

Suggested Reading

Cooke, R. A., and R. H. Ide. 1985. *Principles of fire investigation,* chaps. 8 and 9. Leicester, UK: Institution of Fire Engineers.

DeHaan, J. D. 2007. *Kirk's fire investigation,* 6th ed., chaps. 4 and 5. Upper Saddle River, NJ: Prentice Hall.

Basic Fire Dynamics

2 CHAPTER

There is nothing like first-hand evidence.

—Sir Arthur Conan Doyle,
"A Study in Scarlet"

The body of fire dynamics knowledge as it applies to fire scene reconstruction and analysis is derived from the combined disciplines of physics, thermodynamics, chemistry, heat transfer, and fluid mechanics. Accurate determination of a fire's origin, intensity, growth, direction of travel, duration, and extinguishment requires that investigators rely on and correctly apply all principles of fire dynamics. They must realize that variability in fire growth and development is influenced by a number of factors such as available fuel load, ventilation, and physical configuration of the room.

Fire investigators can benefit from a working knowledge of published research involving the application of the knowledge that exists in the area of fire dynamics. The published research includes several recent textbooks and expert treatises (DeHaan 2007; Drysdale 1999; Karlsson and Quintiere 1999; Quintiere 1997 and 2006), traditional fire protection engineering references (SFPE 2002b), U.S. Government research done primarily by the National Institute of Standards and Technology (NIST) and the U.S. Nuclear Regulatory Commission (NRC) (Iqbal and Salley 2004), and fire research journals. These journals include peer-reviewed fire-related ones such as *Fire Technology*, the *Journal of Fire Protection Engineering*, and *Fire Safety Journal*. Fire-related research also appears in peer-reviewed forensic journals such as the *Journal of Forensic Science* and *Science and Justice*.

Many fire investigators do not routinely access these sources. Bridging this knowledge gap is the primary purpose of this chapter. Many basic fire dynamics concepts are described in *NFPA 921* (NFPA 2004a). These basic concepts and other, more advanced ones will be discussed later in this textbook. Existing real-world examples and training programs illustrate their application and describe how basic fire dynamics principles apply to fuels commonly found at fire scenes. Case examples are offered in this chapter along with discussions on the application of accepted fire protection engineering calculations that can assist investigators in interpreting fire behavior and conducting forensic scene reconstruction.

◆ 2.1 BASIC UNITS OF MEASUREMENT

The common units of measurement and dimensions used in fire scene reconstruction come from two systems: United States (U.S.) and metric. The United States is in the gradual process of converting to the international standard, which is known as the International System of Units (SI from the French, Système International). This textbook uses the SI system as its standard, and in most cases the equivalent English units follow in parentheses for comparison. Table 2.1 lists many of these commonly used fundamental dimensions from fire dynamics along with their typical symbol, unit, and conversion factor.

The most fundamental property of fire dynamics is heat. All objects above absolute zero ($-273°C$, $-460°F$, 0 K) contain heat owing to the motion of their molecules. Sufficient heat transfer (by conduction, convection, radiation) to an object increases its **temperature** and may ignite it, which may spread fire to other neighboring objects.

TABLE 2.1 ◆ Fundamental Dimensions Commonly Used in Fire Dynamics

Dimension	Symbol	SI Unit	Conversion
Length	L	m	1 m = 3.2808 ft
Area	A	m^2	1 m^2 = 10.7639 ft^2
Volume	V	m^3	1 m^3 = 35.314 ft^3
Mass	M	kg	1 kg = 2.2046 lb
Mass density	ρ	kg/m^3	1 kg/m^3 = 0.06243 lb/ft^3
Mass flow rate	$\dot{m}$	kg/s	1 kg/s = 2.2 lb/s
Mass flow rate per unit area	$\dot{m}''$	kg/(s-m^2)	1 kg/(s-m^2) = 0.205 lb/(s-ft^2)
Temperature	T	°C	$T(°C) = [T(°F) - 32]/1.8$
			$T(°C) = T(K) - 273.15$
			$T(°F) = [T(°C) \times 1.8] + 32$
Temperature	T	K	$T(K) = T(°C) + 273.15$
			$T(K) = [T(F) + 459.7]0.56$
Energy, heat	Q, q	J	1 kJ = 0.94738 Btu
Heat release rate	$\dot{Q}, \dot{q}$	W	1 W = 3.4121 Btu/hr = 0.95 Btu/s, 1 kW = 1kJ/s
Heat flux	$\dot{q}''$	W/m^2	1 W/cm^2 = 10 kW/m^2 = 0.317 Btu/(hr-ft^2)

Source: Derived from SFPE 2002a, Quintiere 1998, Karlsson and Quintiere 1999, and Drysdale 1999.

Fire is simply a rapid oxidation reaction or combustion process involving the release of energy. Although a fire's energy release may be visible in the form of flames or glowing combustion, nonvisible forms exist. Nonvisible forms of energy transfer may include heat transfer in the form of radiation or conduction (Quintiere 1997). Oxidation processes like rusting of iron or yellowing of newsprint generate heat but so slowly that their temperatures do not increase perceptively.

FIRE TETRAHEDRON

For fires to exist, four conditions must coexist, which may be represented by what is commonly known as the **fire tetrahedron**, shown in Figure 2.1. The basic components of the tetrahedron are the presence of **fuel** or combustible material, sufficient **heat** to raise the material to its ignition temperature and release fuel vapors, enough of an **oxidizing agent** to sustain combustion, and conditions that allow **uninhibited exothermic chemical chain reactions** to occur. The science of fire prevention and extinguishment is based on the isolation or removal of one or more of these components.

The action of heat on a flammable liquid fuel causes simple evaporation (transforming it into a vapor). The action of heat on most solid fuels causes the chemical breakdown of the molecular structure, called **pyrolysis**, into vapors, gases, and residual solid (char). The pyrolysis also absorbs some heat from the reaction (being an endothermic process).

The flaming combustion of the gases from a liquid or solid takes place within an area above the fuel's surface. Heat and mass transfer processes create a situation in which the fuel generates a region of vapors that can be ignited and sustained with the proper conditions (Figure 2.2).

TYPES OF FIRES

There are four recognized categories of combustion generated by fire (Quintiere 1997). **Diffusion flames** make up the category of most natural fuel-controlled flaming

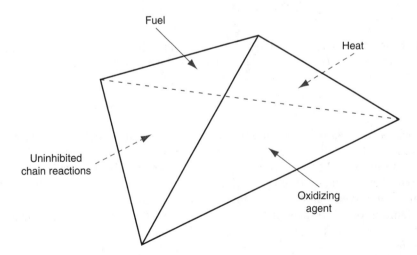

FIGURE 2.1 ◆ The fire tetrahedron.

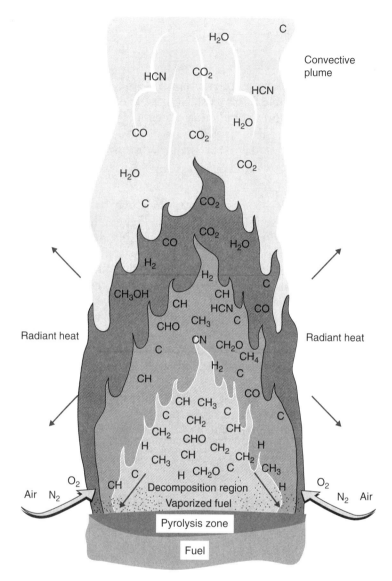

FIGURE 2.2 ◆ Schematic representation of the heat and mass transfer processes associated with a burning surface. *From J. D. DeHaan,* Kirk's Fire Investigation, *5th ed., 2002. Reprinted by permission of Pearson Education, Inc., Upper Saddle River, NJ.*

fires such as candles, campfires, and fireplaces. These occur as fuel gases or vapors diffuse from the fuel surface into the surrounding air. In the zone where appropriate concentrations of fuel and oxygen are present, flaming combustion will occur.

Combining of fuel and oxygen prior to ignition results in a combustion process known as **premixed flames**. These fuels consist of either vaporized liquid fuels or gases. Ignition of these fuels or gas mixtures is possible within regions bounded by upper and lower fuel/oxygen concentrations. In a jet engine, a special type of flames can occur when the fuel vapor is released under pressure into the air and mixes by mechanical action to produce an ignitable mixture.

Smoldering is typically a slow exothermic process in which oxygen combines directly with the fuel on its surface (or within it, if it is porous). If enough heat is being produced, the surface may glow with incandescent reaction zones. Smoldering fires are characterized by visible charring and absence of flame. A discarded or dropped cigarette on the surface of a couch, mattress, or other upholstered furniture can cause a smoldering fire.

The terms **glowing combustion** and **smoldering combustion** are often used interchangeably, but we should note there is a distinction. Both terms refer to flameless oxidative exothermic combustion where oxygen is combining directly with the solid fuel surface creating heat. If this heat is produced at such a rate that the oxidizing surface of the solid can reach temperatures greater than about 550°C (1000°F) (Drysdale 1999, 53), a "glow" visible to the unaided human eye can be observed. If ventilation is aided (by draft or extra oxygen) the glowing can be maintained (as observed when using a bellows on a smoldering wood fire). This is usually not a self-sustaining reaction. A torch flame played against a wood surface can cause it to char and glow where the flame is applied, but as soon as the flame is removed, the glow stops and the combustion diminishes, often to a complete stop. Smoldering combustion can best be described as a self-sustaining flameless combustion where the heat generated by the oxidation heats surrounding fuel to the point where the combustion front advances and continues to grow. Note that the layperson often uses "smoldering" to mean "burning with flames that are not large," but this is not the correct scientific definition.

A similar error creeps in when a target surface heated by radiant or convective heat transfer is observed to darken and emit white vapors. This process is sometimes erroneously described as smoldering. It is not. Rather, it is the heat-induced pyrolysis of the fuel surface causing the formation of wisps of water vapor and other vapors, sometimes called **off-gassing**. It is an endothermic process (absorbing heat), and when the external heating is halted, the vapor or smoke emission stops as the surface temperature drops rapidly.

Like smoldering, **spontaneous combustion** is another slow chemical reaction. Self-heating can produce enough heat in a mass of fuel that results in flaming or sustained smoldering combustion at a critical point known as *thermal runaway*. Spontaneous combustion usually involves naturally occurring vegetable-based fuels, such as those found in peanut and linseed oils (DeHaan 2007, chap. 6; Babrauskas 2003).

◆ 2.3 HEAT TRANSFER

During a fire, heat is traditionally transferred from the fire plume to surrounding objects, surfaces, and humans by three fundamental methods:

- ◆ Conduction
- ◆ Convection
- ◆ Radiation

The relative importance of these three methods that dominate a fire depends on not only the intensity and size of the fire but also the physical environment. The heat transferred increases the surface temperature of the target, which can cause observable and measurable effects.

FIGURE 2.3 ◆ Example of internal conductive heat transfer through a car door from a fire inside the vehicle. *Courtesy of Ross Brogan, NSW Fire Brigades, Greenacre, NSW, Australia.*

CONDUCTION

Through the process of **conduction**, thermal energy passes from a warmer to a cooler area of a solid material. Heat conducted through walls, ceilings, and other adjacent objects often leaves signs of fire damage based on the varying temperature gradients (Figure 2.3). As heat is conducted through a material, there will often be rings or bands of varying thermal damage.

The general equations used in conductive heat transfer through a solid whose faces are at different temperatures, T_1 (cooler) and T_2 (warmer), are

$$\dot{q} = \frac{kA(T_2 - T_1)}{l} = \frac{kA\,(\Delta T)}{l} \tag{2.1}$$

$$\dot{q}'' = \frac{\dot{q}}{A} = \frac{kA(\Delta T)}{Al} = \frac{k\,(\Delta T)}{l} \tag{2.2}$$

where

$\dot{q}$ = conduction heat transfer rate (J/s or W),

$\dot{q}''$ = conductive heat flux (W/m^2),

k = thermal conductivity of material (W/m-K),

A = area through which the heat is being conducted (m^2), and

$\Delta T = (T_2 - T_1)$ wall face temperature differences between the warmer T_2 and cooler T_1.

l = length of material through which heat is being conducted

Conduction requires direct physical contact between the source and the target receiving the heat energy (Quintiere 1997). For example, heat from a fire in a living room may be conducted through a poorly insulated common wall to an adjoining room, producing surface demarcation patterns. In an investigation, examine and

TABLE 2.2 ◆ Thermal Characteristics of Common Materials Found at Fire Scenes

Material	Thermal Conductivity k (W/m-K)	Density ρ (kg/m³)	Specific Heat c_p (J/kg-K)	Thermal Inertia $k\rho c_p$ (W²s/m⁴K²)
Copper	387	8940	380	1.30×10^9
Steel	45.8	7850	460	1.65×10^8
Brick	0.69	1600	840	9.27×10^5
Concrete	0.8–1.4	1900–2300	880	2×10^6
Glass	0.76	2700	840	1.72×10^6
Gypsum plaster	0.48	1440	840	5.81×10^5
PPMA	0.19	1190	1420	3.21×10^5
Oak	0.17	800	2380	3.24×10^5
Yellow pine	0.14	640	2850	2.55×10^5
Asbestos	0.15	577	1050	9.09×10^4
Fiberboard	0.041	229	2090	1.96×10^4
Polyurethane foam	0.034	20	1400	9.52×10^2
Air	0.026	1.1	1040	2.97×10^1

Source: Data derived from Drysdale 1999, 33.

record damage to both sides of interior and exterior walls and surface materials. Interior wall finishing materials such as wooden paneling may also have to be removed to evaluate the effects of heat transfer. See Table 2.2 for the thermal characteristics of typical materials found at fire scenes.

Shown in Figure 2.4 are graphs of temperature during short- and medium-duration conductions and steady-state conduction through a solid material. During short-duration conduction, heat flux events produce high surface temperatures, with the inner core staying much cooler. Increasing the time duration of heat transfer increases the internal temperature of the solid, eventually producing a nearly linear temperature gradient from front (hot) to rear (cool).

CONVECTION

The second process, **convection**, is the transfer of heat energy via movement of liquids or gases from a warmer to a cooler area, also known as **Newton's law of cooling**. An indicator of convective heat transfer damage may be found in the region where a fire plume contacts the ceiling directly above the plume, as shown in Figure 1.7.

The general equations used in convective heat transfer are

$$\dot{q} = hA(T_\infty - T_s) = hA(\Delta T) \tag{2.3}$$

$$\dot{q}'' = \frac{\dot{q}}{A} = \frac{hA(\Delta T)}{A} = h\Delta T \tag{2.4}$$

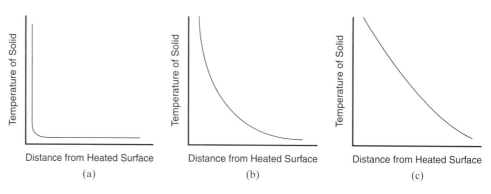

FIGURE 2.4 ◆ Temperature distribution through conduction for (a) short duration, (b) medium duration, and (c) steady state.

where

$$\dot{q} = \text{convective heat transfer rate (J/s or W)},$$

$$\dot{q}'' = \text{convective heat flux (W/m}^2),$$

$$h = \text{convective heat transfer coefficient (W/m}^2\text{K}),$$

$$A = \text{area through which the heat is being convected (m}^2), \text{ and}$$

$$\Delta T = (T_\infty - T_s) \text{ temperature differences of the fluid, } T_\infty, \text{ and the surface, } T_s.$$

The term h is the convective heat transfer coefficient, expressed in W/m²K. It is dependent on the nature of the surface and the velocity of the gas. Typical values for air are 5–25 W/m²K for free convection and 10–500 W/m²K for forced convection (Drysdale 1999, 33).

Closer examination of surface features will usually indicate the direction and intensity of the fire plume and ceiling jet, which contain heated products of combustion such as hot gases, soot, ash, pyrolysates, and burning embers. The most extensive damage normally exists at the plume's impingement on the ceiling directly above the fuel source. Generally, materials farther from the fire plume will exhibit less damage. Figure 2.5 shows the external convective heating of a car door from a gasoline fire of short duration.

RADIATION

Radiation is the emission of heat energy via electromagnetic waves from a surface due to it being at a temperature above absolute zero (0 K). Heat transfer by radiation often damages surfaces facing the fire plume, including immobile objects such as furniture. Convection and radiation are responsible for most victim burn injuries. Radiant heat travels in straight lines. It can be reflected by some materials and transmitted by others. The total thermal effect on a material's target surface may be a combination of conductive, convective, and radiative heat transfer rates.

The equation for **radiative heat flux** received by the target surface is represented in general terms by

$$\dot{q}'' = \epsilon\sigma T_2^4 F_{12}, \tag{2.5}$$

where

$$\dot{q}'' = \text{radiative heat transfer flux (kW/m}^2),$$

$$T_2 = \text{temperature of the emitting body (K)},$$

FIGURE 2.5 ◆ Example of external heating of a car door from a very-short-duration gasoline fire that only scorched the paint on the door. The gasoline was poured outside the door with its window closed. The glass shattered by convected and radiated heat. *Courtesy of J. D. DeHaan.*

ϵ = emissivity of the hot surface (dimensionless),

σ = Stefan–Boltzmann constant = 5.67×10^{-11} (kW/m^2)/K^4 *or* kW/m^{2}K^4, and

F_{12} = configuration factor (depends on the surface characteristics, orientation, and distance).

On heated solid and liquid surfaces the emissivity is commonly 0.8 ± 0.2. The emissivities for gaseous flames depend on their fuel and thickness, and may be very low if the layer is thin. Most are approximately 0.5–0.7 (Quintiere 1998). The incident heat flux falling on an object or other surface is discussed further in a later section.

Guidelines for interpreting the damage due to radiation are simple—the investigator should generally start with the surfaces farthest from the apparent fire plume, systematically advancing toward the apparent area and point of fire origin. Areas of more intense damage relative to these surfaces often were closer to the heat source, as shown in Figure 1.5. By comparing these damaged surfaces, an investigator can often determine the fire's movement and intensity as it burned. After this evaluation, the investigator reverses his/her steps, tracing the damage back away from the point of fire origin.

SUPERPOSITION

Superposition is the combined effects of two or more fires or heat transfer effects, a phenomenon that may generate confusing fire damage indicators. The predominant heat transfer mode through direct flame impingement (flame contact, for example) involves both radiant and convective heat transfer from the flaming gases to the surface as well as some conduction through the surface. Direct flame impingement is a case of superposition in which the high radiant heat flux and high convective transfer of the flame quickly affect the exposed fuels. Such combined effects can quickly ignite fuels and raise others to very high surface temperatures.

Because the goal of an arsonist is often the rapid destruction of a building, multiple-set fires are often encountered. These fires, which initially consist of separate plumes, can confuse even the most experienced investigator, particularly when their sequence of ignition is unknown, and the effects of separate fires overlap. Note also that separate multiple fires can occur when there is a collapse or "falldown" of burning materials, such as hanging draperies or window shades involved as fire spreads from a single origin. If multiple points of fire origin are suspected, the investigator should evaluate the combined heat transfer effects or superposition of these plumes.

EXAMPLE 2.1 Heat Transfer Calculations

Problem. A fire in a storage room in an automobile repair garage raises the temperature of the interior surface of the brick walls and ceiling over a period of time to 500°C (932°F or 773 K). The ambient air temperature on the outside of the brick wall to the garage is 20°C (68°F or 293 K). The thickness of the brick wall is 200 mm (0.2 m), and its convective heat transfer coefficient is 12 W/m²K. Calculate the temperature to which the exterior of the wall will be raised.

Suggested Solution—Conductive-Convective Problem. In this solution, the configuration is considered a plane wall of uniform homogenous material (brick) having a constant thermal conductivity, a given interior surface temperature of $T_s = 500°C$ (932°F or 773 K), and exposed to an external garage ambient air temperature of $T_a = 20°C$ (68°F or 293 K). The external surface temperature, T_E, of the wall is unknown. This temperature might be important later if combustible materials come into contact with this wall surface.

Since the steady-state heat transfer rate of conduction through the brick wall is equal to the heat transfer rate of convection passing into the ambient air outside the garage,

$$\frac{\dot{q}}{A} = \frac{T_s - T_E}{\frac{\Delta x}{K_a}} = \frac{T_E - T_a}{\frac{1}{h_G}}$$

$$\frac{773 \text{ K} - T_E}{\frac{0.2 \text{ m}}{0.69 \text{ W/mK}}} = \frac{T_E - 293 \text{ K}}{\frac{1}{12 \text{ W/m}^2\text{K}}}$$

$$2665 - 2.448 \, T_E = 12.05 \, T_E - 3530$$

$$T_E = \frac{6195}{15.49} = 399 \text{ K} = 126°C (258°F),$$

where

$\dot{q}$ = conduction heat transfer rate (J/s or W),
$\dot{q}''$ = conductive heat flux (W/m²),
h = convective heat transfer coefficient (W/m²K) (brick to air),
k = thermal conductivity of material (W/m-K) (through brick),
A = area through which the heat is being conducted (m²), and
$\Delta T = (T_S - T_E)$ wall face temperature differences between the warmer T_s and cooler T_E.

Suggested Alternative Solution—Electrical Analog. Assuming that the brick maintains its integrity during the fire, determine the steady-state temperature of the wall's surface in the garage (example modified from Drysdale 1999, 72). A one-dimensional electrical circuit analog, shown in Figure 2.6, is used in this solution, because we can look at the way heat transfers through a material in the same way that current flows through resistors.

Solutions by electrical analogs are often used. The steady-state heat transfer solution is best illustrated by a direct-current electrical circuit analogy, where the temperatures are voltages ($V = T$), the thermal resistances of heat convection and conduction are represented as electrical resistances ($R = 1/h_h$ and $R = L_n/k_n$, respectively). The heat flux, $\dot{q}''$, whose units are W/m², is analogous to the current; that is $I = \dot{q}''$.

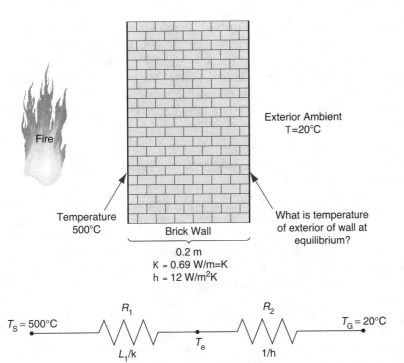

FIGURE 2.6 ◆ (a) Heat transfer by conduction: surface temperature calculation. (b) Heat transfer by conduction: surface temperature using electrical analog.

The problem can be represented as shown in Figure 2.6b, as a steady-state electrical circuit analog, where R_1 and R_2 are the thermal resistances of the brick wall and inner air of the garage.

Temperature in storage room	$T_S = 500°C$ (773K)
Temperature in garage	$T_G = 20°C$ (293 K)
Thermal conductivity	$k = 0.69$ W/mK
Thickness of the brick	$L = 200$ mm (0.2 m)
Conductive thermal resistance	$R_1 = \Delta L/k = (0.2\text{ m})/(0.69\text{ W/mK}) = 0.290\text{ m}^2\text{K/W}$
Convective heat transfer coefficient	$h = 12$ W/m²K
Convective resistance	$R_2 = 1/h = (1)/(12\text{ W/m}^2\text{K}) = 0.083\text{ m}^2\text{K/W}$
Total heat flux	$\dot{q}'' = (T_S - T_G)/(R_1 + R_2) = 1287\text{ W/m}^2$
	$T_E = 773\text{ K} - (\dot{q}'')(R_1)$
Exterior wall temperature T_E	$= 773\text{ K} - 374\text{ K}$
	$= 399\text{ K}$
	$T_E = 399\text{ K} - 273 = 126°C\ (258°F)$

◆ 2.4 HEAT RELEASE RATE

RELEASE RATES

The **heat release rate** (HRR) of a fire is the amount of heat released per unit time by a heat source. The term is often expressed in watts (W), kilowatts (kW), megawatts

(MW), kilojoules per second (kJ/s), or BTUs per second (BTU/s) and is denoted in formulas as $\dot{Q}$ (the dot over the Q means per unit time). For general conversion purposes, 3412 Btu/hr = 0.95 BTU/s = 1 kJ/s = 1000 W = 1 kW.

The HRR is essentially the size or power of the fire. Three important effects of HRR make it the single most important variable in describing a fire and the hazards associated with its interaction with buildings and their occupants (Babrauskas and Peacock 1992):

- Creation of more heat
- Correlation with other effects (smoke production, temperatures)
- Survivability of occupants

First, and most important, heat released by a fire is the driving force for that fire subsequently to produce more heat by creating more fuel by evaporation or pyrolysis. As long as adequate quantities of fuel and oxygen are available, the transfer of heat from the flames of the burning fuel causes the formation of more heat to be released from it and surrounding fuels. It is not guaranteed that all the fuel will burn. As with burning characteristics of fuels, there are physical limits to theoretical and actual heat release rates.

Second, another important role of the heat release rate in forensic fire scene reconstruction is that it directly correlates with many other variables. Examples include the production of smoke and toxic by-products of combustion, room temperatures, heat flux, mass loss rates, and flame height impingement.

Third, the direct correlation of high heat release rates and lethality of a fire is significant. High HRRs produce high mass loss rates of burning materials, some of which may be very toxic. High heat fluxes, large volumes of high-temperature smoke, and toxic gases may overwhelm occupants, preventing their safe escape during fires (Babrauskas and Peacock 1992).

The concept of heat release rates was introduced to fire investigators in 1985, when Drysdale set forth formulas for the calculation of estimated flame heights and onset time for flashover. These equations attempted to answer the following important questions:

- How intense was the fire, and could it ignite nearby combustibles or produce thermal injuries to humans?
- Was an ignitable liquid of sufficient quantity used in the fire, and how high did the flames reach?
- Were conditions right for flashover to occur?
- When did the smoke detectors and/or sprinklers activate?

An investigator's most basic question centers on the peak heat release rate, which can range from several watts to thousands of megawatts. Table 2.3 lists peak HRRs for common burning items of potential interest to fire investigators.

The peak heat release rate is dependent on the fuel present, its surface area, and its physical configuration. The HRRs of many typical fuels are available on the NIST website, *fire.nist.gov.* Other concepts in fire dynamics relate to the HRR. These include mass loss, mass flux, heat flux, and combustion efficiency.

The mass loss rate $\dot{m}$ of a fuel typically depends on three factors: type of fuel, configuration, and area involved. For example, a campfire constructed with small sticks of dry wood arranged in a tent configuration will burn more quickly and yield a higher heat release rate over a shorter period of time than one using wet, large-dimensioned wood placed in a haphazard arrangement.

TABLE 2.3 ◆ Peak Heat Release Rates for Common Objects Found at Fire Scenes

Material	Total Mass (kg)	Peak Heat Release Rate (kW)	Time (s)	Total Heat Released (MJ)
Cigarette	–	.005 (5 W)	–	–
Wooden kitchen match or cigarette lighter	–	.050 (50 W)	–	–
Candle	–	.05–.08 (50–80 W)	–	–
Wastepaper basket	0.94	50	350	5.8
Office wastebasket with paper	–	50–150	–	–
Pillow, latex foam	1.24	117	350	27.5
Small chair (some padding)	–	150–250	–	–
Television set (T1)	39.8	290	670	150
Armchair (modern)	–	350–750 (typical)–1.2 MW	–	–
Recliner (synthetic padding and covering)	–	500–1000 (1 MW)	–	–
Christmas tree (T17)	7.0	650	350	41
Pool of gasoline (2 qt, on concrete)	–	1 MW	–	–
Christmas tree (dry, 6–7 ft)	–	1–2 MW (typical)—5 MW	–	–
Sofa (synthetic padding and covering)	–	1–3 MW	–	–
Living or bedroom (fully involved)	–	3–10 MW	–	–

Source: Derived from Babrauskas (SFPE 2002b, sec. 3-1) and DeHaan 2007, 35.

A material's heat of combustion is the amount of heat released per unit of mass during combustion, expressed in MJ/kg or kJ/g. The term Δh_c denotes the *complete* heat of combustion and is used when all the material combusts leaving no fuel behind. The effective heat of combustion, denoted by Δh_{eff}, is reserved for more realistic cases where combustion is not totally complete and the heat lost to heating the fuel is considered.

MASS LOSS

The heat release rate can be calculated from the **mass loss rate** and heat of combustion of an object. This "burning rate" is expressed in kg/s or g/s and is denoted by $\dot{m}$ in most equations. The heat release rate can be calculated using the general equation (2.6). Mass loss rates can be experimentally determined by simply weighing a fuel package while it burns.

The heat release rate, $\dot{Q}$, can be calculated from

$$\dot{Q} = \dot{m}\,\Delta h_c, \tag{2.6}$$

where

$\quad \dot{Q} \quad =$ heat release rate (kJ/s or kW),
$\quad \dot{m} \quad =$ burning rate or mass loss rate (g/s), and
$\quad \Delta h_c =$ heat of combustion (kJ/g).

MASS FLUX

Another concept related to the heat release rate is a material's **mass flux** or mass burning rate per unit area, expressed in kg/(m^2 s) and denoted by $\dot{m}''$ in equations. It is measured in laboratory tests where the fuel surface area (pool diameter) and orientation are controlled variables, as they can affect $\dot{m}''$ dramatically. When the horizontal burning area of a material is known along with the burning rate per unit area, the heat release rate equation is written

$$\dot{Q} = \dot{m}''\,\Delta h_c A, \tag{2.7}$$

where

$\quad \dot{Q} \quad =$ total heat release rate (kW),
$\quad \dot{m}'' \quad =$ mass burning rate per unit area (g/m^2-s),
$\quad \Delta h_c =$ heat of combustion (kJ/g), and
$\quad A \quad =$ burning area (m^2).

The amount of heat needed to convert a solid or liquid fuel to a combustible form (gas or vapor) is called the latent **heat of vaporization** and represents heat that has to be created to initiate and sustain combustion. The energy needed to evaporate a mass unit of fuel, expressed in kJ/kg, is denoted by H_L in equations. Solids, unlike liquids, have H_L values that are grossly time dependent. It is very hard to obtain accurate H_L values for solids. Values for Δh_c, $\dot{m}''$, and H_L are listed in Table 2.4 and are available in the reference literature for various fuels.

HEAT FLUX

Another fire dynamics concept vital to interpreting ignition, flame spread, and burn injuries is **heat flux**. Heat flux is the rate at which heat is falling on a surface or passing through an area, measured in kW/m^2 and represented by the symbol $\dot{q}''$.

The radiant heat output (emissive power) of a surface at temperature T_2 is $\dot{Q}'' = \epsilon\sigma T^4$ (where $\epsilon =$ emissivity of the source fire, $\sigma =$ Stefan-Boltzmann constant, and $T =$ temperature of the surface in Kelvin), so the higher the temperature of a source, the greater the heat flux from that source. Note that it increases as a function of temperature, T (K), to the fourth power.

Two major pieces of information can be obtained if an investigator can estimate the heat flux and its duration: (1) whether ignition of a secondary or target surface

TABLE 2.4 ◆ Typical Heats of Mass Burning Rates, Vaporization, and Heats of Combustion for Common Materials

Common Material	Mass Flux $\dot{m}''$ (g/m² s)	Heat of Vaporization, H_L (kJ/g)	Heat of Combustion (effective), Δh_{eff} (kJ/g)
Gasoline	44–53	0.33	43.7
Kerosene	49	0.67	43.2
Paper	6.7	2.21–3.55	16.3–19.7
Wood	70–80	0.95–1.82	13–15

Source: Derived from SFPE (2002b) and Quintiere (1997).

can be achieved and (2) whether thermal injury to a victim exposed to a fire is possible. The minimum heat flux needed to produce thermal injuries and to ignite some common fuels is presented in Table 2.5 (NFPA 2004a, Table 5.5.4.2.8).

Heat fluxes from various flames and fires are listed in detail in the *Ignition Handbook* (Babrauskas 2003, chap. 1). Because in most cases these values are difficult to measure and even more difficult to accurately estimate, the investigator should study carefully these published experimental data.

In the simplest of cases, radiant heat flux from a fire plume can be approximated as emerging from a single point source known as the **virtual origin**, a concept discussed in more detail in Chapter 3. A schematic illustration of the concept of radiant heat transfer from a point source of a flame to a target is shown in Figure 2.7 (NFPA 2000a, chaps. 10–11).

TABLE 2.5 ◆ Impact of a Range of Radiant Heat Flux Values

Radiant Heat Flux (kW/m²)	Observed Effect on Humans and Wooden Surfaces
170	Maximum heat flux measured in postflashover fires
80	Thermal protective performance (TPP) test
52	Fiberboard ignites after 5 s
20	Floor of residential family room at flashover
16	Pain, blisters, second-degree burns to skin at 5 s
7.5	Wood ignites after prolonged exposure (piloted or not)
6.4	Pain, blisters, second-degree burns to skin at 18 s
4.5	Blisters, second-degree burns to skin at 30 s
2.5	Exposure while firefighting, pain and burns after prolonged exposure
<1.4	Exposure to sun

Source: Derived from *NFPA 921* (NFPA 2004a, Table 5.5.4.2.8) and Gratkowski, Dembsey, Beyler 2006.

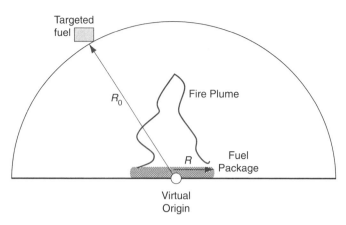

FIGURE 2.7 ◆ Schematic illustration of radiant heat transfer from a point source (virtual origin) of a fire plume to a target.

The equation used to predict the incident heat flux falling on an object or victim from a "point-source" fire, as in Figure 2.7, is

$$\dot{q}_0'' = \frac{P}{4\pi R_0^2} = \frac{X_r \dot{Q}}{4\pi R_0^2}, \tag{2.8}$$

where

$\dot{q}_0''$ = incident radiation flux on target (kW/m²),

P = total radiative power of the flame (kW),

R_0 = distance to the target surface (m), and

X_r = radiative fraction (typically 0.20–0.60), representing the percentage of heat released from the source as radiant heat,

$\dot{Q}$ = total heat release rate of source (kW),

or, by rearranging,

$$R_0 = \left(\frac{X_r \dot{Q}}{4\pi \dot{q}_0''}\right)^{1/2}. \tag{2.9}$$

This equation is valid when the ratio of the distance to the targeted object to the radius (R) of the burning fuel package is

$$\frac{R_0}{R} > 4 \tag{2.10}$$

or when the radius of the burning surface of the fire, R, is small compared with the distance between the target and the fire, R_0, as shown in Figure 2.7.

For cases where

$$\frac{R_0}{R} \leq 4, \tag{2.11}$$

the source cannot be treated as a point, and the geometry of the target surface relative to the source becomes critical. For each geometry, a configuration factor will

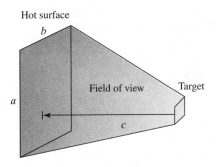

FIGURE 2.8 ◆ A small target facing a large radiant heat source (*a* by *b* in size) a distance *c* away is subject to heating from a wide angle of view.

have to be calculated. The configuration factor (F_{12}) appears in the general radiant heat relation

$$\dot{q}'' = \epsilon\sigma F_{12}T^4 \quad \text{or} \quad \dot{Q}F_{12}$$

Because the calculations can be very complex, F_{12} for a geometry such as that in Figure 2.8 is usually determined from a nomograph such as that in Figure 2.9. Nomographs for various geometries are found in the *SFPE Handbook* (SFPE 2002b, App. D, A43–46).

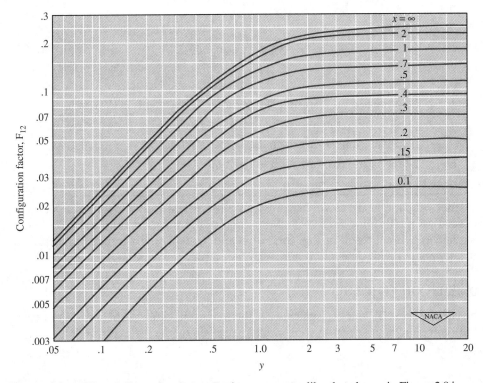

FIGURE 2.9 ◆ The configuration factor F_{12} for a geometry like that shown in Figure 2.8 is taken from this nomograph, where $x = a/c$ and $y = b/c$. The curve with the closest value for x is selected. Where that curve intersects the vertical line equal to the calculated value of y determines F_{12}. *From Hamilton and Morgan, NACA Tech. Note 2836, December 1952.*

EXAMPLE 2.2   Heat Flux Calculation

Problem.   A 500-kW fire represented as a point source as shown in Figure 2.7 has a surface radius of 0.2 m. Assume the radiative fraction to be 0.60. What is the heat flux falling on the target fuel at distances of R_0 equal to 1, 2, and 4 m?

Suggested Solution.   The expected heat release rate can be calculated from equation (2.6).

Heat release rate	$\dot{Q}$ = 500 kW
Radiative fraction	X_r = 0.60
Radius of the burning fuel surface	R = 0.20 m
Equation ratio test	$R_0/R > 4$
Incident radiation flux on target	$\dot{q}_0'' = X_r\dot{Q}/4\pi R_0^2$
At R_0 equal to 1, 2, and 4 m	$\dot{q}_0''$ = 23.87, 5.96, and 1.49 kW/m^2

Note that as the distance is doubled, the radiant heat flux decreases by a factor of one-fourth. This is known as the *inverse square* rule.

STEADY BURNING

Once a material is exposed to sufficient incident heat on the fuel and an ample air supply, the fire may spread evenly, become fully involved, and burn at **steady state**. The formulas for steady-state burning are

$$\dot{m}'' = \dot{q}''/H_L \qquad (2.12)$$

and

$$\dot{Q}'' = \frac{\dot{q}''}{H_L} \Delta h_c, \qquad (2.13)$$

where

$\dot{Q}''$ = heat release rate per unit surface area of burning fuel (kW/m^2),
$\dot{m}''$ = mass burning rate per unit area (g/m^2-*s*),
$\dot{q}''$ = net incident heat flux from the flame (kW/m^2),
H_L = latent heat of vaporization (kJ/g), and
Δh_c = heat of combustion (kJ/g).

These equations are useful because they allow investigators to make *estimates* of the steady-state burning rate. It should be noted that for most solid fuels, H_L is not constant, as a direct result of char formation and heat and mass transfer processes. It can vary greatly with time and is not easily measured.

COMBUSTION EFFICIENCY

The **combustion efficiency** of substance is the ratio of its effective heat of combustion to the complete heat of combustion and is expressed by the term X, where

$$X = \Delta h_{\text{eff}}/\Delta h_c. \qquad (2.14)$$

Thus fuels with more complete combustion produce values of X approaching 1. Fuels with lower combustion efficiencies—characterized by flames containing products of incomplete combustion, which produce sootier and more luminous fires—will result in values of X between 0.6 and 0.8 (Karlsson and Quintiere 1999, 64). For example, gasoline typically has a Δh_c of 46, but its Δh_{eff} is 43.8 owing to incomplete combustion ($X = 0.95$). Examples of soot-producing substances include many flammable liquids and thermoplastics.

To take into account incomplete combustion (for fuels such as plastics and ignitable liquids) the combustion efficiency can be inserted into the equation for the convective heat release rate, $\dot{Q}_c$, where X_r is the energy loss attributable to radiation (typically 0.20 to 0.40).

$$\dot{Q}_c = X\dot{m}''\Delta h_c A(1 - X_r). \qquad (2.15)$$

In calculations of mean flame heights and virtual origins, which are explored in Chapter 3, the total heat release rate, Q, is used. When other properties of fire plumes such as plume radius, centerline temperatures, and velocity are calculated, the convective heat release rate, $\dot{Q}_c$, is used (Karlsson and Quintiere 1999, 64).

EXAMPLE 2.3 Ignitable Liquid Fire

Problem. An arsonist pours a quantity of gasoline onto a concrete floor, creating a 0.46-m (1.5-ft)-diameter pool. The arsonist ignites the gasoline, creating a fire very similar to a classic burning pool. Estimate the heat release rate from the burning pool using historical fire testing results.

Suggested Solution. Assuming an unconstrained pool fire, the expected heat release rate can be calculated from equation (2.8).

Heat release rate	$\dot{Q}$	$= \dot{m}''\Delta h_{eff}A$
Area of the burning pool	A	$= (\pi/4)(D^2) = (3.1415/4)(0.46^2) = 0.164\ m^2$
Mass flux rate for gasoline	$\dot{m}''$	$= 0.036\ kg/m^2\text{-}s$ (reduced by incomplete combustion)
Heat of combustion for gasoline	Δh_{eff}	$= 43.7\ MJ/kg$
Heat release rate	$\dot{Q}$	$= (0.036)(43,700)(0.164) = 258\ kW$

The heat release rate of a real fire under these conditions would be expected to be in the range 200–300 kW; but in actuality, a thin layer of gasoline will show a heat release rate of only approximately 25 percent of the value computed by the preceding equations, which were developed for pools of substantial depth. For detailed discussion, see the work by Ma et al. (2004).

◆ 2.5 FIRE DEVELOPMENT

Fires, particularly those within enclosed structures, burn in a predictable manner over time. For the purpose of fire scene reconstruction, development can be divided into four separate **fire phases**, each containing unique characteristics and its own time frame. These phases are applicable to both single fuel packages as well as compartment fires. These fire development curves are often referred to as the **fire signature**. The phases are as follows:

- ◆ Phase 1—Incipient ignition
- ◆ Phase 2—Growth
- ◆ Phase 3—Fully developed fire
- ◆ Phase 4—Decay

As shown in Figure 2.10, these phases form a characteristic "fire signature" or "design curve" fire, which is often studied for items found as the source of the first

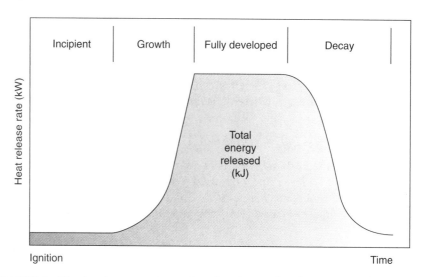

FIGURE 2.10 ◆ Fire development phases that often form a fire signature, consisting of (1) ignition, (2) growth, (3) fully developed fire, and (4) decay. *From the* SFPE Handbook of Fire Protection Engineering, *3rd ed., 2002, by permission of Society of Fire Protection Engineers.*

fuel burned in typical structural fires. These phases are important when predicting a fire's growth, physical effects on the room and its contents, calculating the activation of smoke, heat, and automated sprinkler systems, estimating evacuation times, and determining the risk of heat exposure to its occupants.

The following physical features are associated with each fire development phase:

- ◆ Phase 1—Low heat, some smoke, no detectable flame
- ◆ Phase 2—More heat, lots of smoke, flame spread
- ◆ Phase 3—Massive heat and smoke output, extensive flame
- ◆ Phase 4—Decreased flame and heat, considerable smoke, with increased smoldering

At the transition from phase 2 to phase 3, **flashover** occurs, depending on the amount of heat produced during the initial stages of the fire and the size and configuration of the room of fire origin. In the absence of flashover, the fire may not become fully developed and instead may reach a maximum size because of limited available fuels, flame spread, or a ventilation limit. Flashover is discussed in detail later.

PHASE 1: INCIPIENT IGNITION

The fundamental properties that influence an object's **ignitability** are its density, ρ, heat capacity, c_p, and thermal conductivity, k. Table 2.2 listed these characteristic values for some common materials found at fire scenes. The product of these three characteristics ($k\rho c_p$) is often referred to as an object's **thermal inertia** and the calculated value $k/\rho c_p$ is called the **thermal diffusivity**. Like mechanical inertia, where the higher the inertia, the harder it is to move, the higher the thermal inertia, $k\rho c_p$, the harder a material is to ignite—that is, the more heat has to be applied or the longer the duration of exposure required before ignition.

Table 2.6 lists the ignition properties of some common fuels. Note that T_{ig} is the approximate surface temperature observed at ignition, and $\dot{q}''_{critical}$ is the

TABLE 2.6 ◆ Ignition Properties

Material	$k\rho c$ $(kW/m^2K)s^2$	T_{ig} (°C)	$\dot{q}''_{critical}$ (kW/m^2)
Plywood, plain (0.635 cm)	0.46	390	16
Plywood, plain (1.27 cm)	0.54	390	16
Plywood, FR (1.27 cm)	0.76	620	44
Hardboard (6.35 mm)	1.87	298	10
Hardboard (3.175 mm)	0.88	365	14
Hardboard, gloss paint (3.4 mm)	1.22	400	17
Hardboard, nitrocellulose paint	0.79	400	17
Particleboard (1.27-cm stock)	0.93	412	18
Douglas fir particleboard (1.27 cm)	0.94	382	16
Fiber insulation board	0.46	355	14
Polyisocyanurate (5.08 cm)	0.020	445	21
Foam, rigid (2.54 cm)	0.030	435	20
Foam, flexible (2.54 cm)	0.32	390	16
Polystyrene (5.08 cm)	0.38	630	46
Polycarbonate (1.52 mm)	1.16	528	30
PMMA, type G (1.27 cm)	1.02	378	15
PMMA, polycast (1.59 mm)	0.73	278	9
Carpet #1 (wool, stock)	0.11	465	23
Carpet #2 (wool, untreated)	0.25	435	20
Carpet #2 (wool, treated)	0.24	455	22
Carpet (nylon/wool blend)	0.68	412	18
Carpet (acrylic)	0.42	300	10
Gypsum board, common (1.27 mm)	0.45	565	35
Gypsum board, FR (1.27 cm)	0.40	510	28
Gypsum board, wallpaper	0.57	412	18
Asphalt shingle	0.70	378	15
Fiberglass shingle	0.50	445	21
Glass-reinforced polyester (2.24 mm)	0.32	390	16
Glass-reinforced polyester (1.14 mm)	0.72	400	17
Aircraft panel, epoxy fiberite	0.24	505	28

Source: From Quintiere and Harkleroad 1984.

minimum heat flux needed to bring about ignition (Drysdale 1999, 33). Note that these results are applicable to short-term exposures (10 to 30 min) of fresh, whole products. If heating is prolonged (hours), much lower $\dot{q}''_{critical}$ values may be found. For example, wood products may ignite at approximately 7.5 kW/m² under such conditions. These values do not include long-term heating (months to years), for which the self-heating properties of wood and some other materials will have to be recognized.

The temperature of an object is a measure of the relative amount of heat (energy) contained within it. This heat is determined by the mass of the object and its heat capacity or specific heat. The heat capacity (c_p) of an object is a property that determines how much heat must be added to the object to increase its temperature.

The ignition of an object can be calculated when it is exposed to radiant or convective heat flux, $\dot{q}''$, with no initial contact with flames (autoignition). The calculation assumes that a first burning object (ignition source) provides sufficient radiant heat flux to ignite a second, nonburning fuel object nearby. Experimental fire testing by Babrauskas has established three levels of ignitability: **easy, normally resistant**, and **difficult** (SFPE 2002c). This fire testing included paper, wood, polyurethane, and polyethylene fuels to establish these relationships.

Thin materials such as paper and curtains will easily be ignited when exposed to a threshold radiant flux of 10 kW/m² or greater. Thicker or more massive items with low thermal inertia values such as upholstered furniture will ignite when exposed to a radiant flux of 20 kW/m² or greater. Solid materials with thicknesses greater than 13 mm (0.5 in.) such as plastics or dense woods having large thermal inertia values will ignite when exposed to a radiant flux of 40 kW/m² or greater (SFPE 2002c).

Equations (2.16), (2.17), and (2.18) are derived from data plots from Babrauskas (1982) published in *SFPE Engineering Guide to Piloted Ignition of Solid Materials under Radiant Exposure* (SFPE 2002c). The heat release rates, $\dot{Q}$, needed for a fire source to ignite another fuel package a distance D (measured in meters) away, can be calculated for fuels classified as *easy*, ($\dot{q}''_{crit} \geq 10$ kW/m²), *normally resistant* ($\dot{q}''_{crit} \geq 20$ kW/m²), and *difficult* ($\dot{q}''_{crit} \geq 40$ kW/m²), as shown, respectively, in the following equations:

$$\dot{Q} = 30 \cdot 10^{\left(\frac{D+0.8}{0.89}\right)}, \qquad \dot{q}'' \geq 10\,\text{kW/m}^2, \tag{2.16}$$

$$\dot{Q} = 30\left(\frac{D+0.05}{0.019}\right), \qquad \dot{q}'' \geq 20\,\text{kW/m}^2, \tag{2.17}$$

$$\dot{Q} = 30\left(\frac{D+0.02}{0.0092}\right), \qquad \dot{q}'' \geq 40\,\text{kW/m}^2, \tag{2.18}$$

where

$\dot{Q}$ = heat release rate of the source fire (kW),

D = distance from the radiation source to the second object (m).

Note that these relationships were derived for ignition of a second item from an initially burning piece of furniture. They should not be applied to heat release rates from sources that can burn indefinitely (e.g., gas jets).

The time to ignition for **thermally thin** and **thermally thick** materials can be estimated by several equations contained in engineering practice guides developed by the Society of Fire Protection Engineers (SFPE 2002a, 2002b).

The generally accepted equations for **ignition time** of thermally thin and thermally thick are (2.18) and (2.19), respectively. Under thermally thin conditions, $l_p \leq 1$ mm (l = thickness of the material),

$$t_{ig} = \rho c_p l_p \left(\frac{T_{ig} - T_\infty}{\dot{q}''} \right). \tag{2.19}$$

In thermally thick conditions, $l_p \geq 1$ mm,

$$t_{ig} = \frac{\pi}{4} k \rho c_p \left(\frac{T_{ig} - T_\infty}{\dot{q}''} \right)^2, \tag{2.20}$$

where

t_{ig} = time to ignition (s),
k = thermal conductivity (W/m-K),
ρ = density (kg/m^3),
c_p = specific heat capacity (kJ/kg-K),
l_p = thickness of material (m),
T_{ig} = piloted fuel ignition temperature (K),
T_∞ = initial temperature (K), and
$\dot{q}''$ = radiant heat flux (kW/m^2).

Note that for the thermally thick material, the thickness term, l_p, disappears. The ignition behavior of liquids is different, and the preceding equations for solids should not be used.

PHASE 2: FIRE GROWTH

Fire growth rates, as shown in Figure 2.10, can sometimes be modeled using mathematical relationships that show the flame spread and fire growth estimations using heat release rates. The growth of the flame front across a horizontal fuel surface of a solid fuel is known as the **lateral flame spread** and can be expressed by the equation (Quintiere and Harkelroad 1984)

$$V = \frac{\dot{q}''}{\rho c_p A \, (T_{ig} - T_s)^2} \tag{2.21}$$

or as

$$V = \frac{\phi}{k \rho c_p \, (T_{ig} - T_s)^2}, \tag{2.22}$$

where

V = lateral velocity of flame spread (m/s),
A = cross-sectional area affected by the heating of $\dot{q}''$,
ϕ = ignition factor from flame spread data (kW^2/m^3) (experimentally derived),
k = thermal conductivity (W/m-K),
ρ = density (kg/m^3),
c_p = specific heat capacity (kJ/kg-K),
T_{ig} = piloted fuel ignition temperature (°C), and
T_s = unignited, ambient surface temperature (°C).

TABLE 2.7 ◆ Example Lateral Flame Spread Data

Material	Piloted Ignition Temperature T_{ig} (°C)	Ambient Surface Temperature T_s (°C)	Ignition Factor ϕ (kW²/m³)
Wool carpet			
Treated	435	335	7.3
Untreated	455	365	0.89
Plywood	390	120	13.0
Foam plastic	390	120	11.7
PMMA	378	<90	14.4
Asphalt shingle	356	140	5.4
Acrylic carpet	300	165	9.9

Source: Derived from ASTM 2002.

The values for ϕ and other variables needed for calculating lateral flame spread can be obtained from *ASTM E 1321-97a* (ASTM 2002). Table 2.7 lists the horizontal flame spread data for several common materials. Obviously, an external draft pushes the flame down against the "downwind" side and increases its radiant heat effect on the fuel in front of its path, so the flame front will grow faster if wind aided.

Lateral or downward flame spread on thick solids such as wood is typically of the order of 1×10^{-3} m/s (1 mm/s). In contrast, upward spread on solid fuels is typically 1×10^{-2} to 1 m/s (10–1000 mm/s). Flame spread rates across liquid pools are typically 0.01–1 m/s, depending on the ambient temperature and the flash point of the liquid, and 0.1–10 m/s for premixed vapor/air mixtures (Figure 2.11).

An example of lateral flame spread over a fuel surface is a grass fire that can be affected by wind over a surface. For example, lateral flame spread in forests depends on several key variables: the type of fuel and its configuration, wind, humidity, and the slope of the terrain. The brush or duff along the forest floor is usually involved in the initial phases of the fire. The terrain and radiant and wind-induced convective heat transfer contribute significantly to the flame spread along the forest floor. Flame spread through the canopy of trees or brush is a different process.

The **Thomas** (1971) **formula**, which can be used to approximate flame spread in grass fires on flat ground, is

$$V = \frac{k(1 + V_\infty)}{\rho_b}, \qquad (2.23)$$

where

$\quad V \;=\;$ velocity of flame spread (m/s),

$\quad V_\infty =\;$ wind speed (concurrent) (m/s),

$\quad \rho_b \;=\;$ bulk density of the fuel (kg/m³), and

$\quad k \;\;=\;$ 0.07 wildland fire, 0.05 wood crib (kg/m³).

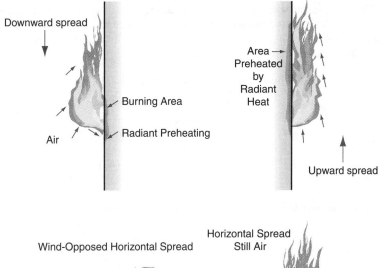

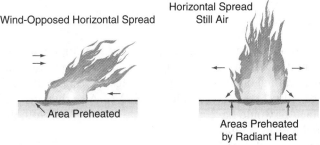

FIGURE 2.11 ◆ Radiant heat ahead of an existing flame determines its rate of spread.

EXAMPLE 2.3   Estimating Lateral Flame Travel Velocity

Problem.   An investigator determines that a fire ignited at the edge of a wool carpet (un-treated). What was the estimated velocity of the lateral flame spread assuming a high radiant heat impact causing an initial ambient surface temperature of 365°C?

Suggested Solution.   Use the Quintiere and Harkelroad equation (equation 2.22) and *ASTM 1321-97a* data, from Table 2.7.

Ignition factor	$\phi = 0.89 \text{ kW}^2/\text{m}^3$
Fuel thermal inertia	$k\rho c_p = 0.25 \, [\text{kW}^2\text{-s}/(\text{m}^4 - \text{K}^2)]$
Piloted fuel ignition temperature	$T_{ig} = 455°C$
Ambient surface temperature	$T_s = 365°C$
Velocity of flame spread	$V = \phi/k\rho c_p(T_{ig} - T_s)^2 = 0.89/0.25(455 - 365)^2$
	$= 0.33 \times 10^{-3} \text{ m/s} = 26 \text{ mm/min}$

Notice that the lower the initial temperature (T_s) is, the larger the denominator of equation (2.22) is, and the slower V will be. If the fuel surface is not preheated, the ambient surface temperature becomes $T_s = T_a = 25°C$, where T_a = ambient air temperature, and the velocity of the flame spread V is calculated as

Velocity of flame spread	$V = \phi/k\rho c_p(T_{ig} - T_s)^2 = 0.89/0.25(455 - 25)^2$
	$= 0.019 \times 10^{-3} \text{ m/s} = 1.14 \text{ mm/min}$

EXAMPLE 2.4 Wildland Arson

Problem. Children playing with matches set a fire during a dry season of the year in a flat-terrain wildland area. The wind is blowing at a velocity of 2 m/s, and the bulk density of the low brush along the forest floor is 0.04 g/cm^3. Determine the velocity of the flame spread from this fire with and without considering the effect of the wind.

Suggested Solution. Use the Thomas equation (equation 2.23).

Velocity of flame spread	$V = k(1 + V_\infty)/\rho_b$
Wind speed	$V_\infty = 2$ m/s
Bulk density	$\rho_b = 0.04$ g/cm$^3 = 40$ kg/m^3
Equation constant	$k = 0.07$ kg/m^3
Flame spread velocity (with $V_\infty = 0$ m/s)	$= (0.07)(1 + 0)/(40)$ $= 0.00175$ m/s $= 1.75$ mm/s $= 0.1$ m/min
Flame spread velocity (with $V_\infty = 2$ m/s)	$= (0.07)(1 + 2)/(40)$ $= 0.00525$ m/s $= 5.25$ mm/s $= 0.315$ m/min

Fire growth rates during the initial phases of development can sometimes be modeled using mathematical relationships. One common relationship assumes that the initial growth rate geometrically approximates the time that the fire has burned squared (also called t^2 **fires**) as shown in Figure 2.12. A perfect fire, unlimited by fuel or ventilation factors, would have an exponential growth rate. Note that this behavior applies only to the growth phase.

This accepted growth rate formula is

$$\dot{Q} = (1055/t_g^2)t^2 = \alpha t^2, \tag{2.24}$$

where

t = time (s),

t_g = time for fire to grow from ignition to 1.055 MW (1000 BTU/s),

$\dot{Q}$ = heat release rate (MW) at time t, and

α = fire growth factor (kW/s^2) for the material being burned.

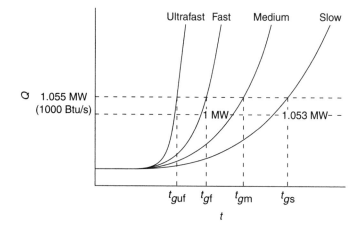

FIGURE 2.12 ◆ The fire growth factor α is derived from the graph of Q versus time, where $\alpha = (1055/t_g^2)$ and t_g is the time required for the heat release rate to grow from its baseline (incipient fire) value to 1.055 MW (1000 Btu/s).

TABLE 2.8 ◆ Four Growth Rates in t^2 Fires

Type of Fire	Example Objects	Time t_g (s)	Fire Growth Factor α (kW/s²)
Slow	Thick wooden objects (tables, cabinets, dressers)	600	0.003
Medium	Lower density objects (furniture)	300	0.012
Fast	Combustible objects (paper, cardboard boxes, drapes)	150	0.047
Ultrafast	Volatile fuels (flammable liquids, synthetic mattresses)	75	0.190

Source: Derived from *NFPA 92B* and *NFPA 72* (NFPA 2000c; 2002).

According to an accepted principle used by *NFPA 72* (detectors) and *NFPA 92B* (smoke control systems), Table 2.8 and Figure 2.13 illustrate four types of growth times used to define fires by the time required to reach 1.055 MW (1000 BTU/s). The time t_g is obtained from repeated calorimetry tests of specific fuel packages (NFPA 2000b, 2002).

Table 2.9 shows typical ranges for the heat release rates per unit area and characteristic times for a fire to reach 1.055 MW (1000 Btu/s) in size. These data were obtained from actual fire testing of common materials stored in warehouses.

The **slow growth curve** (t_g = 600 s) usually involves thick solid wooden objects such as tables, cabinets, and dressers. The **medium growth curve** (t_g = 300 s) applies

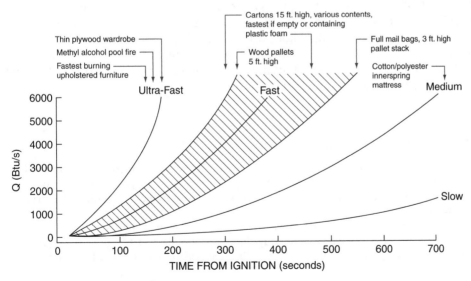

FIGURE 2.13 ◆ Relation of t^2 fires to several individual fire tests. Note that the heat release rate is in BTU/s and that 1000 BTU/s = 1.055 MW. *Courtesy of NIST, from Nelson and Tu [1991].*

TABLE 2.9 ◆ Typical Ranges for Fire Behavior of Materials Stored in Warehouses

Material	Typical Heat Release Rate per Floor Area Covered (MW/m^2)	Characteristic Time for t^2 Fires to Reach 1 MW (s)
Wood pallets,		
Stacked 0.46 m (1.5 ft) high	1.3	155–310
Stacked 1.5 m (5 ft) high	3.7	92–187
Stacked 3.1 m (10 ft) high	6.6	77–115
Stacked 4.6 m (15 ft) high	9.9	72–115
Polyethylene bottles in cartons, stacked		
4.6 m (15 ft) high	1.9	72
Polyethylene letter trays, stacked		
1.5 m (5 ft) high	8.2	189
Mail bags, filled, stored 4.6 m (15 ft) high	0.39	187
Polystyrene jars in cartons, 4.6 m (15 ft) high	14	53

Source: Derived from Quintiere 1998, Tables 6-5 and 6-6.

to lower-density solid fuels such as upholstered and lightweight furniture. The **fast growth curve** ($t_g = 150$ s) involves combustible items such as paper, cardboard boxes, and drapes. The **ultrafast growth curve** ($t_g = 75$ s) is for flammable liquids, synthetic mattresses, and volatile fuels. Of course, the geometry of the objects also has an impact on growth rates.

Note that the t^2 growth rate applies after the initial incipient fire phase, where the growth rate is nearly zero. In some cases involving polyurethane upholstered furniture, the steady-state phase does not exist, and the approximation for the heat release rate resembles a triangular curve. The approach of fitting triangular curves to model the heat release rate signatures of furniture can represent up to 91 percent of the total heat released (Babrauskas and Walton 1986).

EXAMPLE 2.5 Loading Dock Fire

Problem. A fire occurred on the back loading dock of a supply distribution center, which was extinguished by an automatic sprinkler system. A nearby security camera captured the silhouette of a young male running from the loading dock prior to the activation of the sprinkler system.

On examination of the fire scene, the investigator determines that the fire was intentionally set to a cart of polyethylene letter trays in two pallets stacked 1.56 m (5 ft) high on the loading dock. Determine for the investigator the heat release rate 30 s and 120 s after the arsonist set the fire.

Suggested Solution. Use equation (2.24) for t^2 fire growth and data from Table 2.9 on typical ranges for fire behavior of polyethylene letter trays.

Time to reach 1 MW	$t_g = 189$ s
Growth time of fire	$t = 30$ s
Heat release rate	$\dot{Q} = (1055/t_g^2)\, t^2$
	$= (1055/189^2)(30^2) \cong 25$ kW
and for	$t = 120$ s
	$\dot{Q} = (1055/189^2)(120^2) = 425$ kW

PHASE 3: FULLY DEVELOPED FIRE

The fully developed phase is also referred to as the **steady-state phase**, either when the fire reaches its maximum burning rate (based on the amount of fuel available) or when there is insufficient oxygen to continue the growth process (Karlsson and Quintiere 2000, 43). Fuel-controlled fires are associated with steady-state fires, particularly when sufficient oxygen is supplied. Note that "fully developed" does not distinguish postflashover fires from those that die out due to a lack of oxygen.

When insufficient oxygen is available, the fire becomes ventilation controlled. In this case, the fire is usually in an enclosed room, and temperature may become hotter than in fuel-controlled fires. Fully developed fires in compartments are often post-flashover fires in which all the fuel is involved, and the size of the fire becomes ventilation controlled.

PHASE 4: DECAY

The fire decay during phase 4 usually occurs when approximately 20 percent of the original fuel remains (Bukowski 1995b). Although most fire service and fire protection specialists are interested in the first three phases of a fire, there are significant problems associated with the decay phase.

In high-rise buildings, for example, even after the fire is extinguished, trapped occupants may need to be rescued. Furthermore, fire investigators entering the building may also be exposed to residual amounts of toxic by-products of combustion, since a high concentration of CO and other toxic gases is produced during the decay phase (often dominated by smoldering).

EXAMPLE 2.6   Reliable Test Data

Problem.   A fire investigator is trying to develop reliable test data on the length of time a fire may have burned and an estimate of the corresponding heat release rate. The fire was reportedly started by a child playing with a cigarette lighter while sitting at the center of a mattress. The mattress was the only piece of furniture involved in the fire in the room, and a smoke detector in the hallway reportedly activated shortly after the fire began.

Suggested Solution.   The information needed by the investigator is found at the Fire on the Web Internet site of the Building and Fire Research Laboratory, National Institute of Standards and Technology (NIST), Gaithersburg, Maryland (fire.nist.gov). In the fire tests/data section, are collections of actual test data for commonly found items in residential and commercial settings. The data files include still photos, video, graphs, miscellaneous data, and setup data for NIST fire modeling software programs.

Figure 2.14 shows the information needed by the investigator to compare this case with a similar fire under laboratory testing conditions. The fire test data showed that the peak heat

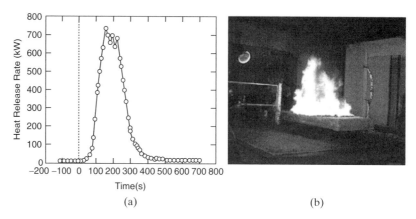

FIGURE 2.14 ◆ Reliable fire test data on (a) the heat release rate and (b) the fire after 120 s to support an investigation of a mattress fire. *Courtesy of NIST, Fire on the Web* [http://fire.nist.gov].

release rate occurred at approximately 150 s (2.5 min) after ignition, resulting in a 750-kW fire. When forming the hypothesis for the fire, the investigator could evaluate whether the fire's timeline is reasonably consistent with witness accounts, damage indicators, and heat release data. The investigator needs to remember that the test mattress may not be the same construction as the one involved in the fire and that even the same mattress will behave differently if ignited at a corner or from underneath. Note that the test was conducted in a large enclosure so as to minimize heat radiation effects (which would increase the rate of burning) and ventilation effects (which could limit the maximum rate). These variables must be recognized by the investigator when testing the relevant hypothesis. Good documentation of the fire scene will be the key to success.

◆ 2.6 ENCLOSURE FIRES

The growth of a fire confined to a room is usually constrained by to the ventilation-limited flow of air, smoke, and hot gases in and out of the enclosure. Confining variables may include the ceiling height, ventilation openings formed by windows and doors, room volume, and location of the fire in the room or compartment. While the fire gases are constrained, a notable event called *flashover* may occur, as a consequence of the heat release rate of the combustible materials contained within the room.

MINIMUM HEAT RELEASE RATE FOR FLASHOVER

In the growth of a compartment (room) fire, **flashover** is defined (per *NFPA 921*) as the transition phase in the development of a compartment fire in which surfaces exposed to thermal radiation reach ignition temperature more or less simultaneously, and fire spreads rapidly throughout the space, resulting in full room involvement or total involvement of the compartment or enclosed space. This transition may occur in just a few seconds in small, heavily fueled rooms, or over a period of minutes in large rooms, or may not take place at all due to insufficient heat release rate of the fire. A useful reference on this subject is found in *NFPA 555—Guide on Methods for Evaluating Potential for Room Flashover* (NFPA 2000a).

Fire researchers have documented other characteristics that occur during this transition phase. Definitions often include one or more of the following criteria that address the effects of flashover: ignition of floor targets, high heat flux (>20 kW/m^2) to the floor, and flame extension out of the compartment's vents (Babrauskas, Peacock, and Reneke 2003; Defining Flashover for Fire Hazard Calculation. Part II. Fire Safety Journal, 38, 613–622; Millke and Mowrer 2001; NFPA 2000a; Peacock et al. 1999).

- Flames are emitted from openings in the compartment.
- Upper-layer gas temperature rises to $\geq 600°$C.
- Heat flux at floor level reaches ≥ 20 kW/m^2.
- Oxygen level in the upper portions of the room decreases to approximately 0–5 percent.
- There is a small, short-term pressure rise of approximately 25 Pa (0.0036 psi).

There are two fundamental definitions for the occurrence of flashover. The first looks at flashover as a "thermal balance" in which critical conditions are produced when the compartment exceeds its ability to lose heat. The second definition views the compartment as being in a mechanical fluid-filling process. In this approach, the point at which the compartment's cool air layer is replaced with hot fire gases is flashover.

Because flashover represents a transition phase, defining the exact moment of occurrence is often a problem. Prior to flashover, the highest average temperatures are found in the upper or "hot gas" layer. If the average temperature of the hot gas layer exceeds $600°$C (with or without the flaming ignition of the hot smoke called **flameover** or **rollover**), the radiant heat from the layer that is reaching all other exposed fuel in the room exceeds the minimum ignition radiant heat flux for exposed fuels, and those fuels ignite. If flashover occurs, temperatures throughout the room reach their maximum ($1000°$C is not uncommon) as the two-layer environment of the room breaks down and the entire room becomes a turbulently mixed combustion zone—from floor to ceiling.

The active mixing promotes very effective combustion, with oxygen concentrations dropping below 3 percent and producing very high temperatures. This environment in turn produces radiant heat fluxes of 120 kW/m^2 or higher, ensuring rapid ignition and burning of all exposed fuel surfaces, including walls, carpets, flooring, and low-lying fuels like baseboards. Also, combustion near ventilation points is enhanced.

Ignition of carpets produces floor-level flames that sweep under chairs, tables, and other surfaces that were initially protected from the downward radiant heat from the hot gas layer alone. Burning then proceeds throughout the room until the fuel supply is exhausted or some attempt is made at extinguishment.

Several variables can influence the minimum heat release rates necessary for flashover to occur. First, the size of the compartment may influence the impact of radiation from the fire to the surfaces of the walls, floor, and ceiling. For small compartments, radiation from the fire enhances rapid temperature increase of exposed surfaces. Also, small compartments may negate the two-zone model.

Vent openings have an impact on flashover, as large vent openings make it necessary to have a very large fire in a room to produce flashover and may produce inaccurate calculations for vent flows. The surface materials on the walls can influence the minimum flashover energy, and some calculations take the heat transfer characteristics of the materials into account (Peacock et al. 1999).

Recent work on flashover calculations along with comparison with experimental data demonstrates that the wall surfaces play a role in heat transfer. Researchers report a trend that the shorter exposure times increase the needed minimum heat release rate for flashover (Babrauskas, Peacock, and Reneke 2003).

An approximation for the minimum heat release rate required for flashover in a room, $\dot{Q}_{fo}$, can be found using the following equation, called the **Thomas correlation** (Thomas 1981):

$$\dot{Q}_{fo} = (378\,A_o)\,\sqrt{h_o} + 7.8A_w, \tag{2.25}$$

where

$\dot{Q}_{fo}$ = heat release rate for flashover (kW),
A_o = area of the opening to the compartment (m^2),
h_o = height of the compartment opening (m), and
A_w = area of the walls, ceiling, and floor minus the opening (m^2).

EXAMPLE 2.7 Calculating Flashover

Problem. Given a 10 × 10-m room, with a 3-m-high ceiling and a 2.5 × 1-m opening (see Figure 2.15), determine the minimum heat release rate needed to cause flashover.

Suggested Solution. Using equation (2.25), the minimum heat release rate for a fire to progress to flashover in this room can be calculated as follows:

Area of compartment opening	A_o	$= (2.5\,\text{m})(1\,\text{m}) = 2.5\,\text{m}^2$
Height of compartment opening	h_o	$= 2.5\,\text{m}$
Area calculations	A_w	$=$ floor + walls + ceiling − compartment opening
	1 floor	$= (10\,\text{m})(10\,\text{m}) = 100\,\text{m}^2$
	4 walls	$= (4)(10\,\text{m})(3\,\text{m}) = 120\,\text{m}^2$
	1 ceiling	$= (10\,\text{m})(10\,\text{m}) = 100\,\text{m}^2$
		$= 320\,\text{m}^2 - 2.5\,\text{m}^2 = 317.5\,\text{m}^2$
Flashover heat release rate	$\dot{Q}_{fo}$	$= (378A_o)\sqrt{h_o} + 7.8A_w$
		$= (378)(2.5)\sqrt{(2.5)} + (7.8)(317.5)$
		$= 1494 + 2477 = 3971\,\text{kW} = 3.97\,\text{MW}$

A realistic answer would be about 4 MW.

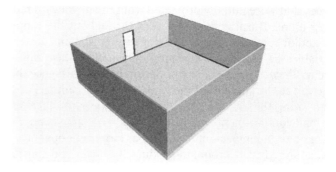

FIGURE 2.15 ◆ Example room with doorway opening for calculation of the minimum heat release rate needed for flashover.

ALTERNATIVE METHODS FOR ESTIMATING FLASHOVER

There are several alternative methods for estimating the minimum heat release rate necessary for flashover. The use of several estimates is necessary when applying the scientific method. Consider alternative approaches, evaluate the results of fire models, and compare them with eyewitness accounts of the fire.

The first alternative approach assumes a flashover criterion temperature of a suitable increase in temperature (ΔT), $\Delta T = 575°C$ (Babrauskas 1980), noting that $\Delta T = T_{fo} - T_{\text{ambient}} = 600 - 25 = 575°C$.

$$\dot{Q}_{fo} = 0.6 A_v \sqrt{H_v},\qquad(2.26)$$

where

$\dot{Q}_{fo}$ = heat release rate for flashover (MW),

A_v = area of the door (m^2), and

H_v = door height (m).

The term $A_v \sqrt{H_v}$ is often referred to as the **ventilation factor** and frequently appears in similar equations. In actual experiments, two-thirds of the cases agree and are bounded by the following equations (Babrauskas 1988):

$$\dot{Q}_{fo} = 0.45 A_v \sqrt{H_v},\qquad(2.27)$$

$$\dot{Q}_{fo} = 1.05 A_v \sqrt{H_v}.\qquad(2.28)$$

EXAMPLE 2.8   Calculating Flashover—Alternative Method 1

Problem.   Repeat the calculation for minimum heat release rate needed to cause flashover as previously presented in Example 2.7 using an alternative method.

Suggested Solution.   Use equations (2.26), (2.27), and (2.28), also known as the *Babrauskas correlation formulas*.

Area of the door opening	$A_v = (2.5 \text{ m})(1 \text{ m}) = 2.5 \text{ m}^2$
Height of door opening	$H_v = 2.5 \text{ m}$
Flashover heat release rate	$\dot{Q}_{fo} = (0.6)(A_v)\sqrt{H_v}$
	$\quad = (0.6)(2.5)\sqrt{(2.5)} = 2.37 \text{ MW}$
Minimum heat release rate	$\dot{Q}_{fo} = (0.45)(A_v)\sqrt{H_v} = 1.78 \text{ MW}$
Maximum heat release rate	$\dot{Q}_{fo} = (1.05)(A_v)\sqrt{H_v} = 4.15 \text{ MW}$

The second alternative approach is based on test data and a flashover criterion temperature of $\Delta T = 500°C$ (Lawson and Quintiere 1985; McCaffrey, Quintiere, and Harkleroad 1981) "Estimating room temperature and likelihood of flashover using fire test data correlation," Fire Technology, vol. 17, No. 2, pp. 98–114.):

$$\dot{Q}_{fo} = 610 \sqrt{h_k A_s A_v \sqrt{h_v}},\qquad(2.29)$$

where

h_k = enclosure conductance to ceiling and walls [kW/m^{2}K or (kW/m^2)/K],

A_s = thermal enclosure surface area, excluding vent or door area (m^2),

A_v = total area of vent or door opening (m^2), and

H_v = height of vent or door opening (m).

The **McCaffrey and Quintiere approach** requires an estimate for the effective heat transfer enclosure **conductance coefficient**, h_k, of the enclosure. Assuming uniform heating of the enclosure with adequate thermal penetration, the conductance coefficient h_k can be estimated as

$$h_k = \frac{k}{\delta}, \qquad (2.30)$$

where

h_k = enclosure conductance to ceiling and walls [(kW/m^2)/K],

k = thermal conductivity of enclosure material [(kW/m)/K], and

δ = enclosure material thickness (m).

EXAMPLE 2.9   Calculating Flashover—Alternative Method 2

Problem. Repeat the calculation for the minimum heat release rate needed to cause flashover given an enclosure lined with 16-mm ($\frac{5}{8}$-in.)-thick gypsum board. Also calculate the estimated enclosure conductance.

Suggested Solution. Assuming uniform heating of the enclosure with adequate thermal penetration of the enclosure material, the typical values are as follows:

Heat release rate	$\dot{Q}_{fo} = 610\sqrt{h_k A_s A_v \sqrt{h_v}}$
Thermal conductivity	$k = 0.00017$ (kW/m)/K
Enclosure thickness	$\delta = 0.016$ m
Enclosure conductance	$h_k = k/\delta$
	$= 0.00017/0.016 = (1.062 \times 10^{-2})[(kW/m^2)/K]$
Total surface area	$A_s = 317.5$ m^2
Total vent area	$A_v = 2.5$ m^2
Height of vent	$h_v = 2.5$ m
Flashover heat release rate	$\dot{Q}_{fo} = 610\sqrt{(0.01062)(317.5)(2.5)(\sqrt{2.5})}$
	$= 2226.9$ kW $= 2.23$ MW

A realistic answer would be about 2.2 MW.

Note that the results 4.0 MW (Example 2.9) and 2.23 MW (Example 2.10) fall within the minimum and maximum calculated range of 1.78 to 4.15 MW, from equations (2.27) and (2.28). Taking a numerical average of the three $\dot{Q}_{fo}$ predictions may offer a useful estimate of the actual requirement.

IMPORTANCE OF RECOGNIZING FLASHOVER IN ROOM FIRES

Those who examine a fire scene after extinguishment or burnout should be aware of the importance of recognizing whether flashover did actually occur and the impact it may have had on the burning of a room's contents. Chapter 3 addresses the interpretation of fire burn patterns, some of which may be created during the **postflashover** burning period.

Postflashover burning in a room produces numerous effects that were once thought to be produced only by accelerated (arson) fires involving flammable liquids such as gasoline. A "fireball" of burning gasoline vapors can sometimes produce floor-to-ceiling damage whose shallow penetration will be the result of the brief duration of such fires. Walls scorched or charred from floor to ceiling can be due to postflashover burning no matter how the fire actually began, as well as from a fireball of burning gasoline vapors (DeHaan 2007).

The combustion of carpets and underlying pads, once considered by fire investigators to be fairly unusual in accidental fires, is quite common in rooms filled with today's high-heat-release materials such as polyurethane foam, and synthetic (thermoplastic) fabrics in upholstery, draperies, and carpets. This problem is exacerbated by the increasing use of highly combustible fibers like polypropylene for the face yarns of economical carpets as well as in the backings of nearly all wall-to-wall carpets. These carpets now melt, shrink, and ignite under radiant heat flux, exposing the combustible urethane foam pad underneath to the radiant heat and flames. This combustion produces intense fires at floor level that in turn can create deep irregular burn patterns on the floor beneath. The involvement of common carpet in corridor fires was the reason that flooring radiant panel test was developed. The intense radiant heat effects in postflashover fires are not uniform owing to the extremely turbulent combustion, so the effects on floors are not uniform (DeHaan 2001).

Other published experiments report the presence and absence of burn-throughs with gasoline poured and ignited on flooring along with various tile and carpet arrangements (Sanderson 1997). The experiments reported that without the presence of carpet padding, burn-throughs did not occur. However, burn-throughs sometimes were reported with carpet padding, but not at the location where the gasoline was poured. Babrauskas (2005) concurs with DeHaan (2007) that the most common reason for burn-throughs of flooring is radiant heat from above and not the burning of combustible liquid on the floor. Extended slow combustion of collapsed furniture and bedding can also produce localized burn-throughs of flooring.

Fire damage under tables and chairs (once thought to be the result of the burning of flammable liquids at floor level) can be produced by the flaming combustion of the carpet and pad as they ignite during flashover. High temperatures and total room involvement, at one time thought to be linked to flammable liquid accelerants, are produced during postflashover burning without accelerants. The extremely high heat fluxes produced in postflashover fires can char wood or burn away other material at rates up to 10 times the rate at lower heat fluxes (Babrauskas 2005; Butler 1971).

Localized patterns of smoke and fire damage that can help locate fuel packages, characterize their flame heights, and reveal the direction of flame spread can be obliterated with prolonged postflashover burning, making the reconstruction of the fire very difficult. The fire investigator, then, should be aware of the possibility of flashover in modern room fires and the effects that postflashover burning can have. Postflashover or full room involvement fires typically exhibit ignition of all exposed fuel surfaces (but not necessarily complete combustion) from floor to ceiling with room corners sometimes spared. Larger rooms have been observed by the authors to sometimes exhibit "progressive" flashover, with floor-to-ceiling ignition at one end and merely undamaged floor covering at the other. Brief exposure (under 5 minutes) to postflashover fires does not necessarily compromise fire patterns on walls (see Chapter 8; Hopkins, Gorbett, and Kennedy 2007).

In preflashover burning, the most intense thermal damage will be in areas immediately around (or above) burning fuel packages. The higher temperatures of the hot gas layer will produce more thermal damage in the upper half or upper third of the room. In postflashover fires, all fuels are involved, the fires may become ventilation limited, and the most efficient (highest temperature) combustion will be occurring in turbulent mixing around the ventilation openings, where the oxygen supply is best. Oxygen-depleted burning occurs in many postflashover rooms where the available fuel exceeds the air supply.

Interviews of first-in firefighters may reveal whether a room was fully involved, floor to ceiling, when entry was first made. Investigators must be aware of the conditions that can lead to flashover and how to recognize when it has or has not occurred. A thorough understanding of heat release rates and fire spread characteristics of modern furniture and the dynamics of flashover will be of great assistance (Babrauskas and Peacock 1992).

POSTFLASHOVER CONDITIONS

As previously stated, the average hot gas temperature *throughout* a room or compartment at flashover rises to $\geq 600°C$. At postflashover conditions, the entire compartment is treated as a single homogenous (turbulent and well-mixed) volume sharing a common temperature and percentages of concentrations of oxygen and other byproducts of combustion.

For **ventilation-controlled fires** during postflashover, the heat release rate of the fuel within the compartment is regulated by the available ventilation. More specifically, the rate of inflow of fresh air and the outflow of gases is determined by a combination of the temperature differential across the vent, and the ventilation factor $(A_v\sqrt{H_v})$.

Understanding this insight, we can use the relationship of mass flow into an opening to estimate the maximum heat release rate of fuels burning within a compartment. The mass flow rate of air through an opening is approximately (Karlsson and Quintiere 1999)

$$\dot{m}_{air} = 0.5 A_v\sqrt{H_v} \tag{2.31}$$

Because the heat release rate is regulated by the amount of available air, the following relationship can be used to estimate this rate:

$$\dot{Q} = \dot{m}_{air}\frac{\Delta h_c}{r} \tag{2.32}$$

$$\dot{Q} = 0.5 A_v\sqrt{H_v}\frac{\Delta h_c}{r} \tag{2.33}$$

where

Δh_c = heat of combustion of the fuel (kJ/kg),
$\dot{m}_{air}$ = mass of air required to burn a mass of fuel (kg(air)/kg(fuel)),
$\dot{Q}$ = heat release rate (kJ/s or kW), and
r = air-to-fuel ratio 5.7 kg (air)/kg (wood).

The term is approximately constant for most fuels. For example, wood is

$$\Delta h_c = 15{,}000 \text{ kJ/kg}$$
$$r = 5.7 \text{ kg (air)/kg (wood)}$$
$$\frac{\Delta H_c}{r} = 2630 \text{ kJ/kg (air)}$$

For a wood-fueled fire in a room, the maximum-sized fire can be estimated by assuming 100 percent efficiency. Substituting into equation (2.33), we obtain

$$\dot{Q} = 1370 A_v \sqrt{H_v} \text{ (kJ/s = kW)} \tag{2.34}$$

In postflashover fires, the temperature of the compartment, which may reach 1100°C (2012°F), can also be estimated. The combined work of Thomas (1974) and Law (1978) produced the following correlation predicting the maximum postflashover temperature, assuming natural ventilation:

$$T_{FO(\text{max})} = 6000 \frac{(1 - e^{-0.1\Omega})}{\sqrt{\Omega}} \tag{2.35}$$

$$\Omega = \frac{A_T - A_v}{A_v \sqrt{h_v}} \tag{2.36}$$

where

$T_{FO(\text{max})}$ = maximum compartment temperature at flashover,
Ω = ventilation factor
A_T = total area of the compartment enclosing surfaces, excluding the area of vent opening (m^2),
A_v = total area of the vent openings (m^2), and
h_v = height of the vent openings (m).

EXAMPLE 2.10   Calculating Maximum Heat Release Rate and Temperature in a Ventilation-Controlled Compartment Fire

Problem.   Given a ventilation-controlled fire consisting of burning wooden furniture in a 10 × 10-m room, with a 3-m-high ceiling and a 2.5 × 1-m opening (see Figure 2.15), determine the minimum heat release rate and maximum compartment temperature, assuming natural ventilation.

Suggested Solution.   The minimum heat release rate for a fire to progress in this room with the given opening can be calculated from equation (2.35) as follows:

Maximum heat release rate $\dot{Q} = 1370 A_v \sqrt{H_v}$
Height of opening $H_v = 2.5$ m
Area of opening $A_v = (1)(2.5) = 2.5$ m^2
Maximum heat release rate $\dot{Q} = 1370 A_v \sqrt{H_v}$
  $= 1370(2.5)\sqrt{2.5} = 5415$ kW $= 5.415$ MW

Maximum compartment temp. at flashover $T_{FO(\text{max})} = 6000\frac{(1 - e^{-0.1\Omega})}{\sqrt{\Omega}}$

Ventilation factor $\Omega = \frac{A_T - A_v}{A_v \sqrt{h_v}}$

Total area of the vent openings $A_v = (1)(2.5) = 2.5$ m^2

Total area of the compartment enclosing surfaces, excluding area of vent opening

$$1 \text{ floor} = (10 \text{ m})(10 \text{ m}) = 100 \text{ m}^2$$

$$A_T = \quad 4 \text{ walls} = (4)(10 \text{ m})(3 \text{ m}) = 120 \text{ m}^2$$

$$1 \text{ ceiling} = (10 \text{ m})(10 \text{ m}) = 100 \text{ m}^2$$

$$= 320 \text{ m}^2 - 2.5 \text{ m}^2 = 317.5 \text{ m}^2$$

Height of the vent openings $\qquad h_v = 2.5 \text{ m}$

Ventilation factor $\qquad \Omega = \dfrac{A_T - A_v}{A_v \sqrt{h_v}} = \dfrac{317.5 - 2.5}{2.5\sqrt{2.5}} = \dfrac{315}{3.95} = 79.75$

Maximum compartment temperature at flashover
$$T_{FO(\text{max})} = 6000 \dfrac{(1 - e^{-0.1\Omega})}{\sqrt{\Omega}}$$

$$= 6000 \dfrac{(1 - e^{-7.975})}{\sqrt{79.75}}$$

$$= 6000 \dfrac{0.999}{8.93} = 671.6\,^\circ\text{C}$$

Discussion: Since we earlier predicted that a fire of 3–4 MW would be at the verge of triggering flashover in this large room, and the ventilation opening will support a fire of 5 MW or better, there is agreement that the maximum predicted temperature 671°C will exceed the threshold for flashover (i.e., 600°C).

◆ 2.7 OTHER ENCLOSURE FIRE EVENTS

DURATION

As part of forensic fire scene reconstruction and analysis, other features of fire behavior in enclosures also become important to investigators. The following are commonly cited features:

- Smoke detector activation
- Heat and sprinkler system operation
- Limited ventilation fires

These enclosure fire features can be used to answer several questions that arise during an investigation, such as: How long did the fire effectively burn? and, When during this time period did the smoke detector or sprinkler head activate? Answers to these questions can be placed on a timeline of information needed to fill the voids, particularly when fire deaths occur.

In tightly closed rooms, a fire consumes available oxygen and pumps oxygen-depleted air into the hot smoke layer. This layer descends until it reaches the level of the fire. The fire then experiences vitiated burning, existing on what oxygen is available to it in the smoke layer, and reducing, accordingly, its HRR. If there is a leak or vent up high in the room, the smoke layer may rise to where the fire is back in "normal" room air and the fire will resume its full HRR, until the layer descends again. This sequence may produce a cyclical behavior of open flame/smolder/open flame/smolder until the fuel is depleted.

NFPA 555 offers two useful relationships for estimating the duration of burning in a tightly closed compartment—one for a fire with a steady HRR of $\dot{Q}$, the second for unsteady t^2 fires (NFPA 2000a).

For steady fires:

$$t = \frac{V_{O_2}}{\dot{Q}(\Delta h_c \rho_{O_2})} \tag{2.37}$$

For growing fires:

$$t = \left[\frac{3V_{O_2}}{\alpha(\Delta h_c \rho_{O_2})} \right]^{1/3} \tag{2.38}$$

where

V_{O_2} = maximum volume of oxygen available to be consumed in the combustion process (m^3),

$\dot{Q}$ = heat release rate from steady fire (kW),

$\Delta h_c \rho_{O_2}$ = heat release rate per unit mass of oxygen consumed (kJ/m^3), and

α = constant governing the speed of fire growth (kJ/s^2) (slow 2.93×10^{-3}, medium 11.72×10^{-3}, fast 46.88×10^{-3})

The quantity V_{O_2} is generally considered to be half of the total available oxygen in the room (so $V_{O_2} = 0.5(0.21\, V_{room})$), since flaming combustion will generally not be sustained once O_2 levels are in the range 8–12 percent (*NFPA 555*, p. 6). This also assumes the fire is at floor level in the room; if the fire is elevated and not so large that it will induce a great deal of localized turbulent mixing, V_{O_2} will be controlled by the volume of the room at and above the level of the fire.

SMOKE DETECTOR ACTIVATION

A common problem faced by fire investigators is determining when on a timeline of events a thermal fire detector, smoke detector, or sprinkler system was activated. Historically, much work has been done on this area by fire protection engineers, such as the DETACT-QS program (Evans and Stroup 1986). These milestone events are usually noticed by a witness or electronically reported by alarm systems and often serve as critical markers in an investigation.

Figure 2.16 shows the measurements needed to calculate these activations. The dimension H is the distance to the ceiling above the fuel surface, and r is the radial distance from the plume centerline to the heat detector or sprinkler head. The centerline is the imaginary vertical line emerging from the center of the fuel source to the ceiling. R is the equivalent radius of the burning fuel package. Virtual origins are discussed in detail in Chapter 3.

Three common methods are used in estimating smoke detector response time, namely, Alpert (1972), Milke (2000), and Mowrer (1990). Note that in some approaches, such as Alpert and Milke, the practice is to use the convective portion of the heat release rate, $\dot{Q}_c$, in calculations. The convective heat release rate fraction, X_c (typically 0.70), is used, where $\dot{Q}_c = X_c \dot{Q}$.

In the Mowrer method (1990), two calculations are necessary to estimate smoke detector activations for steady-state fires. The first calculates the time for the fire gases to reach the ceiling at the plume centerline, or the **plume lag time**. The second

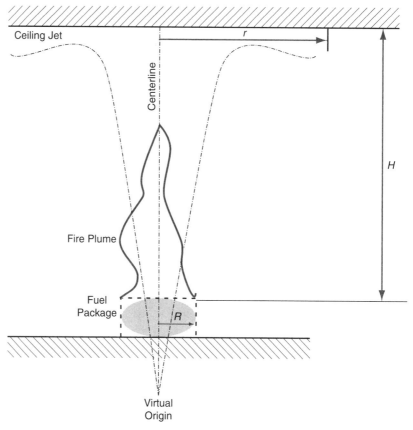

FIGURE 2.16 ◆ Fire plume measurements needed in fire detector, smoke detector, or sprinkler system calculations, consisting of fuel surface to ceiling along centerline (*H*), radius of burning fuel surface (*R*), and radial distance of ceiling jet from centerline (*r*).

calculates the time for the gases to reach the detector from the plume centerline, or the **ceiling jet lag**.

For a steady-state fire of heat release rate $\dot{Q}$, these equations are expressed as

$$t_{\text{pl}} = C_{\text{pl}} \frac{H^{4/3}}{\dot{Q}^{1/3}} \tag{2.39}$$

$$t_{\text{cj}} = \frac{1}{C_{\text{cj}}} \frac{r^{11/6}}{\dot{Q}^{1/3} H^{1/2}} \tag{2.40}$$

where

t_{pl} = transport lag time of plume (s),

t_{cj} = transport lag time of ceiling jet (s),

C_{pl} = plume lag time constant = 0.67 (experimentally determined),

C_{cj} = ceiling jet lag time constant = 1.2 (experimentally determined),

r = radius of distance from plume centerline to the detector (m),

H = height of ceiling above top of fuel (m), and

$\dot{Q}$ = heat release rate of the fire (kW).

EXAMPLE 2.11   Smoke Detector Activation by Wastepaper Basket Fire

Problem.   A business owner closes up his shop and later claims that he may have caused a fire by carelessly discarding a lit match into a wastepaper basket before leaving. The fire ignited a small 0.3-m (0.98-ft)-diameter basket filled with ordinary papers. The fire developed quickly to reach a steady heat release rate of 100 kW.

The initial temperature of the room was 20°C (293 K, 68°F). The distance from the top of the wastepaper basket to the ceiling is 4 m (13.12 ft). A smoke detector attached to the ceiling is located radially 2 m (6.56 ft) from the fire plume's centerline. Assuming that there are negligible radiation losses in a smooth, level ceiling, determine the time of the smoke detector's activation by the heated ceiling jet gases.

Suggested Solution.   Assume that steady-state conditions exist, along with immediate activation.

Height of ceiling above fuel	$H = 4\text{m}$
Radius from plume centerline	$r = 2\text{m}$
Plume lag time constant	$C_{pl} = 0.67$
Ceiling jet lag time constant	$C_{cj} = 1.2$
Heat release rate	$\dot{Q} = 100 \text{ kW}$

Plume transport time lag
$$t_{pl} = C_{pl}\frac{H^{4/3}}{\dot{Q}^{1/3}} = 0.92 \text{ s}$$

Ceiling jet transport lag time
$$t_{cj} = \frac{1}{C_{cj}}\frac{r^{11/6}}{\dot{Q}^{1/3}H^{1/2}} = 0.32 \text{ s}$$

Detector activation time
$$t = t_{pl} + t_{cj} = 0.92 + 0.32 = 1.24 \text{ s}$$

Discussion.   Owing to variables such as fire size, room geometry, the type of detector (photoelectric or ionization), and environmental interactions, the time to activate a smoke detector may vary, especially with ceiling-mounted units. Therefore, the preceding solution may not be accurate because it does not account for the lag time for the smoke to enter the detector's sensing chamber. Note that this solution assumes that an "instantaneous" fire of 100 kW occurs, something that takes a real-life fire 30 or more seconds to achieve. A suggested practice for determining the activation time for ceiling-mounted smoke detectors is to consider the temperature of the smoke layer (Heskestad and Delichatsios 1977). Note however that current detectors have much faster response times than did those in 1977.

Testing by Collier (1996) showed that approximately a 4°C rise at the detector location is sufficient for an activation. In several approaches to predicting the time for smoke, heat, and sprinkler systems to activate, the engineering calculations typically use only the convective heating of the sensing elements by the hot fire gases and do not account for any direct heating by radiation from the flames.

For comparison, the method of Alpert and Milke provides estimates of 5.89 s and 12.48 s, respectively, for this problem. Experimental reenactment of a fire condition may reveal critical variables (e.g., placement/location on a wall, ceiling vent location, or time required for development of the fire to steady state).

SPRINKLER HEAD AND HEAT DETECTOR ACTIVATION

In estimating the time to activate a sprinkler head or heat detector, both the ceiling jet temperature and its velocity must be calculated along with the ratio r/H. The equations for these estimates are based on data from a series of actual tests for 670-kW to 100-MW fires (Alpert 1972; NFPA 2000b, chaps. 11–10).

The first calculation necessary is the centerline temperature directly above the plume produced by the burning fuel source. This equation is valid when $r/H \le 0.18$.

$$T_m = 16.9 \frac{\dot{Q}^{2/3}}{H^{5/3}} + T_\infty \qquad \text{for} \qquad r/H \le 0.18. \qquad (2.41)$$

For r/H ratios greater than 0.18, the detector or sprinkler falls within the ceiling jet portion of the plume.

$$T_{m_\text{jet}} = 5.38 \frac{(\dot{Q}/r)^{2/3}}{H} + T_\infty \qquad \text{for} \qquad r/H > 0.18, \qquad (2.42)$$

where

T_m = plume gas temperature above fire (K),
$T_{m\text{jet}}$ = temperature of ceiling jet (K),
T_∞ = ambient room temperature (K),
$\dot{Q}$ = heat release rate from fire (kW),
$\dot{Q}_c$ = convective heat release rate (kW), where $\dot{Q}_c = X_c Q$
r = radial distance from plume centerline to device (m), and
H = distance above fuel surface (m).

Note that when calculating the time to operation of a sprinkler head the generally accepted engineering practice is to consider only the convective heating of the sensing elements by the hot fire gasses (Iqbal and Salley 2004, 10–12). This practice does not account for direct heating by radiation. Therefore, the value for the convective heat release rate for the fire, $\dot{Q}_c$, is often used in place of the total heat release rate, $\dot{Q}$.

To study detector and sprinkler activation problems further, the maximum velocity of the ceiling jet, U_m, needs to be calculated. The following correlations depend on the value of the r/H ratio.

For maximum jet velocity close to the centerline,

$$U_m = 0.96 \left(\frac{\dot{Q}}{H} \right)^{1/3} \qquad \text{for} \qquad r/H \le 0.15. \qquad (2.43)$$

For jet velocities farther away from the centerline,

$$U_m = 0.195 \left(\frac{\dot{Q}^{1/3} H^{1/2}}{r^{5/6}} \right)^{1/3} \qquad \text{for} \qquad r/H > 0.15, \qquad (2.44)$$

where

$\dot{Q}$ = heat release rate (kW),
H = distance above fuel surface (m),
r = radial distance from plume centerline to device (m), and
U_m = gas velocity (m/s).

Determination of the time to activation of the heat detector or sprinkler during steady-state fires relies on a term called the **response time index** (RTI). This index assesses the ability of a heat detector to activate from an initial condition of ambient room temperature. Because the "detector" of a sprinkler has a finite mass, the *RTI* takes into account the time lag before the temperature of the detector rises.

$$t_\text{operation} = \frac{RTI}{\sqrt{U_m}} \log_e \left(\frac{T_m - T_\infty}{T_m - T_\text{operation}} \right), \qquad (2.45)$$

where

RTI = response time index ($m^{1/2} s^{1/2}$),

U_m = gas velocity (m/s),

T_m = plume gas temperature above fire (K),

T_∞ = ambient room temperature (K), and

$T_{operation}$ = operation temperature (K).

RTI values are specified by the manufacturer for each style or model of sprinkler.

EXAMPLE 2.13   Sprinkler Head Activation to Wastepaper Basket Fire

Problem.   A closer examination of the debris in Example 2.12 shows that the wastebasket actually contains both paper and plastic, producing a steady-state fire of 500 kW. A standard response bulb sprinkler head, whose operation temperature is 74°C (165°F, 347 K) and *RTI* is 235 $m^{1/2}s^{1/2}$, is located directly above the basket. The ambient temperature of the room is 20°C (68°F, 293 K). Estimate the time to activation of the sprinkler head. When calculating the ceiling jet temperature, use the convective heat release rate, $\dot{Q}_c$.

Suggested Solution.   Since the area of interest is directly above the fire plume, the radius from the centerline is zero, producing a ratio $r/H < 0.18$. Also, the conservative assumption is that the convective heat release rate fraction (X_c) of the fire is 0.70. Therefore, use the appropriate equations to calculate the temperature and velocity.

Heat release rate $\quad\quad\quad\quad\quad\quad\quad \dot{Q} = 500 \text{ kW}$

Convective heat release rate $\quad\quad \dot{Q}_c = 350 \text{ kW}$

Distance above fuel surface $\quad\quad\quad H = 4 \text{ m}$

Ambient room temperature $\quad\quad T_\infty = 20°C + 273 = 293 \text{ K}$

Response time index $\quad\quad\quad\quad RTI = 235 \text{ m}^{1/2}\text{s}^{1/2}$

Temperature of operation $\quad\quad T_{operation} = 74°C + 273 = 347 \text{ K}$

Temperature of ceiling jet $\quad\quad\quad T_m = 16.9(\dot{Q}^{2/3}/H^{5/3}) + T_\infty$

$\quad\quad\quad\quad\quad\quad\quad\quad\quad\quad\quad\quad = (16.9)(350^{2/3}/4^{5/3}) + 293$

$\quad\quad\quad\quad\quad\quad\quad\quad\quad\quad\quad\quad = 83 + 293 = 376 \text{ K} = 103°C$

Ceiling jet velocity $\quad\quad\quad\quad U_m = 0.96\left(\dfrac{\dot{Q}}{H}\right)^{1/3}$

$\quad\quad\quad\quad\quad\quad\quad\quad\quad\quad\quad\quad = (0.96)(500/4)^{1/3} = 4.8 \text{ m/s}$

Time to operation $\quad\quad\quad\quad t_{operation} = (RTI/\sqrt{U_m})log_e\,[(T_m - T_\infty)]/$

$\quad\quad\quad\quad\quad\quad\quad\quad\quad\quad\quad\quad\quad\quad [(T_m - T_\infty)/(T_m - T_{operation})]$

$\quad\quad\quad\quad\quad\quad\quad\quad\quad\quad\quad\quad = (235/\sqrt{4.45})log_e[(376 - 293)/(376 - 347)]$

$\quad\quad\quad\quad\quad\quad\quad\quad\quad\quad\quad\quad = 61.8\text{ s} \approx 1 \text{ min}$

Note that the calculations for ceiling jet behaviors are valid only for smooth, level ceilings. Open joists, ceiling ducts, or pitched surfaces will dramatically affect the movement of gases.

Example 2.14   Nursing Home Case Study

Problem.   An elderly woman seated in a wheelchair in a nursing home died when her clothing and lap blanket ignited while she was sitting smoking cigarettes on a covered exterior patio of the facility. When staff members first noticed a problem, the flames were readily visible, and by the time they reached her, the flames had set off a sprinkler immediately above her position. The sprinkler doused the flames but not before they had consumed the plastic seat back of the wheelchair and much of the clothing and lap blanket. The woman succumbed to inhalation injuries (soot and edema in airway) and shock from 70 percent total body surface area (TBSA) burns some 8 hours after the incident. (No postmortem or toxicology tests were conducted.) Witnesses reported they observed no movement and no outcry from the victim who "sat like a mannequin." She had had a stroke some months prior, and it is possible she had suffered another. On the table next to her were found a partial pack of cigarettes, an ashtray, and a partial box of matches. The woman, who was a lifelong smoker, had normal access to smoking materials. Exemplars of a dressing gown and lap blanket fabric were provided. These fabrics were tested using *NFPA 705* methods and were characterized as a cotton/synthetic blend dressing gown and a combustible lightweight quilt with a thin polyester fiberfill pad covered with a combustible cotton/polyester fabric.

Proposed Solution. Here, data from various databases and published sources indicated that a fire in such synthetic materials as the lap blanket or wheelchair seat back could be ignited only by an open flame. The chair was a manual one with no power supply or electrical fittings. A dropped match on such materials could provide competent ignition to ignite a rapidly growing, flaming fire. Data from the NIST website (fire.nist.gov) on synthetic-upholstered chairs indicated a fire of a maximum heat release rate of 300–700 kW would be established within 1 min of flame ignition.

At the location of the fire, the ceiling was finished with painted wood with recessed lighting fixtures. It was sloped with a height of 2.36 m (93 in.) at the fire location as shown in Figure 2.17. The sprinkler head was identified as a 135°F normal pendant fixture whose deflector head was set 2.3 m (90 in.) from the floor. Sprinklers were located on 3-m (10-ft) centers well away from support beams (also on 3-m centers). There was a circular charred area on the ceiling surrounding the sprinkler head approximately 1 m (3 ft) in diameter with a halo of soot deposit outside it. The seat of an exemplar wheelchair was found to be approximately 0.5 m (19 in.) above the floor. Laboratory analysis of samples of clothing and debris found no residues of ignitable liquid.

From the Zukoski (1978) relationship for plume height (plume heights are discussed in Chapter 3) to reach the ceiling at 2.36 m (93 in.) from a fire surface at 0.5 m (19 in.) height,

$$Z = 0.175k\dot{Q}^{0.4}, \tag{2.46}$$

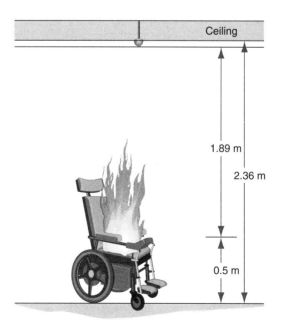

FIGURE 2.17 ◆ A fire in a wheelchair under a sprinkler causes rapid activation of the sprinkler (less than 100 s).

where

$Z = $ fire plume height (m),
$k = $ constant (1), and
$\dot{Q} = $ heat release rate (kW).

Therefore,

$$\begin{aligned}
Z &= 0.175^K \dot{Q}^{0.4} \\
1.86 &= 0.175^1 \dot{Q}^{0.4} \\
\dot{Q}^{0.4} &= 10.69 \\
(\dot{Q}^{0.4})^{2.5} &= (10.69)^{2.5} \\
\dot{Q} &= 375 \, \text{kW}
\end{aligned}$$

Thus, a typical fire from the combustion of a synthetic-upholstered chair (or its equivalent) could easily reach the ceiling and cause ceiling jet damage to the surrounding finish.

Time to activation of a standard-response link sprinkler ($RTI = 130$, temperature rating = 165°F) at a height of 1.86 m above the seat of the chair (immediately below the sprinkler head), would be approximately 12 s, calculated using NRC Fire Dynamics Tools spreadsheet 10 (Iqbal and Salley 2004, ch. 10) for sprinkler response time (later discussed in detail) and assuming a steady-state fire of 375 kW. Since the fire would have to grow from ignition to 375 kW, a fast t^2 fire (given the properties of synthetic blankets and clothing) could be assumed, Then,

$$\dot{Q} = \alpha t^2,$$

and

$$375 = 0.047 \, t^2$$

$$t^2 = \frac{375}{0.047} = 7978$$

$$t = 89.3 \, \text{s}$$

So, realistically, the sprinkler would have activated in less than 100 s from the time of ignition.

It was concluded that the decedent had dropped a match while attempting to light a cigarette (probably owing to a stroke or other medical event), and the match ignited a very rapidly growing fire whose flames reached ceiling height. The pattern of burn injuries to the victim and to the chair was entirely consistent with the finding that her clothing and lap blanket were the major fuel packages. The combustion and collapse of the wheelchair were the result of fire spread from there.

Case study courtesy of John D. DeHaan

◆ **2.8 SUMMARY AND CONCLUSIONS**

The body of fire dynamics knowledge as it applies to fire scene reconstruction and analysis is based on the combined disciplines of thermodynamics, chemistry, heat transfer, and fluid mechanics. We have seen that the growth and development of fires are influenced by a number of variables such as available fuel load, ventilation, and physical configurations of the room. To estimate accurately a fire's origin, intensity, growth, direction of travel, and duration, investigators must rely on and understand the principles of fire dynamics.

Fire investigators can also benefit by applying the wealth of knowledge of fire dynamics contained in textbooks, expert treatises, and authoritatively conducted fire research. Case examples along with discussions on sound fire protection engineering calculations can assist investigators in interpreting fire behavior.

Problems

2.1. Recalculate the heat release rate needed for flashover using several methods for a room measuring 5 × 5 m, with a 3-m ceiling and two 2.5-m-high × 1-m-wide openings.

2.2. Photograph or review a recent fire scene and estimate the heat release rate of the first article ignited.

2.3. Search for references to fires in warehouses in which the fire protection system detected and extinguished the incipient fire with only one sprinkler head activating. Obtain the floor plan to a warehouse and estimate the time to sprinkler head activation.

2.4. Visit the *fire.nist.gov* web site and examine the fire data availability there. What are the maximum heat release rates for upholstered chairs, mattresses, and Christmas trees?

2.5. Test the sensitivity of some of the mathematical calculations described in this chapter by changing a room dimension or ventilation opening by 0.2 m and comparing the results.

Suggested Reading

DeHaan, J. D. 2007. *Kirk's fire investigation,* 6th ed. Upper Saddle River, NJ: Prentice Hall, chap. 3.

Drysdale, D. D. 1999. *An introduction to fire dynamics,* 2nd ed. Chichester, UK: Wiley.

Karlsson, B., and J. G. Quintiere. 1999. *Enclosure fire dynamics.* Boca Raton, FL: CRC Press.

Quintiere, J. G. 1998. *Principles of fire behavior.* Albany, NY: Delmar, chaps. 3–9.

Iqbal, N., and M. H. Salley. 2004. Fire dynamics tools (FDTs): Quantitative fire hazard analysis methods for the U.S. Nuclear Regulatory Commission Fire Protection Inspection Program.

Fire Pattern Analysis

3 **CHAPTER**

There is no branch of detective science which is so important and so much neglected as the art of tracing footsteps.

—Sir Arthur Conan Doyle,
"Study in Scarlet"

The ability to document and interpret fire patterns accurately is essential to investigators reconstructing fire scenes, and they are often the only visible evidence remaining after a fire is extinguished. This chapter describes how these fire patterns are analyzed and used by investigators in assessing fire damage and determining a fire's origin.

Smoke deposits, heat transfer, and flame spread are the major causes of change to the exposed surface and appearance of materials during a fire. Fire patterns are formed by the heating effects of fire plumes on exposed solid surfaces such as floors, ceilings, and walls. These burn patterns are influenced by a number of variables including the available fuel load, ventilation, and the physical configurations of the room. Many common combustible materials and ignitable liquids can produce these fire plumes and their resulting damage.

Case examples are offered here along with discussions on validated fire science and engineering calculations to assist investigators in interpreting fire patterns. Also addressed in this and subsequent chapters are various documentation techniques that can be helpful in fire scene reconstruction using the concepts of fire pattern analysis.

◆ 3.1 FIRE PLUMES

The single most important factor in fire scene reconstruction is the **fire plume**, which is a buoyant column of hot gases produced by the combustion of a fuel source emitting a vertical column of flames and hot products of combustion (Drysdale 1999). Fire plumes can originate from any fuel combusting that has sufficient heat release rate to generate an upward column of flames and smoke. Fire plumes result from any significant fire. They can be produced by floor-level pools of ignitable liquids; however, many combustible solids, including certain types of foam mattresses and plastics, melt and collapse while burning and behave like liquid fuels. Of course, fire plumes can also result from fuel geometries having multiple vertical and horizontal fuel surfaces, complex internal structures, and varying fuel types. For the purpose of the following discussion, we shall treat the effects of fire plumes from a simple flat fuel array as behaving analogously to a pool fire.

The shape of burning pools that produce fire plumes depends on several variables including the geometry of the containment of the pool, the type of substrate on which the pool is resting, and, in some cases, the external winds. Because fire plumes are three-dimensional, their location can often be determined and documented by evaluating the patterns of heat transfer, flame spread, and smoke damage they cause to adjacent flooring, wall, and ceiling surfaces, as illustrated earlier in Figure 1.5. A fire investigator can gain additional insight through a fundamental understanding of the nature, physics, and heat transfer characteristics of fire plumes. The buoyancy of hot gases is the driving force behind their vertical spread and horizontal movement as they encounter obstacles.

V PATTERNS

The impingement, intensity, and direction of the fire plume's travel form lines or areas of demarcation on walls, ceilings, floors, and other materials. As the hot gases and smoke rise from a fire, they mix with surrounding air, with the mixing zone becoming wider as hot gases rise above the fuel. Entrainment mixes and spreads the rising column, so it forms a V approximately 30° in width (i.e., half-angle of 15°) if unconfined in still air from a turbulent diffusion fire (You 1984). Therefore, the total width of the *unconfined* plume in still air is approximately half its height above the fuel surface. As the fire gases mix, they are diluted and cooled. As a result, the diameter of the hottest part of the plume along the centerline becomes *smaller* with increasing height, as seen in Figure 3.1. The rapid cooling of fire gases by mixing with the air limits their thermal damage on noncombustible walls, so the thermal pattern on adjacent walls will not reflect the entire 30° angle width described for the plume.

Analyzing the shape of fire patterns caused by plumes can provide valuable information. For example, the most pronounced ceiling damage is often in the plume impingement area directly above the fuel source (as in Figure 1.5), a point from which a gas-movement vector points directly back to the fire's origin. The temperature of the gases determines the extent of thermal effects on surfaces they encounter. If their temperature is too low, the heat transfer will be insufficient and there may be no thermal effect; but combustion products will still condense on cooler surfaces.

Damage patterns forming lines of demarcation frequently appear in simple two-dimensional views of the location where the fire plume comes into contact with and damages wall surfaces. These lines are often referred to as **V-shaped fire patterns**, based on their characteristic upward-sweeping curves.

NFPA 921 (NFPA 2004) and *Kirk's Fire Investigation* (DeHaan 2007) dispelled past misconceptions regarding the shape and geometry of V patterns, once thought to relate to the rapidity of fire growth. The shape of a V pattern is actually related to the heat release rate, geometry of the fuel, ventilation effects, ignitability and combustibility of affected surfaces, and intersection with horizontal surfaces (NFPA 2004, sec. 6.17.1).

If the V-shaped fire pattern's angle of demarcation is followed downward to its base, the area closest to the virtual origin of the plume can be located. Several

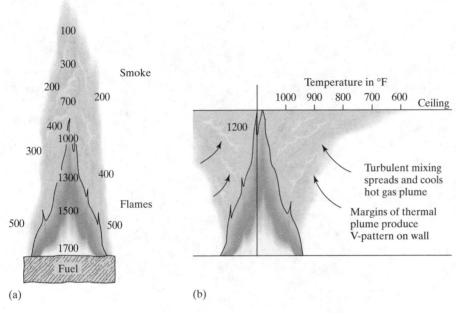

FIGURE 3.1 ◆ The fire plume includes both flames and buoyant smoke. The temperature distribution of gases in the buoyant plume: (a) unrestricted, (b) under ceiling (temperatures in °F).

mathematical relationships aid in documenting and analyzing the features of fire plume height, temperature, velocity, vortex shedding frequency, and virtual origin (DeHaan 2007, Chap. 7; Heskestad 1988; Quintiere 1998).

HOURGLASS PATTERNS

In certain cases, when a fuel package burns in a room adjacent to a wall or corner, the damage appears in the shape of an "hourglass" burn pattern instead of a V, as shown in Figure 3.2 (Icove and DeHaan 2006). Both fire testing and mathematical analysis by the authors have shown that the formation of hourglass burn patterns is a direct function of the fire plume's virtual origin, which is mathematically tied to the heat release rate and surface area of the fuel package. A later section in this chapter will illustrate in full the formation of these hourglass patterns.

FIRE PLUME DAMAGE CORRELATIONS

Testing by the Factory Mutual Research Corporation (FMRC) provided additional insight into the formation of V patterns. In FMRC's testing, a closer examination of a fire plume's lines of demarcation reveals distinct areas of damage to the wall surfaces, also known as the **fire propagation boundary**. This boundary is a visually distinguishable line where the heavy pyrolysis of the surface ends.

Actual fire testing by FMRC has documented the close correlation of the fire propagation boundary with the **critical heat flux boundary**, which is where the minimum heat flux is at or below the point at which a flammable vapor–air mixture is produced by pyrolysis at the surface of the solid (Tewarson 1995). Critical heat fluxes have also been experimentally determined by successive exposures of material

FIGURE 3.2 ◆ Hourglass pattern from fire in a corner. Note demarcation between flame contact area (burned paper on drywall) and radiant heat (scorched paper). *Courtesy of J. D. DeHaan.*

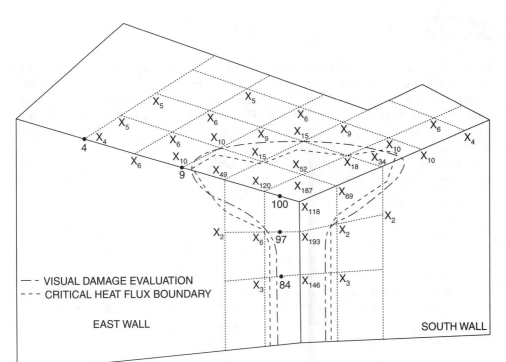

FIGURE 3.3 ◆ The correlation between critical heat flux boundary (dashed line) and visual lines of demarcation (dashed-dotted line) in an FMRC 25-ft corner fire test provides a scientific explanation for the formation of V patterns on walls. *From* The SFPE Handbook of Fire Protection Engineering, *3rd ed., 2002, by permission of Society of Fire Protection Engineers.*

samples to progressively decreasing incident heat fluxes until ignition no longer takes place (Spearpoint and Quintiere 2001).

Figure 3.3 illustrates this close correlation in FMRC's 8-m (25-ft) corner tests of a growing fire peaking at a 3-MW heat release rate (Newman 1993). This test is useful for evaluating fires involving low-density, high–char forming wall and ceiling insulation materials. In Figure 3.3 the critical heat flux boundary as measured by radiometers attached to the surface is denoted by the dashed line, and the visual damage evaluation by the dashed–dotted line. Note how closely these observations correlate. The position and impact of the fire plume on the walls, ceilings, and surface areas are important to interpreting these correlations correctly when evaluating the various types of fire patterns.

At first, the soot and pyrolysis products of the smoke condense on the cooler surfaces, with no chemical or thermal effect. As hot gases from a plume come into contact with the surface, heat is transferred by convective and radiative processes. As the heat is transferred, the temperature of the surface increases. The heat may be conducted into the surface. At some point, the temperature increases to the point at which the surface coating begins to scorch, melt, or char (pyrolyze). At higher temperatures it may actually ignite. These temperatures are reached when the critical heat flux is reached for that material and sustained for a sufficient time. These observations allow us to characterize patterns as surface deposit only, thermal effect to surface only, charring and ignition, penetration, and consumption.

When a fire scene is examined, often the only evidence remaining is of the fire burn pattern indicators reflecting their thermal impact. In cases where a fire was extinguished quickly by fire suppression or has self-extinguished, clearly defined burn patterns from the plume are left on the walls, floors, ceilings, and exterior surfaces of the structure. Many of these patterns are formed by the intersection of fire plumes with structure and targets.

There are four major types of plumes of interest in fire investigation and reconstruction (Milke and Mowrer 2001):

- ◆ Axisymmetric
- ◆ Window
- ◆ Balcony
- ◆ Line

Understanding these plumes provides a keener insight into the dynamics of more complex fires. The investigator examining the damage of these plumes must have a working knowledge of their origin, type of fuel package, and ventilation. Furthermore, empirical testing has established mathematical relationships describing the behavior, shape, and impact of these classes of fire plumes.

AXISYMMETRIC PLUMES

Axisymmetric plumes have uniform radial distribution from a common vertical plume centerline. Generally, they occur in open areas or near centers of rooms where there is no nearby wall. Much research has been conducted on the correlations of the plume height, temperatures generated, gas velocities, and entrainment. Figure 3.4 shows the typical descriptions and measurements used for flame and plume characteristics describing the plume centerline, mean flame height, and virtual origin (NFPA 1997, chap. 11–10). These measurements are used in calculations that show relationships of the fire plume with heat release rates, flame height, temperatures, and smoke production.

Figure 3.5 is an example of an axisymmetric fire plume under the NIST furniture calorimetry hood during research on flammable and combustible liquid spill and burn patterns (Putorti 2001, 15). These tests provide valuable data on heat release rates as well as on burn patterns on various flooring surfaces.

Plumes occurring near walls and corners do not behave as axisymmetric plumes. Their behavior will be discussed in a later section of this chapter.

WINDOW PLUMES

Plumes that emerge from doors and windows into large open spaces are known as **window plumes** (NFPA 2000C, sec. 3.8.3). These plumes emerge from openings during the fire and are commonly ventilation controlled. In Figure 3.6, the plume is emerging from the opening and measures Z_w above the soffit.

Window plumes result when the rate at which combustible vapors and gases are produced by pyrolysis inside the room exceeds the rate at which those products can be burned by combining with air getting into the room and flow out openings to burn outside the compartment. Window plumes are often (but not exclusively) associated with postflashover burning in that room. As such, they may be a visible indicator of a fire's development.

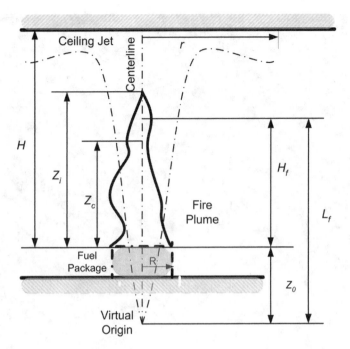

FIGURE 3.4 ◆ A schematic diagram of the typical axisymmetric fire plume and its interaction with a ceiling, where Z_0 is the distance from the fuel surface to the virtual origin, H_f is the mean flame height above the fuel surface, Z_c is the continuous flame height above the fuel surface, Z_i is the intermittent flame height above the fuel surface, H is the distance to the ceiling from the fuel surface, R is the radius of the fuel package, and r is the distance of the ceiling jet from the plume centerline. Note that the statistical mean flame height is the portion of luminous flame that is visible during 50 percent of its total exposure time.

Estimating the heat release rate may be worthwhile when considering the observations of on-scene eyewitnesses and responding firefighters. For example, the description of the window plume may be sufficient to estimate the heat release rate based on the observed area and height of the ventilation opening. Window plumes can be a major factor in vertical fire spread in multistory buildings, since entrainment can push the plume against combustible siding or spandrel panels or windows of stories above. The building experiences very high heat fluxes under such conditions (DeHaan 2007, 56–58).

WINDOW PLUME CALCULATIONS

As we saw in Chapter 2, the area and height of the ventilation openings of window plumes can be used to estimate the maximum heat release rate of the fire in a compartment that is being ventilated by that opening. A mathematical relationship based on actual experimental data for wood and polyurethane can be used to predict the maximum heat release rate, assuming that the single opening is ventilating the fire (Orloff, Modale, and Alpert 1977; Tewarson 1995).

$$\dot{Q}_{\max} = 1260\, A_o \sqrt{H_o}, \tag{3.1}$$

FIGURE 3.5 ◆ An axisymmetric fire plume under the NIST furniture calorimetry hood during research on flammable and combustible liquid spill and burn patterns. *Courtesy of Putorti [2001, 15].*

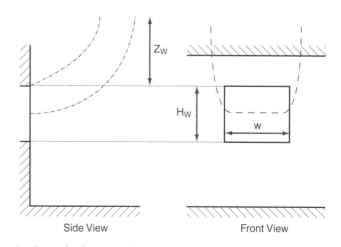

FIGURE 3.6 ◆ A schematic diagram of a window plume showing side and front orientations.

where

$\dot{Q}_{max}$ = maximum heat release rate (kW),

A_o = area of ventilation opening (m^2), and

H_o = height of ventilation opening (m).

This simple relationship for estimating the heat release rate may be worthwhile when considering the observations of on-scene eyewitnesses and responding fire-fighters. For example, the description of the window plume may be sufficient to estimate the heat release rate based on the observed area and height of the ventilation opening. Compare equation (3.1) with equation (2.25) for calculating the $\dot{Q}$ needed to cause flashover (converting the latter to kW units) to see that $\dot{Q}_{fo} \approx 0.5\,\dot{Q}_{max}$. In other words, after flashover, the heat release rate of a fire can increase by a factor of 2 until it reaches its upper limit, where all the available inflowing oxygen is being consumed by combustion. Excess fuel produced by the intense radiant heat flux of the postflashover fire that cannot be combusted in the room escapes through the window openings to burn as the observed window plumes.

BALCONY PLUMES

Plumes that emerge under overhangs and from doors are known as **balcony spill plumes** (NFPA 2000c, sec. 3.8.2). These plumes (Figure 3.7) are characteristic of fires occurring in enclosed rooms and spreading through patio doors or windows to covered porches, patios, or balconies. The buoyancy of these gases causes them to flow along the underside of horizontal surfaces until they can rise vertically.

The pattern of thermal damage or smoke damage to the underside of horizontal surfaces may be used to estimate the dimensions of the plume. The plume spreads laterally as it flows. The width of the plume is estimated by re-creating the horizontal and vertical dimensions of its contact area such that

$$W = w + b, \qquad (3.2)$$

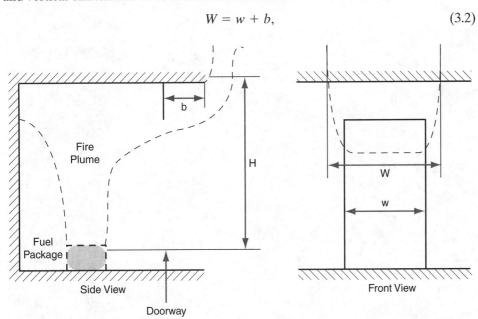

FIGURE 3.7 ◆ A schematic diagram of a balcony spill plume showing side and front orientations.

where

W = maximum width of the plume (m),

w = width of the opening from the area of origin to the balcony (m), and

b = distance from the opening to the balcony edge (m).

This calculation has been used in estimating the width and height of the fire plume to evaluate its properties and assess smoke production. Locations of balcony plumes include exterior doors leading to garden apartments, internal hotel rooms in atrium-style buildings, multistory shopping malls, and multiple-level prison cells that face catwalks.

The characteristic of a balcony plume is that flames are deflected away from the upper floor by the external overhang, as they would be by an open porch or deck in an apartment or condominium. The overhang is an example of the use of passive fire engineering in building construction to reduce the chance that external plumes will spread fires to upper floors above the floor of fire origin.

Figure 3.8 shows an example of a window plume coming into exterior contact with a porch overhang, creating a situation similar to a balcony spill. As the plume exits the window, it travels along the ceiling of the porch overhang, and flames are projected from the side of the overhang. This situation is common in residential fires, particularly when the dwelling has a porch or a deck.

FIGURE 3.8 ◆ Window plume that comes into exterior contact with a porch overhang, creating a situation similar to a balcony spill. *Courtesy of D. J. Icove.*

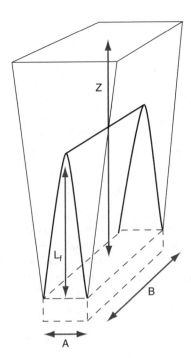

FIGURE 3.9 ◆ Illustration of a line source plume.

LINE PLUMES

Line plumes (Figure 3.9) have elongated geometric shapes that produce narrow, thin, shallow plumes. The relationship of the heat release rate to the flame height is described in a later section. A line plume with a flame plume height, L_f, greater than five times the B measurement more closely resembles an axisymmetric plume (Hasemi and Nishihata 1989).

Possible scenarios for line plumes include exterior fires in ditches where spilled ignitable liquids collect and ignite, a row of townhouses, a long sofa, the advancing front of a forest fire, flame spread over flammable wall linings, and even a balcony spill plume (Quintiere and Grove 1998). Line plumes can also be used to approximate elongated fires in open areas of atria or warehouses.

FLAME HEIGHTS OF LINE PLUMES

The relationship of the heat release rate to the flame height for line plumes (Figure 3.9), where $B > 3A$, is described next. For flame plume heights greater than five times the B measurement in Figure 3.9, the line plume more closely resembles an axisymmetric plume (Hasemi and Nishihata 1989):

$$L_f = 0.035 \left(\frac{\dot{Q}}{B}\right)^{2/3}, \quad \text{for} \quad B > 3A, \tag{3.3}$$

$$\dot{m} = 0.21z \left(\frac{\dot{Q}}{B}\right)^{2/3}, \quad \text{for} \quad B < 3A \quad \text{and} \quad L < z < 5B, \tag{3.4}$$

where

L_f = flame height (m),

$\dot{Q}$ = heat release rate (kW),

$\dot{m}$ = mass flow rate (kg/s),

A = shorter side of line plume base (m),

B = longer side of line plume base (m), and

z = ceiling height above fuel (m).

ALTERNATIVE METHOD

The entrainment of air around a long line plume can be envisioned as air coming from two opposite sides of an axisymmetric plume. It has been modeled in a theoretical study by Quintiere and Grove (1998). For a wall fire where the flame is against a vertical, noncombustible wall, the entrainment is from one side only. If the Q for an equivalent fuel surface can be calculated (from the equations in Chapter 2), the $\dot{Q}$ (heat release rate per unit length of line) can be calculated from the ratio $\dot{Q}/B$.

The *Fire Protection Handbook* gives the relationships

$$L_f = 0.017\dot{Q}^{2/3}, \quad \text{for line fire plumes}$$

$$L_f = 0.034\dot{Q}^{3/4}, \quad \text{for wall fire plumes}$$

(NFPA, 1997).

◆ 3.3 AXISYMMETRIC FIRE PLUME CALCULATIONS

As indicated in Chapter 2, investigators can benefit from information on fire plumes, which are merely the physical manifestations of the combustion process. Sound scientific and engineering principles, based on a combination of theoretical and actual fire testing data, describe fire plume behavior.

Information on fire plume behavior is expressed in terms of five basic calculated measurements often used in hazard and risk analysis (SFPE 2002a, 387). They model the following fire plume characteristics:

- ◆ Equivalent fire diameter
- ◆ Virtual origin
- ◆ Flame height
- ◆ Plume centerline temperatures and velocities
- ◆ Plume air entrainment

These basic characteristics constitute the bare minimum necessary information for fire reconstruction. Several common equations for performing a fire engineering analysis of fire plumes rely on the energy or heat release rate and provide answers to the questions: How tall was the fire plume? What was the placement of the virtual source in relation to the floor?

A fire investigator must be able to use these calculations in applying basic fire science and engineering concepts. These calculations are essential in evaluating the

various working hypotheses, such as how much fuel was burned, the fire's time duration, and the impact of ventilation of the room of origin.

Many of these calculations are found in the various fire modeling software tools,
such as FPETool and CFAST. The calculations are briefly explained here to demonstrate their underlying concepts and potential applications.

EQUIVALENT FIRE DIAMETER

In calculations of the flame height for fire plumes, the base of the fire is assumed to be
circular. This is normally not the case when ignitable liquids are spilled onto a floor or
when the fuel package is rectangular in shape. In these cases, the **equivalent diameter**
must be calculated. From the basic relationship for circular areas ($A = \pi r^2$, where
$D = 2r$), the relationship is

$$D = \sqrt{\frac{4A}{\pi}}, \qquad (3.5)$$

where

$D =$ equivalent diameter (m),

$A =$ total area of burning fuel package (m^2), and

$\pi =$ 3.1416.

Areas of pools can be calculated by re-creating the shape and size of pools from
fire scene photos using wrapping-paper cutouts and obtaining dimensions from fire
scene notes. The cutouts can then be weighed using a postal scale and the area determined from the weight of a reference paper sample of known area. Areas can also be
re-created from photogrammetry of the fire scene photos (see Chapter 4 on scene
documentation).

Rectangular contained spills include those produced when an oil-filled electric
transformer or storage tank is compromised and spills its contents into a rectangular
diked area and ignites. It is then necessary to determine the diameter of a circle
containing the same area as the rectangular spill, using the aspect ratio of the spill
(Figure 3.10).

The *aspect ratio* is the width-to-length ratio of the spill area. The equivalent diameter relationship generally holds true for pools with an aspect ratio ≤ 2.5. Aspect

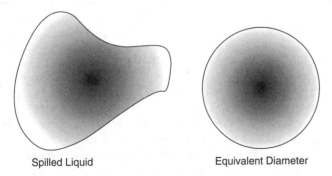

Spilled Liquid   Equivalent Diameter

FIGURE 3.10 ◆ The equivalent-diameter concept used in estimating the diameter of a circle
that represents the same area as a noncircular spill pattern.

ratios >2.5, such as would be seen with trench or line fires, utilize other models (Mudan and Croce 1995).

In cases of short-duration fires, the area of a liquid fuel spill can be measured directly from the burned areas of the floor or covering. In postflashover or prolonged fires where the floor covering is readily combustible, the fire-damaged areas may extend well beyond the margins of the original pool and are therefore not a reliable guideline. DeHaan (1995, 2002, and 2004) has shown a simple relationship among surface area, volume of liquid, and type of surface that controls the equivalent depth of the pool. Putorti (2001) has confirmed this general relationship.

For spills on nonporous surfaces such as sealed concrete, the depth of a free-standing pool of gasoline is about 1 mm (1×10^{-3} m). The area (m²) is then calculated by dividing the volume of the spilled liquid (m³) by the depth (m), where

$$\text{Area (m}^2) = \frac{\text{Volume (m}^3)}{\text{Depth (m)}}. \tag{3.6}$$

For the purposes of conversion, 1 liter (L) = 10^{-3} m³.

On semiporous surfaces such as wood, the equivalent depth is 2–3 mm ($2 - 3 \times 10^{-3}$ m). On porous surfaces such as carpet, the maximum depth will be the thickness of the carpet, assuming full saturation of the carpet.

EXAMPLE 3.1   Equivalent Fire Diameter

Problem.   A container filled with 3.8 L (1 gal or 3800 cm³) of gasoline is accidentally tipped over onto an enclosed patio of sealed concrete. The depth of the spilled pool is 1 mm. Calculate the (1) coverage area and (2) equivalent diameter of a circle of the same area.

Suggested Solution.

$$\text{Area of spill} = \frac{\text{Volume of liquid}}{\text{Depth of pool}} = \frac{3800 \text{ cm}^3}{0.1 \text{ cm}} = 38,000 \text{ cm}^3 = 3.8 \text{ m}^2$$

The equivalent diameter D then is.

Equivalent diameter	$D = \sqrt{4A/\pi}$
Total surface area	$A = 3.8 \text{ m}^2$
Constant	$\pi = 3.1416$
Equivalent diameter	$D = \sqrt{(4)(3.8)/3.1416} = 2.2 \text{ m}$

VIRTUAL ORIGIN

As shown in Figures 2.7, 2.16, and 3.4, the **virtual origin** represents the actual fire as a point source. Its location is a point along the fire plume centerline where flames appear to originate, measured from the top burning surface of the fuel package. Identifying the virtual origin is helpful when evaluating the exposure of other "targets" in the room to the fire. Several assumptions may need to be made depending on the distance to the target and the size of the burning fuel surface.

The virtual origin can be mathematically calculated, and its location can be helpful in reconstructing and documenting the fire's virtual source, point of origin, area, and direction of travel. The virtual origin is also used in some calculations for the measurement of the fire plume flame height.

The virtual origin, Z_0, is calculated from the **Heskestad equation** (1982):

$$Z_0 = 0.083\ \dot{Q}^{2/5} - 1.02D, \qquad\qquad (3.7)$$

where

D = equivalent diameter (m),

$\dot{Q}$ = total heat release rate (kW), and

Z_0 = virtual origin (m)

for the conditions

T_{amb} = 293 K, (20°C, 68°F),

P_{atm} = 101.325 kPa, and

D ≤ 100 m (328 ft)

The virtual origin, depending on the equivalent diameter and the total heat release rate, may fall above or below the fuel's surface. This placement depends primarily on the heat release rate and effective diameter of the fuel. The resulting value can then be used in equations and models of other fire plume relationships.

Small-diameter ignitable liquid spills are more likely to exhibit virtual origins above the fuel surfaces. Inspection of equation (3.7) reveals that the smaller the value of the fire's equivalent diameter, D, and/or the larger the value of the heat release rate, $\dot{Q}$, the greater the chance for a positive virtual origin, Z_0, that is, above the fire surface. Because not all fires burn at floor level, the concept of Z_0 is useful in estimating the effective origin of the fire plume in relation to the fuel surface and the compartment itself.

EXAMPLE 3.2   Virtual Origin

Problem.   A fire starts in a wastepaper basket filled with papers and quickly reaches a steady heat release rate of 100 kW. The basket measures 0.305 m (1 ft) in diameter. Determine the virtual origin of this plume using the Heskestad equation.

Suggested Solution.   Use the equation for the virtual origin.

Virtual origin	$Z_0 = 0.083\ \dot{Q}^{2/5} - 1.02D$
Equivalent diameter	$D\ = 0.305$ m
Total heat release rate	$\dot{Q}\ = 100$ kW
Virtual origin	$Z_0 = (0.083)(100)^{2/5} - (1.02)(0.305)$
	$\quad = 0.524 - 0.311 = 0.213$ m (0.698 ft or 8.9 in.)

Note that in this case the virtual origin is a positive number, indicating that it is above the surface of the fuel. If the same fuel is spread out over the floor such that its equivalent diameter is 1.0 m while its $\dot{Q}$ remains 100 kW, its virtual origin would be $Z_0 = -0.5$ m (1.64 ft, or 19.7 in., below the floor). This means that the impact of the fire on the room would be as if the fire were burning farther away from the ceiling.

Discussion.   If the trash can was filled with gasoline and $\dot{Q} = 500$ kW, recalculate the virtual origin. What can you assume about the virtual origin when the heat release rate is raised?

FLAME HEIGHT

Flames from plumes represent the visible portion of a fire's combustion process. They are visible owing to the luminosity of heated soot particles and pyrolysis products. The buoyancy of these fire gases allows a fire plume to rise vertically. The height of the flames is described by several terms, including continuous, intermittent, and average.

Knowing the **flame height** allows the investigator to examine and confirm what potential damage can be expected based on heat transfer to the ceiling, walls, flooring, and nearby objects. Several equations, based on fitting regression formulas to actual fire testing, are used to estimate the flame height. As with all these equations, users are cautioned that the equations are bounded by the limitations of the testing.

Within the flaming region, the two **McCaffrey flame height measurements** (1979) along the centerline are depicted as continuous, Z_c, and intermittent, Z_i:

$$Z_c = 0.08\ \dot{Q}^{2/5}, \tag{3.8}$$

$$Z_i = 0.20\ \dot{Q}^{2/5}, \tag{3.9}$$

where

Z_c = continuous flame height (m),

Z_i = intermittent flame height (m), and

$\dot{Q}$ = heat release rate (kW).

Crude estimates of heat release rates made by witnesses and responding firefighters can be drawn by algebraically rearranging the McCaffrey flame height equation for intermittent flame heights:

$$\dot{Q} = 56.0\ Z_i^{5/2}. \tag{3.10}$$

EXAMPLE 3.3   Flame Height—McCaffrey Method

Problem.   Using Example 3.2 involving the 100-kW wastepaper basket fire, determine the continuous and intermittent flame heights using the McCaffrey method.

Suggested Solution.   Use the equation for the McCaffrey method.

Heat release rate	$\dot{Q}$ = 100 kW
Continuous flame height	$Z_c = 0.08\ \dot{Q}^{2/5} = (0.08)(100)^{2/5}$
	= 0.505 m (1.66 ft)
Intermittent flame height	$Z_i = 0.20\ \dot{Q}^{2/5} = (0.20)(100)^{2/5}$
	= 1.26 m (4.1 ft) above the fuel surface.

EXAMPLE 3.4   Heat Release Rate from Flame Height—McCaffrey Method

Problem.   A witness to an accident of an overturned lawn mower sees the gasoline ignite and burn, forming an axisymmetric fire plume. He observes intermittent flames reaching 3 m (9.84 ft) into the air from the ground. Estimate the heat release rate, $\dot{Q}$, given off by the burning fuel spill.

Suggested Solution.   Use the equation for the McCaffrey method.

General equation	$Z_i = 0.20\ \dot{Q}^{2/5}$
Intermittent flame height	$Z_i = 3.0$ m

Estimated heat release rate

$$\dot{Q} = 56.0\, Z_i^{2/5}$$
$$= (56.0)(3.0)^{2/5} = 873 \text{ kW}$$

A practical estimate would be 800–900 kW.

EXAMPLE 3.5 Calculating the Area of a Pool Fire

Problem. From the data in Table 2.4, calculate the area of the pool fire at 873 kW from Example 3.4.

Suggested Solution. Use the equation for the heat release rate.

General equation	$\dot{Q}$	$= \dot{m}'' \Delta h_c A$
Heat release rate	$\dot{Q}$	$= 873$ kW
Mass burning rate per unit area	$\dot{m}''$	$= 53$ g/m²s
Heat of combustion	Δh_{eff}	$= 43.7$ kJ/g
Area of pool fire	A	$= \dot{Q}/\dot{m}'' \Delta h_c = 873/(53)(43.7) = 0.38$ m² (4.1 ft²)

One assessment of the flame height is its statistical **mean flame height**, which is the portion of luminous flame that is visible during 50 percent of its total exposure time. When the human eye observes a pulsating flame, it tends to interpolate the variations into a rough approximation of the mean flame height.

A good estimate of the mean or average flame height, L_f, uses the **Heskestad equation:**

$$L_f = 0.235\, \dot{Q}^{2/5} - 1.02D, \tag{3.11}$$

where

$D =$ effective diameter (m),

$\dot{Q} =$ total heat release rate (kW), and

$L_f = Z + Z_0 =$ mean flame height above the virtual origin (m).

Note that this equation includes a diameter term, accounting for the effects of large-diameter fuels on the visible flame. The larger the size of the burning fuel area, the lower the flames will be for the same heat release rate.

EXAMPLE 3.6 Flame Height—Heskestad Method

Problem. From Example 3.2 involving the 100-kW wastepaper basket fire, determine the mean flame height, L_f, above the virtual origin, Z_0.

Suggested Solution. Use equation (3.11) for calculating the mean or average flame height.

Mean flame height	$L_f = 0.235\, \dot{Q}^{2/5} - 1.02 D$
Diameter of basket	$D = 0.305$ m
Total heat release rate	$\dot{Q} = 100$ kW
Mean flame height	$L_f = 0.235\, \dot{Q}^{2/5} - 1.02 D$
	$= (0.235)(100)^{2/5} - (1.0)(0.305)$
	$= 1.17$ m (3.84 ft) above the virtual origin (3–4 ft)

Note that the virtual origin from Example 3.2 was about 0.2 m above the fuel surface.

ENTRAINMENT EFFECTS

When a fire is burning in the center of a room, the buoyant upward flow of the fire gases drags fresh room air into contact with it. This process, called **entrainment** because surrounding cooler air is "mixed" or "entrained" into the upward plume flow, changes the plume's temperature, velocity, and diameter above the fuel source. Entrainment reduces the plume temperature and increases its width.

If there is a horizontal ceiling and the fire plume reaches this height, the plume hits the ceiling and flows horizontally, forming what is called a **ceiling jet**. In this case ceiling entrainment also takes place (Babrauskas 1980b).

When some of the heat is transferred to the air and is removed by convective processes, there is usually turbulent mixing that dilutes the rising column of hot gases and cools it further. Looking down on the fire plume from above as in Figure 3.11a, we can see that the inward air currents are roughly equal around the entire perimeter, so the plume remains symmetrical and upright.

If the fire is against a wall, it cannot draw room air into itself equally from all sides, so the air being drawn from the "free" side acts like a draft to force the plume sideways against the wall (shown in Figures 3.11b and c). The reduced entrainment means that less cooling air is brought in, slowing the cooling process. Thus, the gases have more time to rise farther, stretching out the flame. This process can occur on nearby wall surfaces even if they are not vertical. This is the mechanism responsible for external flame spread between floors of modern high-rise buildings. When the windows of the room first involved fail, the plume exits the building and then entrains

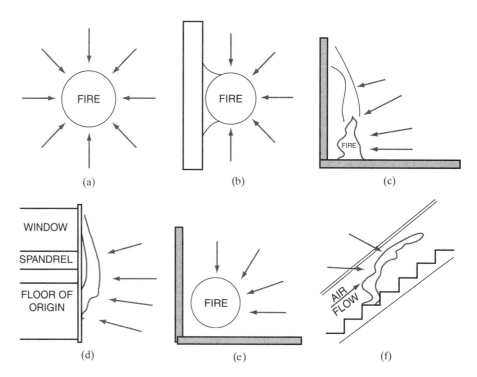

FIGURE 3.11 ◆ (a) Entrained air flow for an axisymmetric plume, from above; (b) entrainment in wall configuration; (c) side view; (d) typical high-rise fire plume; (e) entrainment in corner fire; (f) trench effect, side view.

itself against the side of the building (as in Figure 3.11d). The windows and spandrel panels of the floors above are then exposed to the radiant and convective heat transfer ($>50\,\text{kW/m}^2$) of the plume and quickly fail, allowing the flames to enter the floor above and ignite the contents of that floor. If there is not sufficient fire protection to suppress this newly ignited fire, then fire can leapfrog up the building. In buildings with a ledge or balcony at each floor level, the plume is directed away from the side of the building, so it never gets a chance to attach to the facade. The chances of ignition by window failure of the floors above are much reduced by a ledge flame-guard effect, compared with a (sheer) curtain wall design.

If the fire occurs in a corner, the perimeter through which cooling air can be drawn is reduced to one-fourth its original size (note that there is still plenty of air reaching the combustion zone, so the fire continues to burn at the same rate as represented in Figure 3.11e). In a corner, then, the gases cool even more slowly and have even more time to rise before cooling to the point at which they are no longer incandescent ($500°C$), and the visible flame becomes even taller. This effect is enhanced by the increased radiant heat that reaches the fuel surface after being reflected off the nearby wall or corner surfaces, thus increasing the heat release rate.

Because a nearby wall or corner reduces the cooling entrainment of room air into the rising plume and increases radiant heat feedback to the burning fuel, the height of the visible flame can be affected. *NFPA 921* lists relationships that are very useful to investigators and have been subjected to numerous studies and tests for determining the mean flame height near walls (NFPA 2004, sec. 5.5.6; Alpert and Ward 1984) These relationships are as follows:

$$H_f = 0.174\,(k\dot{Q})^{2/5}, \tag{3.12}$$

$$\dot{Q} = \frac{79.18\,H_f^{5/2}}{k}, \tag{3.13}$$

where

$H_f =$ flame height (m),

$\dot{Q} =$ heat release rate (kW), and

$k =$ wall effect factor, with

$k = 1$: no nearby walls,

$k = 2$: fuel package at wall, and

$k = 4$: fuel package in corner.

EXAMPLE 3.7   Flame Height—*NFPA 921* Method

Problem.   Using Example 3.2 involving the 100-kW wastepaper basket fire, estimate the height of the flame in three configurations—in the middle of the room, near a wall, and near a corner. Assume that this is an unconstrained fire burning in a large room with no ventilation constraints.

Suggested Solution.   From equation (3.12), the flame height H_f measured from the top of the fuel surface is as follows:

Total heat release rate $\qquad\qquad \dot{Q} = 100\,\text{kW}$
Flame height $\qquad\qquad\qquad\quad H_f = 0.174\,(k\dot{Q})^{2/5}$

No nearby walls ($k = 1$)

$$H_f = (0.174)(1 \times 100)^{2/5}$$
$$= (0.174)(100)^{0.4} = 1.10 \text{ m (3.61 ft)}$$

Near a wall ($k = 2$)

$$H_f = (0.174)(2 \times 100)^{2/5}$$
$$= 1.45 \text{ m (4.76 ft)}$$

In a corner ($k = 4$)

$$H_f = (0.174)(4 \times 100)^{2/5}$$
$$= 1.91 \text{ m (6.27 ft)}$$

A special case of this phenomenon is called the **trench effect**, also known as the **Coanda effect**, and is the tendency of a fast-moving stream of air to deflect itself toward and along nearby surfaces. The trench effect can occur where a fire starts on the floor of an inclined surface with walls on both sides, such as a stairway or escalator (shown in Figure 3.11f). In this case, the flow of the air drawn into the fire from below is aided by the confinement of the buoyant flow. When the fire reaches a critical size (i.e., a critical airflow) the plume lies down in the trench and is stretched out by the directional entrainment. If the floor and/or walls of the trench are combustible, they ignite, and the fire can spread the length of the stairway very quickly.

The trench effect was the mechanism that drove a small fire on the floor of an all-wood escalator in the King's Cross underground station in London in 1987 to develop quickly into a massive fireball that engulfed the large ticket hall at the top of the escalator (Moodie and Jagger 1991). Further research by Edinburgh University identified four factors relating to the conditions that produce the trench effect: (1) the slope of the trench, (2) its geometric profile, (3) the combustible construction material, and (4) the fire ignition source (Wu and Drysdale 1996). Researchers determined that the critical angle above horizontal is 21° for a large trench and 26° for a small trench.

PLUME CENTERLINE TEMPERATURES AND VELOCITIES

The temperature distribution along the centerline of a typical plume is shown in Figure 3.12a, and the temperature distribution along a horizontal cross section through the flames is shown in Figure 3.12b. Estimates of the centerline maximum temperature rise, velocity, and mass flow rate of a plume are useful in determining the impact of the fire plume in compartment fires. This impact includes the maximum effect of the plume as it is rising to the ceiling and may come into contact with other fuel packages. There are several formulas for estimating these values.

The **Heskestad method** (1982) used for estimating the centerline maximum temperature rise, velocity, and mass flow rate of a plume is

$$T_0 - T_\infty = 25 \left(\frac{\dot{Q}_c^{2/5}}{(Z - Z_0)} \right)^{5/3}, \tag{3.14}$$

$$U_0 = 1.0 \left(\frac{\dot{Q}_c}{(Z - Z_0)} \right)^{1/3}, \tag{3.15}$$

$$\dot{m} = 0.0056 \, \dot{Q}_c \frac{Z}{L_f} \quad \text{for} \quad Z < L_f, \tag{3.16}$$

(i.e., for points within the flame plume)
where

$$\dot{Q}_c = 0.6 \, \dot{Q} \quad \text{to} \quad \dot{Q}_c = 0.8 \, \dot{Q}. \tag{3.17}$$

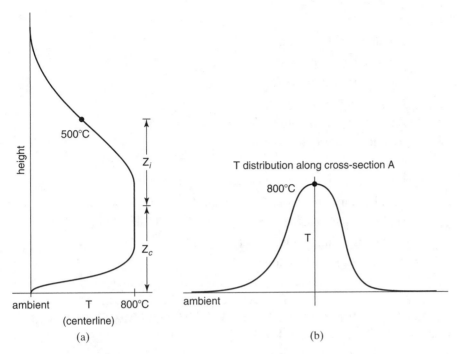

FIGURE 3.12 ◆ (a) Plot of average flame temperature versus height shows the uniform maximum temperature in the center of the continuous flame zone and the decreasing average flame temperature in the intermittent region. (b) Plot of average flame temperature versus horizontal position shows the maximum temperature at the axis of the flame plume.

The (1979) **McCaffrey method** for estimating the centerline maximum temperature rise, velocity, and mass flow rate of a plume is

$$T_0 - T_\infty = 21.6 \; \dot{Q}^{2/3} \; Z^{-5/2}, \tag{3.18}$$

$$U_0 = 1.17 \; \dot{Q}^{1/3} \; Z^{-1/3}, \tag{3.19}$$

$$\dot{m}_p = 0.076 \; \dot{Q}^{0.24} \; Z^{1.895}. \tag{3.20}$$

An alternative method for estimating the maximum temperature rise and velocity of a plume is (Alpert 1972)

$$T_0 - T_\infty = \frac{16.9 \; \dot{Q}^{2/3}}{H^{5/3}} \qquad \text{for} \qquad \frac{r}{H} \le 0.18, \tag{3.21}$$

$$T_0 - T_\infty = \frac{5.38 \; (\dot{Q}/r)^{2/3}}{H^{5/3}} \qquad \text{for} \qquad \frac{r}{H} > 0.18, \tag{3.22}$$

$$U = 0.96 \frac{\dot{Q}^{1/3}}{H} \qquad \text{for} \qquad \frac{r}{H} \le 0.15, \tag{3.23}$$

$$U = \frac{0.195 \; Q^{1/3} \; H^{1/2}}{r^{5/6}} \qquad \text{for} \qquad \frac{r}{H} > 0.15, \tag{3.24}$$

where

T_0 = maximum ceiling temperature (°C),

T_∞ = ambient air temperature (°C),

$\dot{Q}$ = total heat release rate (kW),

$\dot{Q}_c$ = convective heat release rate (kW),

Z = flame height along the centerline (m),

Z_0 = distance to the virtual origin (m),

L_f = $Z + Z_0$.

$\dot{m}$ = total plume mass flow rate (kg/s),

r = radial distance from the plume centerline (m),

H = ceiling height (m),

U = maximum ceiling jet velocity (m/s), and

U_0 = centerline plume velocity (m/s).

EXAMPLE 3.8   Plume Temperature, Velocity, and Mass Flow Rate at Ceiling Level—McCaffrey Method

Problem.   A fire is discovered in a wastepaper basket in a room that has a ceiling height of 2.44 m (8 ft). The heat release rate estimate is 100 kW. Assume that the ambient air temperature is 20°C (68°F). Estimate the plume temperature, velocity, and mass flow rate at the ceiling using the McCaffrey method.

Suggested Solution.   Use the McCaffrey equation.

Maximum plume temperature

$$T_0 = 21.6\,\dot{Q}^{2/3}\,Z^{-5/2} + T_\infty$$
$$= (21.6)\,(100)^{2/3}\,(2.44)^{-5/2} + 20 = 68.5°C$$

Centerline plume velocity

$$U_0 = 1.17\,\dot{Q}^{1/3}\,Z^{-1/3}$$
$$= (1.17)\,(100)^{1/3}\,(2.44)^{-1/3} = 4.033\,\text{m/s}$$

Plume mass flow rate

$$\dot{m}_p = 0.076\,\dot{Q}^{0.24}\,Z^{1.895}$$
$$= (0.076)\,(100)^{0.24}\,(2.44)^{1.895} = 1.244\,\text{kg/s}$$

Discussion.   In Example 3.2, the calculated virtual origin, Z_0, was 0.213 m (0.698 ft) above the burning fuel surface. Solve this same problem using the methods of Heskestad and Alpert.

SMOKE PRODUCTION RATES

The **smoke production rate** is estimated to be close to twice that of the mass flow rate of gas produced by a fire plume. This assumption is based on the theory that the amount of air entrained into the rising plume equals the amount of smoke-filled gas (NFPA 2000b, sec. 11, chap. 10).

   In the **Zukoski** (1978) **method** for determining smoke production rates, the equation for estimating the rate of the plume mass flow above the flame at 20°C (68°F) is

$$\dot{m}_s = 0.065\,\dot{Q}^{1/3}Y^{5/3}, \tag{3.25}$$

where

$\dot{m}_s$ = plume mass flow rate (kg/s),

$\dot{Q}$ = total heat release rate (kW), and

Y = distance from virtual point source to bottom of smoke layer (m).

The best correlations to this theory are found in cases involving circular equivalent pool fires and in which the enclosure is vented in the lower layer.

ENCLOSURE SMOKE FILLING

When smoke from a fire accumulates within an enclosure or compartment, it tends to rise to the ceiling. This accumulation or **enclosure smoke-filling rate** depends on the amount of smoke being produced by the fire and its ability to escape through vents within the compartment. As the smoke accumulates, a layer forms and decends toward the floor of the compartment.

The enclosure smoke-filling rate is important when estimating how much smoke accumulated below the ceiling. Factors contributing to how fast smoke stratifies and descends from the ceiling relate primarily to the amount of smoke produced and the size and location of smoke vents. Fire investigators finding a room with smoke stratification deposits evenly distributed in an enclosed room may find this evidence important when later determining the fire duration or the smoke exposure a victim might have encountered at a particular time.

The relationship for the descent of this upper layer in a closed room is expressed as

$$U_t = \frac{\dot{m}_s}{\rho_l A_r},\qquad(3.26)$$

where

U_t = rate of layer descent (m/s),

$\dot{m}_s$ = plume mass flow rate (kg/s),

ρ_l = density of the upper gas layer (kg/m^3), and

A_r = enclosure floor area (m^3).

Note that the density of smoke varies inversely with temperature but is conservatively estimated to be the same as air at standard temperature and pressure (STP) (1.22 kg/m^3).

EXAMPLE 3.9   Smoke Filling

Problem.   The 100-kW fire discovered in a wastepaper basket in the previous example is in a closed room that has a ceiling height of 2.44 m (8 ft). The room measures 2.44 m (8 ft) square. Estimate the enclosure smoke-filling rate for this fire.

Suggested Solution.   Use the data previously calculated using the McCaffrey method. Assume the lower limit of the velocity of descent by using the ambient density of air as 1.22 kg/m^3 and 17°C.

Rate of layer descent	$U_t = \dfrac{\dot{m}_s}{\rho_l A_p}$
Mass rate of smoke production	$\dot{m}_s = 1.244$ kg/s
Density of upper gas layer	$\rho_l = 1.22$ kg/m^3
Enclosure floor area	$A = (2.44)\,(2.44) = 5.95$ m^2
Rate of layer descent	$U_t = (1.244)/(1.22)\,(5.95) = 0.171$ m/s

---

◆ 3.4 FIRE PATTERNS

TYPOLOGY

When reconstructing fire scenes, skilled fire investigators use numerous indicators to estimate the areas and points of origin, distribution, and behavior of a fire such as direction of spread. These indicators, called **fire patterns**, are characteristically broken down into the following five major groupings summarized in Table 3.1. Some specific patterns are illustrated in Figure 3.13.

- Demarcations
- Surface effects
- Penetrations
- Loss of material
- Victim injuries

Note that the first four of these correlate to the categories of the fire effects discussed previously. Victim injuries can be considered a special case of applying those categories to human skin and tissue.

DEMARCATIONS

Demarcations occur in locations where a combination of smoke, heat, and flames impinge on materials, forming intersections between affected and unaffected areas. Examples include smoke layers deposited on walls, and plume patterns on walls. See item 1 in Figure 3.13. Demarcations can vary depending on the type of exposed material, gas temperature, rate of heat release, and ventilation. Areas of material and non-material loss are helpful in determining demarcations.

Heat and **smoke levels** (also referred to as **horizons**) refer to the height at which smoke or heat has stained or marked the walls and windows of a room. In fire modeling programs, these levels can be predicted throughout the structure being modeled and correlated with the duration and size of the fire. Heat and smoke levels are also important when evaluating victim accounts of whether they could see an exit if standing in a room or whether they would be affected by smoke or heat as they navigated through the building. In cases involving victims injured or killed in fires, the smoke and heat levels become important when evaluating whether victims stood up in a smoke-filled room and sustained their injuries from the cloud of heated toxic gases and smoky by-products of combustion.

As the ceiling jet plume extends throughout the room, demarcations are formed by stratified heat and smoke damage. In Figure 3.13, the ceiling plume extends to the right along the wall, forming the start of a stratified line. Notice the sharp demarcation in the degree of damage between the affected and the unaffected areas on the wall surface.

Note also the acute angle between the demarcation line and with the ceiling. This angle is the result of buoyancy-driven flow along the ceiling and is a very reliable indicator of direction of spread.

Fire suppression efforts such as the application of hose spray to the walls, ceilings, and floors can also affect demarcations. Ventilation efforts during fire suppression, such as the venting of the roof and the opening of doors and windows, also can affect the demarcations formed by heat and smoke throughout the structure. Figure 3.14 illustrates the formation of an upward pattern on the wall surface

TABLE 3.1 ◆ Typology of Common Fire Patterns

Type of Pattern	Description	Variables	Examples
Demarcations	The intersection of affected and unaffected areas on materials where smoke and heat impinge	Type of material Fire temperature, duration, and rate of heat release Fire suppression Ventilation Heat flux reaching surface	V patterns on walls Discoloration patterns Smoke stains
Surface effects	Determined by the boundary shape, areas of demarcation, and nature of heat application to various surface types	Surface texture (rougher surfaces sustain more damage) Surface-to-mass ratio Surface coverings Combustible and noncombustible surfaces Thermal conductivity Temperature of surface	Spalling of floor surfaces Combustible surfaces scorched or charred by pyrolysis and oxidation Calcination Noncombustible surfaces changing color, oxidizing, melting, or burning cleanly smoke stains
Penetrations	Penetration of horizontal and vertical surfaces	Preexisting openings in floors, ceilings, and walls Direct flame impingement Heat flux Duration of fire Suppression	Downward penetrations under furniture Saddle burns on exposed floor joists Failure of walls, ceilings, and floors Internal damage to walls and ceilings
Loss of material	Combustible surfaces with loss of material and mass	Type of materials Construction methods Room contents and fuel loads Duration of fire	Tops of wooden wall studs Fall-down of debris Heat shadowing Isolated areas of consumed carpet Beveling of corners and edges
Victim injuries	Areas, depths, and degrees of burns on victim's clothing and body	Location of victim in relation to other objects or victims Actions taken prior to, during, and after the fire	Absence of burn injuries protected by clothing or furniture Burns on lower torso Heat shadowing

Source: Expanded from Icove. 1995.

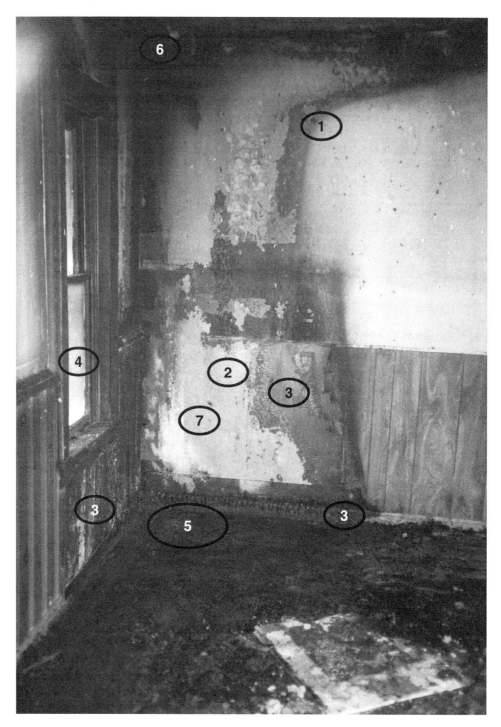

Figure 3.13 ◆ Fire pattern indicators found at fire scenes consisting of (1) demarcation, (2) calcination, (3) loss of material to wooden wainscoting and baseboard, (4) fractured glass, (5) ignitable liquid burn pattern on carpet, (6) penetration into the ceiling, and (7) area of clean burn where paper and pyrolysis products burned completely off wall, and calcination, indicates the most intense heat. *Courtesy of D. J. Icove.*

FIGURE 3.14 ◆ The impact of a hose spray on a wall during fire suppression efforts formed an upward pattern on the wall surface demarcations previously formed by heat and smoke. *Courtesy of D. J. Icove.*

by the impact of a hose spray, which scoured the heat-affected paper and propelled it off the wall.

The cooler the surface, the faster soot and pyrolysates will condense. Airflow will also control depositions. Figure 3.15 shows the effects of airflow on soot deposition and how it can reveal the direction of hot gas spread.

SURFACE EFFECTS

Surface effects occur when heat transfer is sufficient to cause discoloration of masonry surfaces, scorching or charring of combustible surfaces, melting or changes in color, or oxidizing of noncombustible surfaces. Areas not damaged by the heat and flames are called **protected** areas and are significant in cases where objects or bodies have been found or have been removed prior to inspection. Surface effects can also include nonthermal deposits of soot or pyrolysis products that condense on the cooler surfaces. The cooler the surface, the faster these deposits occur. Variables influencing surface effects include the type and smoothness of the surface, the thickness and nature of the coverings, and even the object's surface-to-mass ratio.

Extended postflashover burning in a room exposes surfaces to high temperatures and heat fluxes. This "erases" the demarcations and surface effects (DeHaan 2007). Research at Eastern Kentucky University on burn patterns showed that brief

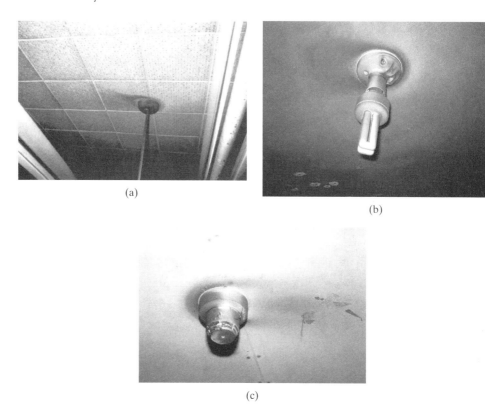

(a)

(b)

(c)

FIGURE 3.15 ◆ Deposits of soot (a and b) around light fixtures and (c) around a heat detector indicate direction of air movement. *(a) Courtesy of Jim Allen, Nipomo, California; (b and c) Courtesy of Ross Brogan, NSW Fire Brigades, Greenacre, NSW, Australia.*

(2–9 min) exposure to postflashover fires did not totally obliterate the fire patterns needed by investigators (Hopkins, Gorbett, and Kennedy 2007). Examples are included in Chapter 8.

An example of a surface effect is the discoloration of paint on the external metallic surfaces of a burned automobile. Heat transfer through conduction often produces discernible fire patterns on the exterior and interior surfaces of the vehicle (as in Figures 2.3 and 2.5).

Discoloration of metallic surfaces can also be caused by **oxidation**, which is a chemical process associated with combustion. Particularly with metal surfaces, the boundaries between higher and lower temperatures can often be documented through photographs and/or drawings.

Gypsum plasterboard or gypsum board is a material used in residential and commercial construction primarily for interior walls, in a process sometimes referred to as drywall construction. The drywall board contains gypsum (calcium sulfate dihydrate) bonded between two layers of paper. Fire-rated or X-type drywall is reinforced with glass fibers. The ASTM C 36 standard is the international standard for its manufacture.

Gypsum is a rock mineral that can be mixed with water and then dried into a hardened paste, resulting in a product that is 21 percent water, by weight. The **calcination** of gypsum wallboard is a process by which heat transfer causes the calcium sulfate to dehydrate in two discrete steps at characteristic temperatures, typically around 100°C (212°F) and 180°C (356°F).

As it dehydrates, the gypsum shrinks and loses the water of hydration, and its mechanical strength degrades. It can fail, exposing combustible wall structure, when it loses so much strength from dehydration, shrinkage, or combustion of paper coverings that it can no longer support its own weight. Eventually, the fully dehydrated gypsum turns powdery and loses its mechanical strength, causing the wall to collapse. This process of calcination can also form lines of demarcation on the wall surface (DeHaan 2007). The effect of calcination on a gypsum board wall is shown as items 2 and 7 in Figure 3.13.

A change in color from white to gray to white is seen in a cross section of affected gypsum. Dehydration alone probably does not cause the gray-to-white change, since this is a colorless transition. During fire exposure other carbonaceous pyrolysis products accumulate that turn the absorbent gypsum gray. Subsequent heating burns away (or evaporates) these deposits, causing whitening to progress through the thickness of the gypsum.

During fire exposure the exposed paper and painted surfaces of the gypsum wallboard burn away. This reduces the overall mechanical strength of the gypsum wallboard. The process in which dehydrated gypsum falls or flakes off the surface during heat exposure is called **ablation**, which occurs at 600°C (1112°F) for ceilings and 800°C (1472°F) for walls (König and Walleij 2000). This process reduces the thickness of the board (Buchanan 2001, 341). Several techniques are available for documenting the measured postfire loss of gypsum from the wall surface and estimating the radiant heat flux exposure time (Kennedy 2000; Schroeder and Williamson 2003). The measurements of depth of calcination of gypsum surfaces are often recorded on a plot of the room.

Research by Schroeder and Williamson (2001) indicates that gypsum wallboard is a viable source for documenting a fire's intensity, spread, and direction. In a series of experiments using an ASTM E 1354 cone calorimeter, they concluded that thermally induced changes in gypsum wallboard, along with X-ray diffraction analysis to document the crystalline structure change, provided a quantifiable method in assessing time, temperature, and heat flux exposure. Additional cited references on the thermal properties of gypsum board can be found in Lawson (1977), Thomas (2002), and McGraw and Mowrer (1999). Also see additional references on calcinations of gypsum board during fire in Mowrer (2001) and Chu Nguong (2004).

Charring is the **pyrolytic** action of heat, which converts the organic material (most often paper or wood) into its volatile fractions, leaving behind carbonaceous char. This char layer may burn off or flake off as the fire exposure continues (depending on the type of wood). The linear depth of the pyrolyzed material affected, including the char layer and lost material, is referred to as the **char depth**.

The rate at which wood chars is not linear. Wood chars quickly at first and then at a much slower rate owing to various factors such as the insulation effect of the new char, reduced air–fuel interactions, ventilation, heat transfer, and physical arrangement. The reported charring behavior of large-section structural lumber has been modeled. A review of 55 specimens of 2 × 4-in. spruce-pine-fir lumber subjected to a constant-temperature exposure of 500°C (932°F) produced a linear model. The model showed a constant rate of charring of 1.628 mm^2/s and 0.45 mm/min (Lau, White, and Van Zeeland 1999). It should be noted that real-world fire exposures do not produce constant temperature or heat flux conditions.

Work was undertaken to numerically simulate pyrolysis. In this research, the movement of the char/virgin front was modeled with good agreement of mass loss with experimental testing (Jia, Galea, and Patel 1999). Recent work by Babrauskas is a comprehensive analysis of charring rate of wood as a tool for fire investigators

FIGURE 3.16 ◆ Varying degrees of damage to a paneled wall and previously protected wood studs within the wall after being exposed to a test fire. *Courtesy of D. J. Icove.*

(Babrauskas 2005; also DeHaan 2007, chap. 7). It includes data illustrating the variability of char rates with type of wood and nature of fire exposure. The char rate is highly dependent on applied heat flux (Babrauskas 2005). During a fire, heat fluxes on a surface may vary from near zero to 50 kW/m^2 for contact by small flames to 120–150 kW/m^2 under postflashover conditions, resulting in a char rate that can vary from zero to several millimeters per minute.

To identify conductive heat transfer through walls, investigators often document the charring of wood studs within walls covered by materials such as wooden paneling or gypsum board. Figure 3.16 shows varying degrees of damage to a gypsum board–covered wall and previously protected wood studs within the wall after exposure to a test fire. The test results agreed with the varying degrees of charring damage predicted by a fire model known as WALL2D (Figure 3.17), which predicts heat transfer and char damage to wood studs protected by gypsum board and exposed to fire. The WALL2D developers reported very good agreement of their heat transfer model

FIGURE 3.17 ◆ The progression of a char front upward through a wood ceiling joist after exposure to a test fire. *Courtesy of D. J. Icove.*

with the results of both small- and full-scale fire tests (Takeda and Mehaffey 1998). WALL2D was the focus of an engineering study that identified it as a viable fire model for users in both the regulatory and fire investigation communities (Richardson et al. 2000).

Char depth is usually documented using a penetrating tool and survey grid diagrams (NFPA 2004a, sec. 17-2.4). In these diagrams, the data from the measurements are converted to lines of equal depth, called **isochar** lines. Each line connects depths of the same value, and investigators should consider recording elevation (height above the floor) measurements for greater precision (Sanderson 2002). Because variations in char rate can be caused by the differences in affected woods, coatings, and flame-retardant treatments, investigators are cautioned to realize that char depths should be measured on similar types of wood (baseboards, door frames, etc.).

Due to the nonlinear nature of charring, the measurement of char depth cannot establish precise times of fire exposure, but deeper relative depths of char may indicate areas of more intense or prolonged burning. Also, the appearance of the char surface often depends on the individual characteristics of the wood and ventilation conditions (DeHaan 2007). However, char depth relationships assuming a constant 50-kW/m^2 incident flux have been suggested for fire protection engineering applications (Silcock and Shields 2001).

Spalling is the chipping or crumbling of a concrete or masonry surface when it is exposed to varying heat, cold, or mechanical pressures. As a fire pattern damage indicator, spalling can indicate steep temperature gradients caused by radiated heat, a significant fuel load of ordinary combustibles, or other sources of localized heating that last long enough for heat to penetrate the concrete.

Spalling occurs during rapidly rising temperatures, typically 20°C–30°C/min (70°F–86°F/min) (Khoury 2000). Several factors influence spalling, but the moisture content and nature of the aggregate used (limestone versus granite) are the leading variables. Spalling can also be caused by suddenly cooling a very hot concrete surface with a hose stream. High-strength concrete tends to spall more than normal concrete owing to its higher density. The pores of the high-density concrete become filled more quickly with the high-pressure water vapor (Buchanan 2001, 228).

Steel reinforcements expand much more rapidly when heated than does concrete. Thus, if reinforcements are not protected deeply in the concrete, they expand and cause tensile failure of the concrete. Lightweight concrete spalls more readily because of the vermiculite used as an aggregate. An example of extreme spalling of a concrete wall and ceiling surfaces is shown in Figure 3.18. Spalling can be reduced if the concrete is made with fire plastic fibers mixed into the slurry.

Testing by the SP Swedish National Testing and Research Institute on spalling reinforced the two classic explanations for concrete spalling. The most common cause is thermal stress, which forces water vapor out of the hotter side of the concrete. Recent experiments documented that water vapor is also forced out of the cold side of the concrete during fire tests. During these fire tests, the internal pressures also started to fluctuate severely after 15 min (Jansson 2006). New material test methods have been proposed for determining whether an actual exposed concrete may suffer from explosive spalling at a specified moisture level. A specimen cylinder is used, which is a cost-effective alternative to full-scale testing (Hertz and Sorensen 2005).

Studies by the UK Building Research Establishment (BRE) in fire damage on natural stone masonry in buildings showed that natural stone can be seriously affected in building fires, with damage concentrated around window openings and doorways (Chakrabarti, Yates, and Lewry 1996). Color changes at 200°C–300°C (392°F–572°F) were followed by localized disintegration at 600°C–800°C (112°F–1472°F). Color changes included the reddening of stones containing iron (which is irreversible and causes significant damage to historical buildings). Other damage noted by BRE included costly smoke staining and salt efflorescence resulting from fire hose streams.

BRE also reported that brown or light-colored limestone containing hydrated iron oxide changes to reddish brown at 250°C–300°C (482°F–572°F), more reddish at 400°C (752°F), and a gray-white powder at 800°C–1000°C (1472°F–1832°F). The depth of calcination of limestone is seldom less than 20 mm, which begins at 600°C (1112°F), reducing the stone's strength. Magnesium limestone is white or buff colored and changes to pale pink at 250°C (482°F) and pink at 300°C (572°F).

Sandstone changes to a brown color owing to the dehydration of iron compounds starting at 250°C–300°C (482°F–572°F) and to reddish brown above 400°C (752°F). Structural weakening of sandstone starts at 573°C (1063°F), when the quartz grains internally rupture. Loss of strength of such stone is particularly critical when solid stone lintels are used above windows, doors, and gateways, as massive structural collapse can occur without warning.

Further research by BRE indicated that granite will show no color changes but will crack or shatter at temperatures above 573°C (1063°F) through quartz expansion, and marble's internal calcite crystals will undergo thermal hysteresis, causing a reduction in flexural strength. Marble has been known to crumble into a powder when exposed to extreme heat (>600°C, 1150°F). This is particularly dangerous for marble staircases in fires.

(a)

(b)

FIGURE 3.18 ◆ Spalling concrete tests: (a) gasoline pool fire, (b) staining but no spalling of concrete after gasoline fire self-extinguished, (c) wood pallet fire on same type concrete, (d) extensive spalling under wood fire. *Courtesy of Jamie Novak, Novak Investigations.*

(c)

(d)

The UK Building Research Station published a historical review of the visible changes in concrete or mortar when exposed to high temperatures (Bessey 1950). The study examined aggregates, concrete and mortar, and reproducibility and interpretation of observed changes in terms of temperature.

Research by Short, Guise, and Purkiss (1996) at the Department of Civil Engineering, Aston University, UK, documents their use of color analysis in the assessment of fire-damaged concrete using optical microscopy.

Flammable liquids themselves very rarely produce significant spalling on bare concrete because a typical flammable liquid pool fire on bare concrete lasts only 1–2 min (and often much less) (DeHaan 1995). This leaves little time for heat to penetrate into the concrete, which has poor conductivity. In addition, the radiant heat from the flame is, in part, absorbed by the liquid fuel during combustion, and the maximum temperature of the surface cannot significantly exceed the boiling point of the liquid in contact with it. Because the boiling point range of gasoline is 40°C–150°C (104°F–302°F), the concrete will not usually be heated to the point at which it will be affected (DeHaan 2007, chap. 7).

If tile or carpet is present on the concrete, the gasoline flames can trigger localized charring, melting, and combustion of the floor covering. Figure 3.19 shows an example of localized burning due to a flammable liquid pour pattern on floor tiles. Because this combustion is more prolonged than the gasoline flame itself, the molten mass has a very high boiling point, and there is no freestanding liquid to protect the concrete, there is more opportunity for heat to affect the concrete and cause it to spall

FIGURE 3.19 ◆ Surface effects due to a flammable liquid pour pattern (center of the photograph) charring the floor tiles. Two parallel pour patterns are documented, characteristic of a back-and-forth spilling of flammable liquids as the arsonist backed out of the building. *Courtesy of D. J. Icove.*

if it is susceptible. The collapse of molten burning plastic or roofing tar onto a concrete surface will produce significant and prolonged heat transfer that will induce spalling. Prolonged burning of structural wood collapsed onto a concrete floor surface is especially likely to cause spalling.

Additional cited references on spalling of concrete include Canfield (1984), Smith (1991), Sanderson (1995), and Bostrom (2005).

Localized combustion of flammable liquids on vinyl- or asphalt-tiled concrete surfaces can produce a type of floor damage called **ghost marks**. Ghost marks are subsurface staining of the concrete beneath the joints of tile floors that appears to be caused by combustion of the cement/mastik dissolved by the applied flammable liquid. Radiant heat in postflashover fires and long-term heating by fall-down of burning materials can produce spalling but do not apparently produce ghost marks (DeHaan 2007, chap. 7). Charring of wood floors along tile seams is not the same.

Fractured glass is another surface effect that usually occurs when glass is stressed by nonuniform heating. Common variables that affect glass breakage include its thickness, the presence of any defects, the rate at which it was heated or cooled, and the method used to secure the glass to the window frame (Shields, Silcock, and Flood 2001).

The potential for glass to fracture and break out has been the topic of several research initiatives. The prediction of the probability of breakage fits a Gaussian statistical correlation, as shown in Figure 3.20 (Babrauskas 2004). This relationship shows that the mean temperature for the breakage of 3-mm-thick window glass is 340°C (644°F), with a standard deviation of 50°C (122°F).

The fire modeling program BREAK1, Berkeley Algorithm for Breaking Window Glass in a Compartment Fire, can calculate the temperature history of a glass window given several fire parameters combined with a physical composition of the glass (NIST 1991). The model also provides the temperature history of the glass normal to the glass surface and the window breakage time. Large-scale glass fracturing by heat is often observed as a room approaches flashover and the temperatures and heat fluxes dramatically increase. Note however, that this program computes only the cracking of glass, whereas the investigator is often interested in knowing when the

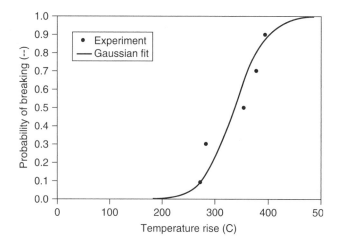

FIGURE 3.20 ◆ The prediction of the probability of the breakage of 3-mm-thick window glass fits a Gaussian statistical correlation with a mean temperature of 340°C (644°F) and a standard deviation of 50°C (122°F). *Courtesy of Babrauskas [2000]; www.doctorfire.com/glass.html.*

glass failed enough so that pieces fell out, rather than the time when cracks first appeared. It is the physical loss of glass from windows that can affect the development of the fire in the room, sometimes dramatically.

Hietaniemi (2005) at the VTT Building and Transport, Finland, has produced a probabilistic simulation model for glass fracture and fallout in a fire. This approach uses two steps: The model first predicts the simulation time for the initial occurrence of a crack using the BREAK1 program, followed by a thermal response model of the glass.

Heat-induced failure of dual-glazed windows is considerably less predictable. Often, the inside pane fails but its fragments protect the exterior glass pane. Dual-glazed windows have been observed to survive postflashover room fires only to break when struck by hose streams. Dual glazing with heat-absorbing film is even more resistant to heat-induced failure.

One common mechanism for thermal fracturing of common window glass is the stress induced by nonuniform heating as a result of "thermal shadow" cast by framing or trim. When a window is exposed to radiant or convected heat, the exposed portions increase in temperature and begin to expand, but the areas shaded by the framing or window putty are not heated. Because glass is a relatively poor conductor of heat, those "shaded" areas remain cool and do not expand. This differentiation induces stress in the glass that can cause the glass to fracture, often in patterns roughly parallel to the edges, as shown in Figure 3.21. The fractures radiating from a single point on the right often are caused by a preexisting nick in the edge of the glass or by stress induced by a glazier point or frame nail (DeHaan 2007, chap. 7).

It should be noted that the observation that the windows "blew out" during a fire is rarely associated with an overpressure event but, rather, with thermally induced massive fracturing. The pressures produced by a flashover transition in a room are very rarely adequate to cause mechanical breakage (Fang and Breese 1980; Mitler and Rockett 1987). However, the development of a "backdraft" condition, in which the fire becomes severely underventilated and is then given additional air, can produce overpressures sufficient to cause windows to break.

The edges of ordinary glass fractured by mechanical stress are characterized by curved, conchoidal ridge lines caused by the propagation of the fracture, as in Figure 3.22a. The fracture pattern on the face of glass when broken by mechanical shock or pressure is typically a spiderweb pattern of mostly straight fractures, as in Figure 3.22b. The pattern usually consists of a number of radial fractures from the point of contact and concentric fractures connecting them.

Mechanical impacts of very brief duration cause minimal radial fractures because the glass does not have time to flex and may produce a domed plug of glass to be ejected from the side opposite the impact. Thermal stress usually results in more random patterns with wavy fractures. These fractures usually have mirror-smooth edges with minimal conchoidal ridges, since they are usually produced at lower speeds than mechanical fractures.

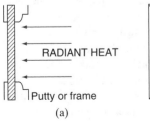

RADIANT HEAT

Putty or frame

(a)

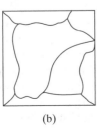

(b)

Figure 3.21 ◆ Heat fractures of a window pane caused by "shadowing" of radiant heat. Fractures radiating out from a point on the right often are caused by a preexisting nick in the edge of the glass or by stress induced by a glazier point or frame nail.

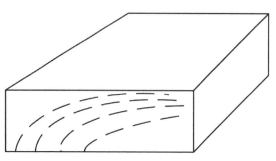

FIGURE 3.22 ◆ (a) Edge of glass fragment showing conchoidal fracture lines. (b) Face of glass sheet broken by mechanical impact near the center, with radial fractures and concentric fractures.

If enough glass can be recovered to jigsaw-fit pieces together for at least part of the pattern, and the inside or outside face can be identified by soil, paint, decals, or lettering, the direction of force can be established by the 3-R rule: the conchoidal lines on **r**adial fractures start at **r**ight angles on the **r**everse (side away from the force). The direction is reversed on concentric fractures. Toughened (safety) glass used in the side and rear windows of vehicles and in structural doors and shower doors fractures into small, rectangular blocks. As a general rule, the cause and direction of failure of safety glass cannot be established, but under some conditions fragments trapped in the frame may reveal the approximate location of a mechanical impact point, as seen in Figure 3.23.

The sequence of impacts can be established from intersecting and ending fractures. Sequence of failure versus time of exposure to fire can be established by the presence or absence of soot on the broken edges. It must be remembered that direct contact with flame can heat glass to the point at which all accumulated soot and smoke deposits are oxidized off, a so-called clean burn. Glass breakage patterns can also reveal a great deal of information about an explosion (DeHaan 2007, chap. 7). The distance to which the glass is blown is related to the thickness of the glass, the size of the window, and the pressures developed (Harris, Marshall, and Moppett 1977; Harris 1983). As a general rule, the thinner the glass or the larger the surface area of the pane, the lower the failure pressure will be.

A distinctive form of glass fracturing called **crazing** is usually produced when fire suppression water spray strikes a heated glass surface. Crazing is characterized by a complex pattern (road map–like) of partial-thickness fractures and concave pitting (Lentini 1992). Field experiments during a fire confirmed that when a water spray was applied to the hot glass panes on one side of a heated window, crazing occurred, as in Figure 3.24. Notice the melted glass in the upper right-hand pane. Common window glass softens when its temperature reaches 700°C–760°C (1300°F–1400°F).

Another descriptive form of surface burn pattern is called **clean burn** (see Kennedy and Kennedy 1985, 451; NFPA 2004a, sec. 6.7.6; DeHaan 2007, 275). When a solid surface is exposed to a fire environment, products of combustion—water vapor, soot, and pyrolysis products—will condense on it. The cooler the surface, the faster products will condense (giving rise to the shadowy outlines of studs, nail heads, and other hidden features of a wall structure that cause a temperature differential). When a sooted surface is exposed to direct flame, it can reach high enough temperatures that the accumulated products can burn away, leaving behind a clean surface. This pattern can be used to identify areas of direct contact between a surface and the flaming portion of the plume. Surfaces

FIGURE 3.23 ◆ Mechanical fractures in tempered glass door (dual glazed) retain the radial fracture pattern to indicate point of breakage (by firefighters). *Courtesy of J. D. DeHaan.*

not in contact with the flame will rarely reach sufficiently high temperatures to allow the complete combustion of condensed soot or pyrolysates. An unpublished study by Ingolf Kotthuff indicates that for a clean burn to occur, the material must reach approximately 700°C (1300°F). A clean burn can also result where charred organic materials are burned away leaving a bare noncombustible substrate exposed.

Melted materials can also reveal much about the temperatures to which surfaces were exposed during the fire. Historical studies by the UK Building Research Station provided an estimation of the maximum temperatures attained in building fires from an examination of the debris. They examined lead and zinc plumbing fixtures, aluminum

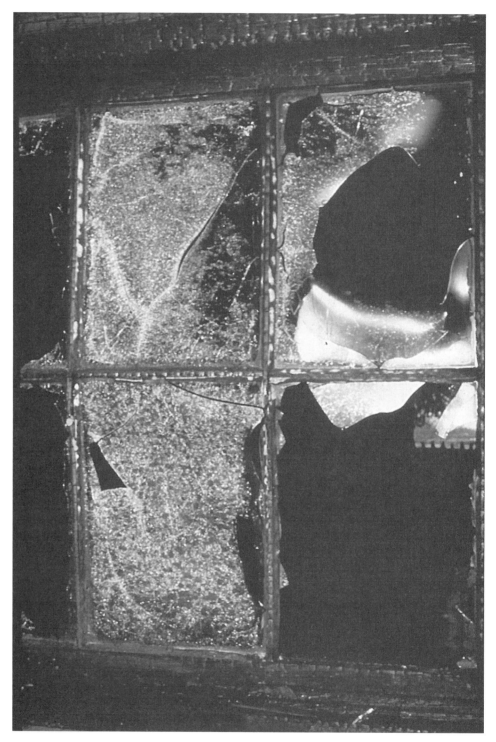

FIGURE 3.24 ◆ Crazing of glass caused by rapid cooling of glass, often during fire suppression, as seen in this field experiment in which a water spray was applied to the window panes. *Courtesy of Lamont "Monty" McGill, McGill Consulting, Gardnerville, NV.*

and alloys used in small machinery, molded glass from windows and jars, sheet glass in window panes, silver jewelry and eating utensils, brass doorknobs and locks, bronze window frames and bells, copper electric wiring, and cast iron pipes and machinery (Parker and Nurse 1950). The findings of this study demonstrated that the analysis of melted materials in a fire could identify and document areas of varying temperature exposure.

Plastics, glass, copper, aluminum, and tin are the most common melted materials found at fire scenes that can document the wide range of temperatures produced by the fire (DeHaan 2007). Note that melting points of thermoplastics can vary depending on the nature of their internal molecular structure. Glass melts, softens, or becomes molten over a range of temperatures. Listed in Table 3.2 are approximate melting temperatures of common materials. One must remember that interactions between melting materials can cause unusual effects. Zinc or aluminum melting onto

TABLE 3.2 ◆ Approximate Melting-Point Temperatures of Common Materials

Material	°C	°F
Aluminum alloys[a]	570–660	1060–1220
Aluminum (pure)[c]	660	1220
Copper[c]	1083	1981
Cast iron (gray)[a]	1150–1200	2100–2200
50/50 solder[a]	183–216	360–420
Carbon steel[a]	1520	2770
Tin[c]	232	450
Zinc[c]	420	787
Magnesium alloy[a]	589–651	1092–1204
Stainless steel[a]	1400–1530	2550–2790
Iron (pure)[c]	1535	2795
Lead (pure)[c]	327	621
Gold (pure)[c]	1065	1950
Paraffin (wax)[d]	50–57	122–135
Human fat[e]	30–50	86–122
Polystyrene[b]	240	465
Polypropylene[d]	165	330
Polyester (Dacron)[d]	dec 250	480
Polyethylene[b,c]	130* (85–110)	185–230
PMMA (acrylic plastic)[b]	160*	320
PVC[b]	dec 250*	480
Nylon 66[b]	250–260	480–500
Polyurethane foam[c]	dec 200	390

Sources: [a]Perry and Green 1984, Table 23-6, 23–40; [b]Drysdale 1999; [c]DeHaan 2007, 274–286; [d]*The Merck Index,* 1989; [e]DeHaan, unpublished.
*Varies with cross linking.

steel can cause the steel to "melt" at ordinary fire temperatures. Zinc can come from pot-metal fittings or galvanized coatings. Aluminum is well known for creating a eutectic alloy with copper with a very low melting point. Calcium sulfate can cause localized melting of black iron under fire conditions.

As we have seen, temperatures in any flame can readily reach 1200°C (2200°F), but the actual values recorded depend greatly on the measuring technique. Temperatures in a growing room fire can vary from ambient room temperature to over 800°C (1470°F). Postflashover fires can produce temperatures of over 1000°C (1800°F) throughout a room. Burning plastics can produce localized flame temperatures of 1100°C—1200°C (1980°F–2200°F). Most other materials whether ignitable liquids or ordinary combustibles, produce maximum flame temperatues in air of 800°C (1470°F). Evidence of high temperatues (800°C (1470°F) or higher) is not proof of the use of accelerants. Plastics can indicate the intensity and distribution of heat exposure as they soften and distort. Figure 3.25 shows the face of a wall clock

FIGURE 3.25 ◆ Softening of plastic clock face indicates temperature of layer (face should be saved and its height from floor measured). *Courtesy of J. D. DeHaan.*

that verify the depth of the hot layer in a room. Polyethylene laundry bags on the floor near it were not affected.

Annealed furniture springs result when their temperature exceeds the typical or annealing point (538°C, 1000°F) and the coiled steel springs lose their temper (Tobin and Monson 1989). **Collapsed springs** are an indicator of this phenomenon, and an investigator should compare the relative degree of collapse to try to ascertain the direction and intensity of the fire (DeHaan 2007). Such collapse is not an indicator that accelerants were used, since smoldering fires of accidental cause can readily produce such effects. The investigator should also examine the framework (usually metal or wood) for other signs of localized thermal damage (NFPA 2004a, sec. 6.14). Modern urethane foam furniture often burns so quickly that even the high-temperature flames cannot raise the steel springs to their annealing temperatures. Steel springs can also corrode from fire exposure.

FIGURE 3.26 ◆ The penetration (burn-through) of carpet, pad, and 12-mm (0.5-in.) plywood floor in a furnished room fire test involving no ignitable liquids. Postflashover fire lasted less than 5 min but produced extensive destruction to exposed areas, especially at the door (foreground). *Courtesy of J. D. DeHaan.*

PENETRATIONS

Another common fire pattern is the occurrence of **penetrations** through horizontal and vertical surfaces as direct flame impingement or intense radiant heat on walls, ceilings, and floors is sustained long enough to affect deeper layers of the material. Ceiling areas directly overhead of fire plumes may burn through or collapse, allowing flames to extend into confined spaces. In some cases, fall-down of combustible debris followed by extended smoldering can lead to localized penetration of wooden floors, sills, and toe plates in wood-framed structures.

Variables influencing penetrations include preexisting openings in floors, ceilings, and walls. Also, direct flame impingement coupled with high heat fluxes play an important role in creating the conditions necessary for penetrations. For example, the area where a fire plume intersects a ceiling, such as item 6 in Figure 3.13, is an example of a penetration into the ceiling directly above the fire plume due to the high heat flux and temperatures that caused the failure of the ceiling materials.

Penetrations are not always directly above the fire plume. During flashover and rollover conditions, downward penetrations are sometimes found. As discussed previously, the most common reason for penetrations through wooden flooring is radiant heat from above and not the combustion of ignitable liquids (DeHaan 1987; Babrauskas 2005). Sustained burning of collapsed furniture, bedding, or clothing also can result in localized penetrations. Fire testing involving no ignitable liquids in furnished rooms has demonstrated that the resulting postflashover conditions can produce penetrations through carpet, pad, and plywood floors, as illustrated in Figure 3.26.

In postflashover fires where ignitable liquids were poured onto carpeted floors, burning penetrations can char supporting joists (from postflashover fire, not the liquid fuel fire). In fire tests conducted by the U.S. Fire Administration, 1.75 L (0.46 gal) of gasoline was poured onto carpeted floor (FEMA 1997a). Flashover occurred at 40 s, with extinguishment at 10.3 min. An examination of the scene after extinguishment showed the fire to have burned through the carpet, carpet underpadding, and 9.5-mm (0.37-in.) plywood floor. As shown in Figure 3.27, there was widespread charring on the 2×8-in. supporting joists. Note that the test involved nearly 10 min. of postflashover burning, which caused most of the observed damage long after the gasoline burned away.

Burning foam mattresses sometimes drop enough burning molten liquid (from pyrolysis) onto the floor directly under the bed to cause burns through the floor.

FIGURE 3.27 ◆ The burn-through of carpet, underpadding and 12-mm (0.5-in.) plywood floor in a room fire test involving 1.75 L (0.5 gal) of gasoline and approximately 10 min of post-flashover burning. *Courtesy of the U.S. Fire Administration.*

Burning thermoplastics such as polyethylene trash containers or thermoplastic appliance liners can do the same. Traditional cotton-stuffed upholstered furniture such as sofas, chairs, and mattresses can support combustion long enough to burn through wooden floors beneath them after they collapse.

Failure of walls, floors, and ceilings during a fire may also contribute to formation of penetrations through which the fire extends throughout a structure. These failures have been known to transport heat, flames, and smoke to other areas of the structure, leading inexperienced investigators to conclude improperly that two or more fires were set within the structure. Such penetrations can occur as gypsum wallboard fails, lath chars, plaster collapses, and metal ceilings or other noncombustible ceiling coverings peel away. Ventilation (drafts) up stairwells can cause very intense fires with rapid failure of doors and ceilings.

LOSS OF MATERIAL

Loss of materials occurs in any combustible material located in the path of a fire's development and spread. This loss of material is helpful in determining the fire's intensity and direction of travel. **Heat shadowing** can affect the loss of materials by shielding the heat transfer and masking the potential lines of demarcation of other fire patterns (Kennedy and Kennedy 1985, 450; NFPA 2004a, sec 6.15.1). Loss of material can be driven by direct flame impingement or radiant heat alone. The rapid charring caused by postflashover radiant heat can be enhanced by ventilation from doors or windows to produce penetrations. These can occur through floors near doors or windows or even in the center of rooms, where postflashover mixing is most efficient (DeHaan 2007) (see Figure 3.26). Extensive destruction around heater vents has been observed in rooms where there was prolonged postflashover burning because of the extra oxygen the vents provided to fuels already burning.

Items 3 and 7 in Figure 3.13 are two examples of loss of materials due to direct impingement of the fire plume. The most significant is the relative loss of the wooden wainscoting material. The closer to the plume centerline, the deeper the charring. Localized damage to the wooden baseboard documents the radiant heat from above.

Loss of material may occur on the tops of wooden studs, on corners and edges of furniture and moldings, and floors and carpeted surfaces. This gives rise to beveling or solid fuels or extended V patterns that can reliably indicate direction of fire spread. Other forms of loss of materials come from fall-down of debris from above. Typical examples include burning drapery or curtains, which ignite other objects in the room and sometimes mislead novice investigators into thinking that multiple fires may have been set.

Loss of material may play an important role in determining the nature and extent of fire damage to upholstered furniture exposed to smoldering versus flaming combustion. Tests published by Ogle and Schumacher (1998) demonstrated that smoldering fire patterns often tended to consist of char zones with a thickness equal to that of the fuel element (i.e., the entire thickness). If the source of the smoldering fire was, for example, a cigarette, they found that the smoldering patterns persisted as long as the cigarette continued to smolder. Fire patterns created by smoldering fires were destroyed if a transition to flaming combustion occurred. In the case of cigarette-initiated fires, the cigarette itself was consumed in the flaming phase.

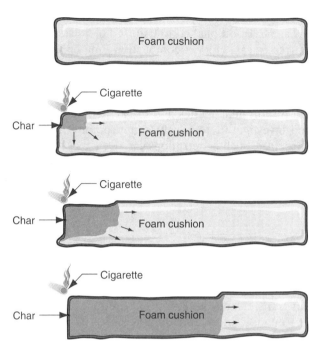

FIGURE 3.28 ◆ The cross-sectional view of the development of smoldering fire patterns in a cotton upholstered foam cushion exposed to a smoldering cigarette at its corner.

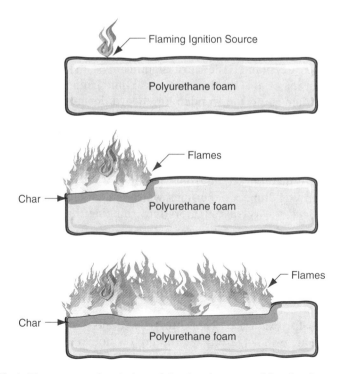

FIGURE 3.29 ◆ The cross-sectional view of the development of flaming fire patterns in a foam cushion exposed to a flaming ignition source at its surface.

Flaming fire patterns had thin char zones with thicknesses smaller than the thickness of the fuel element. Ogle and Schumacher (1998) observed that the flaming fire consumed the cover fabric and underlying padding. When examining flame spread rates with polyurethane foam (PUF) seat cushions, they observed that horizontal burning was faster than vertical downward burning. Also, since the cushions are thermally thick, the depth of char into the foam tended to be thin when compared with the overall cushion thickness.

Shown in Figures 3.28, 3.29, and 3.30 are graphic representations of the burn patterns observed by Ogle and Schumacher in fires involving upholstered polyurethane foam cushions. These figures respectively show the cross-sectional views for smoldering, flaming, and transition. Although polyurethane foam by itself is not usually ignitable by a smoldering cigarette, urethane cushions are more likely to be ignited in combination with cellulosic fabrics and paddings (Holleyhead 1999; Babrauskas and Krasny 1985).

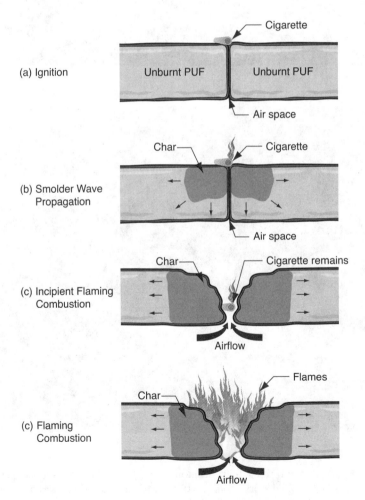

FIGURE 3.30 ◆ The cross-sectional view of the development of a burn-through and the transition from smoldering to flaming combustion in a pair of upholstered foam cushions exposed to a smoldering cigarette at their junction.

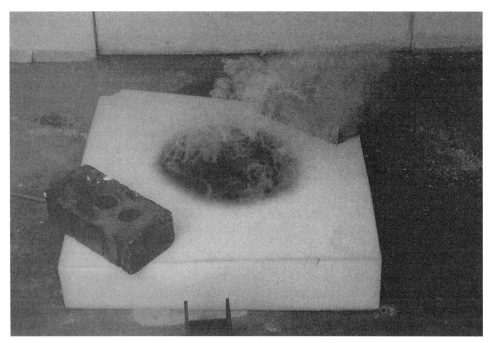

(a)

(b)

FIGURE 3.31 ◆ (a) A polyurethane foam cushion placed over a high wattage lamp. (b) The foam around the lamp formed a dry, rigid, crusty char. When the cushion was ignited by open flame, the residue was the typical sticky liquid and semimolten combustion product.

(c)

FIGURE 3.31 ◆ (c) Postfire residues—dry and rigid where charred near lamp, brown sticky mass where burned in flame. *Test courtesy of Jack Malooly. Photos (a) and (b) courtesy of John Galvin, London Fire Brigade. Photo (c) courtesy of J. D. DeHaan.*

Hot surfaces in contact with a polyurethane foam can cause the formation of a cakelike, rigid char, whereas the flaming combustion of the same foam causes the production of viscous yellow polyol. Postfire examination may reveal the different residues, as seen in Figure 3.31.

VICTIM INJURIES

Victim injuries may occur during discovery of, interaction with, or escape from a fire and may reveal actions taken during critical time periods. This is especially true in situations resulting in fire deaths where the only remaining fire pattern indicators are the areas and degrees of burns on the victim's body and clothing. Human tissues exhibit a variety of behaviors, much like the layers of a tree trunk. Body fat can also contribute to the combustion process in prolonged-exposure fires (DeHaan, Campbell, and Nurbakhsh 1999; DeHaan and Pope 2007).

Radiant heat (infrared radiation) travels in straight lines and is absorbed by most materials. Some thin synthetic fabrics are sufficiently transparent to infrared that first- and second-degree burns can be induced through the fabric. Radiant heat is reflected by most materials to some degree, but it can be reflected from metallic surfaces with sufficient energy to induce thermal damage to secondary target surfaces. The bodies of victims incapacitated during fires may block radiant heat from reaching the surfaces on which they are lying, resulting in a "heat or smoke shadowing" pattern. In these instances, a silhouette of the victim is left on the bed, carpet, or chair, as illustrated in Figure 3.32. The mechanisms of burn injuries are discussed in full later in this text.

FIGURE 3.32 ◆ The bodies of incapacitated victims during fires may block heat from reaching the surfaces on which they are lying, resulting in a "heat or smoke shadowing" fire pattern. In these instances, a silhouette of the victim is left on the bed, carpet, or chair. *Courtesy of D. J. Icove.*

◆ 3.5 INTERPRETING FIRE PLUME BEHAVIOR

FIRE VECTORING

A **vector** is a mathematical pointer that has a direction and a magnitude. An applied technique known as **fire vectoring** provides the investigator with a tool for understanding and interpreting the combined **movement** and **intensity** of plumes that create fire pattern damage. Direction indicators of various types have been used by fire investigators since the earliest investigations.

The concept of heat and flame vector analysis was first seriously set forth by Kennedy and Kennedy (1985). However, few investigators fully understood the application of the concept and systematically applied it. Scientific research into fire scene investigations has addressed the topic only in the last decade.

Heat and flame vectoring, referred to in this textbook simply as fire vectoring, is becoming a standard technique for using fire patterns to determine a fire's area and point of origin (Kennedy 2004; NFPA 2004a, sec. 17.2.3). Several examples throughout this text use this simple yet effective principle for tracing a fire's development, which was fully explained and used to document the 1997 USFA Burn Pattern Study (FEMA 1997a).

Although the underlying concept of fire vectors is that they comprise two characteristics—direction of movement and relative intensity—some investigators successfully use only the vector's directional component.

Movement patterns are caused by flame, heat, and by-products of combustion produced when the fire spreads away from its initial source. Evaluating areas of relative damage from the least to the most damaged areas can reveal the fire's movement, a technique for tracking a fire back to its possible room, area, and point of origin. Beveled edges to burned materials, smoke levels in a series of connected rooms, or protected areas on one side of an object are all examples of movement patterns. As long as the target material's thermal properties are considered, some interpretation of approximate duration may be made by a trained investigator.

Intensity patterns are the result of the fire's sustained impingement on exposed surfaces, such as walls, ceilings, furnishings, wall coverings, and floors. Various intensities of heat transfer result in thermal gradients along these surfaces, sometimes forming lines of demarcation dividing burned and unburned areas.

The fire vector's position and length quantitatively indicate the intensity of the particular burn pattern, and its direction indicates the apparent flow from hottest to coolest. Once the vectors are documented and reviewed, their cumulative effect assists in determining the source of the fire. Each vector is assigned a number that is used for notations in the report. A fire vector diagram used in USFA test 5 is shown in Figure 3.33 (FEMA 1997a, 47).

VIRTUAL ORIGIN

When a majority of the vectors point to the fire plume's base, it is often appropriate to consider this to be the location of the initial fuel package. Vectoring is helpful in estimating the location of the **virtual origin** or source of thermal energy of the fire. The virtual origin is a point located along the plume's centerline that would give the same radiative output as the actual fire. This location is sometimes lower than the floor level. If the physical and thermal properties of the fuel package can be determined from the scene investigation, the virtual origin can be mathematically calculated and its location can be helpful in reconstructing and documenting the fire's virtual source, point of origin, area, and direction of travel. However, if the pattern is clear enough, simple geometric extension of the V sides suffice to establish the virtual origin.

As we saw in Chapter 2, the virtual origin of a fire is located below the floor level when the area of the fuel source is large compared with the energy released by the fuel, as shown in previous examples. Likewise, a fuel releasing a high energy over a small area will produce a virtual origin above the floor level (Karlsson and Quintiere 1999). An example of a high-energy release over a small area is a small gasoline pool fire.

TRACING THE FIRE

The fire investigator's task would be quite easy if he or she were present for the duration of the incident. Unfortunately, fire scenes are often cold and in disarray after the extinguishment and subsequent overhaul to assure that the fire will not rekindle. The investigator's role then becomes one of reconstructing the sequence of the fire backward to its area and point of origin. A firm understanding of fire behavior becomes imperative during this examination and documentation of the fire scene.

The ability to trace the fire's behavior and growth is important to the investigation. Like an experienced tracker, the investigator relies on pattern damage as roadmarks to track the fire's path. Based on experience, the following rules of

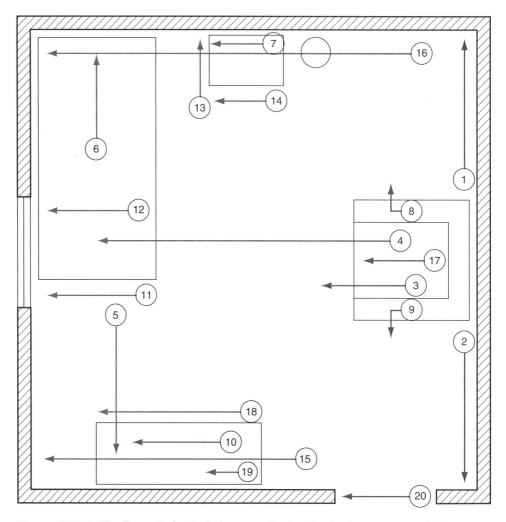

FIGURE 3.33 ◆ The fire vectoring technique uses the length of a drawn vector to indicate quantitatively the direction and intensity of the particular burn pattern. *Courtesy of FEMA [1997a, 47].*

fire behavior are offered to understand and interpret this pattern damage (De-Haan 2007, chap. 7).

- **Fire's Tendency to Burn Upward.** A fire plume's hot gases (including flames) are much lighter than the surrounding air and therefore will rise. In the absence of strong winds or physical barriers (such as noncombustible ceilings) that divert flames, fire will tend to burn upward. Radiation from the plume will cause some downward and outward travel.
- **Ignition of Combustible Materials.** Combustible materials in the path of the plume's flames will be ignited, thereby increasing the extent and intensity of the fire by increasing the heat release rate. The more intense the fire grows, the faster it will rise and spread.
- **Potential for Flashover.** A flame plume that is large enough to reach the ceiling of a compartment is likely to trigger full involvement of a room and increase the chances for

flashover to occur. If there is not more fuel above or beside the initial plume's flame to be ignited by convected or radiated heat, or if the initial fire is too small to create the necessary heat flux on those fuels, the fire will be self-limiting and often will burn itself out.

◆ **Location and Distribution of Fuel Load.** In evaluating a fire's progress through a room, the investigator must establish what fuels were present and where they were located. This fuel load includes not only the structure itself but its furnishings, contents, and wall, floor, and ceiling coverings (as well as combustible roofing materials), which feed a fire and offer it paths and directions of travel.

◆ **Lateral Flame Spread.** Variations on the upward spread of the fire plume will occur when air currents deflect the flame, when horizontal surfaces block the vertical travel, or when radiation from established flames ignites nearby surfaces. If fuel is present in these new areas, it will ignite and spread the flames laterally.

◆ **Vertical Flame Spread.** Upward, vertical spread is enhanced when the fire plume finds chimneylike configurations. Stairways, elevators, utility shafts, air ducts, and interiors of walls all offer openings for carrying flames generated elsewhere. Fires may burn more intensely because of the enhanced draft.

◆ **Downward Flame Spread.** Flames may spread downward whenever there is suitable fuel in the area. Combustible wall coverings, particularly paneling, encourage the travel of fire downward as well as outward. Fire plumes may ignite portions of ceilings, roof coverings, draperies, and lighting fixtures, which can fall onto ignitable fuels below and start new fires that quickly join the main fire overhead, a process called fall-down or drop-down. Radiant heat and momentum flow of hot gases can also cause downward spread under appropriate conditions. An energetic fire can produce radiant heat sufficient to ignite adjacent floor surfaces. The momentum flow of ceiling jets from such fires can extend downward on nearby walls.

◆ **Impact of Radiation.** Fire plumes that are large enough to intersect with ceilings form ceiling jets that extend radially along the ceiling surface. Radiation from overhead ceiling jets or hot gas layers can ignite floor coverings, furniture, and walls even at some distance, creating new points of fire origin. The investigator is cautioned to take into account what fuel packages were present in the room from the standpoint of their potential ignitability and heat release rate contributions.

◆ **Impact of Suppression Efforts.** Suppression efforts can also greatly influence fire spread, so the investigator must remember to check with the fire suppression personnel present about their actions in extinguishing the fire. Positive-pressure ventilation or an active attack on one face of a fire may force it back into other areas that may or may not already have been involved and push the fire down and even under obstructions such as doors and cabinets. The investigator should interview the firefighters and obtain details on how water was applied, since this important information is rarely entered into a fire department's narrative of the fire.

◆ **Behavior of Heat and Smoke Plumes.** Because of their buoyancy, heat and smoke plumes tend to flow through a room or structure much like a liquid, that is, upward in relatively straight paths and outward around barriers.

◆ **Impact of Fire Vectors.** The total fire damage to an object observed after a fire is the result of both the intensity of the heat applied to that object and the duration of that exposure. Both the intensity and the exposure to that heat may vary considerably during the fire.

◆ **Impact of Plume's Radiant Heat Flux.** The highest-temperature area of a plume will produce the highest radiant heat flux and will therefore affect a surface faster and more deeply than cooler areas. This pattern can show the investigator where a flame plume contacted a surface or in which direction it was moving (since it will lose heat to the surface and cool as it moves across).

♦ **Impact of the Plume's Placement.** The contribution a fire makes to the growth process in a room depends not only on its size (heat release rate) and direction of travel but also on its location in the room—in the center, against a wall, in a rear corner, away from ventilation sources, or close to a ventilation opening.

Fire investigation and reconstruction of a fire's growth pattern back to its origin are based on the fact that fire plumes form patterns of damage that are, to a large extent, predictable. With an understanding of fire plume behavior, combustion properties, and general fire behavior, the investigator can prepare to examine a fire scene for these particular indicators. As with the application of the scientific method, each indicator is an independent test for direction of travel, intensity, duration of heat application, or point of origin. The investigator should remember that there is no one indicator that proves the origin or cause of a fire. All indicators must be evaluated together and yet they may not all agree. Finally, the application of fire vectoring can also enrich and buttress the opinion rendered as to the original location of the fire plume.

◆ 3.6 FIRE BURN PATTERN TESTS

Fire testing can produce all types of fire pattern damage phenomena (Table 3.1) including **demarcations, surface effects, penetrations**, and **loss of materials**. Fire testing is one method for expanding the range of knowledge about recognizing and interpreting these fire patterns.

An example of validation testing to explore fire pattern generation was conducted with the assistance of NIST, the Federal Emergency Management Agency's U.S. Fire Administration (FEMA/USFA), and the U.S. Tennessee Valley Authority (TVA) Police in Florence, Alabama. The tests used two forms of structures, a vacant single-family dwelling and a multiple-use assembly building (FEMA 1997a; Icove 1995; NIST 1997). The U.S. TVA Police study concentrated on extending the results of fire pattern burn testing on larger-scale structures while NIST and FEMA worked on the single-family dwelling scenarios.

The primary goals of the project test included demonstrating the production of fire burn patterns, exploring fire scene evidence and documentation techniques, and validating basic fire modeling—all concepts that form the basics of forensic fire scene reconstruction. Owing to the immense amount of diverse information gleaned from the Florence test burns, various aspects of this fire testing appear in examples and problems throughout this text.

FIRE TESTING GOALS

The objectives for conducting the test burns on a multiple-use facility included the simulation of arson fires using plastic containers of ignitable liquids, since arsonists often use these containers when setting fires to structures (Icove et al. 1992). Often when a burned, self-extinguished firebomb is found, investigators ask the questions: How long did the device burn? What damage can be expected? and Where is the best place to collect evidence of the device once it is damaged?

To simulate the arson, a typical firebomb consisting of a 3.8-L (1-gal) plastic container filled with unleaded gasoline was placed in the corner of the multiple-use facility's main assembly room, as shown in the floor plan in Figure 3.34. The main room was carpeted with an institutional-grade synthetic carpeting.

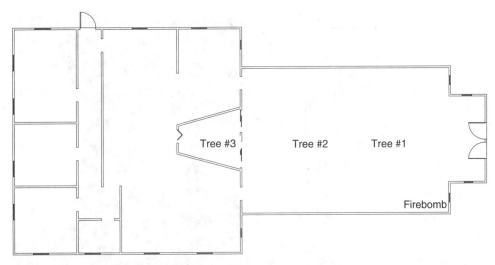

FIGURE 3.34 ◆ Floor plan of structure used to simulate the use of a typical firebomb placed in the corner of the room as indicated. *Courtesy of D. J. Icove.*

As thermocouple data for the test were collected, the firebomb was remotely ignited with an electric match. The resulting fire burned and then self-extinguished, leaving a 0.46-m (1.5-ft)-diameter circular burned area. Three thermocouple trees recorded the room-temperature profiles, as shown in Figure 3.35. Note that the fire lasted approximately 300 s.

The pre- and postfire damage patterns are depicted schematically and photographically in Figures 3.36a, b, and c. The exact location of the firebomb and its subsequent

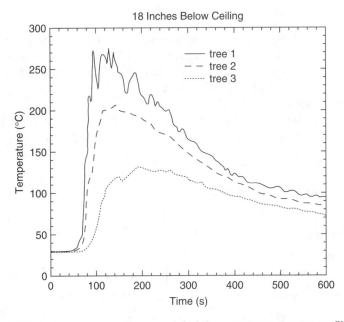

FIGURE 3.35 ◆ Three thermocouple trees recorded the room temperature profiles after the firebomb was remotely ignited with an electric match.

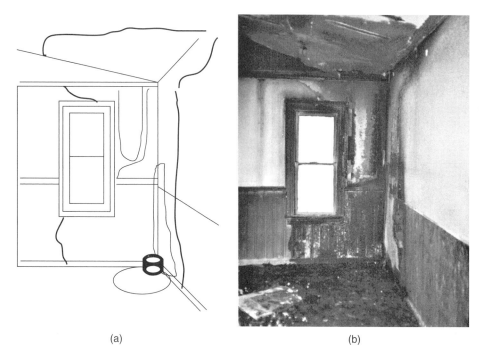

(a)   (b)

FIGURE 3.36 ◆ The postfire pattern damage from the firebomb test depicted both (a) schematically and (b) photographically. (a) Dark lines indicate major demarcations of burned/unburned areas. Light lines indicate isochar demarcations. (b) Note that the wall finish above the wainscoting was painted drywall and played no significant role in vertical flame spread. Colored lines can be used to delineate areas of surface deposit, penetration, and consumption. *Courtesy of D. J. Icove.*

burning pool diameter are indicated. The plastic jug melted quickly, so the majority of the fire occurred when its contents were released. Figure 3.36b illustrates the fire patterns from demarcations, surface effects, penetrations, and loss of material. The schematic uses the relative damages revealed on the exposed surfaces to form isochar lines corresponding to nearly equal char depths.

These isochar lines are useful in assisting in the placement and interpretation of fire vectors. The lines also document the ignitable liquid burn patterns found on carpeted and other floor surfaces.

ESTIMATED HEAT RELEASE RATE

The peak estimated heat release rate, $\dot{Q}$, for the gasoline pool fire can be calculated using equation (2.2).

Heat release rate	$\dot{Q}$	$= \dot{m}'' \, \Delta h_c A$
Area of the burning pool	A	$= (3.1415/4)\,(0.457^2) = 0.164 \text{ m}^2$
Mass flux for gasoline	$\dot{m}''$	$= 0.036 \text{ kg/m}^2\text{-s}$
Heat of combustion for gasoline	Δh_c	$= 43.7 \text{ MJ/kg}$
Heat release rate	$\dot{Q}$	$= (0.036)\,(43700)\,(0.164) = 258 \text{ kW}$

The real-world estimate for the heat release rate would be 200–300 kW.

VIRTUAL ORIGIN

A fire vector analysis of this firebomb test case can be used by the investigator to understand and interpret the combined movement and intensity of plumes that created the fire pattern damage shown in Figure 3.37. Note that the vectors originate from an area above where the firebomb was located on the floor, which corresponds to the theoretical virtual origin of the fire plume.

The calculation for the location of the virtual origin confirms this observation. As previously calculated in Example 2.3, the heat release rate for a 0.46-m (1.5-ft)-diameter gasoline fire is approximately 258 kW. The calculation for virtual origin, as taken from the Heskestad equation, is as follows:

Virtual origin	$Z_0 = 0.083 \, \dot{Q}^{2/5} - 1.02 \, D$
Equivalent diameter	$D = 0.457 \text{ m}$
Total heat release rate	$\dot{Q} = 258 \text{ kW}$
Virtual origin	$Z_0 = (0.083)(258)^{2/5} - (1.02)(0.457)$
	$= 0.766 - 0.466 = 0.3 \text{ m (1 ft)}$

For this case example, Z_0 is determined to be 0.3 m (1 ft), which is above the floor level, as documented in the vector diagram in Figure 3.37. As stated previously, research has shown that a fuel releasing a high energy over a small area, such as in this case example, may produce a virtual origin above the floor level (Karlsson and Quintiere 1999).

A more detailed analysis of the relationship of the virtual origin with changes to the heat release rate and effective diameter of the burning fuel is plotted in Figure 3.38. This plot explores the prediction of the Heskestad equation when comparing fire plumes whose virtual origin may be above or below the burning fuel surface.

For example, a combustible fire with a high heat release rate and small-diameter fuel package will typically have a positive value for the virtual origin, indicating it is located along the centerline above the burning surface of the fuel package. This relationship can be further explored in the graph plotting the relationship of the virtual origin with changes in the heat release rate and effective diameter of the burning fuel.

FLAME HEIGHT

The *NFPA 921* flame height calculation takes into account the placement of the fire within the compartment, particularly in corners. By observation of the damage caused by the fire plume as well as its positive virtual origin above the floor, we can assume that the flame height at least touched the ceiling, which is 3.18 m (10.43 ft) high.

Total heat release rate	$\dot{Q}$	$= 258 \text{ kW}$
Flame height	H_f	$= 0.174 \, (k\dot{Q})^{2/5}$
In a corner ($k = 4$)	H_f	$= (0.174)(4 \times 258)^{2/5}$
		$= 2.79 \text{ m (9.15 ft)}$
Including the virtual origin	H_{total}	$= Z_0 + H_f$
		$= 0.299 + 2.79 = 3.09 \text{ m (10.14 ft)}$
		(about 3 m, or 10 ft)

This value corresponds to a greater-than-ceiling height, thus creating the conditions for a ceiling jet across the ceiling. Therefore, the fire plume was high enough to reach the ceiling and radially disperse ceiling jets extending out from the centerline, a behavior that was observed during the test.

FIGURE 3.37 ◆ The fire vector analysis of the firebomb test case assists in the understanding and interpreting the fire plume and points to an area above the floor that corresponds to the plume's theoretical virtual origin. *Courtesy of D. J. Icove.*

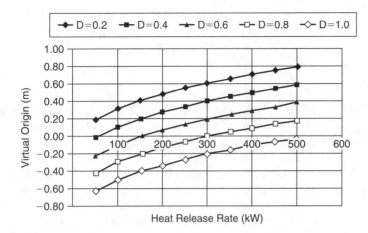

FIGURE 3.38 ◆ Plot of the relationship of the virtual origin with changes to the heat release rate and effective diameter of the burning fuel. *Courtesy of Icove and DeHaan [2006].*

FIRE DURATION

Determining the initial growth period is useful in fire scene reconstruction. In validating estimates of fire duration, it is always fortunate to have historical testing, actual loss history data, and realistic fire models.

When comparing the development of real-life fires with models, the investigator relies on observations, documentation, analysis of historical fire test data, and similar cases that relate to the fire under study. A wealth of information is available from handbooks available in the fire protection engineering field (SFPE 2002b, sec. 3-1).

Fire duration was one of the key questions to be answered in the firebomb test case. After the firebomb ignited, the plastic jug melted, releasing its contents of gasoline into a circular pool of burning liquid. The question then arose: How long could the pool of gasoline be expected to burn? Note that the question of whether the surface under the spill was on a nonporous concrete surface or a carpeted material is addressed in a later section.

The burning duration of pool fires can usually be estimated based on the steady mass burning rate, assuming rapid growth and a large fuel supply. Gasoline has a density of 740 kg/m^3 or 0.74 kg/L. The steady-state mass loss rate (mass flux) of gasoline is 0.036 kg/m^2-s (SFPE 1995a, fig. 3-1.2[a]).

The following calculation is the working hypothesis for the estimated burning duration of the gasoline, assuming that it would self-extinguish after all the available fuel was consumed.

Mass of gasoline	m	$= (1\ \text{gal})\ (3.785\ \text{L/gal})\ (740\ \text{g/L}) = 2.7\ \text{kg}$
Mass burning rate for gasoline	$\dot{m}''$	$= 0.036\ \text{kg/m}^2\text{s}$
Burning area	A	$= \pi r^2 = (3.1416)\ (0.23\ \text{m})^2 = 0.166\ \text{m}^2$
Burn rate of gasoline	$\dot{m}$	$= A\ \dot{m}'' = (0.166\ \text{m}^2)\ (0.036\ \text{kg/m}^2\text{s})$
		$= 0.00598\ \text{kg/s}$
Approximate burning duration		$= \dot{m}/m = (2.7\ \text{kg})/0.00598\ \text{kg/s} = 451\ \text{s}$

The real-world estimate for the average burning duration should be 7 to 8 minutes. If you compare this result with the time–temperature curve (Figure 3.35), this time frame is consistent with the actual test results.

REGRESSION RATES

A flammable liquid pool will burn from the top down at a fairly predictable rate, called its **regression rate**, depending on its chemical composition and physical properties if all other factors (pool diameter and depth) are constant. A thin layer of liquid will burn away very quickly, often so quickly that only the most transient thermal effects will be observed. The depth of a pool is determined by the quantity of liquid, its physical properties (viscosity), and the nature of the surface.

On a level, smooth nonporous surface, a low-viscosity liquid like gasoline will form a pool approximately 1 mm or less deep if not otherwise limited. On a porous surface the liquid will tend to penetrate as far as it can and then spread horizontally by gravity flow and capillary action. The size of the pool will then be controlled by the amount of liquid and the depth of the porous material, the porosity of the substrate, and the rate at which the liquid is poured. On carpeted floors, as a rough guideline, the thickness of carpet can be used as a maximum pool depth, since saturated carpet represents a pool whose depth equals the thickness of the carpet and of any porous pad beneath (DeHaan 1995).

The surface area of the liquid pool is an important variable. A large quantity of gasoline dripping slowly from a leaking gas tank may produce a visible pool on soil or sand not much larger than the diameter of the individual drops but many feet deep. The same quantity dumped quickly onto the same ground may produce a much shallower pool of great diameter. A similar measure of liquid can be broadcast in an arc to generate a shallow layer (film) over a large area. The viscosity and surface tension of the liquid and the speed with which the material is ejected will determine the nature (thickness) of this film. Testing may be needed to establish the bounding conditions of maximum and minimum areas.

Once the liquid is ignited, the rate of burning per unit area (mass flux) is controlled by the amount of heat that can reach the surface of the pool and the size of the perimeter through which air can be entrained. The mass flux depends on the fuel and the size of the pool. For gasoline pools of 0.05–0.2 m in diameter the minimum regression rate is 1–2 mm/min. (Blinov, in Drysdale 1999, 153). For very small pools the rate is much higher (owing to the laminar flame structure). The heat effects (scorching) beneath the burning pool of gasoline are superficial because there is insufficient time to produce them. Because the fire size is driven by the evaporation rate per area (mass flux), the larger the pool in surface area, the larger the fire (the greater the $\dot{Q}$). Gasoline pools over 1 m in diameter have a maximum regression rate of 3–4 mm/min owing to the limitations of turbulent mixing in the plume (Blinov, in Drysdale 1999, 153).

The flame produces a variety of effects on nearby surfaces depending on the geometry and nature of heat application. Because the fuel-covered substrate is being cooled by evaporation of the liquid fuel in contact with it, its temperature cannot be more than a few degrees hotter than the boiling point of the liquid and so will not do much more than scorch the surface directly under the pool. Low-melting-point materials like synthetics can melt as well as scorch. The areas of carpet or floor immediately outside the pool will be exposed to the radiant heat of the plume during the fire. Without the protection of the evaporating liquid, materials will melt and scorch,

FIGURE 3.39 ◆ "Halo" or ring pattern formed at outside of burning methanol pool on carpet. Center of pool is protected by evaporating fuel. *Courtesy of J. D. DeHaan.*

particularly if small or thin (carpet fibers). Materials in direct contact with the flame front will be scorched and ignited or at least charred.

Thermal effects of materials near the flame front often produce what is called a **halo** or **ring effect** around the outside of the pool as shown in tests (Figure 3.39) by DeHaan (2007) and Putorti (2001). If the substrate is ignitable, a ring of damage may extend some distance from the pool. Eventually the center of the protected area may be burned as the protective layer of liquid fuel evaporates. In traditional carpets (wool or nylon) this area will be noticeable, and sufficient residues of the liquid fuel may survive in the center of the burned area to be recovered and identified. If the fire burns itself out or is quickly extinguished by a sprinkler system or oxygen depletion, the pool area can be roughly estimated from the dimensions of the burned margins. If the fire is sustained by the carpet, or if the fire (externally) reaches flashover, the area of burned carpet can grow well beyond the area of the original liquid pool. This is particularly true for the new generation of synthetic fiber carpets with polypropylene backing and polypropylene face yarn pile (DeHaan 2007).

Wool or nylon carpets with jute backing will tend to self-extinguish, but polypropylene will not. Combustion tests have shown that flames on such carpets can propagate at a rate of approximately 0.5–1 m^2/hr (5–11 ft^2/hr), generating very small flames at the margins of the burning area. (Note that such carpets will pass the methenamine tablet test (ASTM 2859), since its igniting flame is about 50 W and of brief duration [similar to a dropped match], but if there is any significant additional heat flux from a larger ignition source [wad of paper] or a continuously burning object like a piece of furniture, these can propagate a fire over a large area if given enough time [DeHaan 2007].)

ADJUSTMENTS TO FIRE DURATION

This test compared the calculated burning duration of 451 s against the actual burning duration shown in Figure 3.35, which showed that the actual duration was less than the estimated duration. This working hypothesis had to be modified to account for the carpet present rather than the concrete floor assumed in the initial calculation.

Experiments on carpet saturated with pentane (a component of gasoline) showed that its rate of evaporation (nonburning) is 1.5 times higher than that for a freestanding pool at the same temperature (DeHaan 1995). Thus, the steady-state mass loss rate by evaporation of gasoline saturated into carpet is approximately 1.5 times the freestanding value of 0.036, or 0.054 kg/m^2-s. Assuming that combustion processes will progress in the same manner as evaporation on this carpet, the recalculated burning duration for a carpeted surface gave a corrected estimate of 316 s and a heat release rate of 387 kW.

Knowledge of the burning characteristics of common flammable and combustible liquids on various surfaces, including those of pool fires such as the one in our test case, is crucial in fire testing and analysis. Synthetic carpets may melt and reduce the mass flux. Also, heat release rate contributions from the molten plastic container pool and carpet were not included. These would be expected to add to the heat release rate and possibly to reduce the duration of the fire.

NIST research on flammable and combustible liquid spill and burn patterns reported that the peak spill fire heat release rates for thin layers of burning liquids on nonporous surfaces were found to be only 12.5 to 25 percent of those from equivalent deep-pool fires. The peak heat release rates for fires on carpeted surfaces were found to be approximately equal to those for equivalent pool fires (Putorti 2001).

In small liquid pool fires, researchers have reported that after the ignition of the fuel, the mass burning rate will increase until a steady rate is reached (Hayasaka 1997). Research and validation experiments by Ma et al. (2004) on burning rates of liquid fuels on carpet indicate that the phenomenon is possibly more complex, particularly in the role the carpet plays in supplying fuel to the fire. Ma notes that several factors affect carpet fires: the capillary or "wick" effect of its fibers; evaporation, combustion, and heat transfer; and mass transfer–limited combustion. He surmises that the carpet has two conflicting roles on the mass burning rate: (1) as an insulating material that blocks heat transfer losses to the depth of the pool and increases the mass burning rate along with resulting higher temperatures and (2) as a porous medium decreasing the mass burning rate owing to an insufficient capillary effect. Note that a thin fuel layer on a hard surface such as concrete or wood, whose conductivity is greater than that of the liquid fuel, will result in greater conduction heat transfer losses. Ma points out that the reports of observations that liquid fires burn more severely on carpets than on smooth uncarpeted floors or ground can be explained. It turns out that carpet enhances the mass burning rate at the earlier stages of burning because it insulates the fabric pile, making the burning liquid fuel appear to behave as a steady deep pool-fire.

FLAME HEIGHT ADJUSTMENT FOR FIRE LOCATION

The flame height is a function of the fire location within the room. As shown in equation (3.12), the flame height, H_f, depends on the fire's placement either in the center ($k = 1$), against the wall ($k = 2$), or in the corner ($k = 4$) of the compartment.

In the initial approach, the corner was assumed to have full impact on the flame height, and a total flame height of 3.1 m (10.1 ft) was predicted. With the new heat release rate at 387 kW and the corner configuration, the new flame height estimate would be recalculated to be 3.28 m plus the virtual origin of 0.43 m, or 3.71 m (12.2 ft).

The damage to the corner of the ceiling above the fire demonstrated that a sufficient flame plume was entrained in the corner to cause such penetration. Note that the additional fire from the combustible wall covering (wainscoting) added an unknown factor to the plume generation.

POOL FIRES AND DAMAGE TO SUBSTRATES

When a flammable liquid burns, it does so by evaporating from the liquid, creating a layer of vapor denser than air. Brownian motion at elevated temperatures causes this layer to have a finite thickness and to diffuse into overlying air to form a steep concentration gradient with distance from the liquid surface. Wherever this gradient is within the flammable range of the vapors, a flame can be supported. A **diffusion flame** is one supported by fuel vapors diffusing from the fuel surface into the surrounding air/oxygen.

The distance between the fuel surface and the flame front varies with the temperature (and thereby the vapor pressure) of the fuel. Radiant heat from the layer of flame travels in all directions. Some of the heat radiated downward is absorbed by the fuel, keeping the temperature high enough that there is a continual supply of vapors to support a flame. Some heat goes through the fuel into the substrate beneath and is absorbed, increasing the temperature of the substrate. Owing to intimate contact between the substrate and the liquid fuel, heat is transferred to the fuel and then distributed through the fuel (if it is a deep enough layer) by convective circulation.

The temperature of the surface under the pool will not be more than a few degrees above the boiling point of the liquid overlying it. If the fuel's boiling-point temperature is low enough ($<200°C$, 390°F), the liquid can burn off without any visible effect if the surface is smooth, with no pores, joints, or seams, and has a relatively high decomposition temperature. The higher the boiling point of the liquid, the greater the chances of thermal damage to the floor, such as pyrolysis (scorching), melting, or both. In a burning pool of fuel, the radiant heat at the edges absorbed by the substrate may produce localized scorching or other heat effects, as illustrated in Figure 3.40. If a fuel contains a mixture of compounds with different boiling points, the low-boiling liquid will tend to burn off first, leaving the higher-boiling compounds to continue to heat. A mixture like gasoline covers the range of boiling points 40°C–150°C (100°F–300°F). As the mixture burns, the boiling point of the residue increases, and therefore the limiting temperature factor increases. At 250°C (450°F), wood surfaces will be only scorched but some synthetic floor coverings can be significantly damaged. Field fire tests demonstrating the absence of significant thermal effects on wood with gasoline pools have recently been published (DeHaan 2007).

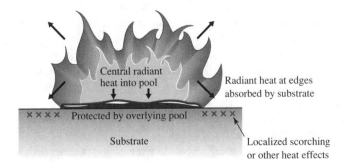

FIGURE 3.40 ◆ Cross section of burning pool of fuel.

FINAL HYPOTHESIS

The final hypothesis noted for this test, which Figure 3.35 reflects, is the consumption of the gasoline in the time–temperature curve, leveling out approximately 400 s after ignition, which is consistent with the preceding estimate. This demonstrates the effects of the use of flammable liquids in arson and other fire-related crimes—a very rapid increase in localized temperature followed by a rapid consumption to self-extinguishment.

Expert conclusions or opinions on the strength of these hypotheses rest partially on the use of accepted and historically proven fire testing techniques that validate this methodology and can be easily replicated. Additionally, this methodology has been peer reviewed and published. There are established error rates for applying this methodology specifically to known variables such as room size and fire development times. Standardized methods are also maintained by independent and unbiased organizations for applying and interpreting these relationships. Finally, this methodology is generally accepted in the scientific community.

◆ 3.7 SUMMARY AND CONCLUSIONS

Fire pattern damage analysis is a vital investigation technique for fire scene reconstruction. The visual interpretation of damage created by fire plumes can isolate and accurately identify the area of fire origin.

Locating and identifying the first fuel package ignited is a critical step in the accurate reconstruction of a fire incident. Careful analysis of fire patterns can significantly aid the scene investigator in this effort. Because effects like charring, melting, ignition, and protection are predictable, their location and distribution offer a sound basis for locating fuel packages, which can be confirmed by interviews or prefire photos.

Systematic steps can be used by fire investigators to support and document thermal damage patterns, identify the fire's direction and intensity, confirm significant witness observations, and verify the results of fire modeling. These systematic steps invoke the scientific method to test and evaluate various hypotheses of the fire's origin and spread.

From these observations, the authors conclude the following:

- As long as there is not too much damage, ample fire pattern heat indicators exist and can be documented.
- Evidence of fire pattern damage at scenes can be found for fire plumes that can indicate a fire's source, area and point of origin, and direction of travel.
- There are only a limited number of scientifically validated fire-related patterns, and they should be evaluated for their intensity, direction, and duration.
- Fire engineering analysis using the scientific method is necessary to perform a comprehensive analysis and may be the major usable tool if there has been substantial destruction to the building.
- Investigators should not overlook signs of physical evidence documenting human activity (hand- or fingerprints, shoeprints, blood spatters, broken glass, discarded hoses/fire extinguishers, and burn injuries).
- Fire analysis and computer-assisted modeling can provide further insights into fire scene behavior.
- Peer review is a healthy and prudent step to ensure that all hypothesis are addressed.

Recommendations include the continued use of *Kirk's Fire Investigation* and *NFPA 921* as the universally accepted guides to documenting all fire and explosion investigations. Continued funding of research into fire pattern analysis and reconstruction technologies, as well as international conferences on the forensic aspects of fire investigation, is an important means by which the body of knowledge in the field may be increased.

Problems

3.1. Find a photograph of a fire plume in a recent news story or article. Estimate the dimensions of the plume and attempt to calculate its heat release rate. What information on the fire can you infer from these calculations that relate to the story or article?

3.2. Photograph a recent fire scene and illustrate as many fire patterns as you can find.

3.3. From the plume information in the photos and the identified first fuel package, estimate the heat release rate and virtual origin of the initial fire.

3.4. Use the equations provided in this chapter to determine the radiant heat fluxes at various distances from the fire's virtual origin.

3.5. A wooden box open on one side can be used to create a small fire with a wad of paper or a small block of urethane foam. Use such a reduced scale test device to observe wall and corner effects, ceiling jets, and rollover events.

Suggested Reading

DeHaan, J. D. 2007. *Kirk's fire investigation,* 6th ed., chap. 7. Upper Saddle River, N.J.: Prentice Hall.

National Fire Protection Association. 2004. *NFPA 921—Guide for fire and explosion investigations,* chaps. 4 and 15. Quincy, MA: NFPA.

Fire Scene Documentation

When you have eliminated the impossible, that which remains, however improbable, must be the truth.

—*Sir Arthur Conan Doyle*

Fire scenes often contain complex information that must be thoroughly documented, a task of paramount importance to the investigator. A single photograph and scene diagram are often not sufficient to capture vital information on fire dynamics, building construction, evidence collection, and avenues of escape for the building's occupants.

Thorough documentation is accomplished through a comprehensive effort comprising forensic photography, sketches, drawings, and analysis. The purposes of forensic fire scene documentation include documenting visual observations, emphasizing development characteristics, and authenticating physical evidence found at the scene.

This chapter provides the fundamental concepts of forensic fire scene documentation, with emphasis placed on a systematic set of guidelines. Reviewed are concepts and techniques that produce viable and legally acceptable documentation for both investigative reports and courtroom presentations. Various computer-assisted photographic and sketching technologies that help ensure accurate and representational diagramming are also explored.

Report writing guidelines are not included in this textbook, only the documentation needed to assemble authoritative reports. For a detailed discussion and examples, see *Kirk's Fire Investigation*, 6th ed. (DeHaan 2007) and *Combating Arson-for-Profit: Advanced Techniques for Investigators*, 2nd ed. (Icove, Wherry, and Schroeder 1998).

◆ 4.1 NATIONAL PROTOCOLS

The **systematic documentation** of a fire scene from its initial stages plays an essential role in the effort to record and preserve the events and evidence for all parties involved, particularly for forensic and other experts who are retained later and may be asked to render their opinion in the case. This systematic approach captures all available information needed for later use in criminal, civil, or administrative matters.

The bottom line is that systematic documentation, when correctly conducted, will be well suited to pass the *Daubert* challenges for courtroom admissibility and ensures that an independent qualified investigator will arrive at the same opinions. Several national interrelated protocols cite the need for systematic documentation of fire scenes. Two of the recent protocols include those by the U.S. Department of Justice and the National Fire Protection Association (NFPA).

U.S. DEPARTMENT OF JUSTICE

One example of a peer-reviewed national protocol for fire investigation was developed and published in 2000 by the National Institute of Justice (NIJ), which serves as the applied research and technology agency of the U.S. Department of Justice. The NIJ's (2000) *Fire and Arson Scene Evidence: A Guide for Public Safety Personnel* is the product of their Technical Working Group on Fire/Arson Investigation, which consisted of 31 national experts from law enforcement, prosecution, defense, and fire investigation communities. This document has become the standard for documenting and recording the collection of fire and arson scene evidence. Both authors participated in the preparation and editing of the NIJ guide.

The expressed intent of the NIJ guide is to expose as many of the public-sector personnel (mainly fire, police, and prosecution) as possible to the process of identifying, documenting, collecting, and preserving critical physical evidence at fire scenes. This document has become the most widely distributed public guide on fire scene processing since the 1980 historic publication by the National Bureau of Standards (NBS; forerunner of the National Institute of Standards and Technology), *Fire Investigation Handbook* (Brannigan, Bright, and Jason 1980).

NATIONAL FIRE PROTECTION ASSOCIATION

The NFPA's *Guide for Fire Incident Field Notes*, first published in 1988 and republished in 1998, serves as a standardized protocol (*NFPA 906*) for recording field notes of fire scenes. The NIJ guide cites *NFPA 906* and includes the data collection forms in its appendix. A synopsis of the forms contained in *NFPA 906* appears in Table 4.1. Similar forms are found in the appendixes to *Kirk's Fire Investigation* (DeHaan 2007). These forms are reprinted in Figures 4.1–4.7, 4.11–4.14, and 4.16–4.18.

The intended users of *NFPA 906* include all persons having responsibility for investigating fires. This includes the fire company officer, the incident commander, the fire marshal, or a private investigator. These data collection forms consist of an organized investigative protocol. The express purpose of these forms is to serve as an input for collecting and recording preliminary information needed in the preparation of a formal incident or investigative report and in constructing a compartment fire model.

The reports cover structure, vehicle, and wildland fires; information on casualties, witnesses, evidence, photographs, and sketches; and documentary data on insurance and public records. A cover case management form is used to track the progress of the investigation. The maintenance of *NFPA 906* is now the responsibility of NFPA's Technical Committee on Fire Investigations, which also oversees *NFPA 921* (NFPA 2004a).

TABLE 4.1 ◆ Forms Used in Assuring the Collection of Uniform and Complete Field Data for Constructing Written Reports

Form	Name	Description
906-0	Case Supervision	Top cover sheet to track the progress of the investigation
906-1	All Fires	Collects general identification and contact information
906-2	Structure Fires	Used for structure fires
906-3	Motor Vehicle Fires	Used for motor vehicle fires
906-4	Wildland Fires	Used for grass, brush, or wildland fires
906-5	Casualties	Records information on persons injured or killed in the fire
906-6	Witness Statements	Identifies and records expected testimonies of witnesses
906-7	Evidence	Documents and records recovered and seized evidence
906-8	Photographs	Logs the descriptions of all photographs taken by the investigators
906-9	Sketches	Contains a scene destruction sketch
906-10	Insurance Information	Records information on insurance coverage, adjustments, and loss
906-11	Records/Documents	Records information on incident, property, business, and personal records that are available
	Fire Modeling	Data for compartment fire modeling

Source: Derived from NFPA 1998, NFPA 2004a, and *Kirk's Fire Investigation,* 6th ed. (DeHaan 2007).

◆ 4.2 SYSTEMATIC DOCUMENTATION

Each fire investigator should adopt a **systematic procedure** or **protocol** for documenting through sketches, photographs, witness statements, and written documentation the examination of a fire scene as well as recording the chain of custody of all evidence removed. A systematic four-phase process (Icove and Gohar 1980) was first introduced with the publication of the NBS (1980) *Fire Investigation Handbook*.

This process recommends a careful documentation of the fire scene in the following four phases.

- Phase 1—Exterior
- Phase 2—Interior
- Phase 3—Investigative
- Phase 4—Panoramic or specialized

Table 4.2 shows the recommended updated guidance and purpose for this systematic documentation philosophy. Note that this approach covers fire investigations without any preconception as to whether the fire was accidental, natural, or incendiary in origin.

	TABLE 4.2 ◆ Systematic Documentation Techniques in Fire Investigation		
Step	*Technique*	*Guidance*	*Purpose*
1	**Exterior**	Photograph perimeter and exterior of property. Sketch exterior. Use GPS to obtain location.	Establishes venue and location of fire scene in relation to surrounding visual landmarks Documents exposure damage to adjacent properties Reveals structural conditions, failures, violations, or deficiencies Establishes extent of fire damage to exterior of scene Establishes egress and condition of doors and windows Documents and preserves possible physical evidence remote from the fire
2	**Interior**	Record and sketch extent of fire damages, potential ignition sources, and data needed for fire reconstruction and modeling.	Traces fire travel and development from exterior to suspected point(s) of fire origin Documents heat transfer damage, heat and smoke stratification levels, and breaches of structural element Documents condition of power utility and distribution, furnace, water heater, and heat-producing appliances Documents position and damage to windows, doors, stairwells, and crawl space access points Documents fire protection equipment locations and operation (sprinklers, heat and smoke detectors, extinguishers) Documents readings on clocks and utility equipment Documents alarm and arc fault information
3	**Investigative**	Document clearing of debris and evidence prior to removal, fire burn and charring patterns, packaged evidence.	Assists in recording fire pattern and plume damage, isochar lines Establishes condition of physical evidence, utilities, distribution, and protection equipment (breakers, relief valves) Documents integrity of the chain of custody of evidence
4	**Panoramic specialized**	Produce multidimensional sketches. Captures evidence using forensic light sources or special recovery techniques.	Provides clearer peripheral views of exterior and interiors Establishes photograph viewpoints of witnesses Documents and preserves critical evidence

Source: Updated from Icove and Gohar (1980).

In an effort to show the utility of the four-phase documentation approach, the use of the *NFPA 906* forms is introduced. This discussion also includes several areas where *NFPA 906* does not capture information needed for full fire scene reconstruction and engineering analysis of the incident.

Agencies that use the *NFPA 906* field notes often copy the blank forms directly from the standard and bind them into packets that are placed into case jackets. Several additional copies of the witness statement form are often included for multiple interviews.

The "Case Supervision Field Notes" (Form 906-0) (Figure 4.1) is the cover sheet for the investigative notes. This cover sheet is to be used as a working document for both investigator and supervisor. Included on the form are check boxes indicating whether a form is contained in the case file, when it was completed, and remarks as to its status. The disposition of evidence, court times, and other major information may be noted in the activity section of the report.

Another high-level case document is Form 906-1, "Any Fire Field Notes" (Figure 4.2). This form records the how an agency was notified of the incident, the conditions on arrival, the owner/occupant of the property, other agencies involved, and an estimated total financial loss. Also documented are the time of arrival, the legal authority to enter the scene, and the time the scene was released.

The **weather conditions** prior to the fire incident sometimes become important, particularly when high winds, temperature fluctuations, or lightning come into play. The National Weather Service, Office of Climate, Water, and Weather Services, has a forensic services program to support investigations, particularly those involved in litigation. Certified climatological records (including radar images, satellite photos, and surface analysis) can be obtained from the National Weather Service Headquarters in Silver Spring, Maryland. Alternative free and low-cost weather notification and historical systems are available in the United States. These services provide radar and severe weather notifications. One popular Web site is the Weather Underground (weatherunderground.com) which can provide historic weather data for many U.S. locations.

The location of **lightning strikes** is an important database when addressing whether to rule in or rule out the possibility that lightning caused the fire. The U.S. National Lightning Detection Network locates strikes across the United States. For a fee, the private firm Vaisala will provide a report on all lightning strikes in a particular area for a given time frame. Vaisala (2003) can also supply data and custom software to its customers at vaisala.com.

◆ 4.3 EXTERIOR

The first major phase of the process is to document the **exterior** of the structure, vehicle, forest, wildland, boat, or object prior to probing into the cause of the fire. While encircling the exterior of the structure or vehicle, the investigator can make a cursory field search for additional evidence.

The purpose of this phase is to establish the location of the fire scene in relation to surrounding visual landmarks. The exterior views should reveal the extent of fire damage, collapse, structural conditions, failures, code violations, deficiencies, or potential safety concerns. This step also documents exposure damage to adjacent properties from the fire. In large investigations, an overhead crane, aerial ladder truck, or aircraft can be used to obtain photographs of the scene.

CASE SUPERVISION FIELD NOTES 906-0	AGENCY	FILE NUMBER

This cover sheet will assist in keeping track of the progress of the investigation. Indicate what has been done, what needs to be done, assignments, dates and so forth, in the Remarks sections. The lower portion should be used to record routine checks or rechecks and other information pertinent to the investigation.

FIELD NOTES FORMS

Form	No.	Complete	N/A	Remarks
ANY FIRE	906-1	☐ COMPLETE ___ DATE	☐ N/A	REMARKS
STRUCTURE	906-2	☐ COMPLETE ___ DATE	☐ N/A	REMARKS
VEHICLE	906-3	☐ COMPLETE ___ DATE	☐ N/A	REMARKS
WILDLAND	906-4	☐ COMPLETE ___ DATE	☐ N/A	REMARKS
CASUALTY	906-5	☐ COMPLETE ___ DATE	☐ N/A	REMARKS
WITNESS	906-6	☐ COMPLETE ___ DATE	☐ N/A	REMARKS
EVIDENCE	906-7	☐ COMPLETE ___ DATE	☐ N/A	REMARKS
PHOTOGRAPH	906-8	☐ COMPLETE ___ DATE	☐ N/A	REMARKS
SKETCH	906-9	☐ COMPLETE ___ DATE	☐ N/A	REMARKS
INSURANCE	906-10	☐ COMPLETE ___ DATE	☐ N/A	REMARKS
RECORDS/DOCUMENT	906-11	☐ COMPLETE ___ DATE	☐ N/A	REMARKS

INCIDENT AND CASUALTY REPORTS UPDATED ☐ YES ___ DATE ☐ NO ☐ NOT NECESSARY

DATE	ACTIVITY	BY

FIGURE 4.1 ◆ NFPA 906-0, "Case Supervision," form is the cover sheet for tracking the progress of the investigation. *Reprinted with permission from NFPA 906—Guide for Fire Incident Field Notes, 1998 Edition. Copyright © 1998, National Fire Protection Association, Quincy, MA 02269. This reprinted material is not the complete and official position of the National Fire Protection Association on the referenced subject, which is represented only by the standard in its entirety.*

ANY FIRE FIELD NOTES 906-1		AGENCY	FILE NUMBER

INCIDENT

ADDRESS/LOCATION			DAY	DATE	TIME	FIRE DEPT. INCIDENT NO.
WEATHER AT TIME OF FIRE	GENERAL CONDITIONS				TEMP.	WIND DIR. / WIND SPEED
PROPERTY DESCRIPTION	STRUCTURE (906-2) ☐	VEHICLE (906-3) ☐	WILDLAND (906-4) ☐	OTHER ☐		

OWNER/OCCUPANT

OWNER'S NAME	PHONE NO.
OWNER'S ADDRESS	
OCCUPANT'S NAME	PHONE NO.
OCCUPANT'S ADDRESS	
DOING BUSINESS AS	PHONE NO.

NOTIFICATION FOR INVESTIGATION

DAY	DATE	TIME	FROM WHOM		
RECEIVED BY				ASSIGNED TO	
ARRIVED AT SCENE	DAY	DATE	TIME	SCENE SECURED ☐ NO ☐ YES	(COMMENT ON CONDITION) (BY WHOM):
AUTHORITY TO ENTER	EMERGENCY	CONSENT ☐ VERBAL ☐ WRITTEN	WARRANT ☐ ADMIN. ☐ CRIM.	OTHER (Describe)	
DEPARTED SCENE	DAY	DATE	TIME	COMMENTS	

OTHER AGENCIES INVOLVED

FIRE DEPT.	INCIDENT NO.	CONTACT PERSON	PHONE NO.
POLICE DEPT.	FILE NO.	CONTACT PERSON	PHONE NO.
OTHER	CASE NO.	CONTACT PERSON	PHONE NO.

ESTIMATED TOTAL LOSS

$	ESTIMATED BY

REMARKS

FIGURE 4.2 ◆ NFPA 906-1, "Any Fire," form collects general dispatch notification, owner, occupant, and investigative and contact information. *Reprinted with permission from* NFPA 906—Guide for Fire Incident Field Notes, 1998 Edition. *Copyright © 1998, National Fire Protection Association, Quincy, MA 02269. This reprinted material is not the complete and official position of the National Fire Protection Association on the referenced subject, which is represented only by the standard in its entirety.*

If there has been any explosion, the distance to which fragments of glass or structure were thrown must be measured and documented via both diagramming and photography. These concepts are discussed in later sections.

The "Structure Fire Field Notes" (Form 906-2a; Figure 4.3) documents the type of property, geographic area, construction techniques, security, alarm protection, and utilities. Documenting the security at the time of the fire is an important consideration in arson cases where the issue of "exclusive opportunity" is raised along with the condition of doors, windows, and protection systems. Also important is documenting the condition of doors, windows, and evidence found as part of the external documentation. The location and distances of window glass displaced from the window or vehicle are documented during the exterior examination.

The condition of the utilities at the time of the fire is important, particularly when considering all other sources of ignition. The status of the utilities may indicate whether the owner/occupant(s) was(were) living in the structure. It is important to document who actually may have given the instructions to disconnect the utilities if that was done prior to the fire. Determining when gas and electric services were cut off (and by whom) during fire suppression is also important.

◆ **4.4 INTERIOR**

The second phase involves the documentation of **interior** damage by showing the extent and progress of the fire, through the room(s), area(s), and suspected point(s) of fire origin. These pictures and sketches are made prior to excavating the fire debris and serve to document the conditions of the scene as found on the investigator's arrival. All undamaged areas of the building should also be examined and documented for comparison and hypothesis formulation (as long as there is legal authority for entry).

The "Structure Fire Field Notes" (Forms 906-2b and-2c; Figures 4.4 and 4.5), "Motor Vehicle Field Notes" (Form 906-3; Figure 4.6), and "Wildland Fire Field Notes" (Form 906-4; Figure 4.7) are used to document further the property, utility service, contents, preliminary area of fire origin, estimated ignition sequences, and fire and smoke spread factors.

DOCUMENTATION OF DAMAGE

While conducting the interior review, the investigator documents **damage** to all rooms, including heat and smoke stratification levels, heat transfer effects, and breaches of structural elements (walls, floors, ceilings, and doors). The investigator may find it useful to map out and delineate the areas of damage corresponding to the pattern types described in Table 3.1, with color-coded chalk or tape on the surfaces themselves, followed by photography. Areas of surface "demarcations," thermal effects, penetrations, and loss of material can also be outlined using colored markers on sketches or photographs. One system might be to use yellow lines to outline areas of surface deposits, green lines for thermal effects, blue lines for penetrations, and red lines to outline areas where the material has been consumed completely.

This phase includes examination and documentation of the condition of power utility and distribution, furnace, water heater, and heat-producing appliances, as well as the contents and conditions of the rooms. The HVAC system should also be documented as to the location and condition of the unit, duct work, and filters. The thickness of window glass, the sizes of windows, and whether they are single or double

STRUCTURE FIRE FIELD NOTES 906-2a	AGENCY	FILE NUMBER

TYPE AND STATUS

PROPERTY USE

STATUS (OCCUPIED, UNOCCUPIED, VACANT)	COMMENTS

AREA DESCRIPTION

☐ RURAL ☐ FARM ☐ URBAN ☐ SUBURBAN ☐ OTHER _____

☐ ZONED ☐ UNZONED ☐ IMPROVING ☐ DECLINING ☐ STABLE ☐ OTHER _____

CONSTRUCTION

FOUNDATION
☐ SLAB ☐ CRAWL SPACE ☐ BASEMENT(S) ☐ OTHER _____

DIMENSIONS
_____ FT LENGTH _____ FT WIDTH _____ FT HEIGHT _____ STORIES _____ NO. UNITS

TYPE OF CONSTRUCTION	EXTERIOR WALLS	INTERIOR WALLS	FLOORS	ROOF

SECURITY (Time of Fire)

DOORS
☐ SECURE ☐ NOT SECURE PER:

WINDOWS
☐ SECURE ☐ NOT SECURE PER:

OTHER
☐ SECURE ☐ NOT SECURE PER:

COMMENTS ON SECURITY

ALARM/PROTECTION SYSTEMS

ALARMS ☐ YES ☐ NO	TYPE ALARM

ALARM COMPANY	CONTACT PERSON	PHONE NO.

COMMENTS

PROTECTION SYSTEMS ☐ YES ☐ NO ☐ OPERATED ☐ DID NOT OPERATE	COMMENTS

DESCRIPTION OF SYSTEM(S)

UTILITIES (Time of Fire)

ELECTRIC	☐ ON ☐ OFF	UTILITY COMPANY NAME	CONTACT	PHONE NO.
GAS	☐ ON ☐ OFF	UTILITY COMPANY NAME	CONTACT	PHONE NO.
WATER	☐ ON ☐ OFF	UTILITY COMPANY NAME	CONTACT	PHONE NO.
PHONE	☐ ON ☐ OFF	UTILITY COMPANY NAME	CONTACT	PHONE NO.
OTHER	☐ ON ☐ OFF	UTILITY COMPANY NAME	CONTACT	PHONE NO.

FIGURE 4.3 ◆ NFPA 906-2a, "Structure Fire," form collects structure description, security, protection, and utility contact information. *Reprinted with permission from* NFPA 906— Guide for Fire Incident Field Notes, 1998 Edition. *Copyright © 1998, National Fire Protection Association, Quincy, MA 02269. This reprinted material is not the complete and official position of the National Fire Protection Association on the referenced subject, which is represented only by the standard in its entirety.*

STRUCTURE FIRE
FIELD NOTES 906-2b

AGENCY	FILE NUMBER

EXTERIOR OBSERVATIONS

INTERIOR OBSERVATIONS

HEATING SYSTEM

TYPE   LOCATION

COMMENTS

ELECTRICAL SERVICE

☐ FUSES   ☐ BREAKERS   ENTRY LOCATION   SERVICE PANEL LOCATION

COMMENTS

OTHER HEATING EQUIPMENT

TYPE(S)   LOCATION

COMMENTS

STRUCTURE CONTENTS

COMMENTS

AREA OF ORIGIN

COMMENTS

FIGURE 4.4 ◆ NFPA 906-2b, "Structure Fire," form collects exterior, interior, heating, electrical service, contents, and area of fire origin information. *Reprinted with permission from NFPA 906—Guide for Fire Incident Field Notes, 1998 Edition. Copyright © 1998, National Fire Protection Association, Quincy, MA 02269. This reprinted material is not the complete and official position of the National Fire Protection Association on the referenced subject, which is represented only by the standard in its entirety.*

STRUCTURE FIRE FIELD NOTES 906-2c	AGENCY	FILE NUMBER

IGNITION SEQUENCE

HEAT SOURCE

MATERIAL IGNITED

IGNITION FACTOR

IF EQUIPMENT INVOLVED
MAKE   MODEL   SERIAL NO.

COMMENTS

FIRE SPREAD

MATERIALS

AVENUES

COMMENTS

SMOKE SPREAD

MATERIALS

AVENUES

COMMENTS

REMARKS

FIGURE 4.5 ◆ NFPA 906-2c, "Structure Fire," form collects ignition sequence and fire and smoke spread information. *Reprinted with permission from* NFPA 906—Guide for Fire Incident Field Notes, 1998 Edition. *Copyright © 1998, National Fire Protection Association, Quincy, MA 02269. This reprinted material is not the complete and official position of the National Fire Protection Association on the referenced subject, which is represented only by the standard in its entirety.*

MOTOR VEHICLE
FIELD NOTES 906-3

	AGENCY	FILE NUMBER

VEHICLE DESCRIPTION

COLOR(S)	YEAR	MAKE	MODEL	LICENSE — NO., STATE, EXPIRES	VIN NO.

OWNER/OPERATOR

OWNER'S NAME	OWNER'S ADDRESS	OWNER'S PHONE NO.
OPERATOR'S NAME/LICENSE NO.	OPERATOR'S ADDRESS	OPERATOR'S PHONE NO.

EXTERIOR

PRIOR DAMAGE	FIRE DAMAGE
TIRES/WHEELS (Missing, Match, Condition)	
PARTS MISSING	

FUEL SYSTEM

PRIOR DAMAGE		FIRE DAMAGE	
TYPE FUEL	CONDITION OF TANK	FILLER CAP CONDITION	FUEL LINE CONDITION

ENGINE COMPARTMENT

PRIOR DAMAGE	FIRE DAMAGE

FLUID LEVELS   OIL _____ TRANSMISSION _____ RADIATOR _____ OTHER _____

PARTS MISSING

INTERIOR

PRIOR DAMAGE	FIRE DAMAGE

IGNITION SYSTEM   KEY IN IGNITION ☐ YES ☐ NO

PERSONAL CONTENTS MISSING

ACCESSORIES MISSING

ODOMETER READING	SERVICE STICKER INFORMATION

VEHICLE SECURITY

ALARM	DOOR AND TRUNK LOCKS	WINDOW POSITIONS

ORIGIN/IGNITION SEQUENCE

AREA

HEAT SOURCE

MATERIAL IGNITED

IGNITION FACTOR

FIGURE 4.6 ◆ *NFPA 906-3*, "Motor Vehicle," form collects vehicle description, owner/ operator, exterior/interior, security, and area of fire origin information. *Reprinted with permission from* NFPA 906—Guide for Fire Incident Field Notes, 1998 Edition. *Copyright © 1998, National Fire Protection Association, Quincy, MA 02269. This reprinted material is not the complete and official position of the National Fire Protection Association on the referenced subject, which is represented only by the standard in its entirety.*

WILDLAND FIRE FIELD NOTES 906-4	AGENCY	FILE NUMBER

PROPERTY DESCRIPTION

FIRE DAMAGE ☐ LESS THAN ACRE _____ NO. ACRES	OTHER PROPERTIES INVOLVED
SECURITY ☐ OPEN ☐ FENCED LOCKED ☐ GATES	COMMENTS

FIRE TRAVEL FACTORS

TYPE FIRE ☐ GROUND ☐ CROWN	FACTORS ☐ WIND ☐ TERRAIN	COMMENTS

AREA OF ORIGIN

PEOPLE IN AREA

AT TIME OF FIRE ☐ YES ☐ NO ☐ UNDETERMINED	COMMENTS

IGNITION SEQUENCE

HEAT OF IGNITION

MATERIAL IGNITED

IGNITION FACTOR

IF EQUIPMENT INVOLVED

MAKE	MODEL	SERIAL NO.

COMMENTS

FIGURE 4.7 ◆ *NFPA 906-4*, "Wildland Fire," form collects property, fire ignition sequence, area of origin, and travel information. *Reprinted with permission from* NFPA 906—Guide for Fire Incident Field Notes, 1998 Edition. *Copyright © 1998, National Fire Protection Association, Quincy, MA 02269. This reprinted material is not the complete and official position of the National Fire Protection Association on the referenced subject, which is represented only by the standard in its entirety.*

glazed must also be noted and documented. The sill height and soffit depth of each door and window in the rooms involved must be noted, as well as their opening dimensions and whether they were open during the fire or broken sometime later.

STRUCTURAL FEATURES

When documenting interior features, the construction of the ceiling may play a key role in spread of smoke or fire, or detection. A smooth, sloping ceiling may direct smoke away from a detector, whereas one with exposed beams or one with deep decorative "bays" may channel smoke or fire in a preferred direction, or prevent it from spreading at all (see Figure 4.8). Because ceilings often suffer major damage during a fire or during overhaul, it is important to document similar undamaged rooms in the building when practical. Otherwise, occupants or maintenance staff should be interviewed about such features. They also may be documented in prefire photos, videos, or building plans. (One of the authors participated in a reconstruction of the 1916 fire that destroyed author Jack London's nearly completed home, Wolf House. There, descriptions in contemporary interviews and features still visible in the masonry remains were compared against builder's plans. It was clear that even though the plans were dated just weeks prior to the fire, there were features in the as-built structure that were different. Wall coverings and stairwell and door locations played a significant role in testing various hypotheses about the fire's area of origin.) (See case example, Chapter 9.)

The type, location, and operation of HVAC components are also important. In one case, an accidental fire in a chair took the life of the sole occupant of a room designated as the "smoking room" for the facility. The midmorning fire was well advanced

FIGURE 4.8 ◆ Complex ceiling structures as in this hotel conference room will affect fire and smoke spread. This photo captures the structural features as well as ventilator, sprinkler, and alarm sensor locations. *Courtesy of John Houde.*

FIGURE 4.9 ◆ An air-return plenum vent opening in this kitchen allowed very rapid extension into the ceiling space and faster-than-expected collapse of the suspended ceiling. *Courtesy of Jamie Novak, Novak Investigations.*

before it was detected by a staff member on the floor above who saw smoke rising past the window of her room and went to investigate. The question ultimately was, Why did the fire get so large and yet the smoke detector in the hall on the ceiling just outside the open door to the room failed to function? (No smoke was seen in the hallway as staff rushed to the room, and the detector was hardwired to an alarm system.) It was ultimately discovered that an exhaust fan in the room's window was left on and its CFM rating was sufficient to draw enough smoke-laden air from the limited fire in the room away from the door, preventing its detection outside. Even passive elements of HVAC systems can play a role. The void space above many suspended ceilings acts as an open air-return plenum, allowing smoke and hot gases to spread quickly. See Figure 4.9 for an example. The filters and condensers on HVAC systems should be checked for soot or pyrolysis products.

FIRE PROTECTION SYSTEMS

It is also important to document for later evaluation the **fire protection equipment** (fire alarms, smoke detectors, automatic suppression systems) in place that would have helped contain the fire as well as alert the occupants. Documentation includes function and location. The time readings on electric and mechanical clocks can serve to document the approximate times that they were damaged and halted by heat or by interruption of power.

The position, type, and operability of smoke, heat, and carbon monoxide (CO) detectors and sprinkler systems should also be documented. Although percentages vary with economic group assessed, between 30 percent and 50 percent of battery-operated smoke detectors are found by inspectors to have dead batteries or no batteries at all. Sadly, this situation is often found during investigations of fatal fires where batteries have been removed for use in toys or remote controls or to prevent false alarms from cooking or even bathroom steam. Hardwired detectors can be disconnected (from unmonitored systems) or covered with tape or plastic film to prevent alarms. The fusing temperature and orifice size of any open sprinkler heads should also be documented and the head(s) preserved as evidence.

The data systems in modern fire alarm systems can be interrogated to capture the time, sequence, and zone of alarms or sprinkler activations. The zones and manner of activation should be established for alarm sensors. Someone knowledgeable about the alarm system should be consulted for assistance. Even systems that are not remotely monitored have a battery-powered memory for recent activations that can be downloaded.

ARC FAULT MAPPING

When a flame or heat attacks an insulated electrical cable, the insulation begins to pyrolyze and degrade. When rubber, cloth, or some plastic insulation chars, the carbonaceous residue becomes conductive, and current can pass if the conductor is energized and a return path established (sometimes called "arcing through char"). In PVC, a temperature of 120°C (248°F) is sufficient to initiate an **arc-tracking** process and allow a current to begin to flow (Babrauskas 2006). The resistance decreases as the insulation gets hotter. In either case, current begins to flow between the hot or energized wire and the neutral or ground (or to a grounded conduit, appliance, or junction box). This increased current creates heat that further chars the insulation, providing even better current flow.

Because of the resistance provided by the charred insulation, this condition is not the same as a direct short or fault (which has nearly zero resistance). The current flow can vary from very low to very high (as high as 150 A has been measured). Because the conductors are not bonded together, they are free to move, so these current flows are usually very short in duration, in many cases too short to cause overcurrent protective devices (OCPD) to function. This means that arcing can occur at many places along a single energized wire and multiple times until the OCPD trips or the wire is severed by melting (sometimes called *fusing*). Once the wire is separated, no current can flow beyond that point.

These phenomena provide a way of locating a possible area of origin of a fire by tracing the wiring from its power source (such as the breaker or fuse box) and mapping the locations of all failures of the wire (since wires can be caused to fail by fire effects even when nonenergized). The failures are then characterized as arcing (energized) or melting (nonenergized) by their gross appearance. The locations of the **arcing failures** farthest downstream along a circuit from the power source are indicators of the point at which the fire first attacked the wiring, as in Figure 4.10. These failures, then, may be very useful in indicating a possible area of origin. This approach was pioneered by Dr. Robert J. Svare (1988) and has been tested in numerous live-burn structure tests.

This procedure has been shown to be effective even in multistory buildings and installations of "ring-mains" wiring (as in the United Kingdom) (Carey 2002). It is not effective in three-phase wiring systems, where there are three energized wires in every circuit. The technique requires diligent and careful tracing of wiring and circuits, since the source of the energy has to be established and the nature of the failure accurately evaluated.

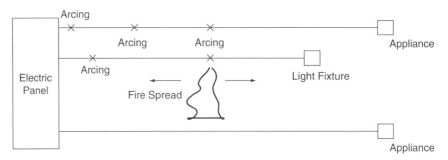

FIGURE 4.10 ◆ Mapping the locations of arcing failures farthest downstream along a circuit from the power source, which are indicators of the point at which the fire first attacked the wiring.

In buildings protected by zone alarm systems, arc fault indicators can be used together with data from the alarm system (alarm, sensor, or sprinkler activation) to estimate areas of possible origin. This approach may not be of use in extensively collapsed or fire-damaged buildings where the relationship and tracing of wires cannot be established accurately.

◆ 4.5 INVESTIGATIVE

The third phase of the systematic investigation concentrates on the debris clearing operations, char and burn patterns, and position of evidence prior to its removal from the fire scene. Evidence of any crimes associated with the fire such as burglary, theft, and homicide should also be documented.

Nothing should be moved, including bodies of victims, until photographically recorded and supported by fire scene notes. Investigative photographs ensure the integrity of the probe and custody of evidence.

CASUALTIES

Information on casualties is recorded on the "Casualty Field Notes" form (906-5; Figure 4.11). This report includes a description of the victim, type of injury, circumstances, treatment received, disposition of the body and its examination, next of kin, and other appropriate remarks. Note that this information should be collected on all injured parties and not just on fatalities, although Health Insurance Portability and Accountability Act (HIPAA) regulations (45 C.F.R. Parts 160 and 164, Subparts A and E) make some information unavailable.

This form does not presently include general information about the victim obtained at autopsies such as burn injuries, tests for blood alcohol, hydrogen cyanide, carboxyhemoglobin levels, and other conditions often documented in fire death investigations. That information is discussed at length in Chapter 7.

WITNESSES

Information from witnesses is documented on the "Witness Statement" form (906-6; Figure 4.12), including identification, home and work addresses, contact information, and expected testimony.

CASUALTY FIELD NOTES 906-5	AGENCY	FILE NUMBER

DESCRIPTION

NAME	ADDRESS	PHONE NO.

RACE	SEX	AGE	DATE OF BIRTH	HEIGHT	WEIGHT	HAIR	EYES	OTHER

DESCRIBE CLOTHING

TYPE OF INJURY

☐ MINOR   ☐ MODERATE   ☐ SEVERE   ☐ FATAL   DESCRIBE INJURY

CIRCUMSTANCES

WHO FOUND VICTIM? WHERE?

VICTIM'S ACTIVITY JUST PRIOR TO AND AT TIME OF IGNITION

VICTIM'S ACTIVITY AFTER TIME OF IGNITION

CASUALTY TREATMENT

☐ TREATED AT SCENE BY?

SENT TO	VIA	TREATED BY

REMARKS

FATALITY

BODY POSITION

BODY REMOVED TO	BODY REMOVED BY	AUTHORITY TO MOVE BODY GIVEN BY

MEDICAL EXAMINER/CORONER	ADDRESS	PHONE NO.

CAUSE OF DEATH

AUTOPSY BY	ADDRESS	PHONE NO.

DATE OF AUTOPSY	CASE NO.	BLOOD TEST ☐ YES ☐ NO	X-RAYS ☐ YES ☐ NO	REPORTS IN POSSESSION ☐ YES ☐ NO

NEXT OF KIN

NAME	RELATIONSHIP	ADDRESS AND PHONE

NOTIFIED BY (How, Date, and Time)

REMARKS

FIGURE 4.11 ◆ *NFPA 906-5*, "Casualty," form collects casualty victim, injury, circumstances, treatment, fatality, and family information. *Reprinted with permission from* NFPA 906—Guide for Fire Incident Field Notes, 1998 Edition. *Copyright © 1998, National Fire Protection Association, Quincy, MA 02269. This reprinted material is not the complete and official position of the National Fire Protection Association on the referenced subject, which is represented only by the standard in its entirety.*

WITNESS STATEMENT FIELD NOTES 906-6	**AGENCY**	**FILE NUMBER**

IDENTIFICATION

NAME			ADDRESS		PHONE NO.
RACE	SEX	AGE	DATE OF BIRTH	SOC. SECURITY NO.	DRIVER'S LIC. NO.
EMPLOYER			ADDRESS		PHONE NO.
RELATIONSHIP TO INCIDENT			CAN BE CONTACTED AT		
STATEMENT TAKEN BY			LOCATION, DATE, AND TIME OF STATEMENT		

STATEMENT

FIGURE 4.12 ◆ *NFPA 906-6,* "Witness Statement," form collects descriptive identification information and expected witness testimony. *Reprinted with permission from* NFPA 906—Guide for Fire Incident Field Notes, 1998 Edition. *Copyright © 1998, National Fire Protection Association, Quincy, MA 02269. This reprinted material is not the complete and official position of the National Fire Protection Association on the referenced subject, which is represented only by the standard in its entirety.*

In cases involving on-scene witnesses to the fire, thorough interviews may reveal additional information relating to the initial stages of the fire and the environmental conditions at the time (rain, wind, extreme cold, etc.). Furthermore, it is important to document details as to the position of the witnesses in relation to the fire scene when recording their visual observations. This process can be enhanced by walking witnesses through the scene (when safe) or back to the location from which they witnessed the fire. This can help establish or confirm lines of sight and prompt more complete statements.

EVIDENCE COLLECTION AND PRESERVATION

The **chain of custody** is intended to trace the item of evidence from its discovery to court. Its purpose is to authenticate the evidence as it is found as well as to prevent its loss or destruction. The documentation includes photographing the evidence at its discovery and preparing a written list itemizing the transfer of the evidence once it leaves the scene, as on the "Evidence Field Notes" form (906-7; Figure 4.13). Such documentation will demonstrate compliance with guidelines such as ASTM E1188.

PHOTOGRAPHY

Photographs should include not only overall views but close-ups of critical evidence and intermediate "establishing" shots when necessary. Photos should record the scene as found, during debris clearance, after clearance, and with furnishings replaced in their prefire positions. An accurate photo log is nearly as important as the photos themselves to ensure that the investigator can correctly reconstruct the scene days, weeks, or months later when called on.

The "Photograph Field Notes" form (906-8; Figure 4.14) is used to record the description, frame, and roll number of each photograph taken at the fire scene. Investigators should follow a similar systematic documentation scheme for fire scene photographic images taken with digital cameras and videotape recorders. The form is designed to be filled out as the photographs are taken. The frame and roll (or photo) numbers are used later on the fire scene sketch to indicate the location and direction from which they were taken. Each roll of film, archived media containing digital images, and videotape should he documented on a separate form. The "Remarks" field is used to document the disposition of the film, archived media, and videotapes. Close-up photos of evidence should be accompanied by overall or orientation shots (as in Figure 4.15).

SKETCHING

Fire scene sketches, whether or not to scale, are important supplements to photographs. A sketch graphically portrays the fire scene and items of evidence as recorded by the investigator. The "Sketch Field Notes" form (906-9; Figure 4.16) is used by investigators to draw simple two-dimensional sketches of fire scenes. Square-grid paper is available in many sizes at stationery supply stores. These sketches may range from rough exterior building outlines or room contents to detailed floor plans. Sketches of the distribution of window glass fragments are critical to later reconstruction of explosion events. They must include distances and angular direction data.

DOCUMENTARY RECORDS

Insurance information (Form 906-10; Figure 4.17) and **documentary records** (Form 906-11; Figure 4.18) are required to ensure that a thorough investigation is conducted, particularly when there are multiple policyholders, as in the case of commercial

EVIDENCE FIELD NOTES 906-7	AGENCY	FILE NUMBER

DESCRIPTION	WHERE FOUND/WHEN	REMOVED TO/BY
1.		
2.		
3.		
4.		
5.		
6.		
7.		
8.		
9.		
10.		
11.		
12.		

REMARKS

FIGURE 4.13 ◆ *NFPA 906-7,* "Evidence," form collects and itemizes evidence, location found, and chain of custody information. *Reprinted with permission from* NFPA 906—Guide for Fire Incident Field Notes, 1998 Edition. *Copyright © 1998, National Fire Protection Association, Quincy, MA 02269. This reprinted material is not the complete and official position of the National Fire Protection Association on the referenced subject, which is represented only by the standard in its entirety.*

PHOTOGRAPH FIELD NOTES 906-8	ROLL NO.	AGENCY	FILE NUMBER

*ONLY ONE ROLL OF FILM PER FORM.

NEG. NO.	DESCRIPTION	NEG. NO.	DESCRIPTION
1		21	
2		22	
3		23	
4		24	
5		25	
6		26	
7		27	
8		28	
9		29	
10		30	
11		31	
12		32	
13		33	
14		34	
15		35	
16		36	
17		37	
18		38	
19		39	
20		40	

REMARKS

FIGURE 4.14 ◆ *NFPA 906-8*, "Photograph," form collects, itemizes, and records descriptions of photographs. *Reprinted with permission from* NFPA 906—Guide for Fire Incident Field Notes, 1998 Edition. *Copyright © 1998, National Fire Protection Association, Quincy, MA 02269. This reprinted material is not the complete and official position of the National Fire Protection Association on the referenced subject, which is represented only by the standard in its entirety.*

(a)

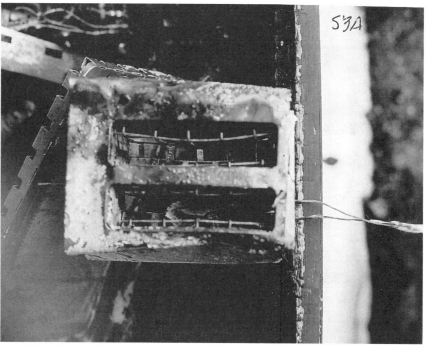

(b)

FIGURE 4.15 ◆ Photographic documentation of critical features such as possible ignition sources should include both (a) overall (orientation) and (b) close-up photos. *Courtesy of J. D. DeHaan.*

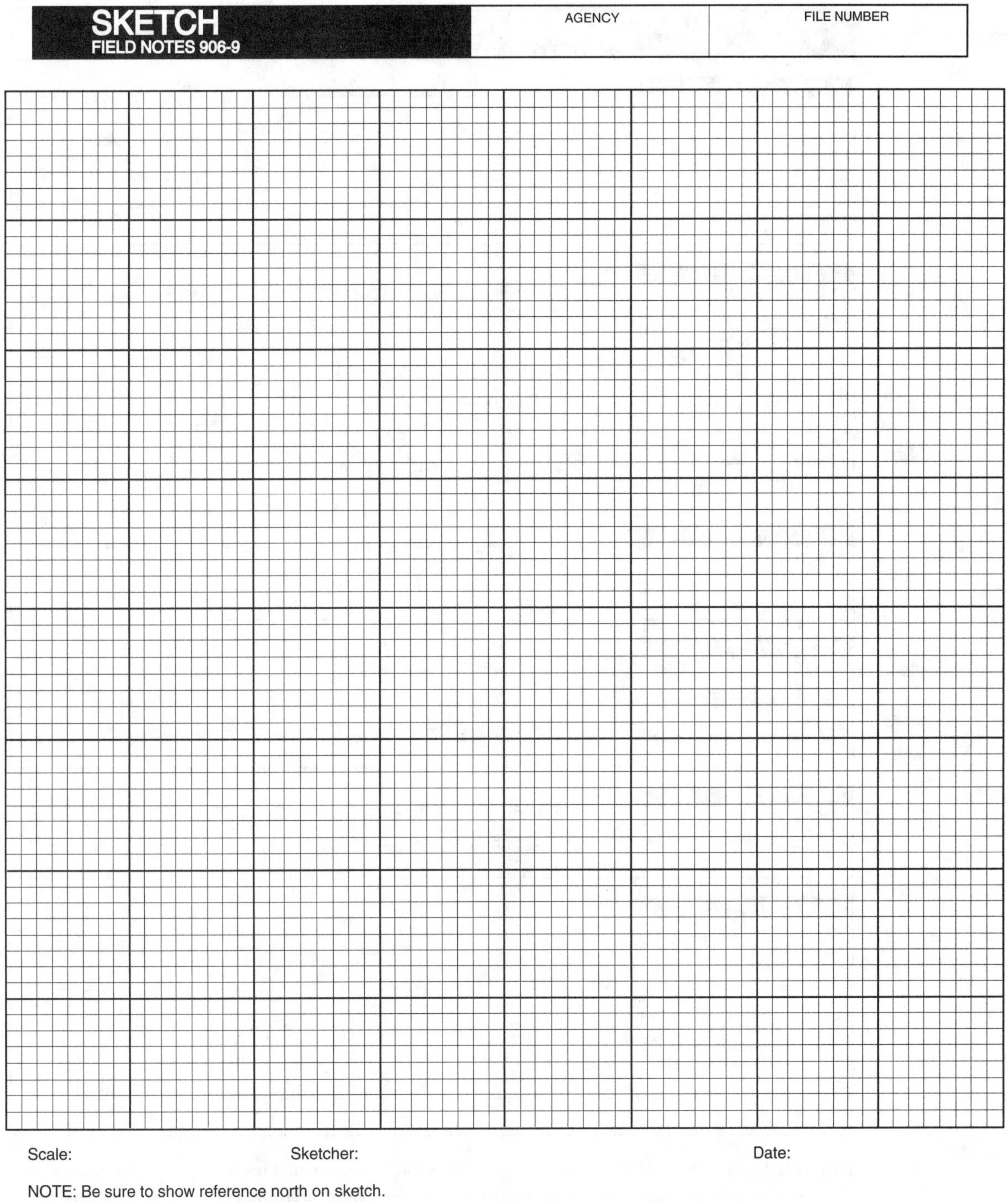

SKETCH FIELD NOTES 906-9	AGENCY	FILE NUMBER

Scale: Sketcher: Date:

NOTE: Be sure to show reference north on sketch.

FIGURE 4.16 ◆ *NFPA 906-9,* "Sketch," form collects on-scene rough sketch of property for use in documenting evidence collected, photographs, and fire pattern damage information. *Reprinted with permission from* NFPA 906— Guide for Fire Incident Field Notes, 1998 Edition. *Copyright © 1998, National Fire Protection Association, Quincy, MA 02269. This reprinted material is not the complete and official position of the National Fire Protection Association on the referenced subject, which is represented only by the standard in its entirety.*

INSURANCE INFORMATION FIELD NOTES 906-10	AGENCY	FILE NUMBER

COMPANY

NAME 1.		ADDRESS		PHONE NO.
POLICY NO.		EFFECTIVE DATE		EXPIRATION DATE
NAME 2.		ADDRESS		PHONE NO.
POLICY NO.		EFFECTIVE DATE		EXPIRATION DATE

COVERAGE

STRUCTURE/VEHICLE	CONTENTS, PERSONAL PROPERTY	BUSINESS INTERRUPTION, LOSS EARNINGS, LIVING EXPENSES
1. ☐ NEW ☐ RENEWAL	NAME OF INSURED	ADDRESS OF INSURED
2. ☐ NEW ☐ RENEWAL	NAME OF INSURED	ADDRESS OF INSURED
PREVIOUS INSURANCE CARRIER NAME	ADDRESS	PHONE NO.

$ _____ STRUCTURE/VEHICLE $ _____ CONTENTS $ _____ OTHER ? _____

PREVIOUS LOSSES, CANCELLATIONS

INSURANCE AGENT

NAME 1.	ADDRESS	PHONE NO.
NAME 2.	ADDRESS	PHONE NO.

ADJUSTER/INVESTIGATOR

NAME OF COMPANY ADJUSTER/INVESTIGATOR 1.	ADDRESS	PHONE NO.
NAME OF COMPANY ADJUSTER/INVESTIGATOR 2.	ADDRESS	PHONE NO.
NAME OF PUBLIC ADJUSTER	ADDRESS	PHONE NO.

TOTAL PAID LOSS

STRUCTURE 1. $	CONTENTS/PERSONAL PROPERTY 1. $	OTHER (Explain) 1. $
STRUCTURE 2. $	CONTENTS/PERSONAL PROPERTY 2. $	OTHER (Explain) 2. $

REMARKS

FIGURE 4.17 ◆ *NFPA 906-10,* "Insurance Information," form collects company, policy coverage, agent, adjuster, investigator, and loss payout information. *Reprinted with permission from NFPA 906—Guide for Fire Incident Field Notes, 1998 Edition. Copyright © 1998, National Fire Protection Association, Quincy, MA 02269. This reprinted material is not the complete and official position of the National Fire Protection Association on the referenced subject, which is represented only by the standard in its entirety.*

RECORDS/DOCUMENTS FIELD NOTES 906-11		AGENCY	FILE NUMBER

Use this form as a checklist to indicate which records have been considered in the investigation. The Remarks sections should be used to note availability, contacts, and so forth.

INCIDENT RELATED

FIRE DEPT. NAME	INCIDENT NO.	REMARKS
POLICE DEPT. NAME	FILE NO.	REMARKS
INSURANCE CO. NAME	CASE NO.	REMARKS
GAS CO. NAME	REMARKS	
ELECTRIC CO. NAME	REMARKS	
MEDIA COVERAGE	REMARKS	
MEDIA COVERAGE	REMARKS	
MEDIA COVERAGE	REMARKS	
OTHER — INCIDENT RELATED	REMARKS	
OTHER — INCIDENT RELATED	REMARKS	

PROPERTY RECORDS

MORTGAGE HOLDER	REMARKS
LIEN HOLDER	REMARKS
TAX RECORDS	REMARKS
CONTRACTS/LEASES	REMARKS
TITLES/REGISTRATIONS	REMARKS
ZONING/CODES	REMARKS
DEEDS	REMARKS
OTHER	REMARKS
OTHER	REMARKS

BUSINESS/PERSONAL

ACCOUNTING	REMARKS
INVENTORY	REMARKS
BANKS/CREDIT UNIONS, ETC.	REMARKS
BUSINESS AND PERSONAL TAX	REMARKS
CRIMINAL HISTORY	REMARKS
CIVIL LITIGATIONS	REMARKS

FIGURE 4.18 ◆ *NFPA 906-11,* "Records/Documents," form collects incident-related property, business, and personal information obtained during the investigation. *Reprinted with permission from NFPA 906—Guide for Fire Incident Field Notes, 1998 Edition. Copyright © 1998, National Fire Protection Association, Quincy, MA 02269. This reprinted material is not the complete and official position of the National Fire Protection Association on the referenced subject, which is represented only by the standard in its entirety.*

buildings. Care should be taken to catalog accurately and to secure incident, property, business, and personal documentary records.

COMPARTMENT FIRE MODELING DATA

The documentation needed for **compartment fire modeling** exceeds the data collected in the *NFPA 906* forms. This additional detail is covered in the fire modeling form shown in Figure 4.19. Included is information on the compartment for the model such as room size, construction, surface materials, openings in doors and windows, HVAC, case timeline, and fuel packages. It is especially important that height information be accurately and fully recorded, not just the floor plan. For example, any places where the ceiling height changes must be clearly identified (slope, deep beams, and architectural features can affect hot gas flow and layer development).

PHYSICAL EVIDENCE

Physical evidence is sometimes called the "silent witness" because it can provide reliable answers to questions other investigative techniques cannot address, fill in details, and corroborate other information. Clearly, the investigative phase of systematic investigations brings together all the essential elements of forensic evidence collected from the fire scene.

Generally accepted forensic guidelines are designed to prevent contamination, loss, or destruction of the evidence and to provide a reliable chain of custody for that evidence. For instance, when there is with dried blood, the object itself should be collected whenever possible and allowed to air-dry before packaging. Bloody objects should be placed into individual sealed paper bags, boxes, or envelopes after drying and kept dry and refrigerated if possible. Plastic bags should not be used, since they will not allow the sample to ventilate. All investigators should be aware of the safety issues involving blood-borne pathogens.

All firearms should be placed into individual manila envelopes, and rifles should be tagged. Firearms should be hand carried to the laboratory. Special handling instructions include the collecting technique, noting the cylinder position in revolvers, and recording the serial number, make, and model of the weapon. Fire debris or containers containing volatile liquids must be sealed in appropriate packaging and kept cool to minimize evaporation. Further details are included in *Kirk's Fire Investigation,* 6th ed. (DeHaan 2007).

◆ 4.6 PANORAMIC PHOTOGRAPHY

PHOTOGRAPHIC STITCHING

A majority of the photographs taken at fire scenes focus on fire patterns and other evidence. In structure fires, an aerial or perspective view of the building can reveal the overall impact of the fire on the structure, including how the building itself was breached by both the fire and efforts to extinguish the blaze. An example of a **panoramic photograph** of a fire scene created by stitching several photographs together is shown in Figure 4.20.

Panoramic cameras, popular in the late 1800s were sometimes used to capture the ravaging effects of fires. Many of the present-day cameras, which simply crop the top and bottom portions of the image, create a false appearance of a panoramic photograph. However, superwide-angle and 360° rotating cameras are now available but are

ROOM FIRE DATA

Room _____   **Room #** _____

Length _____

Width _____ | **Floor Plan** |

Height _____

    Note ceiling height changes _____

Walls:Structure/material _____ Thickness _____ Covering _____ Sample? Y/N

          Structure/material _____ Thickness _____ Covering _____ Sample? Y/N

Ceiling: Structure/material _____ Thickness _____ Covering _____ Sample? Y/N

Floor: Structure/material _____ Thickness _____ Covering _____ Sample? Y/N

Openings (door, window, other vents) into room (number on plan above):

Height (bottom to top of opening)	Sill Height	Soffit Depth (above opening)	Width	Open or Closed? Changes During Fire?
1.				
2.				
3.				
4.				
5.				
6.				
7.				

HVAC System:

    Description _____

    On/off prior to/during fire? _____

Furnishings (descriptions of major fuel items, including floor and wall coverings, draperies):

_____

_____

_____

_____

Time Line: Alarm time _____ FD arr. time _____ Control time _____

Detection: _____

Pre-Fire Events: _____

FIGURE 4.19 ◆ "Room Fire Data" form captures all critical information about each room involved in a fire. *Courtesy of J. D. DeHaan.*

FIGURE 4.20 ◆ Construction of a panoramic view using individual photographs and stitching software. *Courtesy of D. J. Icove.*

costly (Curtin 1999). However, the creation of panoramic images to improve scene review and illustration for court is not new. Panoramic (scanning slit) cameras have been around for over 100 years, but the necessity for heavy, yet delicate, specialized equipment made it rarely used. Simple pasteups of overlapping images are acceptable, but there are always perspective-shift distortions between views that can disorient the viewer. The advent of panoramic view "stitching" as part of various computer photo editing programs has made it possible to create seamless, perspective-corrected panoramic images from a series of simple still photos without specialized cameras. If a series of photos is taken using a normal focal length lens (35–55mm) from a single viewpoint with "overlapping" edges, software easily stitches the images together. If the photos are taken using a leveled rotating-head tripod, with 15–20 photos for a full rotation, distortion is minimized. The Photo Stitch™ program is part of Roxio's PhotoSuite 7 Platinum software that will match features in a series of photos (scanned or digital) to produce a single manipulable panoramic scan (Roxio Corporation, roxio. com, 2004). Photos must be cropped so that all images are exactly the same size and are "manually" overlapped before being merged digitally. The system will produce only end-to-end panoramas. The entire PhotoSuite software requires a 1.2 GHz Pentium III or equivalent computer with 256MB RAM and 1GB of free hard disk space. This system is compatible with Windows XP or Windows 2000. See Figure 4.21 for an example of a Photo Stitch 150° panorama of a large industrial fire scene.

FIGURE 4.21 ◆ Separate sequential photos stitched together using Photo Stitch™ to form a 150° panoramic photo of a large industrial fire scene. *Courtesy of D. J. Icove.*

A more advanced panoramic system is PTGui, based on the successful PT (Panorama Tools) program developed by Bernhard Vogl (ptgui.com, 2004). This system allows the creation of spherical, cylindrical, or flat "interactive" panoramas from any number of source images (as opposed to the flat panoramas from Photo Stitch). This software supports JPEG, TIFF, PNG, and BMP source images. The computer mouse can be used to move the images to change yaw, roll, and pitch with corrected perspective in real time. In a typical application, 16 photos taken of a large complex fire scene using a handheld manual 35mm camera were scanned using Windows XP Picture and Scanner Wizard and then opened into Microsoft Picture It! Photo (v. 6). The files were then opened in PTGui (v. 3.5) and the 360° single row panorama editor was selected. Nine control points were selected on each adjoining photo edge. The created panorama was saved in Windows Picture and Fax Viewer. New HP digital cameras can capture panorama–ready images and produce them immediately.

SCANNING CAMERAS

Another approach to panoramic or three-dimensional photography is the iPIX system (iPIX©, iPIX.com). Its 180° fish-eye lens and tripod mount are compatible with many film or digital 35mm cameras. The camera is mounted on a tripod and a photo is taken, then the camera is rotated 180° to take a second picture in the opposite direction. Fish-eye lenses, of course, produce severely distorted images when viewed directly, but these images are imported into the iPIX software, which corrects the distortion and seamlessly stitches the two images together. The result is a fully immersive, navigable computer image. The iPIX software then permits interactive examination or illustration of fire patterns or explosion damage around the entire periphery of a room. This process requires that the special equipment be available at the scene. It has been used successfully by a number of U.S. police agencies. An equivalent video system called iMOVE uses six video cameras to create an interactive moving virtual reality image. The iMOVE system may provide a navigable computer (virtual reality), three-dimensional tour of large fire scenes but has not been evaluated (iMOVE, Inc., 26 N.W. Second Ave., Portland, OR 97209).

Panoscan of Van Nuys, California, offers a scanning digital camera that can complete a 360° view of a room in an 8-s scan with high resolution and great dynamic range. Its images are readily viewable as a flat panorama or as a virtual reality "movie" on QuickTime VR, iPix, JAVA, and other imaging software. It also offers a photogrammetric capacity using two scans at different heights (Figure 4.22a) with specialized software (PanoMatrix™), allowing accurate measurements of any scene, indoors or out. (See Figure 4.22b for an example.) The system is fully portable. Its data are saved in DXF format for use in AutoCAD, Maya, and other CAD programs. Measurements (with an accuracy of fractions of an inch over a 25-ft radius) can be made or added to the file at any time (panoscan.com). Some advanced digital image systems allow the user to interactively browse around the scene and can even link to views in adjacent rooms. Full 360° immersive images allow viewing of floors, ceilings, and walls in any direction. Images can then be linked together or linked to traditional photographs and renderings and to audio or other file types (iPIX 2003, Panoscan, Spheron VT).

Panoramic photographs can be used to establish **viewpoint photographs.** This type of photography can be used to document what a witness might have seen or not seen from a particular location. The peripheral vision or field of view of the normal human eye is not readily duplicated by a normal camera lens, so panoramic imaging is sometimes the only way to replicate the view a witness may have had.

(a)

(b)

FIGURE 4.22 ◆ (a) Digital scanning cameras can capture panoramic images of entire rooms or exterior scenes in seconds. *Courtesy of Panoscan Inc.* (b) Panoramic digital photo of wildland fire scene. *Courtesy of Panoscan Inc.*

At its simplest constructing a panoramic photograph consists of using a steady tripod, taking a series of overlapping photographs, and then physically overlaying prints to form a mosaic view of the fire scene. Overlaying several photographic shots in the form of a mosaic can compensate for the lack of a true panoramic photograph. For best results a lens with 35- to 55-mm focal length is used, since wide-angle and telephoto lenses introduce unwanted distortions. Once an investigator has become familiar with the results and limitations of the camera, he or she should work on developing an expertise with panoramic photography.

EXAMPLE 4.1   The School Fire

School fires with massive structural destruction are difficult to reconstruct. A fire scene reconstruction was requested of a large school fire that occurred some years ago in the Southeast. A majority of the contents and walls had been removed and placed outside the structure prior to examination of the scene.

FIGURE 4.23 ◆ The panoramic photo overlay from Example 4.1 with graphic chart tape on a plastic overlay that shows missing walls and area of fire origin. Floor plans on right show (bottom to top) the successive fire spread as seen through exterior windows. *Courtesy of D. J. Icove.*

The fire scene was reconstructed using a combination of eyewitness statements, raw news videos, scene analysis, and knowledge of plume geometry. Investigation revealed that a fire started on the school office floor adjacent to filing cabinets where student records were stored.

A methodical examination of the debris, the dynamic time sequence of the fire's movement, and the damage caused by the fire confirmed the office to be the room of fire origin. A careful fire scene analysis of pattern damage revealed a roughly elliptical penetration at the floor of the office where the fire originated. At the center of this elliptical pattern was the apparent point of fire origin.

Yellow chalk was used to sketch on the floor the location of the desk, cabinets, and walls. These chalk lines were photographed from the second floor and displayed as a mosaic overlay forming a panoramic view of the fire scene. After the panoramic photos were pieced together, the walls were marked using graphic chart tape on a plastic overlay. Plastic overlays are important for courtroom presentation, since they can be removed if objections are raised to their use. A photograph of this exhibit with the overlay present is shown to the left of the schematic floor plans in Figure 4.23. Such "ghost" walls can today be inserted using the graphics packages described previously. Such displays can also be helpful in confirming witness observations.

One of the most persuasive visual exhibits offering a perspective view is a scale model for courtroom presentations. The overall dimensions of the building being re-created must be carefully defined and accurately captured. Architectural firms often have technicians on staff who can prepare these models. A removable roof on the model should be included to provide access to the building's interior layout.

Specialized photographic techniques can involve the use of forensic (alternate) light sources and suitable filters, ultraviolet or infrared films or imaging systems, and close-up (macro) photography or other techniques to preserve physical evidence such as fingerprints. Some of these will be discussed in later sections of this chapter.

◆ 4.7 APPLICATION OF CRIMINALISTICS AT FIRE SCENES

The science of criminalistics becomes especially important when the scene is complex or serious (deaths or injuries) and every avenue of information seeking must be explored. Evidence may include physical items such as debris or incendiary devices, witness interviews, casualty injuries or postmortems, and fire scene photographs and sketches. *Criminalistics* is the science of examining all these items of physical evidence to link them to a common origin, identify them, or use them in the reconstruction of events.

CRIMINALISTICS

Criminalistics has been defined as the application of the methods and knowledge of the natural sciences (physics, chemistry, biology, botany, etc.) to legal inquiries. Although not commonly in use in the United States until the 1950s, the term derives from the German word *Kriminalistik* (from the 1880s) for forensic evidence analytical techniques. Criminalistics involves the extensive use of physical science to analyze and identify materials, but more important, to compare, classify, and individualize items of physical evidence, often with the intent of establishing a common origin (or excluding one). This physical evidence can take any form—impressions from shoes, tools, or friction ridge skin (on fingers, palms or feet), tissue, blood or other physiological fluids, glass, paint, soil, grease, oil, dyes, inks, documents, physical matches, or residues of chemicals associated with fires or explosions. Criminalistics often goes further than simple comparison or identification to include the analysis of human behavior and physical dynamics of force, impact, and transfer, and the reconstruction of scenes in both physical and dynamic aspects.

There is a set of generally accepted forensic guidelines for collecting and preserving physical evidence recovered from fires during a fire scene investigation (DeHaan 2007). The application of these best-accepted practices is derived from procedures used by professional criminalists.

As a simple example, an investigator arrives at a crime scene to find a broken window, an unlatched door, and blood smears—all visible evidence and easily documented (Figure 4.24). The reconstruction of this evidence would be that the door was originally locked and its window broken by mechanical force (determined by glass fracture patterns to have come from the outside), and the intruder cut himself/herself on the glass and, while fumbling for the lock and latch, left blood smears behind on the door, as shown in Figure 4.24.

Analysis of the blood (or any fingerprints left on the door, glass, or latch) could identify the individual present. Glass found on the clothing could be compared with window

FIGURE 4.24 ◆ The reconstruction shows that the intruder cut himself/herself on the glass and, while fumbling for the lock and latch, left blood smears behind on the door. Glass fracture patterns document that the entry was made from the outside. *Courtesy of Lamont "Monty" McGill, McGill Consulting, Gardnerville, NV.*

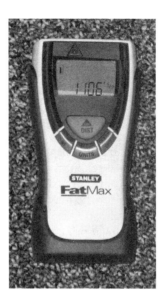

FIGURE 4.25 ◆ New devices such as this FatMax laser measurement tool make taking room measurements easier and faster. *Courtesy of J. D. DeHaan.*

glass from the door to confirm a two-way transfer (suspect blood to scene, scene glass to suspect), and the distribution of cuts or abrasions (and glass/paint fragments) would confirm the method of entry. The services offered by many public and private criminalistics laboratories can be of great help to the fire investigator in reconstructing scenes and events and testing hypotheses about what happened, where, when, and in what sequence. They can also link a person with a scene or a victim, and a scene with a person or a vehicle.

LASER MEASUREMENT TECHNIQUES

Investigators often find themselves documenting the measurements of several rooms within a structure, a cumbersome, time-consuming task that can be inaccurate. Since measuring tapes are most efficient when used by two persons, handheld devices are beneficial when a single person is responsible for measuring distances.

With the introduction of low-cost accurate handheld laser devices, an investigator can now measure not only the distance but also the area and volume of a room. Published specifications for devices sold at less than $100 report an accuracy of ±6.35 mm (1/4 in.) at 30.48 m (100 ft). The typical range of these devices is 0.6–30 m (2–100 ft), can be switched to read English or metric, and operate on low-cost 9-V batteries (see Figure 4.25) (Stanley 2005; DeHaan 2004).

GRIDDING

Methodical techniques for assisting in forensic evidence collection and documentation have been widely used by archaeologists when processing field sites. Fire investigators can adopt similar sketching and photographic techniques (Bailey 1983). There are several accepted methods for **gridding** and documenting scenes (DeHaan 2007, chap. 7). Measurements of the location of recovered physical evidence must

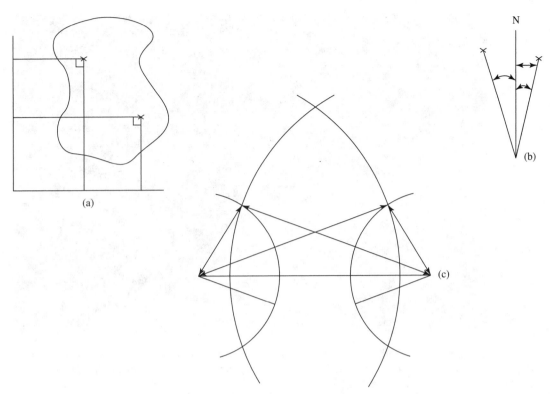

FIGURE 4.26 ◆ There are several measurement methods that are useful for indoor (where walls can serve as baselines) or outdoor scenes. (a) Right-angle transect: baselines at right angles (established with GPS and compass outdoors). (b) Azimuth/baseline: single baseline (in north/south or east/west orientation). Locations are noted by angle and distance from a reference point. (c) Intersecting arcs from two fixed reference points.

be reliable and accurate enough to allow the reconstruction of distances and relationships. Several convenient methods for recording measurements are shown in Figure 4.26.

Coordinate systems aligned perpendicular to major structural walls can aid in the placement of grid lines, especially when the zero point is a corner, usually on the bottom or top left-hand side. Triangulation measures distances to an item of interest within a room from two fixed points. Angular displacement uses one fixed point and a compass direction and is the most appropriate for large outdoor scenes (Wilkinson 2001). A handheld GPS device can be used to establish the location of the fixed reference point as well as compass directions.

A **grid system** is well suited for large-scale scenes such as explosions where shrapnel and other evidence are hurled away from a central location. A grid system can also be used in small scenes, for example, quadrants within a vehicle. Squares within the grid typically use the alphabet along one axis and numerals along the other. Using both alpha and numeric coordinate identifiers simplifies and reduces the possibility of inadvertently switching the coordinates. Figure 4.27 is an example of a grid system at a fire scene. A baseline and distance system can also be used for large outdoor scenes, particularly explosions, where there are no "natural" rectilinear baselines.

FIGURE 4.27 ◆ Methodical techniques using gridding assist in forensic evidence collection and documentation. *Courtesy of Lamont "Monty" McGill, McGill Consulting, Gardnerville, NV.*

Although most scenes can be layered and examined room by room, some scenes require more carefully controlled examination. This is particularly true in fire death scenes where the location of small items is very important, particularly in the vicinity of the body. Using techniques developed in archaeology, the investigator divides the scene into grid squares using the walls as reference base lines. Rope, string, or even chalk can be used to mark out grids, numbered in one direction, lettered in the other, as shown in Figure 4.27. These grids may be 0.5–0.8 m (2–3 ft) square in critical areas and as large as 3×3 m (10×10 ft) in surrounding areas with lighter debris concentrations.

Debris is removed from each grid square, layer by layer, for manual/visual search and sieving. Evidence (or unknown materials) recovered from each grid is kept in a bag, can, or envelope designated by corresponding number/letter. This ensures that at a later stage the evidence can be placed back to within 0.30 m (1 ft) of its original location. Although time-consuming and labor-intensive, this method is the best way of finding and documenting evidence that permits its physical reconstruction after the scene has been completely searched.

Archaeologists use a systematic approach of coordinate systems that account for not only the location but also the depth of where an object is found. Their excavations are carried out layer by layer, so the depth of each is known. This approach can be helpful when layering debris at fire scenes that has fallen onto an area of fire origin. Such cases include multistory buildings, where collapsed floors bury important evidence. In some cases fallen debris often preserves evidence on lower floors.

As previously noted, it is important to place these layers spatially in the documentation process. If forensic evidence collection and documentation are viewed as a scientific endeavor, the use of archaeological techniques will only enhance this effort. The standard of care in the archaeological community is the use of the **Harris Matrix** to show the temporal sequence of layers excavated at a site. The technique is named after Dr. Edward Cecil Harris, who invented it in 1973 (Harris 1975).

Iso-damage curves have been used by archaeologists in documenting the impact of fires on historic monuments. One study placed more intense damage within the interior of a section of the Parthenon due to a fire set by the Celts in 267 (Tassios 2002).

DOCUMENTATION OF WALLS AND CEILINGS

Although most investigators are diligent about identifying floor coverings and assessing their potential contributions to fire spread, the same cannot be said of wall and ceiling materials and coverings. As can be seen from the compartment fire modeling survey form shown in Figure 4.19, the nature and thickness of walls, ceilings, and their coverings are important for accurate reconstruction of the event.

In some cases, noncombustible walls of concrete, masonry, or stucco will not contribute to the fire spread, but their low thermal conductivity and large thermal mass may affect some stages of the fire's development. Plaster (whether with metal or wood lath) will dehydrate and fail, allowing fire to penetrate into ceilings or wall cavities. Modern gypsum board will resist fire spread for some time if properly installed (typically 15–20 min of direct fire exposure) before it collapses. X- or fire-rated gypsum wallboard is 16 mm (⅝ in.) or thicker and contains fiberglass fibers to strengthen the gypsum. As a result, it will withstand 30 min or longer of direct fire contact.

Walls and ceilings can also be made of solid wood, plywood paneling, fiberboard (high density like Masonite or low density like Celotex), particleboard, OSB (oriented-strand board), or even metal. Each makes its own contribution to fire spread. Noncombustible walls can be covered with combustible materials. Thin plywood paneling or low-density cellulose can contribute enormously to fire spread. Karlsson and Quintiere (1999) estimated that lining a room or covering a ceiling with low-density fiberboard would cut the development-to-flashover time in half compared with that for the same room with noncombustible walls or ceilings. These materials will almost guarantee a flashover fire (if adequately ventilated) and can be destroyed so completely as to make their detection difficult.

Fire resistance ratings (1 hr, 2 hr, etc.) are based on laboratory test exposure to a furnace fire whose growth is programmed on a standard time-temperature curve to make the test reproducible, ASTM E 119 (see Figure 4.28). Such tests do not necessarily replicate real-world compartment fires and are not intended to predict failure times in such fires. However, most fires are less severe (on the average, not at a specific time) than those produced on the standard time-temperature curve. This fire testing results can often be used as a "bound" to the actual expected performance.

The investigator must be familiar with these materials and how they are installed. Irregular squiggles or zigzag lines of charred adhesive on cement or plasterboard walls are a sure sign that paneling was glued there at one time. Remnants of the paneling will usually survive at the toe plate or behind baseboards, plumbing, or electrical fixtures. The small nails used to secure paneling or cellulose tiles are very different from those used to secure gypsum board.

Wood or metal furring strips may be used to install tiles on walls or ceilings, and their presence should be taken as a cue to search out and identify the sometimes fragmentary remains of combustible walls or ceilings. Wall or ceiling tiles are sometimes installed only with dabs of cement on the back side, so patterns of large dark dots on remaining surfaces should be carefully examined. Samples of unburned wall covering should be measured, identified, and retained for later confirmation.

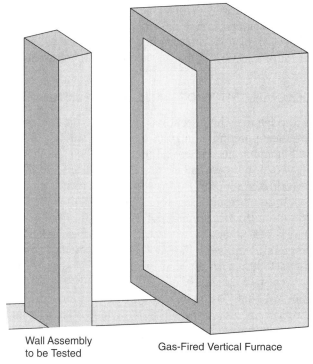

Wall Assembly
to be Tested

Gas-Fired Vertical Furnace

FIGURE 4.28 ◆ The ASTM E 119 test for wall or floor assemblies uses a large vertical or horizontal gas-fired furnace. The test assembly is built against the open face of the furnace, which is operated according to a set time-temperature curve. *Courtesy of NIST.*

In one case of the author's experience, postfire scene photos showed only bare studs, with gypsum board on only one (exterior) face of a common wall. Careful searching revealed that the wall facing the room of origin was covered only in thin plywood veneer paneling that had burned nearly completely and contributed to an extremely intense and fast-growing fire (*Commonwealth of Pennsylvania v. Paul S. Camiolo* 1999).

Although not common in modern structures, the presence of combustible ceilings should not be dismissed. Low-density cellulose ceiling tiles were widely used for many years prior to about 1960. They can be 0.3 × 0.3-m (12 × 12-in.) tiles or 0.61 × 1.22-m (2 × 4-ft) panels. Decorative-edged tongue-and-groove pine or fir boards (sometimes called *beadboard*) were widely used in the late nineteenth to early twentieth centuries for walls and ceilings. Polystyrene ceiling tiles have been widely installed during remodeling of older structures. Such linings will add dramatically to the fuel load and increase the rate of fire development in the room (cutting time to flashover by as much as half, per Karlsson and Quintiere 1999).

The nature and thickness of ceiling materials must be noted. Samples are strongly recommended for later testing. Suspended ceilings are very common in commercial structures, as they can lower ceilings and conceal electrical, plumbing, and HVAC services. These ceilings, although usually noncombustible, allow fire gases to penetrate the large plenum area they offer and spread throughout the concealed space. They typically fail as the steel wires or lightweight steel grid members reach their annealing temperatures and lose their tensile strength. (This can occur in as little as 10 min once the flames reach the ceiling.)

The interior dimensions of all rooms involved in the fire (by flame or smoke penetration) must be recorded (preferably to the nearest ±50 mm (±2 in.). This includes the heights of various rooms (not assuming all rooms in a structure to be the same). Height is often overlooked as a measurement, but it is critical when evaluating the size of fire required for flashover or effects of a given initial fire size, estimating smoke filling rates, visibility, tenability, and the like. Unusual features such as skylights should be measured and photographed.

The nature of the ceiling itself will affect the spread of smoke and fire. A flat (level), smooth ceiling will allow the smoke and hot gases from the ceiling jet to spread uniformly in all directions; a pitched (slanted) ceiling, of course, will direct the majority of the spread upward. Headers above doors, exposed structural beams, even decorative add-on beams, and ceiling-mounted ductwork will limit spread dramatically, in some cases allowing the buildup of a sufficient hot gas layer in the side of the room containing the initial fire to the point of transition to flashover (as seen in Figure 4.8). Open ceiling joists (common in unfinished basements) will direct most of the accumulating hot gases along their length while dramatically reducing (if not preventing) transverse spread into adjacent joist spaces. Such features must be measured and documented.

The action of fire and subsequent extinguishment and overhaul activity may have obliterated signs of wall and ceiling covering, so efforts should be made to find prefire photos or videos that show the character of the ceiling and walls. Failing that, interviews should be taken of owners, occupants, guests, customers, visitors, maintenance personnel, and the like, asking them to describe ceiling and wall finish (as well as type and placement of furniture). Examination and documentation of undamaged areas of the building may reveal the original structural features.

LAYERING

At scenes where there has been significant destruction and collapse of furniture, walls, or ceilings, investigators would be well advised to process at least the most critical portions in **layers.** Most evidence of the critical early stages of a fire will be beneath the sometimes overwhelming overburden of collapsed ceiling and roof structure.

Undamaged areas of the building may be surveyed to establish what sorts of materials to expect. The roof structure can be photographed and its direction of collapse noted and then removed. Roof or ceiling insulation can then be removed noting whether it is loose (blown-in) fiberglass, mineral wool or cellulose, or batts or roll insulation. Samples should be taken for later identification if ignition or spread through the insulation is considered possible.

The ceiling material can then be identified: gypsum wallboard, lath-and-plaster (wire or wood lath makes a difference in manner and time of collapse), ceiling tile, plywood, or wood plank. An investigator should never assume that all insulation, ceiling, or lining materials are consistent throughout even a small residence, since repairs, renovations, or additions will usually be made with materials available at the time.

Light fixtures, furnishings, and victims will be found under the ceiling material. Treatment of fatal fire scenes will be covered in a later chapter. Most of the critical evidence will be found between the ceiling and the floor covering, but even here, the sequence of position may be of importance.

Window glass will tend to break and collapse when the temperatures or heat fluxes become high enough during the fire's development. This usually occurs after the smoke products have condensed on the glass, and the fracture pattern would be expected to be thermal rather than mechanical (unless the glass is toughened safety glass) (DeHaan 2007, chap. 7). Glass falling inward may well fall on furnishings or floors that are already fire damaged. If a window is broken before the fire reaches it, the fracture pattern will be "mechanical" in appearance, and the glass, bearing few or no soot deposits, may fall and protect unburned materials beneath. (Keep in mind that radiant heat in a subsequent postflashover may melt and char even protected material beneath the glass.)

The location of first collapse of ceiling or wall covering may be indicated by careful examination of the debris at this stage, which should be photographed before further excavation. As the furnishings are documented and removed, the nature of floors and floor coverings can be observed. The distribution of carpet, tile, bare wood, composite floor coverings, and vinyl or asbestos (tile or sheet) should be noted. Comparison samples of carpet, floor covering, and underlayment should be taken in areas of suspected origin.

Testing has shown that carpet fiber content cannot accurately be estimated from observation alone, and even using a match or lighter flame will reveal only whether the face yarn is synthetic or natural fiber. Fire tests have further demonstrated that a carpet may not support flame spread alone, but if provided with a particular type of pad, it may burn readily. It is always best to recover and preserve a small (at least 0.15 × 0.15-m [6 × 6-in.]) sample of unburned carpet and pad for later identification. Such samples can be recovered in most scenes from under large furniture or appliances or in protected corners of the room. Such samples can also help forensic testing by being used as a comparison to establish what volatiles they contain or may yield on burning.

SIEVING

In the ashes, all evidence seems to be the same shades of white, gray, or black and not readily distinguishable from fire debris. A manual and visual search may overlook evidence critical to a complete reconstruction. Wet sieving of the debris affords a better chance of detecting small items such as glass fragments, projectiles, keys, or jewelry that may go unnoticed in dry ash. Note that bone fragments that are highly calcined by fire exposure can turn to mush if wetted, so likely areas around hands and feet of burned bodies should be manually searched and sieved dry.

Sieving is most effective when done with at least three sieve frames stacked together using 25, 12.7, and 6.35 mm (1, 0.5, and 0.25-in.) mesh (some examiners will use a fourth sieve with window screen if searching for tiny tooth or bone fragments). Debris is placed in the correct grid and agitated, or a garden hose or hose reel line is used to wash debris through the sieves. The debris should never be pushed through by hand, since that can shatter bones and teeth. Figure 4.29 is an example of a sieve.

Wet sieving for general evidence is preferred, since the water washes off the gray ash and makes small objects visible by color or reflectance. If the objects recovered can be visually identified, they can be placed directly into evidence bags. Items that cannot be readily identified should be kept in a separate container, labeled as to grid of recovery, until they can be properly analyzed.

PRESERVATION

Fire exposure causes materials to become fragile and brittle. This is especially true of copper wiring and electrical insulation and components. Any recovery starts with thorough photographic documentation before the attempt is made to move the object.

Shrink wrap or cling wrap (Saran™ wrap) is very useful for wrapping large items to keep loose pieces together. Wiring can be preserved on 1 × 4 lumber 4–8 ft long, affixed with cable ties or wraps with the ends numbered and distances from a reference point labeled on the wood. Fragile wiring or components can be protected with cling

FIGURE 4.29 ◆ Sieving assists the investigator in collecting evidence that is not readily distinguishable from fire debris. Searching by hand and visual examination only may cause evidence critical to a complete reconstruction to be overlooked. *Courtesy of Lamont "Monty" McGill, McGill Consulting, Gardnerville, NV.*

wrap. Smaller lengths of wiring or wiring from an appliance can be tie wrapped to a piece of corrugated cardboard. Colored or numbered tape can be used around ends of wires to denote the circuit to which they were connected.

When a scene is gridded off for searching, numbered tarps (plastic) can be used for materials recovered from each grid. If the scene is not gridded off, a separate tarp can be used for each room or each sector of a scene. Plastic tarps can be marked off with waterproof tape to produce a full-scale reproduction of the room. This technique has be used to reconstruct positions of furniture and other evidence, and even for court display (Rich 2007).

IMPRESSION EVIDENCE

Impression evidence is a general term for the transfer of pattern or contour information from one surface to another. This transfer may take the form of one harder material deforming a softer material on contact. Examples are a shoe impression in soft clay outside a window or a striated rough cut on a piece of wood made by the ragged edge of a hatchet blade. This transfer can also occur when a transfer medium preserves the shape of contoured surfaces that come into contact with it. For example, a dusty shoe can leave an identifiable print on a clean surface and a clean shoe can remove dust from a dirty surface and leave the same information. Photography is the primary means of documenting most impressions, but care must be taken not to distort

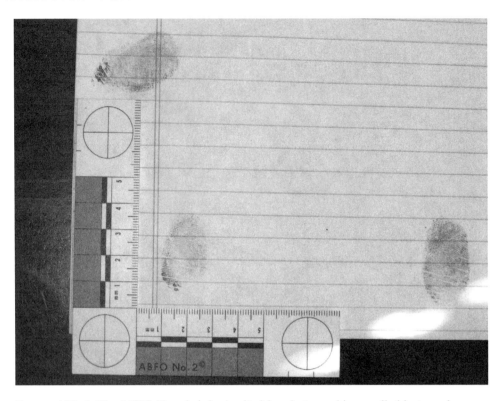

FIGURE 4.30 ◆ The ABFO #2 scale is best suited for photographing small objects, such as fingerprints, toolmarks, or matchbooks. *Courtesy of J. D. DeHaan.*

the image in the photograph. The L-shaped American Board of Forensic Odontology (ABFO) No. 2 scale, measuring 10×10 cm, is the appropriate scale for recording fingerprints, wounds, blood spatters, and similar small evidence, as in Figure 4.30. A larger scale, 15×30 cm as shown in Figure 4.31, is ideal for shoeprints, since it helps prevent distortion (LeMay 2002).

Fingerprints and shoeprints are commonly found forms of impression evidence. When impression evidence is found, it should be documented in place by photograph and note, and then, whenever possible, the object bearing the impression should be recovered and submitted to the lab. The object is always preferable to a photo or cast, but if it is not removable, photos or casting or lifting by adhesive or electrostatic lifter will be necessary.

Items bearing tool marks should also be collected and individually wrapped in packages to protect their surfaces. Similar collection guidelines apply to the suspected tools. Wrap each tool separately to prevent shifting or damage during submission to the laboratory. Place each tool in a separate envelope or box with a folded sheet of paper over the end of the tool to minimize damage and loss of trace evidence adhering to the surfaces and to prevent rusting.

Fingerprints. A finger touching a clean surface like glass or metal can leave a reproduction of its friction ridge contours on the surface by the transfer of the nearly invisible oils, fats, and sweat secretions normally found on the skin. Such patterns are often called **latent** because they are not readily visible to the unaided eye and require

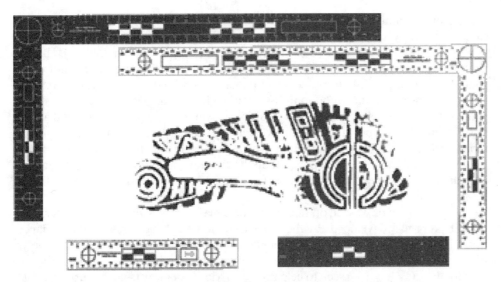

Figure 4.31 ◆ An example of a forensic photographic scale alongside impression evidence for a shoe by the transfer of pattern or contour information from one surface to another. *Courtesy of Armor Forensics.*

some sort of physical, chemical, or optical treatment to make them visible and recordable (and comparable to "inked" record prints).

The skin can also be contaminated with blood, food, grease, or paint that leaves behind a visible or **patent** impression. Even though the skin is pliable and deforms on contact with most other materials, it can deform soft materials such as cheese, chocolate, solvent- or heat-softened plastic or paint, and window putty and leave a three-dimensional or molded "plastic" impression of its ridge detail. The same considerations apply to other deformable materials like rubber shoe soles, gloves, tires, or cloth; each may deform softer materials or leave a pattern on harder ones through transfer of some intermediate medium or even residues of itself.

Impressions of friction ridge features from fingers, palms, and feet can and do survive fires. The multitude of chemical, physical, and optical techniques available today to enhance fingerprints (both latent and patent) has made it possible to recover prints from difficult, textured, or contaminated surfaces. Heat alone can cause the constituents in skin oils to darken or even react with the surface beneath, producing a patent impression from a latent one.

Fingerprints are unique to an individual, they are permanent, everyone has them, and chances for transfer or contact are good (with something at the scene, with an incendiary device, or with a container used to transport or distribute ignitable liquids). In addition, there are vast, long-term repositories of reference prints (which include noncriminals as well as criminals), and the advent of powerful and fast computers has made it possible to scan millions of record prints in a short time to establish a short list of candidate matches. These databases are designed to run single prints, and even partial prints, and some systems can now process palm prints. If fire exposure has not been so severe as to melt or severely char the base material, it is worth considering examination for fingerprints. Handling must be very careful and kept to an absolute minimum. Transport to the lab must be done carefully (preferably by hand) using containers or devices that minimize contact with potential print-bearing surfaces.

Water contamination from condensation, hose stream, or environmental exposure once meant that prints on paper or cardboard would be impossible to develop (since the amino acids detected by ninhydrin processing are water soluble). The advent of physical developer (which reacts with the fatty constituents that are not water soluble) made it possible to process wet paper or cardboard. Small particle reagent (SPR) allows processing on nonporous surfaces like metal, glass, and fiberglass even while still wet. Forensic light sources (multiple wavelength, high intensity) and a variety of chemicals and powders make it possible to recover prints from textured or contaminated surfaces. Fingerprint experts such as Jack Deans have documented how modern optical, chemical, and physical methods are capable of developing latents on all manner of surfaces after fire exposure (Bleay, Bradshaw, and Moore 2006; Deans 2006; DeHaan 2007, chap. 14).

Soot can often be washed off smooth metal or glass surfaces with running water, leaving behind prints "developed" by the soot carbon to be photographed or lifted. Fingerprints in blood can be enhanced with amido black or leucocrystal violet (LCV) sprays, which react chemically with blood to form a dark blue-purple-colored product. The solvent for the LCV has been seen to help rinse overlying soot away from the surface. In one case where two fires had been set in a house, the LCV solution washed soot away, revealing blood spatters on walls where a resident had been bludgeoned to death some 2 years previously (the fires having been subsequently set to simulate drug gang activity).

Charred or burned paper documents are important to a fire scene examination, particularly when business records and important documents are involved. Due to their fragile condition, place fire-damaged documents on soft cotton sheets in rigid containers and hand carry them to the laboratory for examination. Do not treat the papers with any lacquer or coating if they are to be processed for identification or comparison.

Shoe prints. Often compromised by the foot traffic of emergency personnel, tires of vehicles, water, or structural changes, shoe prints can, however, survive fires if they are left on surfaces that have not been destroyed by fire or heat. Shoe prints left behind on doors that are forcibly kicked in may actually be enhanced by the action of the fire. In the example case shown in Figure 4.32, the dusty shoe print was lightened by heat exposure, and the wood around it scorched, increasing its contrast and readability.

Shoe prints are second only to fingerprints in their potential value for linking a person with a scene in that shoes are often (although not always) individualized by the accidental features they acquire through use and damage, they tend not to be changed or discarded for prolonged periods of time, and shoe prints are likely to be left behind at many scenes, since entry (if not approach) to a building is nearly always accomplished on foot. Paper or cardboard can bear latent shoe impressions (developable by physical developer chemical treatment even after water exposure) or patent impressions in soil, dust, or blood.

In addition to identification, shoe prints can offer information about where someone entered or left a room or building, where he/she went, and in what sequence. If two or more people were present, shoe prints have been used to establish which of the perpetrators went where in the building.

TRACE EVIDENCE FOUND ON CLOTHING AND SHOES

Trace evidence such as soil, glass, paint chips, metal fragments, chemicals, and hairs and fibers may cling to the shoes or clothing of a suspect. Liquids may also soak into the clothing or remain on the bottom of the individual's shoes. Glass, plasterboard, sawdust,

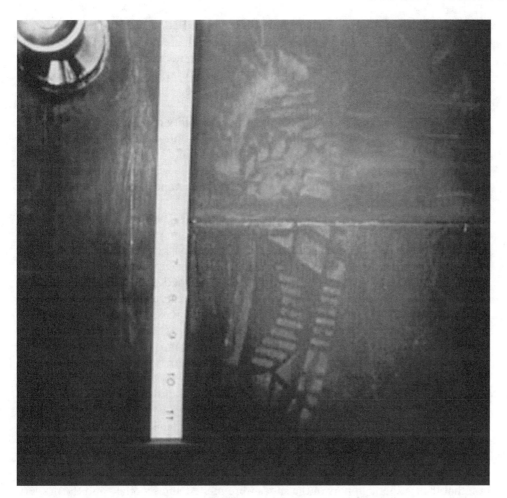

FIGURE 4.32 ◆ Shoeprints left behind on doors that are forcibly "kicked in" may actually be enhanced by the action of the fire. In this case the dusty shoeprint was lightened by heat exposure and the wood around it scorched, increasing its contrast and readability. *Photo by Joe Konebal, by permission.*

and metal shavings have all been found embedded in shoes to offer a link with a scene (as long as comparison samples are recovered). Footwear and clothing can also absorb residues of flammable liquid or chemical accelerants used in arson attacks.

A New Zealand study illustrated the necessity of analyzing clothing and shoes of suspected arsonists for the presence of ignitable liquids. The study found that detectable amounts of petrol were transferred to the clothing and shoes of a person during the action of pouring it around the room, and it addressed the pouring heights and floor surfaces. Results of the study showed that petrol was always transferred to the shoes and often transferred to both the upper and the lower clothing but that seizure and proper packaging must take place quickly (Coulson and Morgan-Smith 2000). In tests using a carbon strip extraction method, it was found that 10 mL (0.34 oz or 2 tsp) of gasoline on clothing worn continuously evaporated much more quickly than on the same clothing left on the lab bench at 20–25°C (68–77°F) and was barely detectable after only 4 hr of wear (Morgan-Smith 2000). Clothing must be recovered and packaged soon after exposure.

The investigator must be aware that the solvents used in glues holding footwear together (particularly athletic shoes) can interfere with identification of ignitable liquid residues, as they can produce false positives (some manufacturers have been known to use gasoline as a substitute glue solvent). Preservation of both shoes separately from other evidence and from each other in vapor-tight containers is essential (Lentini, Dolan, and Cherry 2000).

An Australian transfer study examined the possibility of a natural occurrence of petrol residues tracked inside a vehicle on the shoes of drivers or passengers. The results showed that small quantities of petrol (500 µL) could be detected after 24 hr but evaporation prevented it from being discovered after 1 week. The researchers concluded that it would be significant if fresh or slightly evaporated petrol is found on the car carpet (Cavanagh-Steer et al. 2005).

A separate study by the same Australian researchers found that car carpets, since they may be manufactured from petroleum feed stocks, may decompose when heated and produce volatile organic compounds that produce background interference when testing for the presence of petrol. They found, however, that these components produced chromatographic patterns that were distinguished from those produced by petrol. Both studies underscore the necessity to obtain reference samples to eliminate or account for background interference (Cavanagh, Du Pasquier, and Lennard 2002).

Shoes can also pick up material from scenes. One notable example is a case in which a wad of charred drapery fabric was found melted into the bottom of an athletic shoe. The wearer of the shoe was dumped in the parking lot of a county hospital suffering from extensive second-degree burns from head to foot with his clothing badly burned and still smoldering. His claim of having injured himself when the carburetor of his car engine malfunctioned was not supported by the distribution of burns to himself or his clothing. The charred drapery remnant was matched to drapery material recovered from a nearby apartment fire that had been ignited with a substantial quantity of gasoline poured through several rooms. Clearly, the gasoline had ignited prematurely and the suspect had had to escape through the burning structure. The observation that the synthetic uppers of the shoes were melted and scorched and the drapery fragment was melted into the bottom of the fire-damaged shoe precluded the excuse that he was passing by the fire scene and trod in the debris well after the event.

The general forensic guidelines for handling items of clothing stipulate that they should be marked directly on the waistband, pocket, and collar with the investigator's initials and date found. If being preserved for blood or trace evidence, these items must be packed separately into clean paper bags, with each item wrapped separately in clean paper. If they are damp (water or blood), they must be allowed to air dry before being packaged. Obviously, if flammable liquid residues are possible, the clothing must be sealed separately in vapor-tight cans or sealed nylon or Kapak™ bags.

When dealing with glass, the investigator should collect as many fragments as possible and place them in paper bags, boxes, or envelopes for later physical reconstruction or reassembly. The fragments should be packaged so that their movement within the containers is kept to a minimum. All evidence must be kept separate from any control or comparison samples to avoid cross-contamination. Paint chips, particularly those at least 3.23 sq cm (0.5 in.²) should be placed in pillboxes, paper envelopes, cellophane, or plastic bags and carefully sealed. Smaller glass or paint chips should be placed in a folded paper bindle and then sealed in a paper envelope.

Trace evidence sometimes consists of hairs and fibers. Comparison samples from victims or suspects may reveal a direct or indirect relationship. Loose hair combings and clumps cut from various areas should be packaged separately in pillboxes, paper envelopes, cellophane, or plastic bags. The outside of the container should be sealed and labeled. Hairs from pets in structures may also be transferred to persons, so comparison samples may be suggested.

DEBRIS CONTAINING SUSPECTED VOLATILES

Items containing suspected ignitable liquids should be sealed in clean metal, glass, or specialized polymeric (Kapak™ or nylon) bags that have been developed for fire debris preservation, as shown in Figure 4.33. Common polyethylene plastic or paper bags should not be used, since they are porous and allow evaporation or cross-contamination and may contain volatile chemicals that may contaminate the evidence, resulting in false-positive or false-negative results when subjected to laboratory analysis. The container should be filled no more than three-fourths full, using clean tools to prevent contamination. Clean disposable plastic gloves should be worn when collecting the debris and then discarded after each sample is taken.

Investigators should seek to collect and preserve comparison samples of uncontaminated flammable and combustible liquids from the scene. Samples of suspected liquids measuring up to 1 pt should be sealed in Teflon-sealed glass vials, all-glass bottles, jars with Bakelite or metal tops, or metal cans. Such comparison samples should be carefully labeled as to their source if they are not submitted in their original containers.

The collecting officer should label each container with a brief description of the contents and the location from which the sample was collected (per ASTM E 1459). Collectors are cautioned not to use rubber stoppers or jars with rubber seals, since

FIGURE 4.33 ◆ Clean new airtight metal cans, clean glass jars, or special polymeric bags (such as Kapak™) are all suitable containers for fire debris evidence. *Courtesy of J. D. DeHaan.*

gasoline, paint thinners, and other volatile liquids contain hydrocarbons that can compromise the sample by dissolving the seals. Any odor present or reading obtained on a hydrocarbon detector should be noted prior to sealing the container. Plastic bottles are inappropriate containers for debris, liquid, or comparison samples. Various advanced techniques are available for collecting liquids from concrete floors using adsorbents such as calcium carbonate or "kitty litter" (DeHaan 2007, chap. 14).

Samples containing garden soil should be frozen as soon as possible to minimize possible degradation of hydrocarbons by microbial action and kept frozen until submission to the laboratory. All other samples should be kept as cool as possible and submitted to the laboratory as soon as possible. Further details are included in *Kirk's Fire Investigation,* 6th ed. (DeHaan 2007).

UV DETECTION OF PETROLEUM ACCELERANTS

Ultraviolet (UV) fluorescence of some hydrocarbon liquids has been touted since the 1950s as a way of discovering deposits of such liquids at fire scenes. Unfortunately, simple fluorescence observation (observing the visible light emitted when some materials are exposed to UV) is not reliable for that purpose. Many materials naturally fluoresce, and pyrolysis produces a large number of complex aromatic hydrocarbons as breakdown products that fluoresce and obscure and interfere with any fluorescence from "foreign" accelerants. Some flammable liquids produce no fluorescence at all.

A newly reported innovation promises some success in using UV light, but it involves two electronic manipulations. If the UV source is pulsed rapidly and the electronic detector is "gated" so it observes only the target after a preset delay, the decay time of most pyrolysis products can be discriminated from that of petroleum products. Another avenue is to use an image-intensified charge-coupled device (CCD) detector to look at emissions (fluorescence) at wavelengths and intensities not visible to the unaided eye. These techniques appear promising but await further testing and publication of peer-reviewed results (Takeuchi et al. 2005)

DNA ON MOLOTOV COCKTAILS

Recent work by the Forensic Science Centre at the University of Strathclyde has shown that, thanks to the STR (short tandem repeat) methods now available for DNA analysis, identifiable DNA from saliva can be found on the necks of gasoline-filled Molotov cocktails (petrol bombs). Using both sterilized bottles with 50 μL (2 drops) of saliva deposited on the exterior and a random selection of "real" bottles that had been drunk from, researchers recovered identifiable DNA even after gasoline had been poured in and a wick inserted. When the devices were thrown to explode and burn, identifiable (4–7 loci) DNA was found on about 50 percent of the bottle necks and full DNA profiles identifiable on another 25 percent of the burned bottles (Mann, Nic Daeid, and Linacre 2003).

◆ 4.8 PHOTOGRAPHY

Fire investigators should not fail to photograph and sketch carefully even the seemingly least significant fire scene detail. Some benchmark estimates place the average direct costs of performing a preliminary scene investigation in the range of $500 to $1000 (Icove, Wherry, and Schroeder 1998). With the total cost per color slide amounting to approximately $0.25 in quantity, photographs and sketches become the least expensive investigative tools from a cost-effective standpoint. Fire investigators are usually not required to qualify as expert photographers, yet they must be able to understand and utilize their equipment to the maximum benefit (Berrin 1977). Thus, the following items should be included as minimum equipment for forensic fire scene photography.

DOCUMENTATION AND STORAGE

The "Photograph Field Notes" form (906-8; Figure 4.14) is used to record the description, frame, and roll number of each photograph taken at the fire scene. The form is designed to be filled out as the photographs are taken. The frame and roll numbers are later used on the fire scene sketch to indicate the location and direction from which they were taken. The documentation should include a brief caption that permits an independent reviewer correctly to identify the object of interest and the position from which the photos were taken. Documentation of each roll of film should appear on a separate form. The "remarks" field is used to document the disposition of the film.

At the time of the investigation, a descriptive photographic index should be completed using NFPA 906-8 or a similar photo log from the appendix of *Kirk's Fire Investigation,* 6th ed. (DeHaan 2007). After processing, the original negatives and prints should be stored in a specifically marked envelope. The photographic index or log, complete with narrative, should be included in the investigation report.

FILM CAMERAS

The most versatile cameras available to fire investigators are 35mm single-lens reflex, with focal plane or between-the-lens shutter systems (Berrin 1977). Single-lens reflex cameras are expensive but allow for close-ups and specialized scene photography and introduce little distortion. Viewfinder cameras are low cost but introduce parallax errors and cannot perform many useful photographic functions.

With the introduction of electronically operated shutter systems, the 35mm camera has become the recommended standard for fire investigation photography (NFPA 2004a; Peige and Williams 1977). Competitive pricing has placed the purchase cost of many acceptable units below $100.

FILM FORMATS

Economic considerations play a key role when deciding on a film format. However, standard 35mm color film is the nearly universal choice for fire scene photography. Print film requires development and costly printing of photographs. Some investigators may choose to use color slide films having an American National Standards institute (ASA) rating between 100 and 500 to maintain resolution. In some processing laboratories, the customer can choose prints, negatives, and slides at development time. Black-and-white photography of fire scenes has been all but abandoned owing to its inability to record important colored fire and burn patterns (Eastman Kodak 1968).

For purposes of economy and portability, some investigators choose slides over negatives. Investigators are free to choose only selected slides for printing and inclusion in the report. Advantages of slide film include its ability to display details during courtroom presentations. With the advent of digital slide scanners, the images of selected photographs can also be imported into word-processed documents and electronic presentation programs. Film images can degrade over time if exposed to dampness or high temperatures, so storage sites should be selected accordingly.

DIGITAL CAMERAS

Newer and higher-resolution digital cameras are being introduced each year. Their resolution of image quality is often measured in millions of pixels (megapixels). The present minimum acceptable resolution is upward of 5 megapixels, which is still less than the resolution available using standard film photography. The higher the number of pixels, the better the quality. Many cameras today will capture 6 megapixels per image on a flash card or similar downloadable medium.

The improper use of digital cameras in investigative photography may undermine the viability of a case, owing mainly to evidentiary considerations. Investigators are advised to adhere to a standard protocol for preserving digital images. Some agencies preserve images on their original media (CD or flash cards), whereas others save original images on disk drive memory and make copies for any medium or any digital image processing.

As with all photographs, the courtroom tests for their authentication must be met, each image used being a "true and accurate representation" and "relevant to the testimony" to the fire investigation, particularly under *Federal Rules of Evidence 403* (Lipson 2000). Systematic handling and processing of digital images is the only method for assuring their long-term acceptability (NFPA 2004a, sec. 15.2.3.4).

Computer operator error and unforeseen crashes can cause digital images to be lost forever, so precautions are necessary. It is often impractical and costly to store high-resolution images permanently on some media or flash cards and they must be copied onto another medium. Two copies of the original electronic/magnetic medium should be made and placed in the case file in the same manner as an audio or video recording. A third copy is used as a "working copy." Images should not be compressed, as this causes loss of detail and clarity. Copies should also be placed on external hard drives and stored off site. Operating systems, programs, and file formats change and render some files unreadable. Some investigators use two cameras at critical scenes (print and digital) and take duplicate photos to ensure that some images are always available, even years after the fire.

However, digital cameras and their electronic images are very useful in the production of preliminary photographs of fire scenes or final investigative reports, as the images can easily be processed to lighten, darken, or enhance features and easily transferred to printed documents. They are also helpful in producing panoramic views of fire scenes, as discussed previously. However, file compression and other manipulations may reduce the quality of the images and introduce suspicions of "manipulation" of the photo's content. Digital images can be coded or watermarked while being recorded to demonstrate they are "as taken," an unmodified original. The ease and low per-image cost of digital photography tempts some investigators to take far more photos at scenes than are really necessary or justified. The guidance should be, What information am I trying to capture with this photo?

DIGITAL IMAGING GUIDELINES

The first draft of "Definitions and Guidelines for the Use of Imaging Technologies in the Criminal Justice System" was published in October 1999 in *Forensic Science Communications* (FBI 1999). The guidelines were prepared by the Scientific Working Group on Imaging Technology (SWGIT) and cover the "documentation of policies and procedures of personnel engaged in the capture, storage, processing, analysis, transmission, or output of imagery in the criminal justice system to ensure that their use of images and imaging technologies are governed by documented policies and procedures."

TABLE 4.3 ◆ Recommended Guidelines and Best Practices in Forensic Photography by the Scientific Working Group on Imaging Technology

Section	Description
Section 1	Guidelines for the Use of Imaging Technologies in the Criminal Justice System
Section 2	Considerations for Managers
Section 3	Guidelines for Field Applications of Imaging Technologies in the Criminal Justice System
Section 4	Recommendations and Guidelines for Using Closed-Circuit Television Security Systems in Commercial Institutions
Section 5	Recommendations and Guidelines for the Use of Digital Image Processing in the Criminal Justice System
Section 6	Guidelines and Recommendations for Training in Imaging Technologies in the Criminal Justice System
Section 7	Recommendations and Guidelines for the Use of Forensic Video Processing in the Criminal Justice System
Section 8	General Guidelines for Capturing Latent Impressions Using a Digital Camera
Section 9	General Guidelines for Photographing Tire Impressions
Section 10	General Guidelines for Photographing Footwear Impressions
Section 11	Best Practices for Documenting Image Enhancement
Section 12	Best Practices for Practitioners of Forensic Image Analysis

Source: Forensic Science Communications (FBI 2004).

Table 4.3 lists the present approved guidelines issued by the SWGIT. These guidelines cover imaging technologies, recommended policies, best practices, and techniques for photographing specific impression evidence. These resource documents are available on the FBI (fbi.gov) and International Association for Identification (theiai.org) websites.

Agencies or individuals using photography, both simple and complex, should develop a standard operating procedure and departmental policy. The SWGIT procedures recommend preserving original images by storing and maintaining them in an unaltered state and in their native file formats. Duplicates or copies should be used for working images. Original images should be preserved in one of the following durable formats: silver-based (noninstant) film, write-once compact recordable disks (CDR), or digital versatile recordable disks (DVD-R).

If image processing techniques are used on the original image, they should be documented with standard operating procedures. These procedures should be visually verifiable and include cropping, dodging, burning, color balancing, and contrast adjustment. Advanced techniques include those that increase the visibility of the image through multi-image averaging, integration, or Fourier analysis. Other techniques include cropping, overlaying, and creating panoramic views from individual series of photographs. An image processing log is recommended so that the process could later be replicated.

A chain of custody should be maintained for the media on which original images are recorded. This chain of custody should document the identity of the personnel who had custody and control of the digital image file from the point of capture to archiving.

When compressing images into formats such as JPEGs, investigators should maintain as high a resolution as possible to avoid degradation of the image. Image capture devices should render an accurate representation of the images. Different applications will dictate different standards of accuracy.

Training is important when using imaging technologies. Formal uniform training programs should be established, documented, and maintained. Proficiency testing will assure that camera equipment, software, and media keep pace with hardware or software updates.

LIGHTING

No matter what format camera is used, lighting plays an important role in fire scene photography. The intensely burned areas tend to absorb light from natural or artificial sources. Powerful electronic flashes are useful for lighting char details in dark scenes. Some built-in flashes on film and digital cameras do not have enough power to adequately illuminate large fire scenes, so provisions should be made for external flash connections.

Sometimes, burn patterns can best be photographed using an oblique flash to light the area of interest. Therefore, a camera should be selected that allows for an externally connected flash unit for oblique or remote illumination of the scene (NFPA 2004, pt. 15.2).

ACCESSORIES

Many attachments are available for cameras; however, for the sake of simplicity, non-expert investigators are often recommended to limit their use. The maximum effectiveness of the single lens provided with the camera (typically 50- to 55-mm focal length) is demonstrated later. Furthermore, the use of filters and interchangeable lens configurations might disqualify photographs or slides used in courtroom presentations when investigators cannot authoritatively qualify their uses, advantages, deficiencies, or effects (Icove and Gohar 1980). A clear filter to protect the lens is the only recommended filter for fire scene photography (NFPA 2004a, pt. 15.2.3.6). Some photographers prefer using polarizing or UV filters (which do not affect the color of the image) in place of the clear filter.

The single most recommended accessory for photographing fire scenes is a sturdy tripod. This simple device enables the photographing of scenes with clear detail when low-lighting conditions command longer shutter exposures. Also, a tripod allows the optimization of depth of field according to the available lighting conditions. For example, an aperture setting of f/16 will usually produce a depth of field ranging from 0.91 m (3 ft) to 3.05 m (10 ft) when the camera is focused on a point 1.8 m (6 ft) from the lens. As demonstrated later, panoramic scenes are most easily recorded when using a tripod. Longer exposure times may be needed to accommodate smaller aperture settings, and any exposure longer that 1/30 s is subject to tremors if the camera is handheld.

MEASURING AND IMAGE CALIBRATION DEVICES

Measuring and image calibration devices are used to provide additional information in both processing and interpretation. These devices should be used uniformly

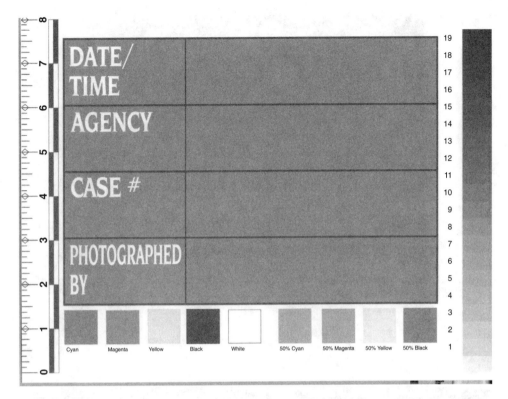

FIGURE 4.34 ◆ Commercially available gray and color calibration chart suitable for use in forensic fire scene documentation. Courtesy of Armor Forensics, by Permission.

from case to case. Even though present fire and explosion investigation guidelines in *NFPA 921* recommend the use of an 18 percent grayscale calibration card (NFPA 2004a, pt. 15.2.3.7.5), there are many advantages to using a color calibration card. With the move toward the universal use of color photography for fire scenes, the investigator should consider the use of a combined color and grayscale calibration card.

A **color and grayscale calibration card** bearing the agency name, case number, date, and time should be photographed on the first frame of each roll of film. Professional film processing laboratories will calibrate their equipment using these cards to ensure that the best color calibrations and gray scales are met. Figure 4.34 is an example of a commercially available card bearing the essential calibration and documentation information.

Armor Forensics produces a 15-cm (6-in.) scale that is printed on 18 percent gray low-reflection plastic so that even close-up digital images can include grayscale calibration. The IAAI also has produced a similar plastic 6-in. ruler as a promotion for their CFITrainer program (Figure 4.35).

Calibration cards also have measuring devices (rulers) on their edges to document distances and sizes of objects photographed. Other measurement devices include longer folding rulers and yellow numbered tent cards to identify the location of individual points of interest or forensic evidence, as shown in Figure 4.36. Bright-colored pointers can be used to indicate important features or document direction of fire travel.

FIGURE 4.35 ◆ A convenient grayscale ruler is available from several sources. *Courtesy of J. D. DeHaan.*

(a)

FIGURE 4.36 ◆ (a) Crime scene measurement devices include yellow numbered tent cards to identify the location of individual points of interest or forensic evidence, such as ignitable liquid containers.

(b)

(c)

FIGURE 4.36 ◆ *Continued* (b) Underside of can bears repeated stabbing penetrations from a knife blade, indicating that it was intentionally made to spread flammable liquid contents. Tool marks may link it to the offender. (c) Numbered evidence markers illustrate the position of evidence. Plastic pointers or arrows indicate positions of positive canine alerts at the scene. They may also be used to indicate the direction of fire spread in the fire scene. *Courtesy of J. D. DeHaan.*

AERIAL PHOTOGRAPHY

In rural areas, the U.S. Soil Conservation Service has **aerial photos** (also known as *orthophotoquads*) of a majority of the farmlands within its area of responsibility. These photos are at various resolutions. Aerial photographs do not always need to be taken from aircraft. Perspective photographs, such as in Example 4.1, can be taken from higher floors of adjacent buildings. Photos taken from the elevated platform of a fire truck can be useful, providing a closer look at the overall pattern of damage.

Satellite and commercial aerial photos are widely available at a modest cost. Many cities and counties use aerial survey to search for nonpermit construction. In the use of commercial databases, care must be taken, as street addresses are not always up-to-date on data files. Users must also confirm that the image was recent enough that it accurately reflects the prefire conditions.

PHOTOGRAMMETRY

Photogrammetry, the science of extracting measurement data from photographs, was once out of the range of expertise of the typical fire investigator but has become more accessible in recent years through affordable, user-friendly computer software. In photogrammetry, two or more photographs are taken and used to extract absolute coordinates and distance measurements from key features in those images (see Figure 4.37). Close-range photogrammetry programs compensate for the shortening of object lines

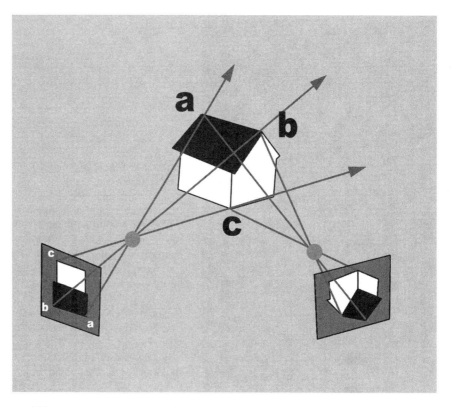

FIGURE 4.37 ◆ In photogrammetry, two or more photographs are used in such a way as to extract absolute coordinates and measurements from key points in those images. *PhotoModeler image courtesy of Eos Systems, Inc.*

FIGURE 4.38 ◆ An example of the use of photogrammetry to combine several two-dimensional images to construct a three-dimensional view. *PhotoModeler image courtesy of Eos Systems, Inc.*

due to perspective that is present in nearly all photographs, thus allowing users to obtain accurate measurements and create realistic models (Eos Systems Inc. 2003).

For example, the actual walking distance from a deceased victim in a bed to the doorway threshold may be crucial in an investigation. Photogrammetry can capture measurement data in large or complex scenes where tape measure and notepad are inadequate or too time-consuming. It also allows investigators to measure evidence after the fact that may have been moved or cleaned up or no longer exists at the scene. Photogrammetry has been in use in North America, Europe, and Australia for many years for accident scenes and major crimes and has been proposed for advanced fire investigations. In one case in the United States, the technique was used forensically to reconstruct a model of a kitchen area to enhance a V burn pattern (King and Ebert 2002).

Some photogrammetry programs allow users to map two-dimensional images over three-dimensional surfaces that have been created, to provide photorealistic models with perspective-corrected phototextures. There is a wide range of software programs for calculating these measurements and constructing three-dimensional models from two-dimensional photographs. Figure 4.38 is an example of the use of photogrammetry to construct a three-dimensional view of the exterior of a building using actual two-dimensional photographs.

One very new innovation that combines the capabilities of both panoramic photography and photogrammetry is the digital scanning camera, such as the Panoscan or Spheron VR model. The core of the system is a specialized digital camera that works much like the moving-slit film cameras of a century ago but uses a high-resolution image converter and a specialized lens-and-shutter system to capture in each vertical "slice" an image of about 170° vertically (see Figure 4.39).

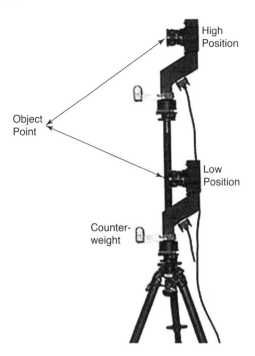

FIGURE 4.39 ◆ A digital scanning camera with an extendable mast and specialized software can be used to capture photogrammetric images. *Courtesy of Panoscan Inc.*

The camera automatically rotates, so that the final image is a spherical, high-resolution image. (The maximum vertical resolution is 5200 pixels and the total image capture can be up to 50 megapixels.) The dynamic range of image capture is up to 26 f-stops, so information from very dark areas to those in full sunlight are captured in a single pass. The camera tripod has a vertically extendable mast that allows the camera to capture an image from the same position in the room but from a view line some 0.5 m higher. The two images are then linked by the software to produce a stereoptic virtual reality where any measurement can then be taken with millimeter precision.

The system uses specialized software that allows fully interactive crime scene documentation to be added to the images. (The original image can be watermarked with GPS location, time, and digital signature by the operator.) Images from different views in a room or adjoining rooms can be linked by "hot spots" so the viewer can "visit" the scene and "view" it from any position used. This eliminates blind spots where the camera was positioned and allows the viewer to see all sides of furniture, walls, and other solid objects as he or she "walks" through the scene. Hot spot links can also be added for still, close-up photos or physical evidence locations. Each "physical evidence" spot can be supplemented with images, notes, or results of comparisons as the forensic examination proceeds. The stand-alone support system has a dedicated PC, 40GB hard drive, and battery for field use. The output of the system can include navigable virtual reality images as computer files (QuickTime VR) or on DVD. The system can also produce fixed images on PowerPoint slides or PDF files of either photo images or line diagrams (plans). See Chapter 10 for an example. Current prices are in the $30,000 range.

Fire scene sketches, whether or not to scale, are important supplements to photographs. A sketch graphically portrays the fire scene and items of evidence as recorded by the investigator. The "Sketch Field Notes" form (906-9; Figure 4.16) is for use by investigators when drawing simple two-dimensional sketches of fire scenes. These sketches may be rough exterior building outlines or detailed floor plans. Sketches allow the investigator to illustrate relationships between objects that cannot be captured via photography such as those in separate rooms, under or behind large furniture, or visible only from overhead or via cross section. See the appendix in *Kirk's Fire Investigation,* 6th ed., for additional details (DeHaan 2007).

GENERAL GUIDELINES

Good scene sketches do not require highly artistic work. Even informal hand-drawn representations can capture locations and relationships between items that cannot be shown in photographs. Forensic fire scene sketches serve many purposes and should depict the following information (DeHaan 2007).

- ◆ The outline and dimensions of the building, room, vehicle, or area(s) of interest
- ◆ The locations of pertinent evidence or critical features, such as fire patterns and plume damage
- ◆ The locations and dimensions of all major fuel packages involved in the fire
- ◆ Locations and travel distances of possible points of entry and exit of victims and suspects using Global Positioning System (GPS) data
- ◆ Conditions and dimensions of windows, doors, floors, ceilings, and wall surfaces including sill and soffit heights for later use in fire scene reconstruction and analysis

In some instances, multiple sketches using the same overall dimensions are necessary. Several separate types of evidence can be preserved in this manner including isochars, plume damage, and sample (evidence collection) sites. A sketch can often reveal characteristics not readily obvious in a photograph.

The necessary elements of a good forensic sketch are as follows (DeHaan 2007).

- ◆ The investigator's full name, rank, and agency, the case number, and the date/time when the sketch was prepared
- ◆ The full names of other individuals involved in constructing the sketch
- ◆ The location and geographic orientation of the fire scene (north arrow). Latitude and longitude using GPS data may be desired for rural scenes to determine jurisdiction boundaries (state, county, township, etc.) or to relocate the scene accurately
- ◆ A legend containing descriptions of all symbols used and their meanings, scale, and other significant information

The areas of surface demarcations, thermal effects, penetrations, and loss of material can also be outlined using colored markers on sketches or photographs. One system, for example, uses yellow lines to outline areas of surface deposits, green lines for thermal effects, blue lines for penetrations, and red lines for areas where the material has been consumed completely. Background patterns (lines, cross-hatch) can be used in computer diagrams.

A sketch is usually initially drawn by hand, sometimes followed up by a computer-assisted drawing. An effective hand sketching tool is a grooved gridpad on which

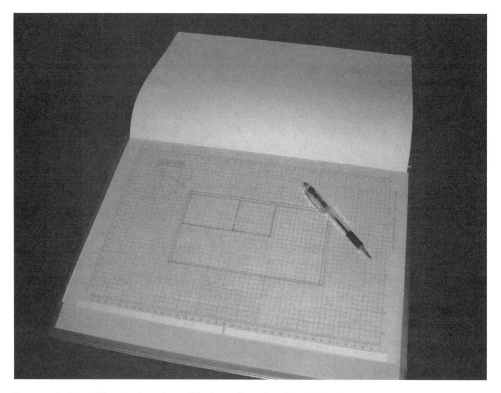

FIGURE 4.40 ◆ New tools such as this Accu-line drawing aid make manual sketching easier. *Courtesy of D. J. Icove.*

the pencil lead follows a slight indentation on a plastic surface (Accu-line 2003) (see Figure 4.40).

A reliable fire scene sketching program called Visio by Microsoft Corporation is sufficient to produce accurate representations of structures. Microsoft provides an add-on crime scene package at no charge. Experienced investigators have developed templates for use in fire scene analysis (Microsoft 2002). Figure 4.41 shows an example of a typical fire scene diagram produced using Visio.

A special application of sketching is to copy an overhead or plan view of the scene onto several clear (transparency) sheets. Each sheet can be used to record a different indicator such as fire travel vectors, char depths, calcination, and furniture patterns. The overlays then can be compared with one another or with the original floor plan to illustrate convergence on a possible area of origin. Putting sketches on separate sheets avoids the problem of too much detail in a single diagram.

TWO- AND THREE-DIMENSIONAL SKETCHES

With the advent of **computer-aided drafting (CAD) tools,** two- and three-dimensional sketches can now be used for courtroom exhibits. Figure 4.42 shows an example of this technique. Plastic overlays can also be placed over the large sketches, which can be annotated in court using either prepared overlays or impromptu markings on the overlays. Using a convenient coordinate system will allow the accurate placement of critical evidence such as the orientation of torsos of victims, fire plumes, and evidence locations.

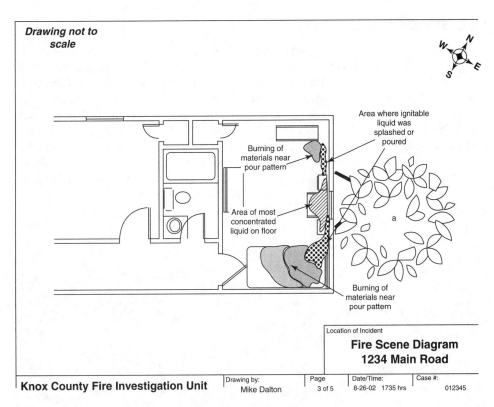

FIGURE 4.41 ◆ A typical fire scene diagram using a sketching program sufficient to produce accurate representations of burn patterns, location of evidence, and other details. *Courtesy of Michael Dalton, Knox County Sheriff's Office.*

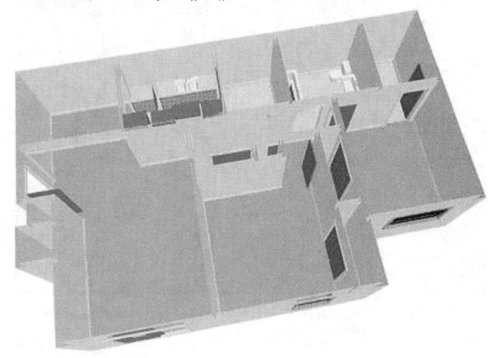

FIGURE 4.42 ◆ Construction of a three-dimensional view using an architectural rendering program. *Courtesy of D. J. Icove.*

TOTAL STATION SURVEY MAPPING

A new generation of computer-driven surveying technologies has come into use for mapping of large interior and exterior scenes. A single reference location is selected and located using a GPS integrated with a **Geographic Information System (GIS).** The laser sighting unit is then trained on individual features—corners of rooms, curb lines, or evidence locations (see Figure 4.43). The computer notes each feature's direction, angle of elevation, and distance. The program then draws a plan of the scene with extremely accurate dimensions (error of ±1 cm [±0.39 in.] at 150 m [492 ft] is typical) even without reflector posts. Systems by Leica, Sokkia, and Topcon that have been used by traffic accident investigation teams in many police jurisdictions for years are being used by some major arson investigation agencies.

FIGURE 4.43 ◆ A Total Station surveying system allows accurate capture of dimensions and three-dimensional relationships. They have been in use by accident investigation teams in large police departments for many years. *Courtesy of Leica Geosystems, Inc.*

Several natural processes can be used to establish one of the most difficult factors in scene investigation—the passage of time. These processes include evaporation, warming, cooling, drying, and melting.

EVAPORATION

One of the most critical factors in fire reconstruction is the use of accelerants. Estimating the **evaporation** of accelerants may provide important insight as to the passage of time and the dynamics of the resulting fire. The evaporation rate of a liquid depends on its vapor pressure (which is temperature dependent), the temperature and nature of the surface on which it is deposited, its surface area, and the circulation of air around it.

Vapor pressure is a fundamental physical property of all liquids and solids. It is the pressure that could be generated by the material if placed in a vacuum and allowed to come to equilibrium. It is a measure of how volatile the material is (i.e., how easily it vaporizes). Because vapor pressure is temperature dependent, the higher the temperature, the higher the vapor pressure. Note that it is the temperature of the liquid (or of the surface on which it is spread) that is critical, not necessarily the ambient temperature of the room. A pan of acetone on a hot stove will evaporate very quickly even in a cold room.

Vapor pressure is also highly dependent on molecular weight. The lower the molecular weight of a substance, the higher its vapor pressure and the faster it will evaporate. The physical form of the material is also critical. A very thin film of volatile liquid on a surface will evaporate more quickly than a deep pool. Volatile liquids on porous surfaces like cloth or carpet will evaporate more quickly than they will from a free-standing pool of the same size. The larger the size of the pool and the more air movement around the pool, the faster the total evaporation (DeHaan 1999).

A complex mixture like gasoline contains more than 200 compounds, some of which, like pentane, are very volatile and evaporate very quickly. Others, like trimethylbenzene, evaporate very slowly and will persist for a long time. When such a liquid is poured out, the lightest, most volatile compounds will represent the entire bulk of the initial vapors generated. Thus, it is the presence of toluene, pentane, and other volatile components that controls the ignitability of the vapors being generated. Components like octane or heavier hydrocarbons hardly evaporate at room temperatures. The vapors created by evaporation are easily distinguished from the residues of the liquid based on their gas chromographic profiles.

This process means that partially evaporated gasoline will have a very different gas chromographic profile than fresh gasoline. Simple evaporation and burning of gasoline result in the same disproportionate loss of the lighter and more volatile components. The process is accelerated by the added heat flux from the flames. This property can be used to estimate the relative time of exposure. Gasoline that is present in fire debris as a contaminant from gas-powered firefighting equipment or intentionally added after the fire will have far more volatiles than would be expected from residues after normal evaporation or combustion.

Changing the physical form of a liquid by placing it on an absorbent wick or aerosolizing it (changing it to a mist by releasing it under pressure) will increase its vapor pressure and make even its heavier components evaporate more quickly or make them more readily ignitable. If evaporation is a concern, from the standpoint of either creating a risk by evaporation of a liquid fuel or reconstructing time factors (time since release), several factors must be documented. These factors include accurate identification of the material itself; the temperature of ambient air, surfaces, or the liquid itself at the time of release; and the nature of surfaces on which it was spread (liquids evaporate much more quickly from thin materials like clothing with free air movement than from solid porous materials like shoes).

The predominant air conditions are also critical: high temperatures, direct sun, wind, mechanical movement by fans, HVAC systems, and movement of equipment, vehicles, or people all increase the evaporation rate. Evaporation of flammable liquids from evidence that could be significant in establishing the cause of the fire or linking a person with the scene needs to be arrested by sealing the evidence in an appropriate vapor-tight container and keeping it as cool as possible to reduce further losses. The time at which it was sealed should be noted on the container.

DRYING

Drying is usually closely related to evaporation (as in the drying of water from clothing), but when blood or other complex liquids are involved, other processes may be involved. Blood, for instance, changes chemistry, color, and viscosity as it dries, becoming darker and stickier (if it is drying as a pool on a nonporous surface). Documentation of the drying state of blood involves more than simple photography, as the texture needs to be assessed, usually using a sterile cotton swab (that would be used in its eventual collection in any event). The time at which this observation is made must also be recorded. The same considerations apply to paint, adhesives, plastic resins, or other materials that change mechanically as they dry. Also, the ambient temperature must be noted if drying is a critical concern.

COOLING

Cooling and **melting** are time indicators that play roles in crime scene investigation (such as still partly frozen ice cream on the counter) but occasionally are useful in fire reconstruction. The cooling (or lack thereof) of materials directly exposed to fire can be assessed by direct contact (ouch!), thermal imaging, or temperature probe (inexpensive digital thermometers that are good from $-45°C$ to $200°C$ ($-50°F$ to $400°F$) are available from kitchen and housewares stores. Once again, ambient conditions of wind, rain, sun, standing or running water, and the like, need to be recorded at the same time.

Cooling of a body is a routine estimate in death investigations but is often ignored in fires. That is unfortunate because the death and the fire need not have occurred at the same time. An adult human body, once cooled to room temperature, must be exposed to heat (even from a fire) for a long time before its internal core temperature rises (ask anyone who's ever tried to cook a 25-lb roast). The thermal inertia of human tissue is similar to that of pine wood or polyethylene plastic. The internal temperature (rectal or liver) of a body at a fire scene should be recorded not for estimation of the time of death interval as much as for exclusion of situations in which the victim has been dead for many hours prior to the fire.

The fire scene is the most important piece of evidence in forensic analysis and reconstruction, particularly when the fire clearly will result in criminal and/or civil litigation. Of major concern to fire investigators is the preservation of evidence to prevent its destruction or alteration before it is submitted for competent examination and analysis. Intentional or negligent destruction or alteration of evidence that will be the subject of pending or future litigation is referred to as ***spoliation*** (NFPA, 2004a, sec. 3.3.144).

Failure to prevent spoliation can result in disallowance of testimony, sanctions, and civil or criminal remedies (Burnette 2000). Published standards (ASTM E 860) establish practices for examining and testing items of evidence that may or may not be involved in products liability litigation. This evidence may include equipment, items, or components that will not be returned to service. In laboratory examinations where the evidence will be altered or destroyed, all persons involved in the present or potential cases should be given the opportunity to make their opinions known and be present at the testing.

ASTM E 860 refers to other ASTM practices, including those concerning the reporting of opinions (ASTM E 620), evaluation of technical data (ASTM E 678), reporting of incidents (ASTM E 1020), and collection/preservation of evidence (ASTM E 1188). For example, ASTM E 1188 (Sec. 4.1) states that the investigator should "obtain statements as early as feasible from all individuals associated with the event and recovery activity. ASTM E 1188 also guides the investigator in maintaining a chain of evidence by identifying and uniquely labeling evidence as to the location, time and date, and name of collector.

When the evidence is subject to in-place or later laboratory examination by a coordinated and informed group of affected parties, these processes must documented and photographed. These tests or examinations may consist of disassembly or destructive testing (ASTM E 1188, Sec. 4.3). Disposal of evidence should also be coordinated with ample notification of all parties and clients.

Common testing standards were listed in Table 1.3. Legal experts provide the following practical advice and pointers for avoiding spoliation issues in potential or pending cases (Hewitt 1997).

- ◆ Recognize the duty to other interested parties.
- ◆ Keep current on your field of expertise and track the law on spoliation.
- ◆ Retain only properly qualified experts and seek their guidance.
- ◆ Have regard for industry standards and recommendations and develop standard procedures for the storage of evidence.
- ◆ Put those in control of evidence on notice of your rights.
- ◆ Raise the opposing party's standard of care if necessary and build a record with the party in control of evidence.
- ◆ Notify other interested parties and provide them with an opportunity to examine evidence.
- ◆ Seek either a voluntary undertaking or a formal agreement to preserve evidence.
- ◆ Consider also obtaining a preservation or protection order.
- ◆ If compelled to destroy or damage evidence, first document it thoroughly.
- ◆ Consult a lawyer.

Guidelines for avoiding sanctions for spoliation of evidence include conducting an immediate investigation, timely notification of potential defendants, and long-term preservation of the evidence collected (Sweeney and Perdew 2005). The realization that the fire scene cannot be preserved throughout litigation is an incentive to allow potential defendants to take steps to identify and preserve other relevant evidence. Suggested strategies other than sanctions for dealing with spoliation include:

◆ Suggest that the judge instruct the jury regarding missing evidence.
◆ Inform the jury of the nature of the plaintiff's conduct.
◆ Show that the destruction of evidence precludes the plaintiff from proving its case.

Summarized in Table 4.4 are relevant fundamental case citations and opinions. Fire investigators should keep up to date on legal issues surrounding the problem of

TABLE 4.4 ◆ Summary of Case Citations and Opinions on Spoliation of Evidence

Principle	*Case Citation*	*Discussion*
Definition	*County of Solano v. Delancy,* 264 Cal. Rptr. 721, 724 (Cal. Ct. App. 1989) *Miller v. Montgomery County,* 494 A.2d 761 (Md. Ct. Spec. App. 1985)	• Failure to preserve property for another's use in a pending or future litigation • Destruction, mutilation, or alteration of evidence by a party to an action
Dismissal	*Allstate Insurance Co. v. Sunbeam Corp.,* 865 F. Supp 1267 (N.D. Ill. 1994) *Transamerica Insurance Group v. Maytag Inc.,* 650 N.E. 3d 169 (Ohio App. 1994)	• Dismissal of action as result of spoliation by deliberate or malicious conduct • Dismissal of action for failure to preserve all evidence prior to suit
Duty to preserve evidence	*California v. Trombetta,* 479, 488–89 (1984)	• Duty to preserve exculpatory evidence that will play a role in suspect's defense
Exclusion of expert testimony	*Bright v. Ford Motor Co.,* 578 N.E. 2nd 547 (Ohio App. 1990) *Cincinnati Insurance Co. v. General Motors Corp.,* 1994 Ohio App. LEXIS 4960 (Ottawa County Oct. 28, 1994) *Travelers Insurance Co. v. Dayton Power and Light Co.,* 663 N.E.2nd 1383 (Ohio Misc, 1996) *Travelers Insurance Co. v. Knight Electric Co.,* 1992 Ohio App. LEXIS 6664 (Stark County Dec. 21, 1992)	• Exclusion of expert testimony for failure to protect evidence prejudicial to innocent party • In products liability actions, expert testimony may be precluded as a sanction for spoliation of evidence • Negligent or inadvertent destruction of evidence sufficient for sanctions excluding deposition testimony of expert witness • Opinion evidence of plaintiff's expert struck because physical evidence no longer available
Evidentiary inferences	*State of Ohio v. Strub,* 355 N.E.2d 819 (Ohio App. 1975) *U.S. v Mendez-Ortiz,* 810 F.2d 76 (6th Cir. 1986)	• Attempts to suppress evidence indicating consciousness of guilt • Inference that evidence unfavorable to spoliator's cause was intentionally spoliated or destroyed
Independent torts	*Continental Insurance Co. v. Herman,* 576 S.2d 313 (Fla. 3rd DCA 1991) *Smith v. Howard Johnson Co. Inc.,* 615 N.E.2d 1037 (Ohio 1993)	• Independent actions for intentional or negligent spoliation of evidence • Existence of cause of action in tort for interference with or destruction of evidence
Criminal statutes	Ohio Statute, Section 2921.32	• "Obstructing justice" by destroying or concealing physical evidence of the crime or act

Source: Derived from information given by Burnette (2000).

spoliation and how to guard against violating its principles and practices. When questions arise, always contact your legal advisor or prosecutor for clarification or advice. Thorough documentation with photography and detailed notes is the best protection against problems arising from accidental damage during storage and transport. A comprehensive review of current spoliation decisions is offered in *Kirk's Fire Investigation*, 6th ed. (DeHaan 2007, chap. 17)

◆ **4.12 SUMMARY AND CONCLUSIONS**

Fire scenes often contain complex information that must be thoroughly documented, since a single photograph and scene diagram are not sufficient to capture vital information on fire dynamics, building construction, evidence collection, and avenues of escape for the building's occupants.

Fire scenes are documented through a comprehensive protocol of forensic photography, sketches, drawings, and evidence collection. Various computer-assisted photographic and sketching technologies now can help ensure accurate and representational diagramming. Comprehensive documentation aids the primary investigation, any subsequent investigations, or inquiries and minimizes charges of spoliation.

The next chapter discusses on accurate fire scene documentation in the analysis of intentionally set fires. From visual observations of the manner of ignition of the fire, the presence or absence of accelerants, and the area of origin, much can be learned about the arsonist. Later chapters demonstrate the need for accurate documentation to construct fire models, to design tests, and to compare case studies.

Problems

4.1. Find an example of a public fire report on the U.S. Fire Administration's Web site. What is your assessment of the completeness of this report? What information would you add to the report?

4.2. Visit the scene of a fire and take exterior survey photographs of the building without crossing onto the property. What information can you glean from the exterior visual information?

4.3. Practice sketching fire scenes by drawing a plan (overhead) view of your living room, showing the location of furnishings, heat sources, and ventilation openings.

4.4. Review the list of case citations on spoliation of evidence. What guidance would you provide to ensure that your local community investigator meets or exceeds this professional conduct?

Suggested Reading

Cooke, R. A., and R. H. Ide. 1985. *Principles of fire investigation,* chaps. 8 and 9. Leicester, U.K.: Institution of Fire Engineers.

DeHaan, J. D. 2007. *Kirk's fire investigation,* 6th ed., chap. 17. Upper Saddle River, N.J.: Prentice Hall.

National Institute of Justice. (June 2000). *Fire and arson scene evidence: A guide for public safety personnel.* Washington, D.C.: NIJ.

CHAPTER 5 ◆ Arson Crime Scene Analysis

You know my methods in such cases, Watson: I put myself in the man's place, and having first gauged his intelligence, I try to imagine how I should myself have proceeded under the same circumstances.

—Sir Arthur Conan Doyle,
"The Adventure of the Musgrave Ritual"

Based on published statistics from the National Fire Protection Association (NFPA), in 2005, U.S. fire departments responded to 1,602,000 fires, or one every 20 seconds. Of these incidents, 32 percent occurred in structures, 16.2 percent in highway vehicles, and 50 percent in outside properties. Loss of life is always a grave concern in fires. The NFPA estimates that 3675 civilians died in 2005, with approximately 82 percent while they were at home. These statistics come from fire departments that responded to the NFPA's national fire experience survey (Karter 2006).

A leading cause of fire in the United States, arson continues to be both an urgent national problem and truly a contemporary crime. Often characterized as a clandestine tool for criminals, the true arson picture is not clearly known. The NFPA statistics for 2004 estimate that arson caused 36,500 structure fires and 36,000 vehicle fires. Intentionally set fires in structure fires resulted in 320 civilian deaths and $714 million in property losses. Vehicle fires caused $165 million in property damage (Karter, 2005). These estimates are almost certainly very low, since many U.S. agencies do not report arson incidents (owing to political or administrative concerns). Many investigators think that the total of intentionally set structure fires is closer to 40 percent.

Economically, arson affects insurance premium rates, removes taxable property assets, and degrades our communities. Historically, the inner areas of our large cities often are the most hard-hit, and the result is that much of the cost of this destructive crime falls on those who can least afford it. The correlation between a healthy economy and a decline in business failure equates to fewer arson-for-profit cases. However, historical trends show repeatedly that a decline in the economy results in an increase in arsons (Decker and Ottley 1999).

If an investigator understands the crime of arson based on its motivations, it may enhance investigative efforts and provide a focus for intervention efforts. Examination of the fire scene and reporting of the results may facilitate dialogue among the various disciplines and investigative units involved in arson study and investigation.

It is intended that the information supplied in this chapter will assist arson investigators in developing skills for reading and interpreting the characteristics of crime scene evidence and applying that evidence to the arsonist's behavior and patterns of thinking.

◆ **5.1 ARSON AS A CRIME**

The definitions for arson range widely, owing primarily to differing statutory terminology. The most accepted definition is that **arson** is the willful and malicious burning of property (Icove et al. 1992).

The criminal act of arson is usually divided into three elements (DeHaan 2007).

- **There has been a burning of property.** This must be shown to the court to be actual destruction, not just scorching or sooting (although some states include any physical or visible impairment of any surface).
- **The burning is incendiary in origin.** Proof of the existence of an effective incendiary device, no matter how simple it may be, is adequate. Proof must be accomplished by showing specifically how all reasonable natural or accidental cases have been considered and ruled out.
- **The burning is shown to be started with malice.** This act of burning requires a specific intent of destroying property.

An arsonist is usually a person apprehended, charged, and convicted of one or more arsons. Arsonists commonly use **accelerants,** which are any type of material or substance added to the targeted materials to enhance the combustion of those materials and to accelerate the burning (Icove et al. 1992). There are many serial arsonists who set fires to available combustibles without using any accelerants.

DEVELOPING THE WORKING HYPOTHESIS

Chapter 1 explored in depth the development of a **working hypothesis** using the **scientific method.** Several factors that may contribute to a working hypothesis for an incendiary fire are discussed in *NFPA 921* (NFPA 2004a, chap. 19). Displayed in Figure 5.1 is a mosaic of several factors that may contribute to a working hypothesis in an incendiary fire or explosion, derived from *NFPA 921*. Note that these factors should not be considered all-inclusive; other indicators may exist. However, one should be cautioned that one or more of these factors are not necessarily sufficient to constitute a finding of an incendiary fire.

MULTIPLE FIRES

When an offender is involved with three or more fires, particular terminology is used to describe that crime (Icove et al. 1992).

- **Mass arson** involves an offender who sets three or more fires at the same site or location during a limited period of time.
- **Spree arson** involves an arsonist who sets three or more fires at separate locations with no emotional cooling-off period between the fires.
- **Serial arson** involves an offender who sets three or more fires with a cooling-off period between the fires.

This chapter includes a detailed discussion of how arson crime scenes are analyzed. Such analysis often leads to the identification of offenders, their method of operation, and their areas of frequent travel.

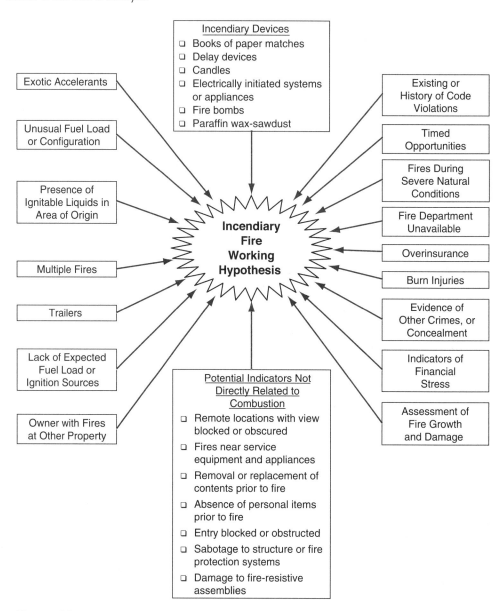

Incendiary Devices
- ❏ Books of paper matches
- ❏ Delay devices
- ❏ Candles
- ❏ Electrically initiated systems or appliances
- ❏ Fire bombs
- ❏ Paraffin wax-sawdust

Exotic Accelerants

Unusual Fuel Load or Configuration

Presence of Ignitable Liquids in Area of Origin

Multiple Fires

Trailers

Lack of Expected Fuel Load or Ignition Sources

Owner with Fires at Other Property

Incendiary Fire Working Hypothesis

Existing or History of Code Violations

Timed Opportunities

Fires During Severe Natural Conditions

Fire Department Unavailable

Overinsurance

Burn Injuries

Evidence of Other Crimes, or Concealment

Indicators of Financial Stress

Assessment of Fire Growth and Damage

Potential Indicators Not Directly Related to Combustion
- ❏ Remote locations with view blocked or obscured
- ❏ Fires near service equipment and appliances
- ❏ Removal or replacement of contents prior to fire
- ❏ Absence of personal items prior to fire
- ❏ Entry blocked or obstructed
- ❏ Sabotage to structure or fire protection systems
- ❏ Damage to fire-resistive assemblies

FIGURE 5.1 ◆ Several factors that may contribute to a working hypothesis for an incendiary fire or explosion. *Derived from NFPA 921* [NFPA 2004a, chap. 19].

◆ 5.2 CLASSIFICATION OF MOTIVE

A jury often wants to have a demonstrable **motive** considered with the evidence so that they can justify the verdict to themselves, even though it is not a legal requirement. Although motive is not essential to establish the crime of arson, and need not be demonstrated in court, the development of a motive frequently leads to the identity of the offender. Establishing motive also provides the prosecution with a vital argument when presented to the judge and jury during trial. It is thought that the motive in an arson case often becomes the mortar that holds together the elements of the crime.

It is in the area of motives that most of the literature on fire setting and arson is concentrated, and this research offers a number of classification schemes and typologies, most often based on motives. Several of the earlier typologies contributed significantly to the current understanding of the motives and psychological profiles of arsonists (Hurley and Monahan 1969; Icove and Estepp 1987; Inciardi 1970; Levin 1976; Lewis and Yarnell 1951; Robbins and Robbins 1967; Steinmetz 1966; Vandersall and Wiener 1970; Wolford 1972). Note that the Lewis and Yarnell (1951) study is cited for historical purposes, since it was based on a very limited sample base. Geller 1992 offers an exhaustive review of that literature and identifies 20 or more attempts to classify arsonists into typologies.

For classification purposes, FBI behavioral science research defines *motive* as an inner drive or impulse that is the cause, reason, or incentive that induces or prompts a specific behavior (Rider 1980). A motive-based method of analysis can be used to identify personal traits and characteristics exhibited by an unknown offender (Douglas et al. 1992 and 1997; Icove et al. 1992). For legal purposes, the motive is often helpful in explaining why an offender committed his or her crime. However, motive is not normally a statutory element of a criminal offense. The motivations discussed in this chapter are also outlined and described in *NFPA 921—Guide for Fire and Explosion Investigations* (NFPA 2004a) and the FBI's *Crime Classification Manual* (Douglas et al. 1992 and 1997; Icove et al. 1992).

OFFENDER-BASED MOTIVES

Law enforcement–oriented studies on arson motives are **offender-based;** that is, they look at the relationship between the behavioral and the crime scene characteristics of the offender as it relates to motive. One of the largest present-day offender-based studies consists of 1016 interviews of both juveniles and adults arrested for arson and fire-related crimes, during the years 1980 through 1984, by the Prince George's County Fire Department (PGFD), Fire Investigations Division (Icove and Estepp 1987). These offenses include 504 arrests for arson, 303 for malicious false alarms, 159 for violations of bombing/explosives/fireworks laws, and 50 for miscellaneous fire-related offenses.

The overall purpose of the PGFD study was to create and promote the use of motive-based offender profiles of individuals who commit incendiary and fire-related crimes. Prior studies failed to address completely the issues confronting modern law enforcement. Of primary concern were the efforts to provide logical, motive-based investigative leads for incendiary crimes.

The study was conducted primarily because fire and law enforcement professionals were entitled to take on themselves the task of conducting their own independent research into violent incendiary crimes. The PGFD study determined that the following motives are most often given by arrested and incarcerated arsonists (Icove and Estepp 1987):

- ◆ Vandalism
- ◆ Excitement
- ◆ Revenge
- ◆ Crime concealment
- ◆ Profit
- ◆ Extremist beliefs

In some cases several of the preceding motives (mixed motives) may be determined. Examples might include a businessperson who murders a partner after work, sets a fire at their store to cover up the crime, removes valuable merchandise, and files an inflated insurance claim. This hypothetical case would exhibit overlapping motives of revenge, crime concealment, and profit. See a later section for a more detailed discussion on mixed motives.

Scientific literature about and research on arsonists have historically been conducted largely from the forensic psychiatric viewpoint (Vreeland and Waller 1978). Many forensic researchers do not necessarily assess the crime from the law enforcement perspective. They may have limited access to complete adult and juvenile criminal records and investigative case files and must often rely on the self-reported interviews of the offenders as being totally truthful. The researchers do this without the capabilities and time to validate the information through follow-up investigations. Other researchers have cited that methodological difficulties including small sample sizes of interviews and skewed databases may also have biased the previous studies (Harmon, Sosner, and Wiederight 1985).

VANDALISM-MOTIVATED ARSON

Vandalism-motivated arson is defined as malicious or mischievous fire setting that results in damage to property (see Table 5.1 and Figure 5.2). Some of the most common targets of these usually juvenile arsonists are schools, school property, and educational facilities. Vandals also frequently target abandoned structures and combustible vegetation. The typical vandalism-motivated arsonist will use available materials to set fires with book matches and cigarette lighters as the ignition devices.

Usually, vandalism-motivated arsonists will leave the scene and not return to it. Their interest is in setting the fire, not watching it or the firefighting activities it generates. On average, vandalism arsonists will be questioned twice before being arrested and charged. They will offer no resistance when arrested but will qualify and minimize their responsibility. After an initial not guilty plea, vandalism-motivated arsonists will typically change the plea to guilty before trial.

EXAMPLE 5.1 Vandalism—Case Study of a Vandalism-Motivated Serial Arsonist

A 19-year-old high-school dropout was responsible for a series of arsons in a northeastern city that were set using available materials and lit by a cigarette lighter. He admitted to setting 31 fires in vacant buildings and garages as well as dumpsters and old vehicles. He was questioned twice before being arrested and formally charged with nine of the house fires. When interviewed he stated, "I just burned them for the hell of it. You know, just to have something to do. Those old houses and that other stuff was not worth anything anyway."

His mother and father had divorced when he was 2 years old, and he had lived alternately with his mother and grandmother. He had no contact with his father and said that his mother had remarried several times. After dropping out of school in the tenth grade, he had worked sporadically at unskilled jobs and continued to live with his grandmother.

Several times, he reported just striking a match, tossing it into dry leaves or grass, and walking away without even seeing if it ignited. "I did not care nothing about watching no fire. It was just something to do. There was not much going on most of the time, you know. Just hanging out. It was just kid stuff, just for the hell of it. Half of the guys I knew that hung out would set a fire for the hell of it" (Sapp et al. 1995).

TABLE 5.1 ◆ Vandalism-Motivated Arson

Characteristics
  Victimology:
    targeted property
	Educational facilities common target
	Residential areas
	Vegetation (grass, brush, woodland, and timber)

Crime scene indicators
  frequently noted
	Multiple offenders acting spontaneously and
	    impulsively
	Crime scenes reflect spontaneous nature of the
	    offense (disorganized)
	Offenders use available materials at the scene and
	    leave physical evidence behind (shoeprints,
	    fingerprints, etc.)
	Flammable liquids occasionally used
	Entrance may be gained through windows of
	    secured structures
	Matchbooks, cigarettes, and spray-paint cans
	    (graffiti) often present
	Materials missing from scene and general destruction
	    of property

Common forensic findings	Flammable liquid analysis Presence of fireworks Glass particles on suspect's clothing if entered by breaking a window

Investigative considerations	Typical offender is a juvenile male with 7–9 years of
	    formal education
	Records of poor school performance
	Not employed
	Single and lives with one or both parents
	Alcohol and drug use usually not associated
	Offender may already be known by police and may
	    have arrest record
	Majority of offenders live more than 1 mi from the
	    crime scene
	Most offenders flee from scene immediately and do
	    not return
	If the offenders return, they view the fire from a safe
	    vantage point
	Investigators should solicit assistance from school,
	    fire, and police

Search warrant suggestions	Spray-paint cans Items from the scene Explosive devices Flammable liquids Clothing: evidence of flammable liquid, glass particles Shoes: shoeprints, flammable liquid traces

Source: Updated from Icove et al. 1992.

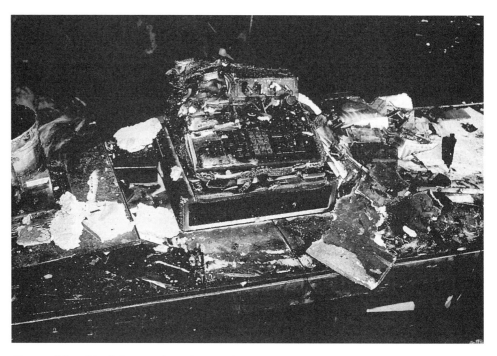

FIGURE 5.2 ◆ Vandalism-motivated arson showing a superficially damaged cash register. *Courtesy of D. J. Icove.*

EXCITEMENT-MOTIVATED ARSON

Excitement-motivated offenders include seekers of thrills, attention, recognition, and, rarely but importantly, sexual gratification (see Table 5.2 and Figure 5.3). The arsonist who sets fires for sexual gratification is quite rare.

Potential targets of the excitement-motivated arsonist run the full spectrum from so-called nuisance fires to fires in occupied apartment houses at nighttime. A limited number of firefighters have been known to set fires so they can engage in the suppression effort (Huff 1994). Security guards have set fires to relieve boredom and gain recognition. Research into this motive category further categorizes excitement-motivated arsonists into several subclassifications, including thrill-, sex-, and recognition/attention-motivated arsonists (Icove et al. 1992).

EXAMPLE 5.2   Excitement—Case History of an Excitement/Recognition-Motivated Serial Arsonist

A 23-year-old volunteer fireman was charged with setting a series of fires shortly after joining the fire department. The fires initially were set in trash cans and dumpsters and then progressed to unoccupied and vacant structures.

He became a suspect in the fires when he consistently arrived first on the scene and frequently reported the fire. He stated that he set the fires so that he could get practice and others would view him as a good fireman. He commented that his father was "real proud of his firefighter son" (Sapp et al. 1995).

TABLE 5.2 ◆ **Excitement-Motivated Arson**	
Characteristics Victimology: targeted property	Dumpsters Vegetation (grass, brush, woodland, timber) Lumber stacks Construction sites Residential property Unoccupied structures Locations that offer a vantage point to observe fire suppression and investigation safely
Crime scene indicators frequently noted	Often adjacent to outdoor "hangouts" Often uses available materials on hand to start small fires If incendiary devices are used, they usually have time- delayed triggering mechanisms Offenders in the 18–30 age group more prone to using accelerants Match/cigarette delay device frequently used to ignite vegetation fires Small group of offenders is motivated by sexual perversions, leaving ejaculate, fecal deposits, pornographic material
Common forensic findings	Fingerprints, vehicle and bicycle tracks Remnants of incendiary devices Ejaculate or fecal materials
Investigative considerations	Typical offender is a juvenile or young adult male with 10+ years of formal education Offender unemployed, single, and living with one or both parents from middle- to lower-class bracket Offender generally is socially inadequate, particularly in heterosexual relationships Use of drugs or alcohol limited usually to older offenders History of nuisance offenses Distance offender lives from the crime scene determined through a cluster analysis Some offenders do not leave, mingling in crowds to watch the fire Offenders who leave usually return later to assess the damage and their handiwork
Search warrant suggestions	Vehicle: material similar to incendiary devices, floor mats, trunk padding, carpeting, cans, matchbooks, cigarettes House: material similar to incendiary devices, clothing, shoes, cans, matchbooks, cigarettes, lighter, diaries, journals, notes, logs, records and maps documenting fires, newspaper articles, souvenirs from the crime scene

Source: Updated from Icove et al. 1992.

FIGURE 5.3 ◆ Excitement-motivated arsonists sometimes target garages, which contain and have easily accessible all the materials and fuels needed for setting the fire. *Courtesy of D. J. Icove.*

REVENGE-MOTIVATED ARSON

Revenge-motivated fires are set in retaliation for some injustice, real or imagined, perceived by the offender (see Table 5.3 and Figure 5.4). Often, revenge is also an element of other motives. This concept of mixed motives is discussed later in this chapter.

What may be of concern to investigators is that the event or circumstance that is perceived as unjust may have occurred months or years before the fire-setting activity (Icove and Horbert 1990). For threat assessment purposes, this time delay may not readily be recognized, and investigators are urged to pursue the historical incidents in which a person or property has been targeted.

Revenge and spite-motivated fires account for more of the serious arson cases owing to the "overkill" reflected in the actions of the offender. In these cases, when one container of a flammable liquid would suffice in setting a building on fire or incinerating a body, the offender may use much larger quantities, reflecting the "rage" or "vengeance" of the act.

The broad classification of revenge-motivated arsonists is further divided into subgroups based on the target of the retaliation. Studies show that serial revenge-motivated arsonists are more likely to direct their retaliation at institutions and society than at individuals or groups (Icove et al. 1992).

**EXAMPLE 5.3   Revenge—Case History of an Institutional Retaliation
                 "Revenge" Serial Arsonist**

John (not his real name) is 31 years old and claimed to have set more than 60 fires at various local government facilities in the city where he lives. His fire-setting activities have taken place since he was 19 years old.

John started setting fires after he was sentenced to 180 days in the local jail facility for a petty theft. He claimed to have set 5 fires while in the jail and then later 20–25 trash can fires in the local city hall. His method of operation consisted of simply walking through and dropping lighted matches into the trash containers.

His stated motive for setting the fires was "to cost the city some trouble and money. They did not treat me fair and I will get even." When asked when his revenge against the city would be satisfied, his response was "when the whole damn city hall burns down and the jail too" (Sapp et al. 1995).

TABLE 5.3 ◆ Revenge-Motivated Arson

Characteristics
  Victimology:
    targeted property
	Victim of revenge fire generally has a history of interpersonal or professional conflict with offender (lovers' triangle, landlord/tenant, employer/employee)
	Tends to be an intraracial offense
	Female offenders usually target something of significance to victim (vehicle, personal effects)
	Ex-lover offender frequently burns clothing, bedding, and/or personal effects
	Societal revenge targets displace aggression to institutions, government facilities, universities, corporations

Crime scene indicators frequently noted	Female offenders usually burn an area of personal significance, using victim's clothing or other personal effects Male offenders begin with an area of personal significance, tend to overkill by using more accelerants or incendiary devices than are necessary
Common forensic findings	Laboratory tests for accelerants, pieces of incendiary bomb, cloth, fingerprints
Investigative considerations	Offender is predominantly an adult male with 10+ years of formal education If employed, the offender is usually a blue-collar worker of low socioeconomic status Resides in rental property; loner, unstable relationships Event happens months or years after precipitating incident Most often, has some periodic law enforcement contact for burglary, theft, and/or vandalism Use of alcohol more prevalent than drugs, with possible increase after the fire Usually alone at scene and seldom returns after fire started to establish an alibi Lives in affected community, with mobility an important factor Revenge-focused analysis assists in determining true victim Investigative investment significant
Search warrant suggestions	If accelerants suspected: shoes, socks, clothing, bottles, flammable liquids, matchbooks

Source: Updated from Icove et al. 1992.

FIGURE 5.4 ◆ A revenge-motivated fire set on a bed, which is characteristic of a focused target having personal significance. *Courtesy of D. J. Icove.*

EXAMPLE 5.4 Revenge—Case History of a Revenge-Motivated Apartment Fire

An apartment fire was reported at 0909 hours (9:09 A.M.) by a neighbor who saw a woman leaving the apartment from which he saw smoke coming seconds later. Fire crews arrived at 0917 hours to find the apartment well involved with flames emanating from two windows. Interior hose stream attack began at 0922 hours, and control of the main fire was reported at 0930. The apartment consisted of two small rooms (as in Figure 5.5) each approximately 3.6 × 3.6 m (12 × 12 ft). The living room/kitchen had two large windows approximately 1.2 × 1 m high—one next to the front door and one over the sink.

The steel-faced entry door was closed at the time of the fire but was burned to failure when fire crews attempted entry. The doorway to the bedroom had no door, and the bedroom window failed during the fire. Both rooms were very heavily damaged by fire, the living room more so than other areas. There were patterns in the doorway and bedroom clearly indicating the fire had extended into that room from the living room. Only the metal frame and springs of the sofa survived, and there was a V pattern of intense damage to the adjoining walls. The top portions of the kitchen cabinets were severely damaged, but they were charred to floor level. There were two large areas where the floor had burned through, allowing debris to fall into the room below. The gypsum board ceiling of the living room had failed, allowing extension of the fire into the attic above.

The registered occupant had left the apartment about half an hour before the fire, after having an argument with his girlfriend (leaving her behind). He reported that no heaters or appliances, no paints or solvents, and no candles were in use prior to his departure. The original scene investigator concluded that the fire was deliberately ignited with an ignitable liquid based on damage to the floor coverings, the floor penetrations, and the alleged rapidity of fire spread (using an 0909 hours alarm time and 0917 hours fire department arrival time as his time

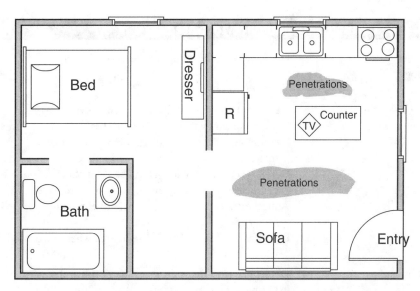

FIGURE 5.5 ◆ Floor plan shows sofa and large windows in small two-room apartment. Sofa was ignited for revenge and drove the living room to flashover, which burned holes through floor. *Courtesy of Mike Dalton.*

frame (not 0930 hours or later as the actual time of suppression). No samples of debris were collected for lab analysis, and no odors of ignitable liquids were detected during the scene examination. An unused "tea light" candle was recovered from the debris in the room below, as were closed, intact cans of paint that the occupant reportedly had in storage.

Unfortunately, no comparison samples of sofa or carpet were collected. The photos of the sofa showed that it was a modern all-synthetic upholstered unit that had burned nearly completely away. Such units are not susceptible to smoldering ignition, but once ignited by an open flame, will produce a very large fire in less than 5 min. In such a small apartment, the ventilation limit would be reached very quickly, but as the large windows failed in the room, more than enough air would be admitted to support a fire (1500 kW per window) far in excess of that needed to achieve and sustain flashover (700–1700 kW, according to NRC spreadsheet calculations). The fire was (at least) approaching flashover when fire crews arrived and observed fire venting from the windows at 0917 hours. Control was reported at approximately 0930 hours; this left at least 13 min (if not more) for a postflashover fire to cause the observed damage without the use of an accelerant. With the elimination of competent accidental ignition sources (by testimony of the occupant), the conclusion was that the girlfriend set fire to the sofa (in a spite-motivated fire) just before leaving the apartment. She eventually pled guilty.

Case study courtesy of John D. DeHaan

CRIME CONCEALMENT–MOTIVATED ARSON

Arson is the secondary criminal activity in this motivational category of **crime concealment-motivated arson** (see Table 5.4 and Figure 5.6). Examples of crime concealment-motivated arsons are fires set for the purpose of covering up a murder or burglary or to eliminate evidence left at a crime scene.

Other examples include fires set to destroy business records to conceal cases of embezzlement and fires set to destroy evidence of an auto theft. In these cases, the arsonist may set fire to the structure to destroy the evidence of the initial crime, to obliterate latent fingerprints and shoeprints, and sometimes to attempt to render useless DNA or serological evidence linking them to a victim left behind to die in the fire.

TABLE 5.4 ◆ Crime Concealment-Motivated Arson

Characteristics
    Victimology:
        targeted property
	Dependent on the nature of the concealment

Crime scene indicators frequently noted	Murder: attempts to obliterate forensic evidence of potential lead value or conceal victim's identity, commonly using an accelerant in a disorganized manner Burglary: uses available materials to start the fire, characteristic of involvement multiple offenders Auto theft: strips and burns vehicle to eliminate fingerprints Destruction of records: sets fire in area where records usually kept
Common forensic findings	Determine if victim is alive at time of fire and why he/she did not escape Document injuries, particularly if concentrated around genitals
Investigative considerations	Common to find alcohol and recreational drug use Offender expected to have history of contacts of arrests by police and fire departments Offender is likely a young adult who lives in the surrounding community and is highly mobile Crime concealment suggests accompanied to scene by coconspirators Murder concealment is usually a one-time event
Search warrant suggestions	Refer to other category of primary motive Gasoline containers Clothing, shoes, glass fragments, burned paper documents

Source: Updated from Icove et al. 1992.

EXAMPLE 5.5 Case History of a Crime-Concealment Serial Arsonist

A 31-year-old unemployed laborer, 3 weeks after being released from prison, admitted responsibility for burglarizing and setting fires to 12 houses in a residential area over a period of 7 months. All but one of the fires occurred while the owners were out of town.

The offender stated, "I just drove through and looked for newspapers and other things where the people were gone for a while. I just took money and jewelry and stuff that would get lost or burn up in the fire. I would pour gasoline on everything and use a candle to light it off."

The fires were all identified as arson fires because of their rapid development and the detected presence of flammable liquids in the debris. The extensive damage made it difficult for the owners to determine if valuables in the homes had been removed. The offender stated that he was identified when "one of the ladies spotted a ring in a pawn shop in another town. I would always take stuff somewhere else to sell it. What happened was she recognized the ring and called the cops."

His criminal history record consisted of a burglary conviction when stolen property was recovered and traced back to him. The offender's rationalization was, "I figured it out that if

FIGURE 5.6 ◆ A crime concealment–motivated arson where valuable inventory was removed prior to the fire. *Courtesy of D. J. Icove.*

the fire burned everything nobody would know anything was gone. Last time, I got caught on a serial number, so this time I decided I would not leave any way for them to know what was gone" (Sapp et al. 1995).

PROFIT-MOTIVATED ARSON

Offenders in the category of **profit-motivated arson** expect to profit from their fire setting, either directly for monetary gain or more indirectly to profit from a goal other than money (see Table 5.5 and Figure 5.7). Examples of direct monetary gain include insurance fraud, liquidation of property, dissolution of businesses, destruction of inventory, parcel clearance, and gaining employment. The latter is exemplified by a construction worker's wanting to rebuild an apartment complex he destroyed or an unemployed laborer's seeking employment as a forest firefighter or as a logger to salvage burned timber.

Arson for profit may have interesting twists when the offender benefits directly or indirectly. Arsonists have set fire to western forests to have their equipment rented out to support part of the suppression effort. What may be the most disturbing of all are cases in which parents murder their own children for profit, with fire used to cover the intentional death of the child. Although this motive is uncommon, it is by no means rare (Huff 1997). Cases have been documented in which an insured child is murdered, but more commonly the parents wish to profit from getting rid of a

TABLE 5.5 ◆ Profit-Motivated Arson

Characteristics Victimology: targeted property	Property targeted includes residential, business, and transportation (vehicles, boats, etc.)
Crime scene indicators frequently noted	Usually well-planned and methodical approach, with crime scene more organized because it contains less physical evidence With large businesses, multiple offenders may be involved Intent is complete destruction with excessive use of accelerants, incendiary devices, multiple points of fire origin, trailers Lack of forced entry Removal or substitution of items of value prior to fire
Common forensic findings	Use of sophisticated accelerants (water-soluble) or mixtures (gasoline and diesel fuel) Components of incendiary devices
Investigative considerations	Primary offender is an adult male with 10+ years of formal education Secondary offender is sometimes the "torch," who is usually a male, 25–40 years of age, and unemployed Offender generally lives more than 1 mi from the crime scene, may be accompanied to the scene, leaves, and usually does not return Indicators of financial difficulty Decreasing revenue with increasing and unprofitable production costs Technology outdates processes Costly lease or rental arrangements Personal expenses paid with corporate funds Hypothetical assets, overstated inventory levels Pending litigation, bankruptcy Prior fire losses and claims Frequent changes in property ownership, back taxes, multiple liens
Search warrant suggestions	Check financial records If evidence of fuel/air explosion at scene, check local emergency rooms for patients with burn injuries Determine condition of utilities as soon as possible

Source: Updated from Icove et al. 1992.

perceived nuisance or hindrance—their own child. This is particularly true in a single-parent or divorce situation in which the child is viewed as an impediment to freedom or marriage. Filicide by fire is discussed in a later section of this chapter.

Other nonmonetary reasons from which arsonists may profit range from setting brush fires to enhance the availability of animals for hunting to burning adjacent properties to improve the view. Also, fires have been set to escape an undesirable environment, such as in the case of seamen who do not wish to set sail (Sapp et al. 1993, 1994).

FIGURE 5.7 ◆ Profit-motivated arsons to leverage overinflated insurance sometimes involve well-planned fires set in vacant dwellings fueled with excessive amounts of accelerants. *Courtesy of D. J. Icove.*

EXAMPLE 5.6   Case History of a Profit-Motivated Serial Arsonist

Arnold (not his real name) was a professional arsonist (a "torch") and was serving 2 years in prison for one of his arsons. He admitted having burned down 35 to 40 vacant houses over his career. He became involved in setting fires for profit when talking with a real estate agent who was trying to find him an apartment. "He asked me if I would find somebody to burn a place. Did I know someone who was involved in demolition? I told him yeah and it started from there."

Arnold then partnered with the real estate agent and the demolition specialist. The real estate agent identified the targeted properties and made sure they were vacant. The demolition specialist taught Arnold how to set the fires, even accompanying him on his first job.

Arnold claimed that the fire-setting technique he used never resulted in a confirmed arson. "They could not tell. They would say it is under investigation. When they say it is under investigation, they are pretty sure it is arson but they cannot prove it." His method of operation was to pour 19 to 38 L (5 to 10 gal) of odorless white gas in the attic and leave a time-delayed chemical timer, literally burning the house from the top down. The ignited gasoline resulted in a fast-developing fire that collapsed the roof and overwhelmed the suspicions of the fire investigators, who found it hard to determine that an accelerant was used.

Arnold was concerned that no one would be injured in the fires he set. "I never burned any place where anybody was inside. I made sure I was not hurting anybody. That was important, it really was."

After being implicated with the real estate agent for an attempted arson, Arnold served 2 years in prison. "It was wrong, but at the time I was making good money and I was not hurting anybody. It was easy, but it is a dangerous job. I was afraid every minute while I was doing it."

As an afterthought, Arnold claimed that he was also a victim of his arsons. "I am the one that got hurt. I was underpaid. I was making $700 or $800 a fire, the real estate man was making about $10,000 on his share of the insurance policy, and I am the one that got sent away and everything" (Sapp et al. 1995).

EXTREMIST-MOTIVATED ARSON

Offenders involved in **extremist-motivated arson** may set fires to further social, political, or religious causes (see Table 5.6). Examples of extremist-motivated targets include abortion clinics, slaughterhouses, animal laboratories, fur farms, furrier outlets, and even, now, sport-utility vehicle (SUV) dealers. The targets of political terrorists reflect the focus of the terrorists' wrath. Random target selection also simply

TABLE 5.6 ◆ Extremist-Motivated Arson

Characteristics Victimology: targeted property	Analysis of targeted property essential in determining specific motive and represents the antithesis of the offender's belief
	Targets include research laboratories, abortion clinics, businesses, religious institutions
Crime scene indicators frequently noted	Crime scene reflects organized and focused attack by the offender(s)
	Frequently employ incendiary devices, leaving a nonverbal warning or message
	Overkill when setting fire
Common forensic findings	Extremist arsonists are more sophisticated offenders and often use incendiary devices with remote or time-delay ignition
Investigative considerations	Offender is frequently identified with cause or group in question
	May have previous police contact or an arrest record with crimes such as trespassing, criminal mischief, or civil rights violations
	Postoffense claims should undergo threat assessment examination
Search warrant suggestions	Literature: writings, paraphernalia pertaining to a group or cause, manuals, diagrams
	Incendiary device components, travel records, sales receipts, credit card statements, bank records indicating purchases
	Flammable materials: materials, liquids

Source: Updated from Icove et al. 1992.

generates fear and confusion. Self-immolation has also been carried out as an extremist act.

EXAMPLE 5.7   Extremist-Motivated Arson

This is a fictional account of an actual extremist-motivated arson targeting a U.S. Government agency (ADL 2003).

The former head of a former antigovernment extremist group was convicted by a federal jury for setting fire to an Internal Revenue Service office. The jury found the 48-year-old leader guilty of destruction of government property and interference with IRS employees.

Prosecutors said he and two other accomplices used 19 L (5 gal) of gasoline and a timing device to start a fire that destroyed the IRS office. The fire caused $2.5 million in damage, and a firefighter was seriously injured while battling the blaze. The leader was also convicted of witness tampering and suborning perjury for asking a witness to lie when testifying before the federal grand jury. He reportedly also threatened a witness to prevent their cooperation with law enforcement officers. Prosecutors stated that the three men were motivated by antigovernment sentiments and set the fire as a protest against paying taxes.

◆ 5.3 OTHER MOTIVE-RELATED CONSIDERATIONS

Other factors involving motives appear in the literature and are important to address. This information is provided to clarify inaccuracies and misconceptions.

PYROMANIA

Perhaps most conspicuous by its absence is any mention of **pyromania** in this discussion of motivations. For an authoritative definition of the term, refer to the American Psychological Association's (APA 1994) *Diagnostic and Statistical Manual of Mental Disorders,* 4th ed. (DSM-IV), the standard for psychological and psychiatric diagnoses for over 40 years.

A review reveals that each edition of the DSM has treated this topic differently. The current edition, DSM-IV, does not list pyromania as a diagnosed personality disorder. The definition of this disorder has cycled through the years owing to varying opinions and the lack of a solid definition.

DSM-IV's diagnostic criteria for pyromania include one or more of the following conditions.

A. Deliberate and purposeful fire setting on more than one occasion.
B. Tension or affective arousal before the act.
C. Fascination with, interest in, curiosity about, or attraction to fire and its situational contexts (e.g., paraphernalia, uses, consequences).
D. Pleasure, gratification, or relief when setting fires, or when witnessing or participating in their aftermath.
E. The fire setting is not done for monetary gain, as an expression of sociopolitical ideology, to conceal criminal activity, to express anger or vengeance, to improve one's living circumstances, in response to a delusion or a hallucination, or as a result of impaired judgment (e.g., in Dementia, Mental Retardation, Substance Intoxication).
F. The fire setting is not better accounted for by Conduct Disorder, a Manic Episode, or Antisocial Personality Disorder.

Researchers ask if the fire-setting impulses—characteristic of the various definitions of pyromania—could be a manifestation of some other disorder. Some report that the "irresistible impulse" to set fires may actually just be an impulse not resisted (Geller, Erlen, and Pinkas 1986).

Sexual fantasies or desires have often been linked to pyromania in the popular literature, but this is not borne out by the interview research. Sex as a motive is highly overrated. In fact, experiments using penile response as an indicator of sexual arousal showed no correlation that sexual motivation is commonly involved in arson (Quinsey, Chaplin, and Upfold 1989). In the study, the penile responses of 26 fire setters and 15 non–fire setters were recorded and compared when the subjects were exposed to audiotaped narratives of neutral, heterosexual activity, and fire-setting activities of several motives (sexual, excitement, insurance, revenge, heroism, and power). There were no significant differences in the responses of fire setters and non–fire setters to any of the narratives they heard.

The so-called motiveless arsonist in many cases knows his or her motive for setting fires, but it does not necessarily make sense to normal outsiders. The arsonist may also lack the capacity to express such concerns.

Fire investigators are cautioned not to label a subject a pyromaniac, since this is a diagnosis to be made by a mental health professional. Each field has a bias or different slant on its view of pyromania, so investigators, criminologists, psychologists, and psychiatrists are working from slightly different definitions.

MIXED MOTIVES

Interviews conducted with incarcerated arsonists underscore the complexity of human behavior, particularly when **mixed motives** arise. When questioned about the motives for their arsons, they gave responses indicating that there were often secondary and supplementary motives in addition to their primary motive (vandalism, excitement, revenge, crime concealment, profit, or extremism).

It should also be noted that arson can be used directly as a weapon, with the intent only of killing the "target." Such homicides can be linked to a wide variety of motives, including self-defense. A thorough reconstruction of the entire incident, rather than just the fire-setting event, may reveal the offender's actual intent for the homicide.

Researchers, most notably Lewis and Yarnell (1951), assert that revenge is present as a motive in all arsons to a greater or lesser degree. Motives for arson, like other aspects of human behavior, often defy a structured, unbending definition. Another rationale is that arsonists may lack the social and communicative skills necessary to articulate their motives clearly. Embarrassment about the true motive may also lead to providing an alternative or false motive to investigators or mental health professionals.

Adding the elements of power and revenge reveals the problem with strict, unyielding classification. Fire investigators also should be aware that since arson is a criminal tool, motivations may change based on the target or situation.

The best examples of many of the factors that arise in a serial arson case are given in an actual account of an arson from an offender's point of view: the well-articulated account by a female serial arsonist who used the pen name Sarah Wheaton (2001). Included in her published article are edited excerpts from her discharge summary, which included a diagnosis of borderline personality disorder. Her treatment therapy included biofeedback, social skills training, and clomipramine, a psychotropic drug. At the time of writing the article, Ms. Wheaton stated that she was fire-thinking-free for 8 months.

EXAMPLE 5.8   Memoirs of a Serial Fire Setter

Sarah Wheaton (not her real name), a former self-admitted compulsive serial fire setter, is now working on a master's degree in psychology. At the end of her college freshman year, during the summer of 1993, she was involuntarily admitted to a psychiatric hospital for 2 weeks for treatment for fire setting.

The following material in italics is a direct quote from Ms. Wheaton's (2001) description of her treatment and perception of what she experienced as a serial arsonist. Her viewpoint was published as "Personal Accounts: Memoirs of a Compulsive Firesetter" in *Psychiatric Services* 52:1035–36. Copyright 2001, the American Psychiatric Association; http://PS.psychiatryonline. org. Reprinted by permission.

__Reason for admission:__ This 19-year-old single female was living in a dorm at the University of California at the time of admission. The patient had been involved in bizarre activity, including lighting five fires on campus that did not remain lit.

__History of present illness:__ This young lady is a highly intelligent, highly active young woman who had been the president of her class in her high school for all four years. She had been going full-time to the university and working full-time at a pizzeria. She had called the police threatening suicide. When she was brought in, she was in absolute and complete denial. Patient claimed that everything was wonderful and she did not need to be here. She was admitted for 72-hour treatment and evaluation.

__Hospital course:__ The hospital course initially was quite tempestuous. It was difficult to determine what the diagnosis was, as the patient was on the one hand very bright and very endearing to the staff and on the other hand was very unpredictable. She jumped over the wall on the patio (AWOL). The police were called and ultimately brought her back. She was subsequently certified to remain in custody for up to 14 days for intensive treatment because she attempted to cut herself with plastic and/or glass. She became more open after this, more tearful and at times more vulnerable. She tended to run from issues, to try to help everyone else, and not look at herself. Her father was seen in family therapy with her by a social worker.

   Mother is reported by father to be both an alcoholic and have a history of bipolar illness. The patient herself reported sexual abuse by an older stepbrother when she was about age nine to age 11.

   Initially I intended to use antimanic medication with her, either carbamazepine or lithium, but opted not to as she was adamant against medication. The patient did seem to stabilize without medication. She showed dramatic improvement, although the staff and I still have concern. The stable environment she has been able to pull together may not exist after discharge. She is scheduled to go to Washington, D.C., on July 1 to work as an intern in the office of one of the congressional representatives. She had done this two years ago working as a page.

__Aftercare instructions:__ No follow-up appointment is scheduled because she is discharged today and will be leaving for Washington on the first.

__Prognosis given by psychiatrist:__ Prognosis is very guarded given the severity of her condition.

__Mental status:__ She firmly denies . . . destructive ideation, including firesetting at this time.

__Discharge diagnosis__ [33 hospitalizations after the initial hospitalization]: Axis I. Major depressive disorder, recurrent, with psychosis. Axis II. Obsessive-compulsive personality disorder; history of pyromania; Borderline personality disorder. Axis III. Asthma. Axis V. GAF 45.

As a student of psychology Ms. Wheaton's article accurately traces many of the traits and characteristics of serial fire setters, particularly those reported in the literature and supported by interviews of law enforcement agents specializing in arson. Quoted next are her observations on how fire dominated her life from preschool to college.

Fire became a part of my vocabulary in my preschool days. During the summers our home would be evacuated because the local mountains were ablaze. I would watch in awe.

Below I have listed some of my thoughts and behaviors eight years after the onset of deviant behavior involving fire. I have also included suggestions for helping a firesetter.

Firesetting behaviors on a continuum. *Each summer I look forward to the beginning of fire season as well as the fall—the dry and windy season. I set my fires alone. I am also very impulsive, which makes my behavior unpredictable. I exhibit paranoid characteristics when I am alone, always looking around me to see if someone is following me. I picture everything burnable around me on fire.*

I watch the local news broadcasts for fires that have been set each day and read the local newspapers in search of articles dealing with suspicious fires. I read literature about fires, firesetters, pyromania, pyromaniacs, arson, and arsonists. I contact government agencies about fire information and keep up-to-date on the arson detection methods investigators use. I watch movies and listen to music about fires. My dreams are about fires that I have set, want to set, or wish I had set.

I like to investigate fires that are not my own, and I may call to confess to fires that I did not set. I love to drive back and forth in front of fire stations, and I have the desire to pull every fire alarm I see. I am self-critical and defensive, I fear failure, and I sometimes behave suicidally.

Before a fire is set. *I may feel abandoned, lonely, or bored, which triggers feelings of anxiety or emotional arousal before the fire. I sometimes experience severe headaches, a rapid heartbeat, uncontrollable motor movements in my hands, and tingling pain in my right arm. I never plan my fire, but typically drive back and forth or around the block or park and walk by the scene I am about to light on fire. I may do this to become familiar with the area and plan escape routes or to wait for the perfect moment to light the fire. This behavior may last anywhere from a few minutes to several hours.*

At the time of lighting the fire. *I never light a fire in the exact place other fires have occurred. I set fires at random, using material I have just bought or asked for at a gas station—matches, cigarettes, or small amounts of gasoline. I do not leave signatures to claim my fires. I set fires only in places that are secluded, such as roadsides, back canyons, cul-de-sacs, and parking lots. I usually set fires after nightfall because my chances of being caught are much lower then. I may set several small fires or one big fire, depending on my desires and needs at the time. It is at the time of lighting the fire that I experience an intense emotional response like tension release, excitement, or even panic.*

Leaving the fire scene. *I am well aware of the risks of being at the fire scene. When I leave a fire scene, I drive normally so that I do not look suspicious if another car or other people are nearby. Often I pass in the opposite direction of the fire truck called to the fire.*

During the fire. *Watching the fire from a perfect vantage point is important to me. I want to see the chaos as well as the destruction that I or others have caused. Talking to authorities on the phone or in person while the action is going on can be part of the thrill. I enjoy hearing about the fire on the radio or watching it on television, learning about all the possible motives and theories that officials have about why and how the fire started.*

After the fire is out. *At this time I feel sadness and anguish and a desire to set another fire. Overall it seems that the fire has created a temporary solution to a permanent problem.*

Within 24 hours after the fire. *I revisit the scene of the fire. I may also experience feelings of remorse as well as anger and rage at myself. Fortunately, no one has ever been physically harmed by the fires I have set.*

Several days after the fire. I revel in the notoriety of the unknown firesetter, even if I did not set the fire. I also return again to see the damage and note areas of destruction on an area map.

Fire anniversaries. I always revisit the scene on anniversary days of fires that I or others set in the area.

Fires not my own. A fire not my own offers excitement and some tension relief. However, any fire set by someone else is one I wish I had set. The knowledge that there is another firesetter in the area may spark feelings of competition or envy in me and increase my desire to set bigger and better fires. I am just as interested in knowing the other firesetters' interests or motives for lighting their fires.

Suggestions for helping a firesetter. The likelihood of recidivism is high for a firesetter. The firesetter should be able to count on someone always being there to talk to about wanting to set fires. Firesetting may be such a big part of the person's life that he or she cannot imagine giving it up. This habit in all aspects fosters many emotions that become normal for the firesetter, including love, happiness, excitement, fear, rage, boredom, sadness, and pain.

A firesetter should be taught appropriate problem-solving skills and breathing and relaxation techniques. Exposure to burn units and disastrous fire scenes may be therapeutic and may enable the firesetter to talk openly about physical and emotional reactions. Doing so will not only help the firesetter but also give mental health professionals a deeper understanding of the firesetter's obsession.

The self-reported insight of this individual clearly confirms the findings of scientific, psychiatric, and law enforcement research. Of particular note are the pre- and postoffense behaviors at the fire scene, the significance of anniversaries, and the highly suggestive impact of the publicity of fires set by other arsonists. Astute fire investigators can use this information to assist them in identifying and solving cases set by compulsive serial arsonists.

EXAMPLE 5.9   The Case of a Mother–Son Serial Arson Team

For more than 2 years a residential upper-middle-class suburb of some 59,000 population was plagued by a series of "small" fires and false alarms. Most of the fires involved roadside grass and median-strip ground cover and garbage cans, but several were in large commercial dumpsters alongside wood-sided commercial and residential structures (with possibly serious consequences). Nearly all the false alarms (phone calls and pull stations) targeted one school. After a review of approximately 800 incidents queried from the District's NFIRS database (by "Incident Type" and by "Location"), investigators compiled a list of some 50 similar incidents. When these were plotted on a local street map (Figure 5.8), a pattern focused on the school and extending the length of the city along a major north-south corridor (Novato Blvd.) was discernible. The concentration around the school led investigators to suspect the arsonist lived nearby.

The actual ignitions were not observed, although a witness in one incident reported seeing a dark-colored SUV with two suspects leaving the scene. The scenes were sufficiently dispersed to suggest the arsonist was not on foot but most probably in a vehicle. At one grass fire, a matchbook cover found near the roadside bore an identifiable latent fingerprint, but it did not have any hits in the database. (The book of matches itself was found burned up in the grass several feet away.) Sunflower seed hulls found near one fire suggested a lead, but no good suspects. Temporal profiling revealed that most of the similar fires occurred between 4:00 and 8:00 P.M. on weeknights, with a higher frequency during the summer (May–July). See Figure 5.9.

When a fire seriously damaged a house (1320 Monte Maria) on August 15, 2004, investigators found several leads. Interviews with neighbors revealed that two of the residents—a 42-year-old mother (widow) and her 24-year-old developmentally disabled son—had been seen going

Street Atlas USA ® 2004 Plus

101

San Marin Dr.

San Marin Dr.

101

Novato Blvd.

02-1158
03-3294

02-2325 83
02-4068 25

03-2517

Novato Blvd.

03-2123

03-2338
02-4058
03-3030
02-0999

02-0999

03-0957
02-3273

4th St.

02-0999

DeLong Ave.

03-3596

101

03-2024

04-0468

04-0867

Novato Blvd.

Lauren Ave.

Hill Rd.

02-2546
02-1771

Garfield St.

Redwood Blvd.

Rowland Blvd.

04-1595
02-1331

02-0

02-1879

Lynwood Dr.

Monte Maria

© 2003 DeLorme
www.delorme.com

★
MN (15.2° E)

ft
0   800 1600 2400 3200 4000
Data Zoom 13-0

FIGURE 5.8 ◆ Street map of city showing locations of school, grass, and dumpster fires over a 2-year period.
Note the pattern along a major thoroughfare near the offenders' house (in cluster of fires in lower right corner).
Based on public records. Case courtesy of Novato Fire Protection District.

256

TEMPORAL ANALYSIS OF SERIAL ARSON INCIDENTS, 2002

Variable	Jan.	Feb.	Mar.	Apr.	May	June	July	Aug.	Sept.	Oct.	Nov.	Dec.	Total
0800 – 1200											O		1
1200 – 1600					*	◈					O		3
1600 – 2000				◈			O**	□					5
2000 – 2400				O						×			2
2400 – 0400				◈							□		2
0400 – 0800	O		□										2
Monday				◈									1
Tuesday	O			O							O		3
Wednesday							O	□			□		3
Thursday						*							1
Friday			□		*		*			×			4
Saturday				◈							O		2
Sunday						◈							1
Total	1	0	1	3	1	1	3	1	0	1	3	0	15

Fire Type:

O Trash
□ Dumpster
◈ Pull Alarm
✳ Roadside
▲ Structure
♠ Wildland
× Railroad

TEMPORAL ANALYSIS OF SERIAL ARSON INCIDENTS, 2003

Variable	Jan.	Feb.	Mar.	Apr.	May	June	July	Aug.	Sept.	Oct.	Nov.	Dec.	Total
0800 – 1200											□		1
1200 – 1600			◈	◈◈◈		◈**	□						8
1600 – 2000			□◈	O◈	◈	◈*	**♠	♠	*		V		12
2000 – 2400				◈◈		□□	□♠♠		*		□		9
2400 – 0400										O			1
0400 – 0800			□		◈		O						3
Monday			□					♠			□		3
Tuesday				◈			O		*				3
Wednesday						*	**♠♠						5
Thursday			□	O	◈	◈							4
Friday				◈◈		□**	□□						7
Saturday			◈◈	◈◈	◈	□◈	♠				□		9
Sunday				◈						O	V		2
Total	0	0	4	7	2	7	8	1	1	1	3	0	34

Fire Type:

O Trash
□ Dumpster
◈ Pull Alarm
✳ Roadside
♠ Wildland
× Railroad
V Vehicle

FIGURE 5.9 ◆ Target and temporal analysis (time, date, day) of fire events for 2002–04. (Persons responsible were arrested August 2004.) *Based on public records. Case courtesy of Novato Fire Protection District.*

TEMPORAL ANALYSIS OF SERIAL ARSON INCIDENTS, 2004

Variable	Jan.	Feb.	Mar.	Apr.	May	June	July	Aug.	Sept.	Oct.	Nov.	Dec.	Total
0800 – 1200		*											*
1200 – 1600					✳								1
1600 – 2000			☽					◇✳▲▲					5
2000 – 2400													
2400 – 0400		□											1
0400 – 0800	□												1
Monday													
Tuesday		□											1
Wednesday								✳◇					2
Thursday		□	☽										2
Friday													
Saturday	□												1
Sunday		□			✳			▲▲					4
Total	1	3	1	0	1	0	0	4	0	0	0	0	10

* Time not available
  for two of the
  dumpster fires.

Fire Type:

O Trash
□ Dumpster
◇ Pull Alarm
✳ Roadside
▲ Structure
☽ Porta-john

FIGURE 5.9 ◆ (*Continued*)

into and out of the house just minutes prior to the fire. A drug-addicted daughter who lived at the house most of the time had confronted her brother during the fire suppression. In view of the neighbors and firefighters, she had slapped him and demanded, "Now what have you done?" The son, who did not drive, had a police–fire scanner in his possession. He had reportedly been out with his mother driving when he heard the fire call on the scanner and returned to find the garage and an interior bedroom ablaze. Investigators determined that the fire had been ignited in furniture in both the garage and the adjacent bedroom. The fire patterns and duration (based on fire damage) of both fires indicated the fires were started separately at about the same time and could not have been the result of fire spread of a single fire through the connecting door (which was very badly burned from both sides).

The next day, a remarkable letter was posted on the craigslist.com website accusing the mother of drug use, endangerment, and arson. The person who posted this notice was identified as an acquaintance of the family who admitted she had done it to draw attention to the problems in the family. During interviews with police investigators, the son admitted frequently burning materials in the barbecue in the backyard and listening on his scanner for reports of smoke. If the smoke was reported, he would extinguish the fire in the barbecue. Examination of the scene revealed burned twigs, leaves, and paper in the barbecue. The son admitted to being driven to various locations by his mother so he could start fires. While she drove down the main street, he would toss lighted matches from the windows of the vehicle. When the vehicle was examined, burned paper matches were found in the door assembly (accounting for a fire in the vehicle's door on November 9, 2003). He admitted repeatedly pulling the alarms on the school (or calling in alarms) and then listening on the scanner or watching the fire department response from his backyard, which overlooked the school. He did not ride a bicycle.

When the son was fingerprinted, the latent print on the matchbook cover from a wildland fire (July 23, 2003) was found to be his. He explained that when he got depressed, setting fires made him feel better. His mother said she helped him because she wanted him to feel better and be happy. She admitted to setting the fires in their home on August 15, 2004, and four other fires. (She was a recovering drug addict using Methadone at the time of the fire.)

Except for a pull alarm at the school just 4 days before the house fire, all other false alarms were generated in 2002. Dumpster and median fires were also taking place during 2002, and they continued on into 2003 and 2004. Wildland or grass fires became dominant in 2003, and a mixture of all types took place in 2004. After their arrests in August 2004 the mother and son negotiated a plea agreement, the mother pleading quickly to two counts, the son to four counts. Both were sent to jail. Dumpster and roadside fires in town reportedly dropped to zero after August 2004.

Case study courtesy of Novato (CA) Fire Protection District.

FAKED DEATHS BY FIRE

Insurance fraud by burning a substituted body in an attempt to obliterate the true identity is a crime unique to profit-motivated arson. This form of arson often involves detailed planning, particularly in acquiring the body and evading suspicion by the investigating authorities.

Case studies have shown that the warning signs for **faked deaths** include, but are not limited to, the following (Reardon 2002).

- ◆ The death occurs shortly after filing of the insurance application and/or during the contestability period.
- ◆ The death occurs abroad.
- ◆ The insured has financial problems.
- ◆ Large policies or multiple small policies that do not require medical examinations exist. The policies have misrepresentations or omissions.
- ◆ The insured uses aliases.
- ◆ Previous policies have been canceled.
- ◆ The levels of insurance are inappropriate for the actual earned income.
- ◆ There is no body, or the body is in an unidentifiable condition.

Reardon emphasizes that the fundamental component in the payment of any life insurance claim is proving that the insured is actually dead. He also recommends thoroughly investigating these types of claims and litigating when appropriate to reduce the likelihood of payment to the claimant.

EXAMPLE 5.10   Out-of-Country Faked Death by Fire Scheme

In 1998, Donny Jones (not his real name) completed an application for a term life policy bearing a face value of $4 million. Jones already had a $3 million policy with another insurer. Six months after the $4 million policy was written, the insurance company received the news that Jones had died in a car accident outside the United States.

The insured prepared his plan carefully. While out of the country with a friend, Jones rented a large SUV, placed his bicycle in the car and, drove away at 10 P.M., supposedly to take a 3-hr drive through the desert to an adjoining town. Early the next morning, the SUV was found burning at the side of a desert highway. Local authorities found no traces of the mountain bike in the SUV, no signs of collision, and, initially, no body. Later, a body was found in the vehicle at an impound yard, but the remains were quickly claimed by a traveling companion of Mr. Jones and cremated.

Owing to the unique circumstances, the insurer initiated an immediate investigation into the claim by arranging for a forensic anthropologist to examine the skeletal bones and for a fire protection engineer to document and impartially determine the cause of the vehicle fire. The forensic anthropologist determined that the remains were those of an elderly man of Native American heritage, not of the insured, who was a 33-year-old Caucasian. The engineer also interpreted for the insurance company technical information on the incendiary nature of the fire, the staged accident, and forensic evidence collected during the investigation by the foreign authorities.

The insurance company denied the claim and filed suit in the U.S. District Court for declaratory judgment, seeking rescission of the policy. Initiating the litigation permitted the insurance company to begin to gather and preserve evidence from the authorities, police, and public and private sectors. The investigation outside the United States required the involvement of the U.S. consulate followed by litigation, for which the Hague Convention generally requires approval from a U.S. court and the appropriate foreign authority.

Eventually, the insured was discovered working in the United States through a routine background investigation by a firm for which the insured had been working, under an alias. The insurance policy was rescinded, and the insured later pled guilty in federal court to criminal charges of wire fraud and was ordered to pay full restitution (Reardon 2002).

FILICIDE

Filicide by fire is a crime in which the murderer is the parent of the victim and has used fire to mask the death, often making it appear to be an accident. Even though cases of filicide appear to be rare, these events are becoming more frequently studied in academics (Stanton and Simpson 2002).

Studies by the FBI, although based on limited case samples, still provide an insight into this crime. This agency speculates that owing to the thoroughness needed to conduct such investigations, this phenomenon may be widespread and underreported (Huff 1997). One of the present authors has been involved in the investigation of at least eight incidents in which a fire was set deliberately to kill one or more of the fire setter's children. One such case was even the subject of a best-selling book (Rule 1999).

Multiple motives are found for filicide, as for any crime, including those in the following examples:

- **Unwanted child.** Falsely believing that she and her spouse or lover can then live unburdened, a mother commits child murder to remove the perceived obstacle or nuisance child(ren).
- **Acute psychosis.** The parent is psychotic, such as the severely depressed single mother who stabbed her three small children to death and then set their apartment on fire.
- **Spousal revenge.** An estranged husband cruelly decides to deprive his wife of her most cherished treasure, her child.
- **Murder-for-profit.** Parents take out large life insurance policies on their children shortly before killing them by fire.

In any of these cases, the authorities may not be suspicious initially, for various reasons, including the lack of a thorough investigation of the fire; the presumption that the fatal fire was accidental, caused by the child's playing with matches; masking by overwhelming compassion for the parents; and sadness for the children. These reactions may override red flags of suspicion (Huff 1997).

Although there is no single indicator, several common factors appear in cases of filicide, including the following:

Victimology. The children are often young, preschool age; have little training of how to behave in a fire; and are thus less likely to escape. The perception is that younger children are more disposable and more easily "replaced" later.

Preoffense behavior. There is unusual behavior just before the fire, such as preparation of the children's favorite meal or a visit to a special place. For example, in one case a fast-food meal was served to the two teenage sons to deliver a dose of common decongestant to make them sleepy when the fire was set (it was identified in the postmortem blood of the decedent and in dried blood stains on the shirt of the survivor). One of the boys died in the fire; the other survived to testify about the prefire events. A young daughter was not targeted and was kept out of the fire area of the home.

Temporal. The fires take place at night or early in the morning when the children are most likely to be in bed asleep. This gives the parents time to plan the event and, in some cases, lock the children in their rooms during the fire setting.

Crime scene characteristics. Children are asked to change their sleeping arrangements the night of the fire and sleep in a room not normally used. Staging of scenes takes place in some cases when the children have already been shot, stabbed, or strangled to death and a fire set to cover the crime, frequently with a flammable liquid such as gasoline. In some cases escape routes are blocked and doors locked. Drugs may be administered to make the child sleepy or unconscious. A full forensic postmortem including X-ray and toxicology assays is essential on any child victim.

Offender characteristics. Halfhearted rescue attempts by adults are reported, with no appreciable element of danger encountered. Therefore, they do not display smoke-filled clothing, burns, or watery eyes, as would be expected. The parents most commonly live in manufactured housing and are no older than their mid-30s.

Postoffense behavior. After the incident, the adults exhibit inappropriate behavior, such as little or no grief; seldom speak about the victims, favoring the discussion of material losses (including insurance coverage); and are too quick to return to life as usual.

There will be one or a combination of two or more of these indicators. However, investigators are cautioned not to allow one factor alone necessarily to raise suspicion. They should move forward in a cautious yet professional manner and not yield to the overwhelming sympathy for the parent

---

EXAMPLE 5.11   Mother Involved in Filicide

Firefighters responding to a late-summer fire found a 22-month-old female hidden under a pile of clothes in a locked bedroom closet, dead from smoke inhalation. The child's mother told firefighters that her young daughter had been known to play with the cigarette lighter, which was found in the kitchen.

Firefighters doubted the story, but it became believable when the deceased child's brother, who was about 4 years old, demonstrated that he was able to operate the lighter. Investigators then ruled the fire accidental, initially concluding that the deceased girl had lit some paper napkins in the kitchen, taken them into the bedroom, and gone into the closet, with the door locking behind her.

All the facts surrounding this fire were consistent with a child's playing with a cigarette lighter and causing her own death. There was no reason to suspect foul play until 5 years later, when the son, now 9 years old, suffered third-degree burns in another fire. When firefighters arrived at the home, they found the child unconscious behind a locked bedroom door. The mother fled immediately after the fire.

One week later, fire investigators located and interviewed the mother, who allegedly confessed to setting both fires, telling them that she was angry with her husband. In each of these fires, individual factors indicated filicide by arson (Huff 1997).

◆ 5.4 GEOGRAPHY OF SERIAL ARSON

Offender-based classification of arson extends beyond merely identifying the motive. Investigators should also closely examine the **geographic locations** selected by the arsonist. An analysis of these sites may reveal much about the intended target, add insight into the arsonist's motive, and provide a potential surveillance schedule for attempting to identify and apprehend the arsonist.

The results of a joint research effort on the geography of serial violent crime illustrate the patterns found in serial arson cases. Results indicate that criminal offenders who repeatedly set fires exhibit temporal, target-specific, and spatial patterns that are often associated with their modus operandi (method of operation; m.o.) (Icove, Escowitz, and Huff 1993; Fritzon, 2001).

Many geographers, criminologists, and law enforcement professionals are concerned with the rising tide of serial violent crimes, including arson, that is plaguing the United States and other Free World countries. Many of these violent offenders purposely use jurisdictional boundaries to evade detection from law enforcement officials.

Serial arsonists often create a climate of fear in entire communities. Community leaders compound this problem by pressuring law enforcement agencies to identify and apprehend the fire setter quickly. Often, the arsonist evades apprehension for months, frustrating even the most experienced investigators. Unpredictable gaps often occur between incidents, leaving law enforcement authorities questioning whether the arsonist has stopped his or her fire setting, left the area, or been apprehended for another offense.

PREVIOUS RESEARCH FINDINGS

Few studies have been conducted on the geographic distribution of serial violent crimes, including arson. The studies that do exist focus on specific characteristics found in particular crimes (Icove 1979; Rossmo 1999).

A keynote analysis of violent crime measured it in terms of a spatial and ecological perspective (Georges 1978). The social-ecological view is that there is a direct relationship between crime and the environment. In this view, an absence of community controls, such as found in inner cities, correlates with high crime rates. Several historical approaches trace methods to interpret the geography of crime, particularly as it affects arson and fire-related crimes.

Transition Zones. The Chicago School of Sociology approach first proposed that crime was related to the environment (Park and Burgess 1921). This ecological approach assumes that high crime rates flourish in urban areas called **transition zones**. These zones have mixed land uses, high population fluxes, and a lack of opportunities for middle and upper economic classes to succeed.

In many cities, concentric rings form zones of transition, representing diffusion between the central business district and residential housing. These zones often contain rooming houses, ghettos, red-light districts, and diverse ethnic groups.

The transition zone theory was illustrated in a study of 15 fire bombings in Knoxville, Tennessee, during 1981 (Icove, Keith, and Shipley 1981). An analysis of the fire bombings investigated by the Knoxville Police Department's Arson Task Force showed that 66 percent of the incidents occurred within specific census tracts having significant population decline or growth.

The Knoxville study also examined a census tract in a low-income area where three fire bombings occurred during the reporting period. U.S. census data indicated that this tract had the highest minority population, the highest percentage of poverty, and the second-lowest mean income of all the tracts where bombings occurred.

Centrography. The technique of **centrography** uses descriptive statistics to measure the central tendency of crime on a two-dimensional spatial plane. In certain crimes, the offender's residence may be close to the site of the crime commission. Another hypothesis is that the older the offender, the greater the mean distance from the crime scenes to his or her residence. This mobility is due, in part, to the offender's access to bicycles and vehicles, as well as being over the curfew age.

A study for the U.S. Department of Justice explored the concept of centrography and described geographic **anchor points** for criminals (Rengert and Wasilchick 1990). Examining the geographic locations of buildings targeted by burglars, the study concluded that criminals select targets close to an anchor point that minimizes the travel and time required to commit their crimes. Often, the dominant anchor point is close to the offender's residence. Other anchor points away from the criminal's residence include bars and video arcades.

Other geographic analysis approaches take place on both the micro and the macro levels. The microlevel approach examines the exact physical location of the crime, such as the type of building, vehicle, or street. The macrolevel approach tends to aggregate the data into zones. These zones can be census tracts, police beats, or other geographic areas. This approach tends to increase the scale of the analysis.

Centrography has been used to examine spatial–temporal relationships of other violent crimes (LeBeau 1987). Observations of the spatial distribution of crime use the *x*- and *y*-axes of a Cartesian coordinate system overlaid on a street-block map. The locations of numerous incidents are analyzed using the **mean center** approach to measure the central tendency by tracking the movement of these mean centers and to determine the standard deviational ellipse to describe the distribution of incidents.

The **spatial and temporal trends** of serial arsonists and bombers have been examined on a local level (Icove 1979; Rossmo 1999). Patterns not previously documented among serial incendiary crimes were found using cluster analysis techniques—individual and group serial arsonists and bombers tend to be geographically bounded by natural and artificial boundaries. (See Figure 5.8 for an example.) When the offender moves his or her residence, the cluster of activity normally follows.

Studies on the historical changes of locations of arsons in the city of Buffalo, New York, have also confirmed many of these observations using cluster analysis. The concept of spatial surveillance brought new meaning to the monitoring of the locations of fires over time. The researchers noted that changes in these spatial patterns may occur for many differing reasons, and conclusions should not be reached merely on visual interpretations (Rogerson and Sun 2001).

Temporal trends have also been examined by Icove (1979) using two- and three-dimensional cluster analysis techniques. This form of analysis detects subtle time-of-day and day-of-week trends correlated with geographic changes in the cluster centers. Other temporal trends such as increases in arsons with lunar activity have also been documented in New York City (Lieber 1978). A research thesis at the University of California, Berkeley, School of Criminology, explored the relationship between lunar phenomena (not necessarily full moon) and the crime of arson (Netherwood 1966).

EXAMPLE 5.12 Incendiary Crime Detection

This is an example of the application of geographic analysis to detect local clusters of activity of serial arsonists. After plotting the locations of arsons in an eastern city from February through September 1974, a major cluster center of activity was detected. Three comparative techniques are shown in Figures 5.10 and 5.11, illustrating the use of two-dimensional gridded incident, shaded, and three-dimensional contour maps by Icove (1979).

A further analysis showed that a major cluster center was formed by two grids totaling 12 arson fires. Investigation revealed that during that time period, a gang of juvenile boys engaged in starting incendiary fires within a section of the large city. Properties targeted by these youths included garages and wooded areas.

When plotted on a map along with other arsons in the area, the incidents formed a major cluster. Data from these fires were collected and analyzed to provide a surveillance schedule for local law enforcement. A temporal analysis of the time of day and day of the week for the incidents contained in this major cluster was performed and revealed that the juveniles set these incendiary fires during breaks in the school day. In the summer months, fires occurred in the evening hours, but the trend reverted to the original pattern at the start of school in the fall. With few exceptions, the fires were on weekdays. During this step, lunar patterns are often detected. In this example, 75 percent of the fires occurred within plus or minus 2 days of either the new or the full moon phase.

Urban Morphology. Georges (1978) examined the concept of **urban morphology**, which posits that crime tends to concentrate within certain zones or within areas with close access to transportation routes. For example, in his earlier study on the geographic distribution of arsons during the 1967 Newark riots, fires were found to have been set along main commercial routes (Georges 1967). The study further documented earlier theories that as cities grow, crime tends to spread along main transportation routes (Hurd 1903). The mother–son case described previously illustrates this type of analysis, where fires were predominantly set along a major traffic route in the city.

CRIME PATTERN ANALYSIS

Crime pattern analysis is a methodology for detecting recurring patterns, trends, and periodic events in the times, dates, and locations of incidents. Once patterns are detected in a serial arson case, much can be learned to help predict the next event and classify the behavioral patterns exhibited by the offender.

Authoritative studies continue to reinforce that co-occurrence of offense characteristics also can be tied to both the geographic selection of targeted properties and the residence/workplace of the offender (Canter and Fritzon 1998). A systematic

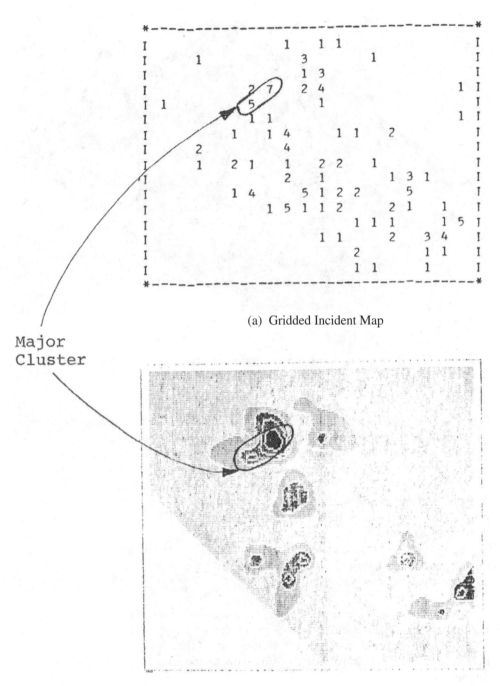

(a) Gridded Incident Map

Major
Cluster

(b) Contour Map of Incidents

FIGURE 5.10 ◆ Gridded incident and contour maps with major cluster centers of arson fires.
Source: Icove 1979. *Courtesy of D. J. Icove.*

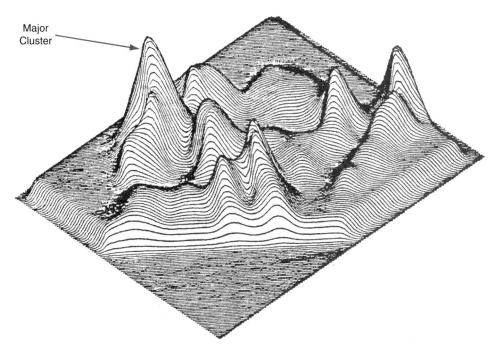

Major Cluster

FIGURE 5.11 ◆ The detection of geographic clusters of arsons in three dimensions with a 35° counterclockwise rotation. Source: Icove 1979. *Courtesy of D. J. Icove.*

algorithm has also been patented that performs crime site analysis in an area of criminal activity to determine a likely center of such activity (Rossmo 1998).

A **criminal investigative analysis** (also referred to as a **profile**) is usually prepared by an expert, based on exposure to numerous case histories, personal experiences, educational background, and research performed. Suspects developed by crime analysis do not always fit every prediction, since the predictions are made based on similar historical cases, and the offender may often modify his or her m.o. in response to an active investigation.

Several key points are developed during a criminal investigative analysis involving the geography of serial incidents: temporal analysis, target selection, spatial analysis of the cluster centers, and standard distance.

Temporal Analysis.   The first type of pattern analysis in serial arson cases is temporal analysis, which determines if any time-of-day or day-of-week trends exist. (See Figure 5.9 for an example.) These trends may vary if the offender modifies his or her m.o. in response to an active investigation. This modification may be a conscious act by the suspect.

Target Selection.   Another form of pattern analysis deals with the type and location of the target selected by the offender. A commonly observed characteristic in cases involving serial arsonists is that the offender often escalates the targets selected, and situations become more life-threatening with time. A typical scenario is that the arsonist starts with small grass and brush fires, then moves to outbuildings and vacant structures and, finally, to occupied structures. Distance traveled is also a factor in target selection (Fritzon 2001).

Spatial Analysis.   Using **spatial analysis,** an investigator can determine that a series of arson fires set repeatedly within the same geographic area forms a cluster of activity (as in Figure 5.8). Finding this center of activity often reveals information about future target selections or where the offender may live or work. As indicated previously, serial arsonists and bombers often tend to be geographically bounded by natural and artificial constraints. Therefore, the offenders either consciously or unknowingly maintain their activities within a bounded area, not often crossing major highways, rivers, or railroad tracks. Spatial monitoring of geographic patterns of 1996 arsons in Buffalo, New York, showed the value of searching for changes in spatial patterns (Rogerson and Sun 2001).

   Geographic profiling has proved to be a trainable and heuristic (rule-of-thumb) talent. In a recent research study on geographic profiling, 215 individuals participated in an experiment to predict the residential locations of serial offenders based on information about where their crimes were committed. In the study, these individuals were pre- and posttested when given formal training on one actuarial profiling technique (Snook, Taylor, and Bennell 2004). Analysis of the study participants' performance showed that 50 percent of them used heuristics that led to accurate predictions before receiving the formal training. Almost 75 percent of the study participants improved in their predictive ability after receiving the formal training.

   Long-term analysis of these **cluster centers** is useful for determining whether the offender has changed his or her residence or place of work. In cases spanning 3 years or more, a cluster center should be calculated for each year and plotted on a map (Icove 1979). If the cluster centers do not move significantly, or if they hover in a specific area, the offender has not likely changed residence or job. When a marked change is detected, this information is conveyed to the law enforcement agency, which compares this knowledge with the potential movement of the suspects.

   Studies continue to suggest a correlation between motivation and distance traveled by arsonists (Fritzon 2001). Those offenders studied whose behavior had a strong emotional component, such as a revenge motivation, tended to travel greater distances.

Cluster Centers.   The **mean center** is the primary measurement of a spatial distribution of a cluster of data points. Given that the columns on a map equate to *x*-values on a Cartesian coordinate system, and the rows to *y*-values, the equations for calculating the mean center (Ebdon 1983) are

$$\bar{x} = \frac{1}{n}\sum_{i=1}^{n} x_i,$$

$$\bar{y} = \frac{1}{n}\sum_{i=1}^{n} y_i.$$

For the general case of cluster center z, the formula is

$$z_i = \frac{1}{n}\sum_{i=1}^{n} x.$$

   A more suitable center of activity for crime pattern analysis is known as the *center of minimum travel* or what is often referred to as the **centroid**. This center is the point or coordinate at which the total sum of the squared Euclidean distances to all other points on the map is the lowest.

The centroid is more meaningful than a mean center when examining the geographic locations of crimes committed by a single offender. The reasoning behind this assumption is that an offender who walks to his or her crime scenes will choose the minimum travel distance. The mean center does not always coincide with the centroid.

The calculation of the centroid is not a simple formula but requires an iterative process of minimizing a performance index. The best clustering procedure using this technique is referred to as the *K*-means algorithm (MacQueen 1967). The application of this algorithm (Tou and Gonzalez 1974) is best broken down into logical steps.

Standard Distance. Once a cluster center is calculated for the geographic locations of a series of crimes, law enforcement officials can use this knowledge to concentrate their investigative and surveillance efforts within that neighborhood. This search radius from the cluster center is useful in focusing a criminal investigation within a reasonable distance from the cluster center of activity. For example, in a detailed crime analysis of a serial arson case profilers often recommend a surveillance schedule and area of concentration within the area bounded by a radius of one standard deviation from the cluster center (Icove 1979). This area is known as the **standard distance**.

Several case examples are presented to illustrate the phenomena associated with the geography of crime. The following cases were submitted for crime pattern analysis after an intensive investigation by a local law enforcement agency had exhausted all logical leads (Icove, Escowitz, and Huff 1983).

TABLE 5.7 ◆ Temporal and Target Analysis of Geographically Shifting Cluster Centers

Variable	Aug.	Sept.	Oct.	Nov.	Dec.	Total
8 A.M.–4 P.M.					1	1
4 P.M.–12 A.M.	1		4	3		8
12 A.M.–8 A.M.		1	4	6	3	14
Unknown			1			1
Monday					1	1
Tuesday	1					1
Wednesday					2	2
Thursday			1	2	1	4
Friday				2		2
Saturday		1	4	6		11
Sunday				2	1	3
Field				1	1	2
Vehicle			2			2
Mobile home				2	1	3
House, vacant	1	1	1			5
House, occupied					1	1
Structure		1	3	5	2	11
Total	1	1	9	9	4	24

Source: Icove, Escowitz, and Huff (1993).

EXAMPLE 5.13   Shifting Cluster Centers

During the fall and winter months in a southwestern state, an unknown arsonist was suspected of setting 24 fires involving fields, vehicles, mobile homes, residences, and other structures. Some of the fires were set after the offender broke into the structures.

A temporal analysis of the 24 incidents revealed that the arsonist favored the late evening and early morning hours. The majority of the fires occurred on the weekends, as shown in Table 5.7. An analysis of the targets selected by the arsonist showed that the structures were either unoccupied, vacant, or closed for business. The arsonist escalated the severity of his fire setting, first using available materials and later turning to flammable liquids to accelerate the fires.

A geographic cluster analysis showed that the first three fires clustered tightly on the west side of the town. The remaining fires concentrated in a larger cluster of activity on the east side (Figure 5.12).

A crime analysis of this case was conducted and the results were forwarded to the agency that requested the assistance. The investigators were told to concentrate their efforts on any suspects who first lived near the west cluster and then changed their residence to close to the center of the east-side cluster.

Armed with this information, the law enforcement agency later charged and convicted a 19-year-old single white male who matched the characteristics determined by the crime analysis. A key element in the case was that the offender lived near both cluster centers of the arsons. The geographic center of the activity shifted when the offender moved from the west to the east side of town (Icove, Escowitz, and Huff 1993).

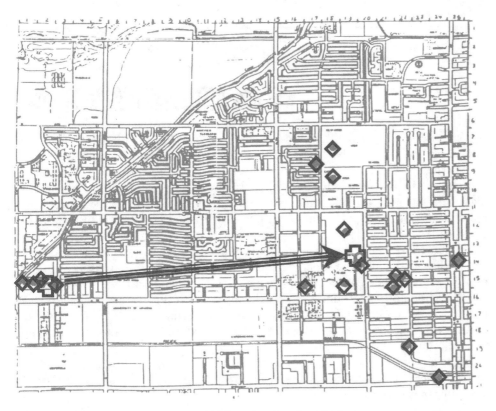

FIGURE 5.12 ◆ Geographic display of 24 fires showing the shift from west to east of centers of activity of two clusters. Source: Icove, Escowitz, and Huff 1993. *Courtesy of D. J. Icove.*

EXAMPLE 5.14 Temporal Clustering

A southeastern city was plagued with 52 arson fires in vacant buildings, vehicles, commercial businesses, residences, and garages over a 2-year period. The law enforcement agency investigating the case was stumped and turned to crime analysis after exhausting all traditional leads.

A crime analysis revealed temporal and geographic patterns in the time and locations of the arsons. The majority of the fires were set during the afternoon and evening weekday hours. There were two 6-month gaps in activity. An overwhelming majority of the arsons occurred within a 1-mi radius of the downtown area of the city, with two additional clusters adjacent to two lakes. The crime analysis returned to the agency stated that the offender lived close to the cluster center of activity of his fires. When later apprehended, the arsonist admitted in most cases walking to the scenes of his fires and using available materials and matches or a lighter carried to the scenes. Even though it is a high-risk scenario for the offender to set a fire in the afternoon and evening hours, his intimate knowledge of the geographic area diminished his possibility of being detected and followed.

Four months after the department received the crime analysis, a 29-year-old white male was arrested after fleeing the scene of a fire he had just set. The suspect later told police that he first started setting fires when his relationship with a girlfriend failed. The different clusters of fires were the result of the subject's changing his residence. Periods of inactivity in his fire setting correlated directly with times when the offender had ongoing social relationships. Only when these relationships failed did the offender return to setting fires.

When arrested, the suspect had in his possession a map that marked the locations of the fires he had set. The offender's prior arrest history included charges for disorderly conduct, criminal mischief, harassment, and filing a false report to law enforcement (Icove, Escowitz, and Huff 1993).

◆ 5.5 SUMMARY AND CONCLUSIONS

Arson is defined as the willful and deliberate destruction of property by fire. Although establishing a motive for fire setting is not a legal requirement of the elements of the criminal offense, it can help focus investigative efforts and aid in the prosecution of the arsonist. Motives for fire setting usually include one or more of the following general categories: vandalism, excitement, revenge, crime concealment, profit, and political terrorism. Motiveless arson or pyromania is not considered an identifiable classification.

Arson crime scene analysis is still in its infancy. Problems associated with its use and application include the broad degree of knowledge required to assess and interpret the actual scenes, particularly using a motive-based approach.

A combined background in the principles of fire protection engineering as well as in behavioral and forensic sciences can certainly enhance the analysis. Clearly, the concept that the "ashes can speak" is paramount in the application of this technique.

Future work in this area should combine the knowledge of fire pattern analysis with advanced photographic techniques to assist in the visualization of fire scenes long after they occur and deteriorate.

Problems

5.1. Research a local serial arson case that appears in the media. Plot the locations of these fires and conduct a temporal analysis. What can you learn from this analysis?

5.2. For the case from problem 5.1, determine what motive was developed. Is the assessment by the media correct? What was the motive stated by the prosecution? By the defense?

5.3. What published studies support the popular notion that sex is a major motive for arson? What is the reliability of such conclusions today?

Suggested Reading

Allen, Douglas H. 2007. *Multiple fire setters—analysis: Pattern recognition.* dhallen2@verizon.net.

Canter, D. 2003. *Mapping murder.* London: Virgin Books.

Canter, D., and K. Fritzon. 1998. Differentiating arsonists: A model of firesetting actions and characteristics. *Legal and Criminological Psychology* 3: 73–96.

Danforth, J. C. 2000. Interim report to the deputy attorney general concerning the 1993 confrontation at the Mt. Carmel complex, Waco, Texas. U.S. Government Publication. July 21.

Gardiner, M. 1992. Arson and the arsonist: A need for further research. Project Report, Polytechnic of Central London, June.

Huff, T. G. 1993. Filicide by fire—The worst crime? *Fire and Arson Investigator* (June): 24–26.

Icove, D. J., V. B. Wherry, and J. D. Schroeder. 1998. *Combating arson-for-profit: Advanced techniques for investigators.* Columbus, OH: Battelle Press.

Rossmo, D. K. 2000. *Geographic profiling.* Boca Raton, FL: CRC Press.

Sapp, A. D., T. G. Huff, G. P. Gary, and D. J. Icove. 1995. *A motive-based offender analysis of serial arsonists* (monograph). Quantico, VA: National Center for the Analysis of Violent Crime, FBI Academy.

Stauss, E. 1993. Paul Kenneth Keller. *National Fire and Arson Report* 2: 4–6.

Wood, B. 1997. Arson profiling. *Fire Engineering Journal* 60: 208.

CHAPTER 6

Fire Modeling

Data! Data! Data! I can make no bricks without clay.

Sir Arthur Conan Doyle,
"The Adventure of the Copper Beeches"

The concept of fire modeling as it applies to forensic fire investigation is unique to the last decade, although models have existed since the 1960s. Previously, fire modeling was centered on explaining the physical phenomena of fires, particularly when applied to verifying existing experimental data.

It was the effort of a few fire scientists and engineers working at and associated with the National Institute of Standards and Technology (NIST) and the Building Research Establishment, Fire Research Station (BRE, FRS, UK) that pushed the acceptability and application of fire modeling out of laboratory conditions and into the world of forensic fire scene reconstruction. These early successes of fire modeling applied to the field of fire litigation and reconstruction further underscored its usefulness (Bukowski 1991; Babrauskas 1996). Several of the keynote studies that contributed significantly to this effort are described in this text.

The purpose of this chapter is not to make the reader an expert in fire modeling but to allow him/her to gain a better appreciation for its added value to an investigation. Ample references are provided should more information be needed. This chapter also answers the following questions that often arise when fire investigators are confronted with a fire of exceptional scale or impact: (1) What exactly is a fire model? (2) In which aspect of my investigation can fire modeling help? (3) What are the realistic and reliable results of a fire model? (4) Should more than one type of model be run to increase confidence in the results? and (5) What is the future of fire modeling?

◆ 6.1 HISTORY OF FIRE MODELING

Research by Mitler (1991) at NIST best establishes the historical framework for the application of fire scene modeling. Work in 1927 at the National Bureau of Standards (NBS), the predecessor agency of NIST, was the first attempt to understand and explain in scientific terms the issues surrounding postflashover compartment fires by relating the room gas temperature to the available mass of fuel being consumed (Inberg 1927).

The first fire model was developed through work in Japan in 1958 to relate the ventilation factor to steady-state fire development (Kawagoe 1958). The second model was constructed in Sweden (Magnusson and Thelandersson 1970), followed by Babrauskas at the University of California, Berkeley (Babrauskas 1975).

Historical reasons for developing and applying mathematical fire modeling were also accurately articulated and predicted by NIST (Mitler 1991). Suggested are six major ways in which a good mathematical fire model of a structure can render assistance:

- Avoid repetitious full-scale testing,
- Help designers and architects,
- Establish flammability of materials,
- Increase the flexibility and reliability of fire codes,
- Identify needed fire research, and
- Help in fire investigations and litigation.

These NIST areas of concentration are as on point today as they were over a decade ago. With the impact of environmental limitations on testing, factors such as the placement and combination of certain fuel items, changes in venting, and thickness of materials can now be evaluated without the need for repeated full-scale testing.

Designers and architects have already found modeling useful in assessing the flammability of materials as they specify new construction. New flexible fire performance codes are becoming more generally accepted as modeling introduces alternatives or new designs (such as atrium construction) that traditionally have not been addressed in historical prescriptive codes. Fire modeling can now also address the optimum designs for evacuation of people from buildings during emergencies.

As the science of fire dynamics advances, computational modeling identifies needed new areas of fire research into the phases of growth, flame spread, and flashover. Modeling can aid in gap assessments of certain areas that previously were not fully understood by designers, engineers, researchers, and even fire investigators.

Finally, fire modeling has had its greatest impact on forensic fire investigations and litigation (DeHaan 2005). Advancements in these areas are discussed in this chapter.

◆ 6.2 FIRE MODELS

Computer fire models have a broad range of applications in the area of fire science and engineering. The fire model works to supplement the information gleaned from forensic evidence, witness interviews, media film footage, and preliminary fire scene examinations. Historically, these fire models, as shown in Table 6.1, have included eight major overlapping categories (Hunt 2000).

Fire models usually emulate the impact of fires, not the physical fire itself. There are two recognized approaches to modeling fires: probabilistic and deterministic. **Probabilistic** models usually center on the application of stochastic mathematics to estimate or predict an outcome (within certain likelihoods), such as human behavior (SFPE 2002b, Chap. 3-12). In **deterministic** models, the investigator relies on mathematical relationships that form the underpinnings of the physics and chemistry of fire science. Deterministic models can range from a simple straight-line approximation to correlating test data with complex fire models solving hundreds of simultaneous equations.

TABLE 6.1 ◆ Classes of Computer Fire Models and Commonly Cited Examples

Class of Model	Description	Example(s)
Spreadsheet	Calculates mathematical solutions for interpretations of actual case data	FiREDSHEETS, NRC spreadsheets
Zone	Calculates fire environment through two homogeneous zones	FPETool, CFAST, ASET-B, BRANZFIRE, FireMD
Field	Calculates fire environment by solving conservation equations, usually with finite-element mathematics	FDS, JASMINE, FLOW3D, SMARTFIRE, PHOENICS, SOFIE
Postflashover	Calculates time–temperature history for energy, mass, and species and is useful in evaluating structural integrity in fire exposure	COMPF2, OZone, SFIRE-4
Fire protection performance	Calculates sprinkler and detector response times for specific fire exposures based on the response time index (RTI)	DETACT-QS, DETECT-T2, LAVENT
Thermal and structural response	Calculates structural fire endurance of a building using finite-element calculations	FIRES-T3, HEATING7, TASEF
Smoke movement	Calculates the dispersion of smoke and gaseous species	CONTAM96, Airnet, MFIRE
Egress	Calculates the evacuation times using stochastic modeling using smoke conditions, occupants, and egress variables	Allsafe, buildingEXODUS, EESCAPE, ELVAC, EVACNET, EXIT89, EXITT, EVACS, EXITT, Simplex, SIMULEX, WAYOUT

Source: Updated from Bailey (2006); Friedman (1992); and Hunt (2000).

Mathematical models are usually formulated from interpretations of actual test data. These types of models are usually calculated by hand using formulas and a scientific calculator, a spreadsheet, or a simple computer program. Smoke filling rates, flame heights, virtual origin, and other approximations can usually be calculated by hand in several minutes. When based on sound mathematical representations and properly applied, fire models can often assure that the scientific method has been satisfied.

Because mathematical models are discussed throughout this text in the form of quick problem solutions, the zone and field models are emphasized here. The discussion is limited to a survey of the concepts and is by no means intended to serve as an instruction manual for their use. In fact, most guidance issued with fire models cautions that they are for use only by those having significant competency in the fire engineering field and that the results should supplement the user's professional judgment.

One of the better known spreadsheets for solving simple fire protection engineering relationships emerged from a May 1992 study, "Methods of Quantitative Fire Hazard Analysis," prepared for the Electric Power Research Institute (EPRI) by the University of Maryland, Department of Fire Protection Engineering. The study was distributed through the Society of Fire Protection Engineers (Mowrer 1992).

The study describes fire hazard analysis models used in the **Fire-Induced Vulnerability Evaluation (FIVE) methodology,** which is used by the Nuclear Regulatory Commission (NRC). The FIVE analysis uses realistic industrial loss histories to evaluate hazards quantitatively, including those losses produced in fires resulting from ignitable liquid spills, cable trays, and electrical cabinets. Many of these examples predict the impact of the fire plumes and ceiling jet temperatures, hot gas layers, thermal radiation to targets, and critical heat fluxes. Many of these concepts are applicable to problems previously posed in fire scene reconstruction and analysis (Mowrer 1992).

From this study came **FiREDSHEETS** from the University of Maryland, Department of Fire Protection Engineering (Milke and Mowrer 2001). These spreadsheets concentrated on the specifics of compartment fire analysis and included extensive property tables for typical materials first ignited.

Mowrer has continued to develop spreadsheet templates for fire dynamics calculations, which are presently posted along with ample documentation on the **Fire Risk Forum** website (Mowrer 2003). These templates incorporate a number of enclosure fire dynamics calculations used by **FPETool** (Nelson 1990) and other suites, as exemplified in Figure 6.1. Mowrer spreadsheets also include key calculations used in the fire and

PLUMETMP.XLS: Estimate of temperature rise in a fire plume

INPUT PARAMETERS		
FIRE HEAT RELEASE RATE (Q)	500	kW
CONVECTIVE FRACTION (Xc)	0.7	–
HEIGHT ABOVE FIRE (Z)	5	m
FIRE LOCATION FACTOR (kLF)	1	–
CALCULATED PARAMETERS		
CONVECTIVE HRR (Qc)	350	kW
FLAME HEIGHT (Zfl)	2.4	m
PLUME TEMPERATURE (dTpl)	95	C

FIRE LOCATION FACTORS	kLF
FIRE IN OPEN	1
FIRE ALONG WALL	2
FIRE IN CORNER	4

LAYERTMP.XLS: Estimate of upper layer temperature using modified MQH correlation

INPUT PARAMETERS		
ROOM LENGTH (L)	6.16	m
ROOM WIDTH (W)	6.7	m
ROOM HEIGHT (H)	2.8	m
OPENING WIDTH (Wo)	1.8	m
OPENING HEIGHT (Ho)	2.7	m
BOUNDARY CONDUCTIVITY (k)	0.00017	kW/m.K
BOUNDARY DENSITY (p)	960	kg/m3
BOUNDARY SPEC. HEAT (cp)	1.1	kJ/kg.K
BOUNDARY THICKNESS (d)	0.0125	m
FIRE HRR (Q)	2000	kW
FIRE LOCATION FACTOR (kLF)	4	–
CALCULATION TIME (t)	400	s
CALCULATED PARAMETERS		
HEAT TRANSFER COEFF (hk)	0.0211849	kW/m2.K
BOUNDARY AREA (At)	154.56	m2
VENTILATION FACTOR	7.99	m^5/2
TEMPERATURE RISE (dT)	582	C

FIGURE 6.1 ◆ Examples of 2 of the 21 fire dynamics calculations that can be performed using the Mowrer spreadsheets. *www.fireriskforum.com.*

TABLE 6.2 ◆ Mowrer's Fire Risk Forum Spreadsheets	
Template	*Spreadsheet Description*
ATRIATMP	◆ Estimates the approximate average temperature rise in the hot gas layer that develops in a large open space such as an atrium
BUOYHEAD	◆ Estimates the pressure differential, gas velocity, and unit mass flow rate caused by the buoyancy the of hot gases beneath a ceiling
BURNRATE	◆ Estimates the burning rate history of a flammable liquid fire
CJTEMP	◆ Estimates the temperature rise in an unconfined ceiling jet
DETACT	◆ Estimates the response time of ceiling-mounted fire detectors
FLAMSPRED	◆ Estimates the lateral flame spread rates on solid materials
FLASHOVR	◆ Estimates the heat release rate needed to cause flashover in a compartment with a single rectangular wall vent
FUELDATA	◆ Contains thermophysical and burning rate data for common fuels
GASCONQS	◆ Estimates gas species concentrations
IGNTIME	◆ Estimates the time to ignite a thermally thick solid exposed to a constant heat flux
LAYDSCNT	◆ Estimates the smoke layer interface position in a closed room due to entrainment
LAYERTMP	◆ Estimates the average hot gas layer temperature in an enclosure with a single rectangular wall opening
MASSBAL	◆ Estimates the mass flow rate through an enclosure with a single wall opening
MECHVENT	◆ Estimates the fire conditions in a mechanically ventilated space without natural ventilation
PLUMEFIL	◆ Estimates the volumetric rate of smoke flow in a fire plume
PLUMETMP	◆ Estimates the temperature rise in an axisymmetric fire plume
RADIGN	◆ Estimates the potential for radiant ignition of a combustible target
STACK	◆ Estimates the mass flow rate through an enclosure
TEMPRISE	◆ Estimates the average temperature rise in a closed room
THERMPRP	◆ Contains thermal property data for 15 types of boundary materials

Source: F. W. Mowrer, Spreadsheet Templates for Fire Dynamics Calculations, University of Maryland, Department of Fire Protection Engineering, September 2003. Spreadsheets and documentation are posted on the Fire Risk Forum website, fireriskforum.com.

arson investigation training course for agents of the Bureau of Alcohol, Tobacco, and Firearms given by the University of Maryland's Department of Fire Protection Engineering. Listed in Table 6.2 are the calculations contained in the Mowrer spreadsheets.

A later and more industry-oriented set of spreadsheets known as the **Fire Dynamics Tools** (FDTs) was developed for the NRC fire protection inspection program (Iqbal and Salley 2002, 2004). The format of the spreadsheet output is more intuitive in its approach, as with, for example, financial spreadsheets. Table 6.3 is a synopsis of the available FDTs spreadsheets.

TABLE 6.3 ◆ NRC Fire Dynamics Spreadsheets

Chapter	FDTs Spreadsheet Description
2	• Predicting Hot Gas Layer Temperature and Smoke Layer Height in a Room Fire with Natural Ventilation • Predicting Hot Gas Layer Temperature in a Room Fire with Forced Ventilation • Predicting Hot Gas Layer Temperature in a Room Fire with Door Closed
3	• Estimating Burning Characteristics of Liquid Pool Fire, Heat Release Rate, Burning Duration, and Flame Height
4	• Estimating Wall Fire Flame Height
5	• Estimating Radiant Heat Flux from Fire to a Target Fuel at Ground Level under Wind-Free Condition Point Source Radiation Model • Estimating Radiant Heat Flux from Fire to a Target Fuel at Ground Level in Presence of Wind (Tilted Flame) Solid Flame Radiation Model • Estimating Thermal Radiation from Hydrocarbon Fireballs
6	• Estimating the Ignition Time of a Target Fuel Exposed to a Constant Radiative Heat Flux
7	• Estimating the Full-Scale Heat Release Rate of a Cable Tray Fire
8	• Estimating Burning Duration of Solid Combustibles
9	• Estimating Centerline Temperature of a Buoyant Fire Plume
10	• Estimating Sprinkler Response Time
13	• Calculating Fire Severity
14	• Estimating Pressure Rise Due to a Fire in a Closed Compartment
15	• Estimating Pressure Increase and Explosive Energy Release Associated with Explosions
16	• Calculating the Rate of Hydrogen Gas Generation in Battery Rooms
17	• Estimating Thickness of Fire Protection Spray-Applied Coating for Structural Steel Beams (Substitution Correlation) • Estimating Fire Resistance Time of Steel Beams Protected by Fire Protection Insulation (Quasi-Steady-State Approach) • Estimating Fire Resistance Time of Unprotected Steel Beams (Quasi-Steady-State Approach)
18	• Estimating Visibility Through Smoke

Source: Iqbal and Salley 2004.

◆ 6.4 ZONE MODELS

Two-zone models are based on the concept that a fire in a room or enclosure supports two unique zones and predict the conditions within each zone. The two zones refer to two individual control volumes known as the upper and lower zone. The upper zone of heated gases and by-products of combustion is penetrated only by the fire plume. These hot gases and smoke fill the ceiling layer, which then slowly descends. A boundary layer marks the intersection of the upper and the lower areas. The only exchange between the upper and the lower layers is due to the action of the fire plume.

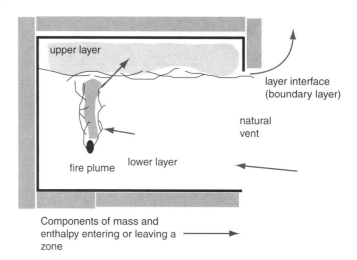

FIGURE 6.2 ◆ Representation of the two-zone fire model showing the location of the upper and lower layers, layer interface, fire plume, and compartment vent. *Courtesy of NIST, from Forney and Moss 1992.*

Zone models use a set of differential equations to solve within each zone for pressure, temperature, carbon monoxide, oxygen, and soot production. The accuracy of these calculations varies with not only the model but also the reliability of the input data describing the circumstances and room dimensions and, very importantly, the assumptions as to the heat release rates of the materials burned in the fire. Complex two-zone models of up to 10 rooms typically take a few minutes to run using Pentium-class computer workstations or laptops.

Figure 6.2 shows a schematic representation of the concepts behind a two-zone model. A survey published in 1992 by the Society of Fire Protection Engineers showed **ASET-B** and **DETACT-QS** to be the most widely used zone models at that time (Friedman 1992).

Today, **CFAST,** the **Consolidated Model of Fire Growth and Smoke Transport** has become the most widely used zone model. CFAST combines the engineering calculations from the **FIREFORM** model and the user interface from **FASTlite** to construct a unique comprehensive zone modeling and fire analysis tool (Peacock et al. 2005).

CFAST OVERVIEW

The CFAST two-zone model is a heat and mass-balance model based on not only the physics and chemistry but also the results of experimental and visual observations in actual fires. CFAST computes

- ◆ Production of enthalpy and mass by burning objects;
- ◆ Buoyancy and forced transport of enthalpy and mass through horizontal and vertical vents; and
- ◆ Temperatures, smoke optical densities, and species concentrations.

CFAST 3.1.7 has a graphical user interface (GUI) that allows the user to enter and modify characteristics about the physical form of the rooms, the fire signature, and graphics output. Enhancements in CFAST 5.0.1 and, most recently, CFAST 6.0.10

TABLE 6.4 ◆ Summary of Numerical Limits in the Software Implementation of the CFAST Model, Version 6

Feature	Maximum
Simulation time (s)	86,400
Compartments	30
Object fires	31
Fire definitions in database file	30
Material thermal property definitions	125
Slabs in a single surface material	3
Fans in mechanical ventilation systems	5
Ducts in mechanical ventilation systems	60
Connections in compartments and mechanical ventilation systems	62
Independent mechanical ventilation systems	15
Targets	90
Data points in a history or spreadsheet file	900

Source: Peacock, Jones, Reneke, and Forney 2005.

allow a graphical windowing interface with NIST's Fire Dynamics Simulator program. Table 6.4 summarizes of the present capabilities in the software implementation of CFAST version 6.

CFAST has been used since its first public release in June 1990, with many enhancements over the years. CFAST Version 6.0.10 includes a newly designed user interface and incorporates improvements in heat transfer, smoke flow through corridors, and more accurate combustion chemistry. NIST added the ability to define a general t^2 growth rate fire allowing the user to select growth rate, peak heat release rate, steady burning time, and decay time, including predefined constants for slow, medium, fast, and ultrafast t^2 fires. The latest CFAST user's guide and documentation follows the guidelines set forth in ASTM E 1355a, "Standard Guide for Evaluating the Predictive Capability of Deterministic Fire Models" (Peacock, Jones, Reneke, and Forney 2005).

Since CFAST does not have a pyrolysis model to predict fire growth, a fuel source is entered or described by its fire signature (heat release rate). The program converts this fuel information into two characteristics: enthalpy (heat) and mass. In an unconstrained fire, the burning of this fuel takes place within the fire plume. In constrained fires, the pyrolyzed fuel may burn in the fire plume where there is sufficient oxygen, but it may also burn in the upper or lower layer of the room of fire origin, the plume in the doorway leading to an adjoining room, or even the layers or plumes in adjacent rooms.

FIRE PLUMES AND LAYERS

As in previous mathematical models in which calculations of heat flux used the fire plume's virtual origin, the CFAST model instructs the fire plume to act as a "pump" to move enthalpy and mass into the upper layer from the lower layer. Mixed with this enthalpy and mass are inflows and outflows from horizontal or vertical vents (doors, windows, etc.), which are modeled as plumes.

Some assumptions of the model do not always hold in real fires. Some mixing of the upper and lower layers does take place at their interfaces. At cool wall surfaces, gases will flow downward as they lose heat and buoyancy. Also, heating and air-conditioning systems will cause mixing between the layers.

Horizontal flow of the fire from one room to the next occurs when the upper layer descends below the opening of an open vent. As the upper layer descends, pressure differentials between the lower layer may cause air to flow in the opposite direction, resulting in the two flow conditions. Upward vertical flow may occur when the roof or ceiling of a room is opened.

HEAT TRANSFER

In the CFAST model, unique material properties of up to three layers can be defined for surfaces of the room (ceiling, walls, flooring). This design consideration is useful, since in CFAST heat transfers to the surface via convective heat transfer and through the surfaces via conductive heat transfer.

Radiative heat transfer takes place among fire plumes, gas layers, and surfaces. In radiative heat transfer, emissivity is dominated primarily by species contributions (smoke, carbon dioxide, and water) within the gas layers. CFAST applies a combustion chemistry scheme that balances carbon, hydrogen, and oxygen in the room of fire origin among the lower-layer portion of the burning fire plume, the upper layer, and air entrained in the lower layer that is absorbed into the upper layer of the next connecting room.

LIMITATIONS

Zone models have their limitations. Six specific aspects of physics and chemistry either are not included or have very limited implementations in zone models: flame spread, heat release rate, fire chemistry, smoke chemistry, realistic layer mixing, and suppression (Babrauskas 1996). Some errors in species concentrations can result in errors in the distribution of enthalpy among the layers, a phenomenon that affects the accuracy of temperature and flow calculations. CFAST has also been known to overpredict the upper-layer temperatures, partly because heat losses due to window radiation are not presently incorporated into the model.

Even with these known limitations, zone models have been extremely successful in forensic fire reconstruction and litigation (Bukowski 1992 and 1996; DeWitt and Goff 2000). CFAST zone models have been successfully introduced in federal court litigation (see Chapter 1, Example 1.2). These successes are a result of their correct application by knowledgeable scientists and engineers, continued revalidation in actual fire testing, and applied research efforts, include verification and validation studies by the NRC (2006).

◆ 6.5 FIELD MODELS

Field models, the newest and most sophisticated of all deterministic models, rely on **computational fluid dynamics (CFD)** technology. These models are attractive, especially as a tool in litigation support and fire scene reconstruction, owing to their ability to display the impact of information visually in three dimensions.

However, the downside of field models is that creating input data is typically very time consuming and often requires powerful computer workstations to compute and display the results. The computation time to run field models may be days or weeks depending on the complexity of the model and the relative computational power of the computer system available.

COMPUTATIONAL FLUID DYNAMICS

CFD models estimate the fire environment by dividing the compartment into uniform small cells instead of two zones. The program then simultaneously solves the conservation equations for combustion, radiation, and mass transport to and from each cell surface. The present shortcomings include the level of background and training needed to codify a model, particularly for complex layouts.

There are many advantages of using CFD models over zone models such as CFAST. The higher geometric resolution of CFD models makes the solutions more refined. Higher-speed laptops and workstations allow CFD models to be run, whereas in the past larger mainframe and minicomputers were required. Multiple workstations have been successfully linked together in parallel processing arrays to perform complex CFD calculations more quickly.

Because CFD models are used in companion areas such as fluid flow, combustion, and heat transfer, their underlying technology is generally more accepted over simpler and coarser models. Once these models become more broadly validated, they will gain further acceptance in situations where scientifically based testimony is sought, such as in forensic fire scene reconstructions.

CFD has a long history of use in forensic fire scene reconstructions, and it is gaining in popularity owing to commercially available and intuitive graphical interfaces [e.g., PyroSim (2007) by Thunderhead Engineering]. CFD programs now outpace the capabilities of the zone models, which, for the most part, operate under maintenance modes only. Moreover, CFD programs such as FDS make the technology economically feasible owing to its wide use, testing, and growing acceptance.

NIST CFD TECHNOLOGY–BASED MODELS

The recommended field model using CFD technology, due to its availability at no cost, focus of research, and acceptance in the community is the **Fire Dynamics Simulator (FDS).** This model is available from NIST's Building and Fire Research Laboratory (McGrattan et al. 2007). The first version of FDS was publicly released in February 2000, and it has become a mainstay of the fire research and forensic communities.

FDS is a fire-driven fluid flow model and numerically solves a form of the Navier–Stokes equations for thermally driven smoke and heat transport. The companion program to FDS is **Smokeview,** which is a graphical interface that produces a variety of visual records of FDS predictions. These include concentration of species, temperature, and heat flux in various animations. User guides for both FDS (McGrattan 2006; McGrattan and Forney 2006) and Smokeview (Forney and McGrattan 2006) are available at NIST's website (fire.nist.gov).

Traditional usage of FDS to date has been divided among the evaluation of smoke handling systems, sprinkler and detector activation, and fire reconstructions. It is also being used to study fundamental fire dynamics problems encountered in both academic and industrial settings. FDS uses a combination of models to handle the computational problems: a hydrodynamic model to solve the Navier–Stokes equations, a Smagorinsky form of large eddy simulation (LES) for smoke and heat

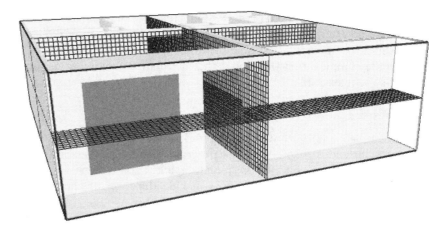

FIGURE 6.3 ◆ CFD (field) models break up the room (or fire area) into cells and calculate the mass heat energy and species transport (fluxes) into and out of each of the six faces of each cell.

transport, a numerical grid simulation to handle turbulence, a mixture fraction combustion model, and a finite volume method approach for radiation heat transfer.

Unfortunately, FDS uses a rectilinear grid to describe the computational cells. This makes it slightly difficult to model sloping roofs, rounded tunnels, and curved walls without some form of approximation. The boundary conditions for material surfaces consist of assigned thermal constants that include information on its burning behavior. FDS, like other CFD models, breaks up the room (or fire area) into cells and calculates the mass heat energy and species transport (fluxes) into and out of each of the six faces of each cell. Figure 6.3 shows the output grids from a typical FDS problem.

The latest release of FDS at the time of this publication is version 5, which is a significant enhancement over previous versions. These enhancements include the ability to evaluate heat transfer through walls and the impact of water suppression and initial conditions, and to add image texturing to solid objects and surfaces. The latest companion release of Smokeview is version 5. Its enhanced features include controls over viewing the rectilinear mesh used in constructing the FDS model. Scene clipping allows the visualization of obstructed boundary surfaces in complex models having numerous walls. A more robust feature allows more control over the movement and orientation of the scene as rendered by Smokeview.

Other features include several visualization modes such as tracer particle flow, animated contour slices of computed variables such as vector heat flux plots, animated surface data, and multiple isocontours. New visualization tools can allow the user to compare fire burn patterns with computed uniform surface contour planes. The option for animated flow vectors and particle animations allows observed heat flow vectors to be compared with computed vectors.

A horizontal time bar displays the duration of the run, and a vertical color-coded bar displays other variables, such as temperature and velocity. Graphic options include the ability to capture the output of the Smokeview program on a screen-by-screen basis. Screens may be exported and made into a movie, but this requires additional (non-NIST) software and can be tedious to produce.

VALIDATION

All fire models must be **validated** by comparing their predictions against the results of real fire tests if their results are to be relied on. This testing will sometimes reveal flaws or "blind spots" where some types of enclosures or conditions produce predicted values that differ from observed real fire behavior. If a model has not been shown to give valid results for a particular type of fire, its predictions in an unknown scenario should not be taken as gospel. For instance, FDS has been shown to be very accurate in predicting growth of fires in large compartments, but data on how FDS predictions will relate to real fires in very small compartments are just now being gathered and published.

It should be noted that FDS was derived from earlier work on modeling the growth and movement of smoke plumes from large stationary fires such as oil pools. This work, based on large eddy simulation (LES) focused on atmospheric interaction with the smoke plumes and not on movement of the fire itself. More recently, FDS has been applied to studying wildland-urban interface (WUI) fires, considering the impact of vegetation fires on nearby structures. These developments depend on validation by comparison with actual incidents, since planned, real-world tests of such fires would be prohibitively expensive and dangerous. Sadly, each year in the United States there are numerous opportunities to collect data from such fires after they occur. Because FDS predicts fire growth and spread based on the thermal characteristics of solid fuel surfaces, when fuel masses are "porous" (i.e., where fire can spread quickly *through* them as well as across them), the physics of heat transfer and ignition for that growth is too complex for FDS to accurately model such fires. The effects of wind-driven fire spread through such porous arrays complicate the process even further. Wildland fires in heavy brush or timber are almost always "crown fires" in their most destructive phase; that is, the fire spreads upward and through porous fuel arrays like leaves or needles, often driven by wind—either atmospheric (external) wind or fire-induced drafts. Although FDS has been validated against tests on flat grasslands (no vertical spread factors) with wind-aided spread only, it has not been shown to give accurate predictions for crown fires through heavier brush or timber.

◆ 6.6 FIRE MODELING CASE STUDIES

Traditionally, fire models have been used in two separate yet important areas:

- Building fire safety and codes analysis and
- Fire reconstruction and analysis.

Specific fire modeling evaluations have looked at the impact or performance of smoke handling systems, sprinkler and detector activation, and fire/burn patterns in reconstructions. Fire modeling found its way into forensic reconstruction through a long history of applications to real-world events. Table 6.5 is a partial listing of cases in which NIST-produced fire models have provided valuable insight into fire investigation. These and other reports can be obtained from the NIST fire publications website (fire.nist.gov).

Science and litigation are the two driving factors behind these applications. Fire protection engineering principles developed from correlating tests to actual case histories stretch the knowledge of fire science. Attorneys seeking answers to liability for damages due to fires also have a reason to seek the answers to similar questions. Therefore, science and law meet in the courtroom.

TABLE 6.5 ◆ Representative NIST Fire Investigations Using Fire Modeling

NIST Report Number or Citation	Case Title and Author
NBSIR 87-3560 May 1987	*Engineering Analysis of the Early Stages of Fire Development—The Fire at the Dupont Plaza Hotel and Casino—December 31, 1986* H. E. Nelson
NISTIR 4665 September 1991	*Engineering Analysis of the Fire Development in the Hillhaven Nursing Home Fire, October 5, 1989* H. E. Nelson and K. M. Tu
NISTIR 4489 June 1994	*Fire Growth Analysis of the Fire of March 20, 1990, Pulaski Building, 20 Massachusetts Avenue, NW, Washington, DC* H. E. Nelson
Fire Engineers Journal 56, no. 185 (November 1996): 14–17	*Modeling a Backdraft Incident: The 62 Watts Street (New York) Fire.* R. W. Bukowski
NISTIR 6030 June 1997	*Fire Investigation: An Analysis of the Waldbaum Fire, Brooklyn, New York, August 3, 1978* J. G. Quintiere
NISTIR 6510 April 2000	*Simulation of the Dynamics of the Fire at 3146 Cherry Road NE, Washington, DC, May 30, 1999* D. Madrzykowski and R. L. Vettori
NISTIR 6923 October 2002	*Simulation of the Dynamics of a Fire in a One-Story Restaurant, Texas, February 14, 2000* R. L. Vettori, D. Madrzykowski, and W. D. Walton
NISTIR 6854 January 2002	*Simulation of the Dynamics of a Fire in a Two-Story Duplex, Iowa, December 22, 1999* D. Madrzykowski, G. P. Forney, and W. D. Walton
NIST SP 995 March 2003	*Flame Heights and Heat Release Rates of 1991 Kuwait Oil Field Fires* D. Evans, D. Madrzykowski, and G. A. Haynes
NIST Special Pub. 1021 July 2004	*Cook County Administration Building Fire, 69 West Washington, Chicago, Illinois, October 17, 2003: Heat Release Rate Experiments and FDS Simulations* D. Madrzykowski and W. D. Walton
NISTIR 7137 2004	*Simulation of the Dynamics of a Fire in the Basement of a Hardware Store, New York, June 17, 2001* N. P. Bryner and S. Kerber
NIST NCSTAR September 2005	*Reconstruction of the Fires in the World Trade Center Towers. Federal Building and Fire Safety Investigation of the World Trade Center Disaster.* R. G. Gann, A. Hamins, K. B. McGrattan, G. W. Mulholland, H. E. Nelson, T. J. Ohlemiller, W. M. Pitts, and K. R. Prasad
Fire Technology 42, no. 4, (October 2006): 273–81	*Numerical Simulation of the Howard Street Tunnel Fire* K. B. McGrattan and A. Hamins
Fire Protection Engineering no. 31 (Summer 2006): 34–36,38,40,42,44,46	*NIST Station Nightclub Fire Investigation: Physical Simulation of the Fire* D. Madrzykowski, N. P. Bryner, and S. I. Kerber

Note: Many of these and other similar reports can be obtained on the NIST website (fire.nist.gov).

Moreover, the public demands answers to why and how incidents occur and whether fire codes perform as anticipated during actual fires. People are not only curious but they have a legitimate concern for answers to questions such as, Is it safe for my child to live in a nonsprinklered dormitory?

Several keynote case histories are worth mentioning to show the development of fire modeling in forensic evaluations of fire scenes. These historically significant cases are discussed along with the important concepts gleaned from the referenced and published studies.

UK FIRE RESEARCH STATION

The UK government operated the outstanding **Fire Research Station (FRS)** located in Borehamwood until the facility was privatized in 1997 and moved to Garston. Now a modest engineering consultancy, the facility has a 50-year history of assisting with fire investigations for the UK government. FRS investigations and reconstructions have included fire incidents at the Stardust Disco in Dublin, at Windsor Castle, and in the Channel Tunnel, and the Paddington–Ladbroke Grove train accident. The FRS uses a combination of a statistical database of loss histories, scene investigations, a fire testing laboratory, and fire models to test hypotheses or investigate unusual phenomenon (FRS 2002).

Early work at the FRS developed a compartment fire model to address smoke and heat venting when its researchers examined the similarities of the Livonia, Michigan, and Jaguar auto plant fire losses (Nelson 2002). These fires at two separate parts manufacturing centers had devastating impacts on the production of automobiles. What began as a search for means to vent hot by-products of combustion through the ceiling formed the basis for the zone fire model, which was based on actual fire testing.

DUPONT PLAZA HOTEL AND CASINO FIRE

The December 31, 1986, fire at the Dupont Plaza Hotel and Casino, San Juan, Puerto Rico, killed 98 persons. Owing to the extreme and needless loss of life, this case has become a textbook example of the application of fire reconstruction of the initial development of what became a large, complex fire. NIST fire researcher Harold "Bud" Nelson worked in cooperation with ATF investigators to collect and analyze data during the scene investigation. This led to a successful prosecution of the individuals who started the fire simply to harass hotel management. The analysis combined technical data and fire growth models to explain the dynamics exhibited by the fire. The analysis looked at mass burning rate, heat release rate, smoke temperatures, smoke layer, oxygen concentration, visibility, flame extension and spread, sprinkler response, smoke detector response, and fire duration. Fire models used in this analysis included FIRST, ROOMFIR, ASETB, and HOTVENT.

The analysis confirmed the area of origin to be the south ballroom. The fire spread next to a foyer, then to the lobby and casino area. Figure 6.4 documents the conditions at 60 s after the fire started.

KING'S CROSS UNDERGROUND STATION FIRE

One of the first important cases using CFD models for fire investigation was the analysis of the King's Cross Underground Station fire, which occurred on November 18, 1987. The CFD model FLOW3D tracked the fire upward along a 30° incline of the wooden sides and treads of the escalators to the ticket concourse level. The theory of the "trench effect" was verified by the CFD model that described this unique phenomenon. Thirty-one persons died in the fire, including a senior station officer (Moodie and Jagger 1991).

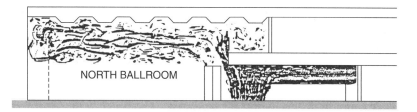

FIGURE 6.4 ◆ The analysis combined technical data and NIST fire growth models to explain the dynamics exhibited by the fire at the Dupont Plaza. The analysis looked at mass burning rate, heat release rate, smoke temperatures, smoke layer, oxygen concentration, visibility, flame extension and spread, sprinkler response, smoke detector response, and fire duration. *Courtesy of NIST, from Nelson 1987.*

In determining the area of origin for this fire, investigators homed in on the area where the fire first started as observed by witnesses, some 21 m (68 ft) below the top of escalator No. 4. This area was at the lowest point of fire pattern damage to wooden handrails, treads, and risers. An analysis of the paint blistering and ceiling damage was also used to locate the fire origin. The physical indicators were corroborated by witnesses who saw the fire partway up the escalator from below.

The investigation concluded that the fire started beneath the wooden steps on the moving staircase from discarded smoking materials. The fire consumed much of the available combustible material in the escalator. Fire damage also extended to the other escalators and the upper-level ticket hall. Conditions in the latter were made untenable by the fact that a temporary plywood construction partition had been erected that was able to burn despite a coating of allegedly fire-retardant paint.

In addition to using a CFD model, government health and safety personnel obtained ignition characteristics of the samples taken from nearby the area of fire origin. Investigators were surprised to determine that escalator lubricating grease was able to be ignited only if combined with fibrous debris (paper, hair, lint) that acted as a wick.

To assess and validate the results of the CFD model, three types of tests were run. Full-scale fire growth tests determined that the initial fire was about 1 MW in size. Small-scale models implied that above a certain fire size, the flame in the channel of the escalator did not rise vertically but was pushed down into the trench by entrained air, forcing rapid spread. Scaled open-channel tests examined the 30° inclined plane and used plywood with various surface coverings to evaluate flame spread. The hot gases moved quickly up the tunnel to fill the ticket lobby. When the gases ignited (in a flameover), the fire engulfed the lobby.

Lessons learned from this fire included the influence of the trench effect on a fire and the speed at which such a fire can develop and spread. Witnesses underestimated the speed of the smoke spread, the growing intensity of the fire, and the need for rapid evacuation. Smoking on underground rail services has since been banned, and combustible escalators, enclosures, and signage eliminated.

FIRST INTERSTATE BANK BUILDING FIRE

Another NIST engineering analysis was conducted on the May 4, 1988, fire at the 62-story, steel-framed high-rise known as the First Interstate Bank Building, in Los

Angeles, California (Nelson 1989). The fire involved an after-hours accidental ignition in an office cubicle on the twelfth floor that spread across the floor of origin. The fire propagated to the thirteenth, fourteenth, fifteenth, and a section of the sixteenth floors as exterior windows failed, allowing large flame plumes to entrain against the sheer sides of the building, causing windows above to fail. The fire burned for over 2 hr before being suppressed by firefighters' hose streams inside the building. One of the maintenance personnel was killed after taking an elevator to the fire floor.

The NIST analysis of the fire used the Available Safe Egress Time (ASET) program to predict the smoke layer's temperatures, smoke level, and oxygen content after the calculated time for flashover (smoke temperature of 600°C [1150°F]). The DETACT-QS model was used to estimate the response of smoke detectors and sprinklers to the fire growth. Because there were no sprinklers on the floor of origin, these studies revealed how much damage could have been avoided.

This analysis was one of the first to evaluate burn patterns and flame spread on combustible furniture, glass breakage, communication of the fire to upper floors by flame extension from broken windows, and compartment burning rates in reconstructing the fire. The study also advanced the theory of a "transfire triangle," which demonstrates the important interdependent relationships among the mass burning (pyrolysis) rate, the heat release rate, and the available air (oxygen) (Nelson 1989).

HILLHAVEN NURSING HOME FIRE

A fire at 10 P.M. on October 5, 1989, at the Hillhaven Rehabilitation and Convalescent Center, Norfolk, Virginia, claimed the lives of 13 persons. None of these individuals were burned but all breathed sufficient carbon monoxide (CO) to produce fatally high carboxyhemoglobin (COHb) levels.

NIST fire researcher Harold E. "Bud" Nelson recognized the importance of evaluating the fire dynamics when reconstructing this incident. His analysis of the room of fire origin in the nursing home showed that the temperature rose to at least 1000°C (1800°F), with a CO concentration of 40,000 ppm.

Using the fire model FIRE SIMULATOR contained in its engineering package, FPETool predicted the smoke temperatures, smoke layer, toxic gas concentrations, and velocity of the smoke front. An important aspect learned in the NIST analysis was the impact of the "what if" questions that are posed in fire reconstructions (Figure 6.5). These questions centered on the hypothetical impact of different fire activation devices on fire suppression as well as the predicted time of flashover.

PULASKI BUILDING FIRE

A fire in Suite 5127 of the Pulaski Building, 20 Massachusetts Avenue, NW, Washington, DC, on March 23, 1990, damaged the offices of the U.S. Army Battlefield Monuments Commission (ABMC). A NIST engineering review of the fire was able to apply sound fire protection engineering principles to estimate the rate of fire growth (Nelson 1989).

At 11:24 A.M., an alert ABMC staff member noticed smoke coming from the open door of the unoccupied audio/visual conference room. The room contained an estimated 4530 kg (10,000 lb) of combustible materials consisting of publications and photos stored in corrugated board boxes and cardboard mailing tubes stacked about 1.5 m (5 ft) high. The fire was thought to have started when cartons of mailing tubes were exposed to a frayed energized electrical lamp cord that passed through a small hole from a podium into the projection room.

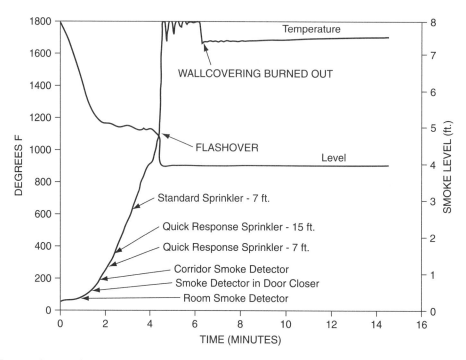

FIGURE 6.5 ◆ The impact at the Hillhaven Nursing Home of the "what if" questions that are posed in fire reconstructions, centering on the hypothetical impact of different fire activation devices on fire suppression as well as the predicted time of flashover. *Courtesy of NIST, from Nelson 1991.*

FPETOOL was used to analyze the fire and arrive at a scenario based on witness testimony and observations. FIRE SIMULATOR estimated the environmental conditions, temperature, depth of smoke, thickness of smoke, and energy vented from the room. Also included was the impact of the failure of the suspended ceiling.

The analysis showed that the fire reached flashover approximately 268 s into the fire. A unique aspect of this fire engineering analysis was the introduction of a timeline of events (Figure 6.6) that compared the approximate time with the observed fire conditions, behavior of the occupants, impact of existing fire protection systems, and "what if" scenarios of the impact of potential fire protection systems.

HAPPYLAND SOCIAL CLUB FIRE

On the morning of March 25, 1990, an arsonist set a fire in the entryway of a neighborhood club using 2.8 L (0.75 gal) of gasoline. The fire killed 87 occupants of the two-story club. A NIST fire engineer used the HAZARD I fire model to examine the fire development as well as potential mitigation strategies that could have influenced the outcome of the fire (Bukowski 1991).

These mitigation strategies included the use of automatic sprinkler protection, solid wooden doors at the base of the stairway, a fire escape and enclosed exit stairwell, and a noncombustible interior finish. The analysis also weighed the economic cost estimates to implement these strategies. The study concluded that the combustible wall coverings and the opening of an entry door, which allowed the entry of gasoline-fed flames, caused the fire to grow and combustion gases spread so rapidly that everyone in the upstairs club was affected within 1–2 min, precluding their escape.

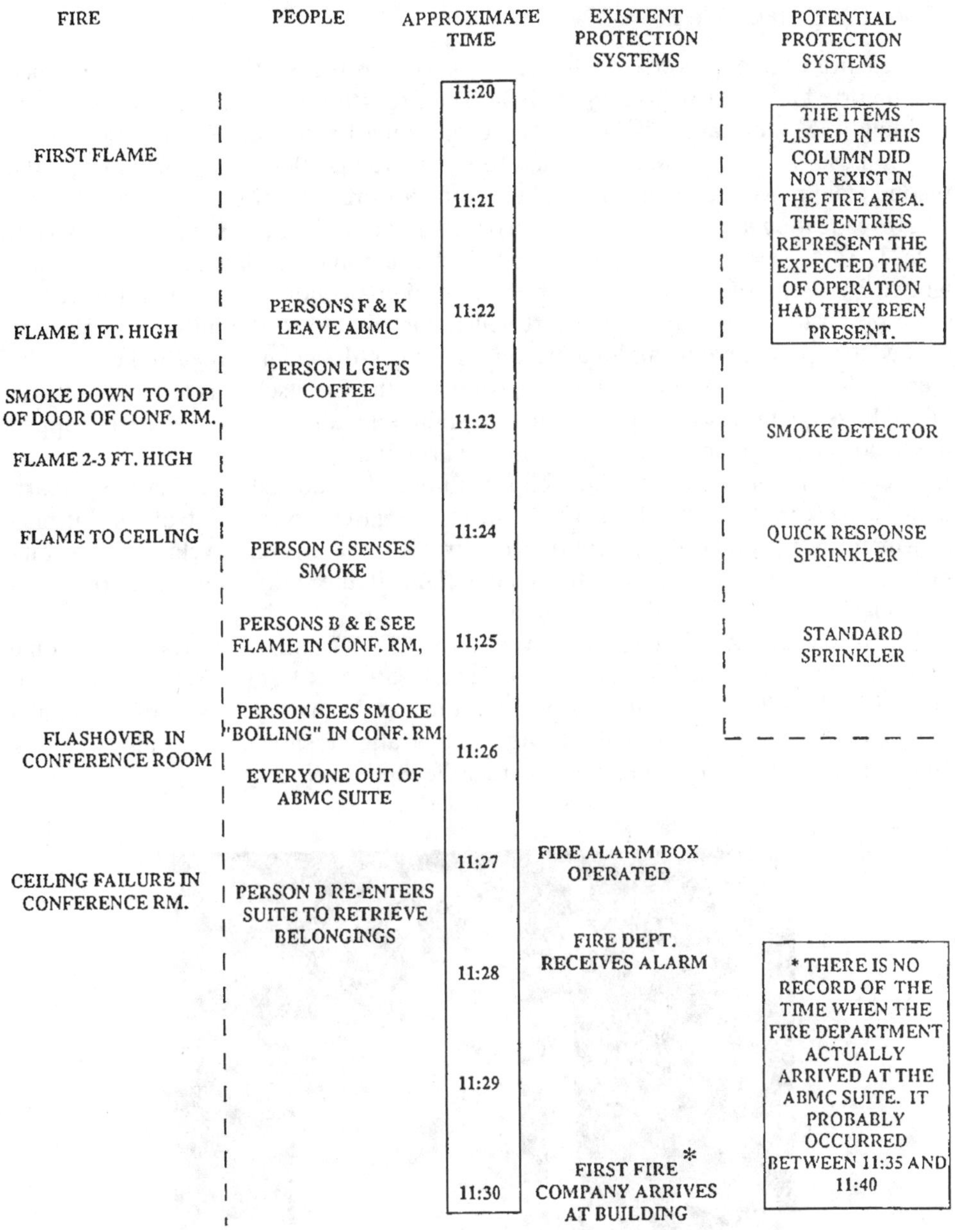

FIRE	PEOPLE	APPROXIMATE TIME	EXISTENT PROTECTION SYSTEMS	POTENTIAL PROTECTION SYSTEMS
		11:20		THE ITEMS LISTED IN THIS COLUMN DID NOT EXIST IN THE FIRE AREA. THE ENTRIES REPRESENT THE EXPECTED TIME OF OPERATION HAD THEY BEEN PRESENT.
FIRST FLAME				
		11:21		
FLAME 1 FT. HIGH	PERSONS F & K LEAVE ABMC	11:22		
	PERSON L GETS COFFEE			
SMOKE DOWN TO TOP OF DOOR OF CONF. RM.		11:23		SMOKE DETECTOR
FLAME 2-3 FT. HIGH				
FLAME TO CEILING	PERSON G SENSES SMOKE	11:24		QUICK RESPONSE SPRINKLER
	PERSONS B & E SEE FLAME IN CONF. RM,	11;25		STANDARD SPRINKLER
FLASHOVER IN CONFERENCE ROOM	PERSON SEES SMOKE "BOILING" IN CONF. RM	11:26		
	EVERYONE OUT OF ABMC SUITE			
CEILING FAILURE IN CONFERENCE RM.	PERSON B RE-ENTERS SUITE TO RETRIEVE BELONGINGS	11:27	FIRE ALARM BOX OPERATED	
		11:28	FIRE DEPT. RECEIVES ALARM	* THERE IS NO RECORD OF THE TIME WHEN THE FIRE DEPARTMENT ACTUALLY ARRIVED AT THE ABMC SUITE. IT PROBABLY OCCURRED BETWEEN 11:35 AND 11:40
		11:29		
		11:30	FIRST FIRE * COMPANY ARRIVES AT BUILDING	

FIGURE 6.6 ◆ The timeline of events in the Pulaski Fire, which compares the approximate time with the observed fire conditions, behavior of the occupants, impact of existing fire protection systems, and "what if" scenarios of the impact of potential fire protection systems. *Courtesy of NIST, from Nelson 1989.*

The NIST analysis also took into account occupant tenability at various locations, based on the output of the HAZARD I fire model (which incorporates CFAST version 2.0). The tenability study looked at heat flux that could cause second-degree burns, temperatures, and fractional exposure doses based on both the NIST and the Purser toxicity models (Bukowski 1991). This analysis showed that even if a fire escape had been installed, there would have been many casualties.

62 WATTS STREET FIRE

A NIST fire engineer modeled what is now known as the 62 Watts Street Fire, which was reported at 7:36 P.M., March 28, 1994. The fire involved a three-story apartment building in Manhattan, New York. The responding firefighters formed two three-person hose teams. One team planned to enter the first-floor apartments while the second team proceeded up a stairwell to search for fire extension.

Investigation later revealed that the occupant of the first-floor apartment had left at 6:25 P.M. A plastic trash bag inadvertently left on top of the kitchen gas range was ignited by the pilot flame. Several bottles of high-alcohol-content liquid were left close by, on the adjoining countertop, which contributed to the initial fire. The fire burned for approximately an hour, becoming oxygen-deficient as the smoke layer descended to the level of the countertop. Although the fire itself was not large, it produced large quantities of carbon monoxide, smoke, and other unburned fuels trapped in the hot smoke layer in the small, closed apartment.

As the firefighters in the stairwell forced open the door to the first-floor apartment, a backdraft occurred, in which warm fire gases flowing out of the burning apartment were replaced by cool outside air. With the ensuing exchange, the combustible gas mixture ignited, creating a large flame in the stairwell for a period of approximately 6 min, killing the firefighters on the stairs.

Bukowski at NIST modeled this event using CFAST, and the analysis has become a classic study into backdrafts and firefighter safety. Recently, McGill (2003), at Seneca College, Toronto, Canada, modeled this case using the NIST Fire Dynamics Simulator. A view of McGill's Smokeview visualization is shown in Figure 6.7. It took approximately 20 hr to define and construct the model.

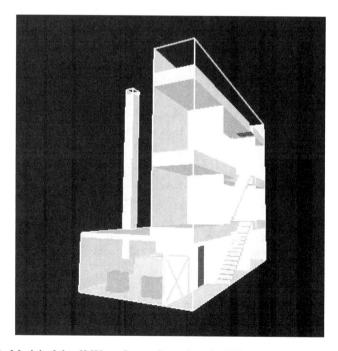

FIGURE 6.7 ◆ Model of the 62 Watts Street fire using the NIST Fire Dynamics Simulator. Room of origin at lower left. The fire plume extending from the apartment door (center) into the hallway trapped and killed three firefighters on the stairs. *Courtesy of D. McGill [2003].*

CHERRY ROAD FIRE

In this case, NIST engineers were called on to demonstrate the results of calculations using the NIST FDS to evaluate the thermal and tenability conditions of a fire that may have led to the deaths of two firefighters and burned four others (Madrzykowski and Vettori 2000). The fire started in a ceiling electrical fixture on the lower level of a townhouse. Responding firefighters entered the first and basement floors of the structure. Both on-scene observations and subsequent modeling showed that the opening of the lower-level sliding glass doors increased the heat release rate of the fire, so that 820°C (1500°F) gases moved up the stairwell. The fire may have smoldered for several hours before flame ignition, producing fuel-rich gases that ignited while firefighters were inside the building searching for the fire and any victims in the heavy smoke.

The Cherry Road fire FDS model may become one of the best examples of application of the NIST FDS. As the technology expands, so does the ability to address harder questions concerning transient heat and ventilation, rapid fire growth, and stratified or localized conditions. Furthermore, the fire pattern damage observed correlated with the model's predictions.

◆ 6.7 CASE STUDY 1—*WESTCHASE HILTON HOTEL FIRE*

A March 6, 1982, fire at the Westchase Hilton Hotel, Houston, Texas, resulted in the deaths of 12 people and serious injuries to three others. Hot gases and smoke progressed from a fire in a guest room on the fourth floor down the corridor and, to some degree, throughout the building as illustrated in Figure 6.8. Records indicate that 137 guests were registered at the hotel the evening of the fire.

The fire started in a fourth-floor hotel room (Figure 6.9), with the first odor of smoke being reported at 2:00 A.M. on the tenth floor, followed by light smoke reported at 2:10 A.M. on the eighth floor; the fire was actually discovered at 2:20 A.M. by an occupant returning to his room.

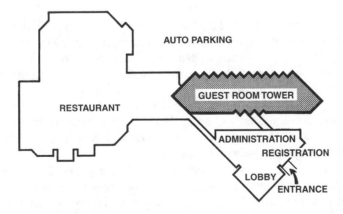

FIGURE 6.8 ◆ The March 6, 1982, fire at the Westchase Hilton Hotel resulted in the deaths of 12 people and serious injuries to 3 others. *Reprinted with permission from NFPA Westchase Hilton Fire Investigation. Copyright © 1982 and 1983, National Fire Protection Association, Quincy, MA 02269.*

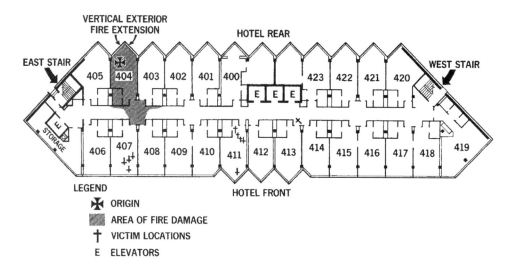

FIGURE 6.9 ◆ The Westchase Hilton Hotel fire started on the fourth floor, Room 404. *Reprinted with permission from NFPA Westchase Hilton Fire Investigation. Copyright © 1982 and 1983, National Fire Protection Association, Quincy, MA 02269.*

At 2:31 A.M., the district fire chief arrived and saw a fire plume projecting out the exterior window of the room of fire origin (Figures 6.10 and 6.11). Extinguishment of the property started at 2:38 A.M. and ended at 2:41 A.M. Fire investigating officials concluded from these times that flashover in the room occurred between 2:20 and 2:31 A.M.

An investigation revealed that a cigarette smoldered in an upholstered chair in the room of origin. The U.S. Consumer Products Safety Commission (CPSC) later had fire tests performed on the furniture. Heat release rate data from these tests along with assumptions about how wide the door was ajar formed the basis for the input to the model (Table 6.6). Data from previous NIST tests of smoldering ignition of furniture were also used in evaluating the time factors between detection of smoke odors and observation of a flaming fire.

A zone model called FIRM (Fire Investigation and Reconstruction Model) was later used to evaluate the single compartment of fire origin using data from the CPSC tests (Janssens 2000). The input screen using these data is shown in Figure 6.12. A sensitivity analysis using FIRM was conducted to examine the impact of the position of a door ajar on the room of fire origin.

The FIRM model also was used to predict the fire development within the room of origin, tenability conditions, and an evaluation of the time to flashover. Table 6.6 lists the observed and predicted data for the room of fire origin, and the output file from the FIRM program is shown in Figure 6.13.

Later studies on the behavior of persons in this fire used a data collection interview form reproduced in Figure 6.14 (Bryan 1983). Approximately a month after the fire this form was mailed to 130 guests registered at the time of the fire by the National Fire Protection Association (NFPA) in coordination with the Houston Fire Department. A total of 55 persons responded, which was 27 percent of the registered guests, representing 42 males and 13 females.

FIGURE 6.10 ◆ At 2:31 A.M., the district fire chief arrived and saw a fire plume projecting out the exterior window of the room of fire origin on the south side of the Westchase Hilton high-rise tower. Note the areas of external, horizontal, and vertical fire extension. *Reprinted with permission from NFPA Westchase Hilton Fire Investigation. Copyright © 1982 and 1983, National Fire Protection Association, Quincy, MA 02269.*

The study showed that some of the guests were aware of the fire for as long as almost 2 hr before the fire department was notified. The majority of the notification was by way of people yelling. The most frequent first actions after notification of the fire were dressing, calling the front desk, and attempting to exit. The study failed to reveal any nonproductive behavior by those persons escaping the fire, and some spent time knocking on doors to alert other occupants. Persons escaping the fire did not reenter the building. Movement through the smoke is always a question in research studies. More than 50 percent of the guests passed through the smoke for distances of from 4.67 to 182 m (15 to 600 ft).

Lessons learned from the analysis of this fire by several sets of experts included the need for better enforcement of the life safety code, education of the public, training of

FIGURE 6.11 ◆ Exterior view, from the hallway, of Room 404, Westchase Hilton Hotel, whose combustible contents were almost totally consumed. *Reprinted with permission from NFPA Westchase Hilton Fire Investigation. Copyright © 1982 and 1983, National Fire Protection Association, Quincy, MA 02269.*

TABLE 6.6 ◆ Numerical Constants in the Fire Investigation and Reconstruction Model

Feature	
Room floor area	24.51 m² (263.8 ft²)
Room ceiling height	2.44 m (8.00 ft)
Door (vent) soffit	1.99 m (6.56 ft)
Vent width (varied)	0.152–0.457 m (6–18 in.)
Fire height (estimated)	0.914 m (3.0 ft)
Heat loss fraction	0.66
Flashover	
0.152-m (0.50-ft) opening[a]	199 s
0.076-m (0.25-ft) opening[a]	299 s
Untenability, 1.2–1.5 m (4–5 ft)	183°C (361°F)

Source: Derived from Janssens 2000.
[a]Predicted value.

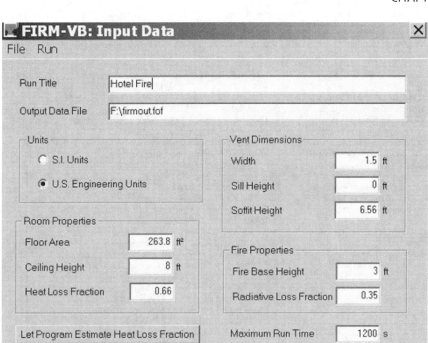

FIGURE 6.12 ◆ Input screen for the Fire Investigation and Reconstruction Model (FIRM).

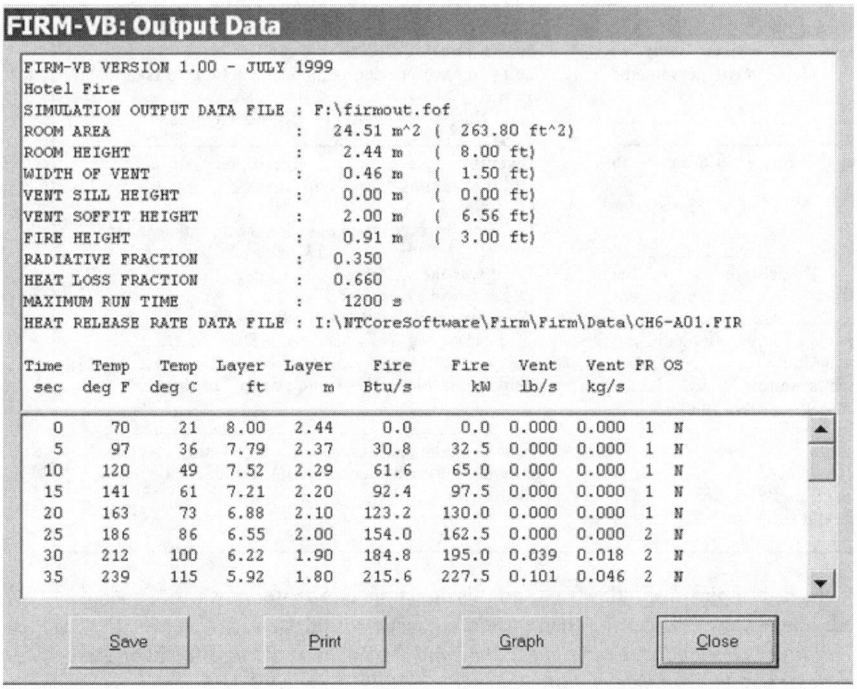

FIGURE 6.13 ◆ Output screen for the Fire Investigation and Reconstruction Model (FIRM).

FIRE EXPERIENCE SURVEY
Westchase Hilton Hotel, Houston, Texas
March 6, 1982

We are grateful for your willingness to share your fire experience with us. Your information will help us find ways to avoid recurrences of such tragedies.

Occupation _____ Sex ___ Age ___
Room No. ___

1) How did you first become aware that there was something unusual occurring in the Hotel? _____
2) What time was it? _____ How did you determine the time? _____
3) When did you realize that what was occurring was a fire? What time was it? _____ How did you determine the time? _____
4) How serious did you believe the fire to be at first?
   ( ) Not at all serious, ( ) Only slightly serious, ( ) Moderately serious, ( ) Extremely serious
5) Were you alone when you became aware of the fire?
   ( ) No ( ) Yes
6) How many persons were with you? ___ They were ( ) relatives ( ) others.
7) Were you injured? ( ) No ( ) Yes:
   If Yes, What was the nature and cause of the injury?
   _____

8) What did you do when you realized there was a fire? (state exact sequence of actions)
   First _____
   Second _____
   Third _____
   Fourth _____
   Fifth _____
9) Once aware of the fire, did you:
   (a) Call or attempt to call the hotel operator (switchboard) ( ) No ( ) Yes; If Yes, At what time? _____
   (b) Call or attempt to call the Fire Department directly ( ) No ( ) Yes;
       If Yes, at what time? _____
   (c) Operate a manual fire alarm pull station?
       ( ) No
       ( ) Yes; If Yes, At what time? _____
10) Did you voluntarily leave (i.e., without being requested by hotel or Fire Department personnel)?
    ( ) No ( ) Yes
    If No, Why not? _____
    If Yes, At what time? _____
    How? ( ) Stairway to ground ( ) Stairway to roof
         ( ) Elevator ( ) Window ( ) Exterior Door
         ( ) Other (Specify) _____
11) I left the building: ( ) Unassisted; Assisted by:
    ( ) Hotel staff ( ) Guest ( ) Fire Department
    ( ) Other (Specify) _____
12) After you realized there was a fire, how long did you wait before leaving the building? ___ minutes.
    What were you doing while waiting? _____
13) In your leaving, was there any visible smoke?
    ( ) No ( ) Yes
    Any smoke odor? ( ) No ( ) Yes
    Any flames? ( ) No ( ) Yes
    Did you try to move through smoke? ( ) No
    ( ) Yes: If Yes, How far did you move? ___ feet

(approx.). How far could you see at the time? ___ feet.
Did you turn back? ( ) No ( ) Yes. If Yes, Why did you turn back? _____
14) Did you notice any lighted exit signs? ( ) No ( ) Yes
15) During your escape, was there any difficulty in following the direction marked by exit signs?
    ( ) No ( ) Yes:
    If Yes, What was difficulty? _____
16) Any obstructions to escape? _____
17) What aids helped you to escape? _____
18) Did smoke enter your room? ( ) No ( ) Yes:
    If Yes, How? _____
    ( ) Heating/cooling unit ( ) Bathroom vent
    ( ) Around door ( ) Window
    ( ) Don't know ( ) Other (Specify) _____
19) Beginning with your first action, number the sequence of events you took while in your room.
    ( ) Put materials over the heating/cooling unit vent
    ( ) Around door ( ) Around window
    ( ) Turned on TV ( ) Turned off radio
    ( ) Took no action ( ) Other _____
20) Did the smoke detector in your room sound an alarm? ( ) No ( ) Yes
21) Did you hear the building fire alarm?
    ( ) No ( ) Yes: If Yes, At what time? ___
    How long did it operate? _____
22) Did you receive any instructions from hotel staff during the fire emergency? ( ) No ( ) Yes: If Yes, What were they? _____
    Did you receive fire safety instructions from hotel employees prior to the fire? ( ) No ( ) Yes:
    If Yes, What were they? _____
    Did you observe fire safety information within your room? ( ) No ( ) Yes. If Yes, What was the information? _____
23) Did you have previous training on actions to take in a fire?
    ( ) No ( ) Yes: If Yes, Number of times ___
    Type _____
    Given by _____ Last time: Date ___
    Did the training help in this fire?
    ( ) No ( ) Yes
24) Did you receive previous fire safety information from: ( ) Radio ( ) TV ( ) Publication.
    What was the message? _____
    Did it help in this fire? ( ) No ( ) Yes
25) Were you ever involved in a fire before?
    ( ) No ( ) Yes Last time: Date _____
26) Please report any additional comments that you think might help others in a similar fire situation.
    _____
    _____

27) Please mark your escape route on the diagram provided. Please identify room number and floor. Thank you.

FIGURE 6.14 ◆ Later studies on the behavior of persons in this fire used this data collection interview form, mailed to 130 guests. A total of 55 persons responded, which was 27 percent of the registered guests, 42 males and 13 females. *Reprinted with permission from NFPA Westchase Hilton Fire Investigation. Copyright © 1982 and 1983, National Fire Protection Association, Quincy, MA 02269. This reprinted material is not the complete and official position of the National Fire Protection Association on the referenced subject, which is represented only by the standard in its entirety.*

hotel staff, and emergency preparedness. Other lessons showed the utility of scientific review of fire scenes, development of a list of behaviorally related questions and answers, and application of a fire model.

◆ 6.8 CASE STUDY 2: *MODELING USED TO ASSESS DISPUTED FIRE ORIGINS AND CAUSES*

The use of FDS in the forensic engineering analysis of fires, particularly ones where the origin and cause is disputed, has begun to gain further popularity and acceptance in the fire investigation community. This case study consists of a series of fires for which FDS simulations provided compelling arguments to match fire burn pattern damage along with witness accounts to prove or disprove hypotheses (Vasudevan 2004).

EXAMPLE 1   Fire in a Minimart: Accident or Incendiary?

A fire was discovered soon after the owner-manager of a minimarket had locked up and left the building. A fire some time after the owner left raised questions as to origin and cause. Two disputed hypotheses were that (1) the fire was a result of an electrical malfunction in the attic area over the storage/display area or (2) the fire was intentionally set in the floor-level liquor storage area.

FDS was used to model the hypothesis that the fire did not originate in the attic space but was an intentionally set fire in the liquor storage area. The fire patterns at the scene (Figure 6.15) were found to be consistent in extent and shape with the prediction of the FDS/Smokeview results for the incendiary fire hypothesis (Figure 6.16).

FIGURE 6.15 ◆ Photo of actual fire burn pattern damage to a minimart. *Courtesy of R. Vasudevan, Sidhi Consultants, Inc.*

FIGURE 6.16 ◆ NIST Fire Dynamics Simulator model results for the incendiary fire hypothesis of the minimart. *Courtesy of R. Vasudevan, Sidhi Consultants, Inc.*

FIGURE 6.17 ◆ Photo of actual fire burn pattern damage to the living room of the one-bedroom apartment. *Courtesy of R. Vasudevan, Sidhi Consultants, Inc.*

EXAMPLE 2   Placement of Smoke Detector

A small fire in a living room of a one-bedroom apartment generated sufficient carbon monoxide to cause the death of one person and serious injuries to a second. It was disputed whether the lack or failure of a smoke detector affected the occupants' ability to respond appropriately to the fire. Although the placement of the smoke detector complied with the local applicable fire code, it did not provide adequate notification time for the occupants to respond to the toxic by-products of combustion.

(a)

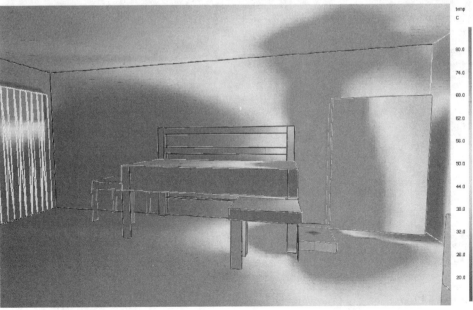

(b)

FIGURE 6.18 ◆ NIST Fire Dynamics Simulator model results for the living room of the one-bedroom apartment showing (a) fire and smoke obscuration and (b) surface boundary conditions. *Courtesy of R. Vasudevan, Sidhi Consultants, Inc.*

EXAMPLE 3 Deadly Kitchen Cooking Fire

A kitchen cooking fire in a two-bedroom apartment resulted in the death of a child and carbon monoxide poisoning to two other persons. The issue concerned (1) the location of the smoke detector, (2) its ability to detect the fire, and (3) tenability levels of carbon monoxide within the children's rooms. The FDS analysis showed that the close location of the smoke detector to the

(a)

(b)

FIGURE 6.19 ◆ Photos of (a) rebuilt kitchen and (b) actual fire burn pattern damage. *Courtesy of R. Vasudevan, Sidhi Consultants, Inc.*

FIGURE 6.20 ◆ NIST Fire Dynamics Simulator model results for the kitchen in the smoke detector placement and carbon monoxide hypothesis tests. Compare these results to the burn patterns in Figure 6.19b. *Courtesy of R. Vasudevan, Sidhi Consultants, Inc.*

incipient fire would damage the unit within seconds of its alarming. The analysis also showed that unsafe carbon monoxide levels would rapidly be reached. Finally, the model results closely matched the fire pattern damage (see Figures 6.19 and 6.20).

EXAMPLE 4   Wildland Case Study

In a recent case, a proponent in a large civil lawsuit attempted to use FDS to predict fire spread on a steeply sloped hillside that was covered in heavy brush [2–3 m (6–10 ft) tall]. This hillside [some 100 m × 200 m (330 × 660 ft) in area] was known to be the general area of origin of a larger fire. Extensive calorimetry tests were carried out on arrays of the same type brush to collect data on the rates of heat release and mass loss for such fuels. The steep hillside was modeled in FDS as a series of rectilinear steps 1 m (3.3 ft) high and 1–2 m (3.3–6 ft) deep, with computational cells of the same proportions. Because FDS does not recognize porous fuel arrays, the fuel load on the hillside was inputted as a series of thin horizontal and vertical panels (in steplike array) whose heat release curves approximated the calorimetry data for those woody shrubs collected by testing. (The calorimeter testing, however, included no wind-driven growth factors so did not duplicate a real-world fire environment.) A mild, cross-slope wind was also inputted. The resulting Smokeview animation showed horizontal (cross-slope) propagation and downslope propagation in a stop-start fashion roughly equal in spread velocity to the upslope propagation. It was proposed that this result supported an origin of the fire very different (well upslope) from the origin identified by the original scene investigators.

When this model was offered to the court in a Frye admissibility hearing, the proponents argued that FDS had been developed and studied for exterior fires (smoke plumes and WUI fires) and therefore should yield valid results for exterior fires of all types. Although proponents could offer peer-reviewed papers discussing FDS modeling for flat grasslands (two-dimensional spread) and WUI (established fires in trees that spread to structures), they could not offer a single reference regarding FDS and fire spread in porous (three-dimensional) fuels on hillsides. Ultimately, the judge ruled that FDS modeling was not generally accepted for this situation and therefore its predictions could not be used as evidence.

Because the predictions of the model (downslope propagation in low-wind conditions) were contrary to logic and observations of countless wildland fires, the proponents should have acknowledged that FDS was not yielding reliable results. This is one of the instances that ASTM E 1355, which evaluates the predictive capacity of fire models by defining scenarios (see referenced guidelines in this chapter), suggests carrying out. That is, if the model results run contrary to normal behavior, there is something wrong with the model or with the data input. Models should be used only for the applications for which they were designed and developed and have been demonstrated to yield results in agreement with actual fires of that type.

Modeling carried out by an opposing consultant using FARSITE, the most widely used and tested computer model for wildland fires, predicted upslope growth whose features paralleled the observations of the first responding fire crews. These crews had staged near the top of the slope and watched the fire progress rapidly upslope (with a mild cross-slope atmospheric wind) toward them from an area well down the slope. These crews were forced to abandon the house being used as an observation post when it was overrun by the growing fire. Growth of the fire was aided by the draft produced by the fire, which quickly developed into a wind-driven crown fire. This wind-aided growth was not duplicated in the calorimeter tests conducted.

(Case study courtesy of John DeHaan).

◆ 6.9 SUMMARY AND CONCLUSIONS

This chapter introduced the concept and application of fire modeling to forensic fire investigation. Although models have existed since the 1960s, fire modeling has centered on explaining the physical phenomena of fires, particularly when applied to verifying existing experimental data.

It was the effort of a few fire scientists and engineers working at NIST that pushed the acceptability and application of fire modeling out of the laboratory and into the world of forensic fire scene reconstruction. As indicated, the purpose of this chapter was not to make the reader an expert in fire modeling but to allow him or her to gain a better appreciation of its value in an investigation.

IMPACT OF ASTM GUIDELINES AND STANDARDS

ASTM Subcommittee E05.33 presently maintains and continues to work on improving four guidelines that standardize the evaluation and use of computer fire models. The following is a synopsis of the current standard guides (Janssens 2002):

- *ASTM E 1355*—evaluates the predictive capacity of fire models by defining scenarios, validating assumptions, verifying the mathematical underpinnings of the model, and evaluating its accuracy.
- *ASTM E 1472*—describes how a fire model is to be documented, including a user's manual, a programmer's guide, mathematical routines, and installation and operation of the software.
- *ASTM E 1591*—covers and documents available literature and data that are beneficial to modelers.

• ***ASTM E 1895***—examines the uses and limitations of fire models and addresses how to choose the most appropriate model for the situation.

A task group from the Society of Fire Protection Engineers (SFPE) has used *ASTM E 1355* since 1995 to evaluate several models. This group is continuing to develop accepted engineering practice guides for each of the models.

IMPACT OF NRC VERIFICATION AND VALIDATION STUDIES

Recent work sponsored in part by the NRC examined the **verification and validation (V&V)** of fire models, including CFAST and FDS (Salley et al. 2007). Although the NRC's study centered on fire hazards particular to nuclear power plants, it addressed many of the concerns raised by those who think fire models are not accurate or appropriate for forensic fire scene reconstruction. These concerns include the ability of the models to accurately predict common features of fires, such as upper-layer temperatures and heat fluxes. One of the features of this report was a comparison between actual fire test results and predictions of hand calculations, zone models, and field models. As shown in Figure 6.21, there is generally good agreement among all the models and the variability of real-world fires.

The V&V reports are presented in seven publications. Volume 1, the main report, provides general background information, programmatic and technical overviews,

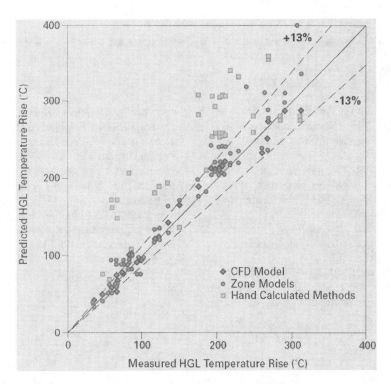

FIGURE 6.21 ◆ Comparison of hot gas layer temperatures measured in full-scale tests compared with predictions of hand calculations (□), zone models (●), and FDS models (◆). A predicted ±13% variability range is included. Note that the hand calculations tended to over-predict layer temperatures whereas both zone and field model predictions were generally within variability limits.

and project insights and conclusions. Volumes 2 through 6 provide detailed discussions of the V&V of the Fire Dynamics Tools (FDTs); Fire-Induced Vulnerability Evaluation, Revision 1 (FIVE-Rev1); Consolidated Model of Fire Growth and Smoke Transport (CFAST); MAGIC; and the Fire Dynamics Simulator (FDS). Volume 7 discusses in detail the uncertainty of the experiments used in the V&V study of these five fire models.

The next chapter explores in more detail the issues of tenability as it applies to the modeling of toxicity of fires.

Problems

6.1. Select one of the cases mentioned in this chapter and model it using the selected program. Obtain a similar fire modeling program and compare and contrast the results derived using it for this analysis.

6.2. Many computer fire models contain a database of materials and their burning

properties. Examine the database of two fire models. Compare and contrast the results obtained using each model.

6.3. Using the FDT spreadsheets, conduct a sensitivity analysis of several of its calculations by varying the input data (such as room dimensions or ventilation openings).

Suggested Reading

DeHaan, J. D. 2005. Reliability of Computer Modeling in Fire Scene Reconstruction. *Fire & Arson Investigator* (January): 40–47.

Hume, B. T. 1993. *Fire models: A guide for fire prevention officers.* Report No. 6/93. London: Home Office Fire Research and Development Group.

Janssens, M. 2000. *Introduction to mathematical fire modeling*, 2nd ed. Lancaster, PA: Technomic.

Madrzykowski, D. 2002. Fire research: Providing new tools for fire investigation. *Fire & Arson Investigator* 52 (4): 43–46.

Salley, M. H., J. Dreisbach, K. Hill, R. Kassawara, B. Najafi, F. Joglar, A. Hammins, K. McGrattan, R. Peacock, and B. Gautier. 2007. "Verification and Validation—How to Determine the Accuracy of Fire Models." *Fire Protection Engineering* 34 (Spring).

Urbas, J. 1997. Use of modern test methods in fire engineering. *Fire & Arson Investigator* 47 (December): 12–15.

Fire Deaths and Injuries

7 **CHAPTER**

There's a scarlet thread of murder running through the colorless skein of life, and our duty is to unravel it, and isolate it, and expose every inch of it.

—*Sir Arthur Conan Doyle,*
"A Study in Scarlet"

In every country, particularly in highly industrialized ones, fire kills a significant number of people. In the United States, it is one of the five leading causes of accidental death. The involvement of the investigator or forensic specialist in fatal fires can come in any form, from any sector, and challenge one's talents and knowledge to come to just and accurate conclusions. These cases require the highest degree of cooperation among the investigators, who all have contributions to make toward a successful investigation. When deaths occur in a fire, the event becomes the focus of the press and the public as well as police, fire, insurance, and forensic professionals. When problems occur, they can have far-reaching consequences.

◆ 7.1 PROBLEMS AND PITFALLS

There are several problem areas that can complicate fatal fire investigations and compromise the accuracy and reliability of the conclusions reached.

- **Linkage between the fire and death investigations.** Prejudging the fire and its attendant death as an accident and automatically treating the scene investigation accordingly is a major problem. Fires can be intentional, natural, or accidental in their cause, and deaths can be accidental, homicidal, suicidal, or natural. The linkage between the two events can be direct, indirect, or simple coincidence. The responsibility of the investigation team in fatal fire cases is to establish the cause of both the fire and the death and to determine the connection (if any) between the two.
- **Time interval to death.** Sudden violent deaths are assumed to be instantaneous exposure to insult followed by immediate collapse and death (a shot is fired and the victim collapses to die shortly afterward), and many forensic investigations are considered (and successfully concluded) in this light. Fires, however, occur over a period of time, creating dangerous environments that vary greatly with time and can kill by a variety of mechanisms. A person

may be killed nearly instantaneously by exposure to a flash fire or only after hours of exposure to toxic gases. Investigators must have an appreciation for the nature of fire and its lethal products and not treat the event as a single exposure to a single set of conditions at a precise moment in time that results in instant collapse.

◆ **Understanding heat intensity and duration.** There is little accurate information available to detectives and pathologists about the temperatures and intensities of heat exposure that occur in a fire as it develops. Misunderstandings and misapprehensions can lead investigators seriously astray when they try to assess injuries or postmortem damage.

◆ **Fire-related human behavior.** In most violent deaths, the victim offers a "fight or flight" response to the threat, suffers an injury, collapses, and dies. In fires, the potential responses include going to investigate; simply observing; failing to notice or appreciate the danger; failing to respond due to infirmity or incapacitation from drugs or alcohol; returning to the fire or delaying escape to rescue pets, family, or personal or valuable property; and fighting or attempting to fight the fire. This variability of responses can vastly complicate the process of solving the critical problem of why the victim failed to escape the fire (and perhaps why other people escaped).

◆ **Time interval between fire and death.** Fires can kill in seconds, or death can occur minutes, hours, days, or even months after the victim is removed from the scene. The longer the time interval between the fire and the death, the harder it is to keep track of the actual cause (the fire) and the result (the death). Evidence is lost when a living victim is removed from a scene, and when the victim dies later, away from the scene, it may be too late to recover or document that evidence. Fatalities that occur after the victim was hospitalized are inevitably omitted from the NFIRS statistics and may also be omitted from the national vital statistics.

◆ **Conflicts among investigating agencies.** There can be conflicts regarding the perceived or mandated responsibilities of police, fire, medicolegal, and forensic personnel that are often involved in fire death scenes.

◆ **Understanding postmortem effects.** After death, there can be severe postmortem effects on the body that can vastly complicate the investigation by obliteration of evidence. The body can bear fire patterns of heat effects and smoke deposits that can be masked by exposure to fire after death. The body can be incinerated by exposure to flames such that evidence of prefire wounds or even clinical evidence such as blood samples is destroyed. There can be structural collapse and effects of firefighting hose streams and overhaul that induce additional damage.

◆ **Premature removal of the body.** A major problem is the premature removal of a deceased victim from the fire scene. The compulsion to rescue and remove every fire victim is a very strong one, particularly among dedicated firefighters. However, once the fire is under control and unable to inflict further damage to the body of a confirmed deceased, there is nothing to be gained and much to be lost in the way of burn pattern analysis, body fragments (especially dental evidence), projectiles, clothing and associated artifacts (keys, flashlight, dog leash, etc.), and even trace evidence by the undocumented and hurried removal of the remains.

◆ 7.2 TENABILITY—WHAT KILLS PEOPLE IN FIRES?

Fire investigators should evaluate the common problems and pitfalls when conducting forensic reconstructions, particularly when death occurs. The purpose of this chapter is to review these human tenability factors, present some analytical approaches, and provide illustrative case histories.

Structural fires can achieve their deadly result in a number of ways—heat, smoke, flames, soot, and others—but fire conditions change continually as a fire grows and evolves, and the conditions of exposure of a would-be victim can vary from "no threat or injury" to almost instantly lethal. Although they can and usually do act in combination, the major lethal agents of fires are as follows: heat, smoke, inhalation of smoke or toxic gases, anoxia, flames, and blunt trauma. These are discussed in detail later in this chapter.

The ability of humans to escape a fire is measured by the time frame during which their environment remains survivable (**tenability**). Fires can produce incapacitating effects on humans when they are exposed to heat and smoke. These physiological effects are generally categorized into the following areas (Purser 2002).

- **Toxic gases.** Toxic gas inhalation causes confusion, respiratory tract injuries, loss of consciousness, or asphyxiation.
- **Heat transfer.** Heat irritates exposed skin and respiratory tracts, causing pain and varying degrees of burn injuries or hyperthermia.
- **Visibility.** Optical opacity of the smoke and irritants produce impaired vision as the distribution of thick smoke descends toward the floor through rooms, stairwells, and hallways.

Of primary concern is the point at which exposure to one or more of the preceding variables would cause injury or block the individual from successfully escaping the fire, resulting in death. The psychological behavior of people in fires when exposed to these variables affects their decisions and the time required to travel via safe escape routes.

Critical limits to human tenability include a limit of visibility to 5 m (16.4 ft), an accumulated dose of carbon monoxide of 30,000 parts per million minute (ppm-min), and a critical temperature of 150°C (302°F). The synergistic effects of two or more of these factors may override these individual limits (Jensen 1998). See Figure 7.1 for a graphic illustration of the exposure of humans walking upright through smoke layers (Bukowski 1995b).

The major goal of the investigator when conducting a **tenability analysis** is to determine how the individual who is escaping a burning structure becomes impaired and how the fire changes his or her environment and perceptions. These techniques are grounded in both experimental and forensic data, giving a balanced and practical approach.

FIGURE 7.1 ◆ Illustration of human tenability in varying smoke layer levels. *Source: R. W. Bukowski, "Predicting the Fire Performance of Buildings: Establishing Appropriate Calculation Methods of Regulatory Applications," National Institute of Standards and Technology, Gaithersburg, MD, 1995.*

The physiological and toxicological correlations of how heat transfer and toxic smoke affect animals and humans are based on experiments. For example, studies correlating carbon monoxide exposure to carboxyhemoglobin levels used subjects ranging from laboratory rats to volunteer medical students (Nelson 1998). Some of these data are extrapolated to model the results at higher levels of exposure. These variables must take into account variations in age, health, and stature of the individual. Note that an individual's height affects his or her exposure to the stratified upper smoke layer, where nearly all the toxic gases reside.

Forensic evaluations of incapacitation also come from forensic data derived from actual case histories and investigations. The major work on the behavior of people in fires came from subject interviews of persons surviving large fires and explosions (Bryan and Icove 1977).

◆ 7.3 TOXIC GASES

The previous section concerned a majority of visibility and irritant effects on humans as they try to navigate through fires. **Toxic gases** contained within smoke can also have a narcotic effect that asphyxiates victims. The dominant narcotic gases in smoke that affect the nervous and cardiovascular systems are carbon monoxide (CO) and hydrogen cyanide (HCN). Carbon dioxide and reduced oxygen levels, although not toxic themselves, may have severe effects on tenability.

Increased exposure to toxic gases may cause confusion, loss of consciousness, and eventually asphyxial death. The prediction of asphyxiation to incapacitation and death can be modeled (Purser in SFPE 2002b). Toxic products of combustion can include a wide variety of chemicals depending on what is burning and how efficiently it is burning (temperature, mixing, and oxygen concentration are all important variables in determining what species are created).

Toxic gases can generally be classified into three basic categories:

- **Nonirritant Gases:** (sometimes referred to as "narcotic gases"): CO, HCN, H_2S (hydrogen sulfide), and phosgene (CCl_2O).
- **Acidic:** HCl (hydrogen chloride)—produced during the combustion of polyvinyl chloride plastics (PVC); sulfur oxides (SO_x), which form H_2SO_3 (sulfurous acid) and H_2SO_4 (sulfuric acid)—produced by oxidation of sulfur-containing fuels; and nitrogen oxides (NO_x), which form HNO_2 (nitrous acid) and HNO_3 (nitric acid)—from nitrogen-containing fuels.
- **Organic Irritants:** Formaldehyde (CH_2O) and acrolein (2-propenal, C_3H_4O)—produced by the combustion of cellulosic fuels, and isocyanates—produced by the combustion of polyurethanes.

Acidic gases dissolve in the water of the mucous membranes and generate the corrosive acids listed. These acids cause the epithelial cell membranes to dissolve and release their fluids, causing edema. Hydrogen sulfide combines with water to form sodium sulfide, which destroys epithelial cell membranes and also inhibits cytochrome a_3 (Cya_3) oxidase, which is necessary for cellular function (Feld 2002).

Exposure to hydrogen chloride (HCl) starting at concentrations of 50 ppm usually causes respiratory or visual impairment sufficient to affect walking movement. Total cessation of movement occurs at a concentration approaching 300 ppm. The effects of concentrations of HCl above the 1000 ppm level are most likely severe enough to prevent escape (Purser 2001).

TABLE 7.1 ◆ Irritant Concentrations of Fire Gases Predicted to Cause 50 Percent Impaired Escape or Incapacitation in the Human Population

Common Fire Gases	Impaired Escape (ppm)	Incapacitation (ppm)
Hydrogen chloride (HCl)	200	900
Hydrogen bromine (HBr)	200	900
Hydrogen fluoride (HF)	200	900
Sulfur dioxide (SO_2)	24	120
Nitrogen dioxide (NO_2)	70	350
Formaldehyde (CH_2O)	6	30
Acrolein (C_3H_4O)	4	20

Source: Derived from Purser 2001.

Table 7.1 shows the effects of various fire gases on both the escape and the incapacitation of humans.

HCl is a major combustion/decomposition product of vinyl plastics, in both the flaming and the smoldering modes. Hydrogen bromide (HBr) or hydrogen fluoride (HF) is produced when some synthetic rubbers are burned. Acrolein is created when wood or cardboard is burned.

The term **fractional effective concentration** (FEC) was developed to assess the impact of smoke on visual obscuration on a subject (Purser 2001). The FEC is expressed in general terms as

$$\text{FEC} = \frac{\text{dose received at time } t(C_t)}{\text{effective } C_t \text{ dose to cause incapacitation or death}} \tag{7.1}$$

The FEC, in special situations, is also referred to as the **fractional incapacitation dose** (FID) or the **fractional lethal dose** (FLD).

CARBON MONOXIDE

Carbon monoxide is produced in fires by the incomplete combustion of any carbon-containing fuel. It is not produced at the same rate in all fires. In free-burning (well-ventilated) fires, it can be as little as 0.02 percent (200 ppm) of the total gaseous product. The CO concentrations in smoldering, postflashover, or underventilated fires range from 1 to 10 percent in the smoke stream. Note that there is no measurable diffusion from an external atmosphere rich in CO into the blood or tissues of a dead body.

When inhaled and absorbed into the bloodstream, CO forms the complex carboxyhemoglobin (COHb) with the heme portion of the hemoglobin molecule. CO has an affinity for hemoglobin that is 200–300 times stronger than that of O_2. It also binds with the heme group in myoglobin (which is the "red" in red muscle). Its affinity for myoglobin is about 60 times that of O_2. Myoglobin stores and transports O_2 in muscle tissue, particularly in cardiac muscle. Under hypoxic conditions, CO shifts from the blood into the muscle with a higher affinity for cardiac than for striated muscle (Myers and Cowley 1979). This may explain why low concentrations of COHb are sometimes found in deceased victims with heart conditions. Carbon monoxide also affects Cya_3 oxidase, an enzyme that catalyzes production of ATP in the cell.

The stability of the COHb complex reduces the O_2-carrying capacity of the blood. Without O_2 and water, adenosine triphosphate (ATP) cannot be produced in the mitochondria of the cell, and the cell dies (Feld 2002).

Goldbaum, Orellano, and Dergari (1976) reported that merely reducing the hematocrit (the blood-carrying capacity) of dogs by as much as 75 percent did not result in death. Even replacing blood with blood containing 60 percent COHb by transfusion or infusion of CO through the peritoneal cavity did not result in death. Only when the CO was inhaled did deaths occur. This suggests that respiration of CO plays a critical role in causing death (not just its presence) (Goldbaum, Orellano, and Dergari 1976).

The mere presence of CO in the blood is not a sign of breathing fire gases. The normal body has COHb saturations of 0.5–1 percent as a result of degradation of heme in the blood. Higher concentrations (up to 3 percent) may be found in nonfire victims with anemia or other blood disorders (Penney 2000). Smokers can have levels of 4–10 percent, since tobacco smoke contains a high concentration of CO. People in confined spaces with emergency generators, pumps, and compressors can have elevated, sometimes dangerous, COHb concentrations.

When a victim is removed from a CO-rich environment to fresh air, the CO is gradually eliminated from the blood. The higher the partial pressure of O_2, the faster the elimination. In fresh air the initial concentration of COHb will be reduced by 50 percent in 250–320 min (approx. 4–5 hr). In 100 percent O_2 via mask, a 50 percent reduction can be achieved in 65–85 min (approx. 1 to 1.5 hr). In O_2 at hyperbaric pressures (3–4 atm), there is a 50 percent reduction of COHb in 20 min (Penney 2000).

The time at which a blood sample is drawn from a subject must be noted, as well as the nature of any medical treatment (such as the antemortem administration of O_2). The COHb saturation of blood in a dead body is very stable, even after decomposition has begun. CO poisoning kills many fire victims before they are ever exposed to fire. It can kill victims even some distance from a fire when they are not exposed to any heat or flames, but it is not the only factor in many fire deaths.

Carbon dioxide is a product of nearly all fires. Levels of 4–5 percent CO_2 in air cause the adult respiratory rate to double. Levels of 10 percent cause it to quadruple. This increases the rate at which CO and other toxic gases are inhaled. High concentrations of CO_2 may also dilute the concentration of breathable oxygen to the point of inducing hypoxic collapse. Carbon dioxide concentration in the blood can be measured in living subjects, but blood chemistry begins to change after death. As a result, CO_2 (and O_2) saturation cannot be measured postmortem.

HYDROGEN CYANIDE

Hydrogen cyanide (HCN) is readily soluble in water (of blood plasma, cells, and organs, forming the CN radical. The CN radical also combines with Cya_3 oxidase, inhibiting its action in the cells. The inhibition of Cya_3 oxidase prevents the formation of water and ATP, the basic route of respiration in the cell (Feld 2002).

Table 7.2 lists the tenability limits for incapacitation or death from exposure to CO, HCN, low O_2, and CO_2. The periods 5 and 30 min are common benchmarks for narcotic products of combustion.

TABLE 7.2 ◆ Tenability Limits for Incapacitation or Death from Exposures to Common Toxic Products of Combustion

	5 min		30 min	
	Incapacitation	*Death*	*Incapacitation*	*Death*
CO (ppm)	6000–8000	12,000–16,000	1400–1700	2500–4000
HCN (ppm)	150–200	250–400	90–120	170–230
Low O_2 (%)	10–13	<5	<12	6–7
CO_2 (%)	7–8	>10	6–7	>9

Source: SFPE 1995a, Table 2-8B[a].

PREDICTING THE TIME TO INCAPACITATION BY CARBON MONOXIDE

The estimation of dosage levels for predicting times to incapacitation is an important concept. Under **Haber's rule** the dosage of toxic gases assimilated by an individual is assumed to be equivalent to the concentration. For example, a 1-hr exposure to a toxic gas at one concentration would be equivalent to a 2-hr exposure to half that concentration.

The Coburn–Forster–Kane (CFK) Equation.   In certain cases, Haber's rule does not hold exactly true for exposure to CO. The relationship between concentration and uptake is linear only for high CO concentrations and is not valid at extremely high concentrations. For lower concentrations, the time to incapacitation is an exponential relationship and is described by the CFK equation. The CFK equation also predicts that the half-time elimination of CO is a hyperbolic function of the ventilation rate (Peterson and Stewart 1975).

$$\frac{A[\text{HbCO}]_t - BV_{CO} - PI_{CO}}{A[\text{HbCO}]_0 - BV_{CO} - PI_{CO}} = e^{-tAV_bB}, \tag{7.2}$$

where

$[\text{HbCO}]_t$ = concentration of CO per blood at time t (mL/mL),

$[\text{HbCO}]_0$ = concentration of CO per blood at beginning of exposure (mL/mL),

PI_{CO} = partial pressure of CO in inhaled air (mm Hg),

V_{CO} = rate of CO production (mL/min),

A = derived constant,

B = derived constant, and

V_b = derived constant.

An obvious disadvantage of using the CFK equation is the number of variables needed. The CFK equation is appropriately used for exposure to CO concentrations less than 2000 ppm (0.2 percent), exposure durations greater than 1 hr, or estimation of time to death where COHb is 50 percent (Purser, in SFPE 2002b, 2-160).

The Stewart Equation.   When dealing with predictions of time to incapacitation where CO concentrations are higher than 2000 ppm (0.2 percent) and COHb is

TABLE 7.3 ◆ Standard Inhalation Values (RMV; Liters per Minute)

Activity	Man	Woman	Child	Infant	Newborn
Resting	7.5	6.0	4.8	1.5	0.5
Light activity	20.0	19.0	13.0	4.2	1.5

Source: Derived from Health Canada 1995.

less than 50 percent, a simpler equation known as the Stewart equation, applies, where

$$\% \, \mathrm{COHb} = (3.317 \times 10^{-5})(\mathrm{ppm \, CO})^{1.036}(\mathrm{RMV})(t), \tag{7.3}$$

where

CO = CO concentration (ppm),

RMV = respirations per minute of volume of air breathed (L/min), and

t = exposure time (min).

Solving the Stewart equation for exposure time gives

$$t = \frac{(3.015 \times 10^4)(\% \, \mathrm{COHb})}{(\mathrm{ppm \, CO})^{1.036}(\mathrm{RMV})}. \tag{7.4}$$

The standard inhalation values (RMV) are listed in Table 7.3, providing typical data for a man, woman, child, infant, and newborn (Health Canada 1995).

According to Stewart, a few breaths of CO in concentrations of 1 to 10 percent (10,000 to 100,000 ppm) extremely elevate the COHb level in the blood. For example, a 120-s exposure at 1 percent (10,000 ppm) CO results in a 30 percent COHb, and a 30-s exposure at 10 percent (100,000 ppm) CO results in a 75 percent COHb (Spitz and Spitz 2006).

The standard approach for evaluating incapacitation from CO is to calculate the fraction of the CO per minute over 1 hr. During moderate activity, human RMV value is approximately 25 L/min, and loss of consciousness occurs at 30 percent COHb (SFPE 2002b, 2-160). The formula for the fractional incapacitating dose (FID) valid for up to 1 hr is

$$F_{I_{\mathrm{CO}}} = \frac{K(\mathrm{ppm \, CO}^{1.036})(t)}{D}, \tag{7.5}$$

where

$$F_{I_{\mathrm{CO}}} = \text{fractional incapacitating dose (FID)} = \frac{\text{concentration of irritant to which subject is exposed at time } t}{\text{concentration of irritant required to cause impairment of escape efficiency}},$$

t = exposure time (min),

$K = 8.2925 \times 10^{-4}$ for 25 L/min RMV during moderate activity, D = 30 percent COHb, and

$K = 2.8195 \times 10^{-4}$ for 8.5 L/min RMV at rest, D = 40 percent COHb.

The calculation for the fractional incapacitating dose (FID) for moderate activity is found by substituting these variables into equation (7.5), where

$$F_{I_{CO}} = \frac{(8.2925 \times 10^{-4})(\text{ppm CO}^{1.036})(t)}{30} \tag{7.6}$$

EXAMPLE 7.1   CO Incapacitation

An adult female is found unconscious in her bed by her rescuers at the scene of a house fire. Rough estimates suggest that she was exposed to a CO concentration of approximately 5000 ppm. Calculate the time to incapacitation and fractional incapacitating dose, assuming that the victim was at rest. Use Table 7.3 for RMV data.

Volume of air breathed	RMV = 6.0 L/min (resting)
Loss of consciousness	COHb = 40 percent (resting)
CO concentration	CO = 5000 ppm
Time to incapacitation	$t = \dfrac{(3.015 \times 10^4)(40)}{(5000)^{1.036}(6.0)} = 30 \text{ min}$
Fractional incapacitating dose	$F_{I_{CO}} = \dfrac{(8.2925 \times 10^{-4})(5000^{1.036})}{30} = 0.188$

EXAMPLE 7.2   CO Incapacitation and Death of Firefighters

Computer fire modeling was used to reevaluate a U.S. Fire Administration investigation (Routley 1995) into a reported Pittsburgh fire that killed three firefighters (Christensen and Icove 2004). NIST's Fire Dynamics Simulator (FDS) was employed to model the fire to estimate the concentration of carbon monoxide present in the dwelling, which was the immediate cause of death of two of the firefighters, who were unable to escape the interior of a burning dwelling.

The fire occurred in a four-story townhouse when an arson fire was ignited using gasoline in a room on the ground floor. Firefighters entered on the street level and were attempting to locate the seat of the smoky fire. Details of the minutes prior to the deaths of the firefighters were unclear, but it appeared that at some point they realized they were running short of air in their self-contained breathing apparatus (SCBA), needed to leave, were unable to find an exit, and exhausted their air supplies. Two of the firefighters were believed to have removed or loosened their face pieces and made attempts to share the air that was available by "buddy breathing," or alternating the use of the breathing apparatus. It was concluded that both had been rendered unconscious owing to toxic gas inhalation. They were found to have COHb saturations of 44 percent and 49 percent, respectively, at autopsy. The third firefighter was found with his face piece in place, and his COHb was 10 percent, indicating death from oxygen deficiency.

This estimate, along with an assumed respiration volume and known blood COHb levels, was used with the Stewart equation to estimate the time of exposure. The FDS model, as shown in Figure 7.2, indicated that 27 min into the fire, the CO concentration had already reached approximately 3600 ppm at the location where the firefighters were found. At this concentration, with a respiration rate of 70 L/min, an estimated 3 to 8 min of exposure would have been required to accumulate the concentrations of COHb measured in the firefighters at autopsy. Note that in calculating time to incapacitation, the calculation uses the average COHb value measured in two of the firefighters at autopsy, or 47%.

Solving this problem using the Stewart equation gives

$$\%\text{COHb} = (3.317 \times 10^{-5})(\text{ppm CO})^{1.036}(\text{RMV})(t), \tag{7.7}$$

EAST SIDE OF DWELLING - 8361 Bricelyn Street

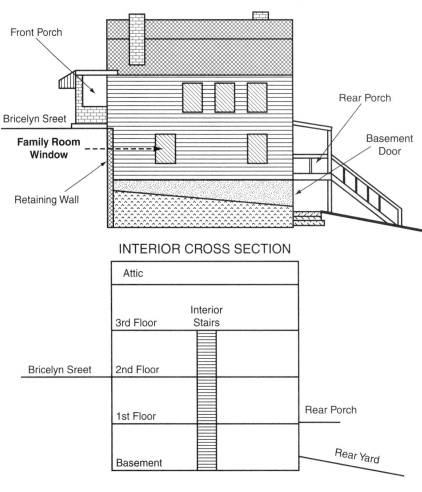

INTERIOR CROSS SECTION

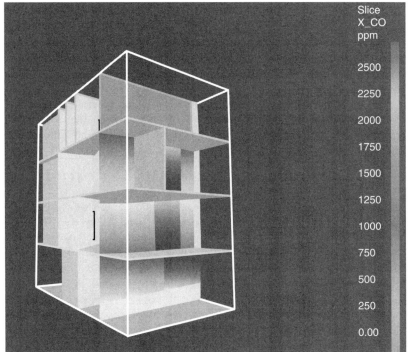

FIGURE 7.2 ◆ (a) Plan of town house where three firefighters were trapped. (b) FDS model of the distribution of carbon monoxide in town house at 27 min. *Courtesy of D. J. Icove.*

where

$$CO = CO \text{ concentration} = 3600 \text{ ppm}$$
$$\%COHb = \text{carboxyhemoglobin} = 47\%$$
$$RMV = \text{respirations per minute of volume of air breathed} = 70 \text{ L/min},$$
$$t = \text{exposure time (min).}$$

Solving for time of exposure, we obtain

$$47\% = (3.317 \times 10^{-5})(3600)^{1.036} (70)(t)$$
$$t = 4.2 \text{ min}$$

Estimated range for $t = 3$ to 8 min

A total exposure time of 4.2 min suggests that following removal of the face pieces, only a few minutes of exposure with no air from the SCBA would have been required to produce the lethal concentrations of COHb observed at autopsy.

INCAPACITATION BY LOW OXYGEN LEVELS

Anoxia (absence of oxygen) or **hypoxia** (low concentration of oxygen) is the condition of inadequate oxygen to support life. This can occur when oxygen is displaced by another inert gas, such as nitrogen or carbon dioxide, by a fuel gas such as methane, or even by benign products of combustion such as CO_2 and water vapor. Normal air contains 20.9 percent O_2. At concentrations down to 15 percent O_2, there are no readily observable effects. At concentrations between 10 and 15 percent, disorientation (similar to intoxication) occurs and judgment is affected. At levels below 10 percent, unconsciousness and death may occur. Anoxia is aggravated by high levels of CO_2, which accelerate breathing rates. The effects of low-oxygen hypoxia include problems with memory and mental concentration, loss of consciousness, and death (see Table 7.4).

TABLE 7.4 ◆ The Impact of Exposure to Low Oxygen Levels

Percent Oxygen	*Reported Effects on Healthy Adults*
14.14–20.9	No significant effects, slight loss of exercise tolerance
11.18–14.14	Slight effects on memory and mental task performance, reduced exercise tolerance
9.6–11.8	Severe incapacitation, lethargy, euphoria, loss of consciousness
7.8–9.6	Loss of consciousness, death

Source: Derived from SFPE 2002b, 2-161.

PREDICTING THE TIME TO INCAPACITATION BY HYDROGEN CYANIDE

Hydrogen cyanide (HCN) is another toxic gas in fires that incapacitates through biochemical asphyxia. As with CO, the time to incapacitation depends on the uptake rate and dosage (Purser in SFPE 2002b).

The formula for time to incapacitation for 80–180 ppm HCN concentrations is

$$t_{I_{CN}}(\text{min}) = \frac{185 - \text{ppm HCN}}{4.4}, \tag{7.8}$$

and the formula for time to incapacitation for HCN concentrations above 180 ppm is

$$t_{I_{CN}}(\text{min}) = \exp[5.396 - (0.023)(\text{ppm HCN})], \qquad (7.9)$$

and the fractional incapacitating dose per minute (FID/min) is

$$F'_{I_{CN}} = \frac{1}{\exp[5.396 - (0.023)(\text{ppm HCN})]}. \qquad (7.10)$$

Note that the term "exp" in equations (7.9) and (7.10) denotes the exponential form. As shown in Table 7.3, the HCN dosages required for incapacitation are much lower than the CO dosages.

EXAMPLE 7.3 HCN Incapacitation

An adult male is found unconscious in the waiting room after being exposed to toxic by-products from the fire in Example 7.1. A burning polyurethane plastic mattress cushion had produced an HCN concentration of approximately 200 ppm. Calculate the time to incapacitation and fractional incapacitating dose per minute (FID/min).

HCN concentration	HCN = 200 ppm
Time to incapacitation	$t = \exp[5.396 - (0.023)(200)] = 2.2$ min
Fractional incapacitating dose per minute	$F_{I_{CN}} = \left(\dfrac{1}{\exp[5.396 - (0.023)(200)]}\right) = 0.45$ FID/min

---

PREDICTING THE TIME TO INCAPACITATION BY CARBON DIOXIDE

Exposure to carbon dioxide can produce a wide range of effects, ranging from respiratory distress to loss of consciousness (see Table 7.5).

Along with being an asphyxiant displacing oxygen, carbon dioxide also increases the RMV, which in turn causes the individual to increase the uptake of the other toxic gases (Purser, in SFPE 2002b). This formula for the multiplication factor VCO_2 is

$$VCO_2 = \exp\left(\frac{CO_2}{5}\right), \qquad (7.11)$$

$$VCO_2 = \left(\frac{\exp[(1.903)(\%CO_2) + 2.0004]}{7.1}\right). \qquad (7.12)$$

The formula for the time to unconsciousness from carbon dioxide is

$$t_{I_{CO_2}} = \exp[6.1623 - (0.5189)(\%CO_2)] \qquad (7.13)$$

TABLE 7.5 ◆ The Impact of Exposure to Carbon Dioxide

Percent Carbon Dioxide	Reported Effect(s)
7–10	Loss of consciousness
6–7	Severe respiratory distress, dizziness, possible loss of consciousness
3–6	Respiratory distress increasing with concentration

Source: Derived from SFPE 2002b, 2-161.

and the fractional incapacitating dose per minute (FID/min) is

$$F'_{I_{CO_2}} = \left(\frac{1}{\exp[6.1623 - (0.5189)(\%CO_2)]} \right) \tag{7.14}$$

The human body is capable of surviving exposure to external heat as long as it can moderate its temperature by radiant cooling of the blood through the skin and, more important, by evaporative cooling. This process occurs internally via evaporation of water from the mucosal linings of the mouth, nose, throat, and lungs and externally via evaporation of sweat from the skin. If the core body temperature exceeds 43°C (109°F), death is likely to occur.

Prolonged exposure to high external temperatures, 80–120°C (175–250°F), with low humidity can trigger fatal hyperthermia. Exposure to lower temperatures accompanied by high humidity (which reduces the cooling evaporation rate of the water from the skin or mucosa) can also be lethal. Fire victims can die of exposure to heat alone even if they are protected from CO, smoke, and flames. Victims may have minimal postmortem changes, although skin blistering and sloughing can occur after death.

PREDICTING THE TIME TO INCAPACITATION BY HEAT

For exposure to convected heat in a fire environment, the fractional incapacitating dose per minute (FID/min) is

$$F'_{I_h} = \left(\frac{1}{\exp[5.1849 - (0.0273)(T[°C])]} \right) \tag{7.15}$$

Assimilation of various data formed the basis for the *Toxic and Physical Hazard Assessment Model* (Purser, in SFPE 2002b, 2-159). Data such as species concentration levels generated by fire models can serve as input to this hazard assessment model. According to this model, the normally accepted threshold for tolerance of radiant heat is 2.5 kW/m² for only a few minutes. Besides the burns to the skin (from both radiant and convected heat), thermal damage to the upper respiratory tract can also occur when dry gases over 120°C (250°F) are inhaled.

Information needed to evaluate this hazard model come from two sets of information: concentration and time profiles of major toxic products, and time, concentration, and toxicity relationships. The estimated toxic products within the victim's breathing zone include concentrations of carbon monoxide, hydrogen cyanide, carbon dioxide, oxygen, radiant heat flux, air temperature, and optical smoke density. Some of these values can be calculated by sophisticated computer models discussed previously.

INHALATION OF HOT GASES

Inhalation of very hot gases causes edema (swelling and inflammation) of mucosal tissues. This edema can be severe enough to cause blockage of the trachea and physical asphyxia. Inhalation of hot gases may also trigger **laryngospasm**, in which the larynx involuntarily closes up to prevent entry of foreign material or **vagal inhibition**, in which the breathing stops and the heart rate drops.

Rapid cooling of the inhaled hot gases occurs as the water evaporates from mucosal tissues, so thermal damage usually does not extend below the larynx if the inhaled gases are dry. If the hot gases include steam or are otherwise water saturated, evaporative cooling is minimized and burns/edema can extend to the major bronchi. If inhaled gases are hot enough to damage the trachea, they will usually be hot enough to burn the facial skin and mouth and singe the facial or nasal hair.

EFFECTS OF HEAT AND FLAME

The human body is a complex target when being affected by heat. Skin consists of two basic layers. The thin layer of **epidermis** (dead, keratinized skin cells) overlays a thicker dermal layer of actively growing cells in which are embedded the nerve endings, hair follicles, and blood capillaries that supply nutrients to the growing skin. Beneath the dermal layer is a layer of tough elastic connective tissue, subcutaneous fat, and, finally, muscle and bone.

Each of these components is affected differently by heat and flames. Application of heat can cause the epidermis to separate from the underlying **dermis** and form blisters in much the way paint or wallpaper blisters away from the wood or plaster beneath when heated. Blistering occurs when the tissue reaches temperatures in excess of 54°C (130°F). The raised epidermal layer is very thin and is more easily affected by continuing heat, sometimes causing it to char. The epidermis can also separate in larger areas and form generalized skin slippage. Exposure of the denuded dermal layer to heat can cause pain when its temperatures exceeds 43–44°C (110–112°F) (Purser, in SFPE 2002b, 2-159).

More prolonged exposure can destroy the proteins of the dermal layer and cause desiccation and discoloration. Higher heat fluxes can cause higher temperatures that cook and even char the tissues. As it desiccates, the skin shrinks, eliminating wrinkles and changing facial contours (making visual identifications of victims very risky). If the skin continues to shrink, it can split, leaving jagged, irregular, torn surfaces (as opposed to the sharply defined surfaces of knife cuts), as shown in Figure 7.3 (Smith and Pope 2003). Heat-split skin often demonstrates subcutaneous bridging of underlying tissue, whereas cut skin does not.

If a victim survives for some time, this shrinkage can constrict blood vessels, so incisions are made (called *escharotomies*) through the damaged dermal layer to relieve the pressure and maintain circulation. The investigator must recognize the effects of escharotomy or skin graft harvesting to distinguish them from fire effects if a fire victim has survived for some time after the fire.

Owing to its small individual dimensions and low thermal mass, hair is affected very quickly by heat. Colors will change (typically changing to darker or redder colors or completely to gray). The hair shaft will bubble, shrink, and fracture as it singes. This shrinkage causes the curling seen as singeing. Its microscopic appearance is very distinctive (as opposed to cut or broken hair shafts). If hair is burned in large masses, it can form a black puffy mass.

If the body continues to be exposed to heat, the shrinkage can affect muscles. When the skin and muscles of the neck shrink, they can force the tongue out of the mouth. Shrinkage of muscle and tendons can cause the joints to flex, causing what is called **pugilistic posturing**, as shown in Figure 7.4. This flexing can cause bodies to move during fire exposure. If the body is on an irregular or unstable surface, this movement can cause the body to fall from a bed or chair and, possibly, change the direction of heat application, eliminating or obscuring previously protected areas.

FIGURE 7.3 ◆ Effects of heat and flame shrink skin, eliminating wrinkles and changing facial contours. If the skin continues to shrink, it can split, leaving jagged, irregular torn surfaces, as opposed to the sharply defined surfaces of knife cuts. This photo illustrates postfire recognition of knife cuts to the chest versus splitting of the skin on the arm. *Courtesy of University of Tennessee Regional Forensic Center.*

FIGURE 7.4 ◆ Shrinkage of muscle and tendons can cause the joints to flex, causing what is called *pugilistic posturing*. *Courtesy of University of Tennessee Regional Forensic Center.*

During legal cremations, pugilistic posturing has been observed after 10 min of exposure to flames at temperatures of 670–810°C (1240–1490°F) (Bohnert, Rost, and Pollak 1998).

Direct flame impingement, with its high-temperature gases of 500–900°C (930–1650°F) and high heat fluxes (55 kW/m²), produces effects very quickly. Blisters will form in about 5 s, with charring induced some seconds later. Skin will be charred away in 5–10 min of direct flame contact, especially where stretched over the bone (joints, nose, forehead, skull) (Pope and Smith 2003). Very short but intense (flash) fire exposure can cause blistering of the epidermal layer without the sensation of pain (because the pain sensors are in the dermal layer beneath) and the heat takes longer to penetrate deeply.

Even in the absence of fire, prolonged exposure to high temperatures (over 50°C) causes desiccation and shrinkage of muscle tissue that causes flexion of joints (pugilistic posturing). Exposure to flames can cause combustion of muscle and fracturing of major limb bones as they degrade where the bone is exposed to flames. Extreme heat causes bones to twist and fracture, with continued flame exposure (30 min or longer in observed cremations) causing calcination (in which the charred organic matter is burned away). As the bones calcine, they become very fragile and may disintegrate of their own accord. The thin bones of the skull may delaminate, with the inner and outer layers failing separately. This delamination has given rise to thermally caused holes in the skull that have beveled edges similar to those produced by gunshots (Pope and Smith 2003). The lower-density major bones from elderly victims of osteoporosis have been seen to disintegrate under fire exposure more quickly and completely than bones of normal density (Christensen 2002). Tests have revealed that fire exposure can cause fluid to leak through fissures in the exposed skull but not an actual explosion of the skull. Internal organs shrink as they desiccate and char, this requiring 30–40 min of cremation (Bohnert, Rost, and Pollak 1998). Bones can also shrink in length (giving rise to potential errors in antemortem height estimation).

Recent tests showed that failure of the skull can occur from fire damage whether or not there was preexisting mechanical trauma. Some thermally induced failures can mimic bullet wounds, requiring extreme care in interpretation (Pope and Smith 2004).

Heat exposure to the head can cause blood and fluids to accumulate, forming a hematoma in the **extradural** or **epidural** space between the skull and the tough bag of tissue (the *dura mater*) that surrounds the brain. With further heating, these fluids desiccate and char, producing a rigid, foamy, blackened mass. Physical trauma can cause subdural hematomas, in which the blood collects between the dura and the brain. Fractures of the base of the skull have not been observed to be produced by fire exposure (Bohnert, Rost, and Pollak 1998). Fire exposure chars and seals exposed tissue, so as a rule, bodies exposed to an enveloping fire do not bleed. When the body is moved, however, the fragile layer of char can be broken, allowing body fluids to seep out.

Within certain considerations, the amount of fire damage on a body can be related to the duration of exposure if the intensity of the fire can be estimated from the fuel, ventilation, and distribution factors described elsewhere in this text. Factors such as combustion rates, thermal inertia, and relative combustibility of body components have been explored in recent studies (DeHaan, Campbell, and Nurbakhsh 1999). Some aspects of postmortem fire damage are discussed later in this chapter.

FLAMES (INCINERATION)

When heat is applied to a surface, the rate at which it penetrates that surface is determined by the thermal inertia of the material (the numerical product of thermal capacity, density, and thermal conductivity). The thermal inertia of skin is not much different from that of a block of wood or polyethylene plastic (see Table 2.4). The pain sensors for human skin are in the dermis, about 2 mm (0.1 in.) below the surface. If heat is applied very briefly, there may not be any sensation of discomfort or pain. The longer the heat is applied, the deeper it will penetrate. The higher the intensity of the heat applied, the faster it will penetrate. Pain is triggered when skin cells reach a temperature of about 48°C (120°F), and cells are damaged if their temperature exceeds 54°C (130°F) (Stoll and Greene 1959).

Exposing skin to 2–4 kW/m² radiant heat for 30 s causes pain. Cellular damage triggers blisters and skin slippage. Exposing skin to 4–6 kW/m² radiant heat for 8 s produces blisters (second-degree burns). Exposing skin to 10 kW/m² radiant heat for 5 s causes deeper, partial-thickness injuries. Exposing skin to 50–60 kW/m² radiant heat for 5 s produces third-degree burns including destruction of the dermis (Stoll and Greene 1959).

BURNS

Even in the absence of fire or flames, prolonged exposure of body parts to heat that raises their temperature over 54°C (130°F) will cause desiccation, sloughing, and blistering, which can also be caused by exposure to caustic chemicals. It is often very difficult, if not impossible, to distinguish between burns suffered near the time of death (perimortem) and those inflicted after death (postmortem).

Most, but not all, medical personnel agree that **first-degree burns** involve only reddening of the skin, whereas **second-degree burns** involve damage to the epidermis with blistering and sloughing. These are sometimes referred to as **partial-thickness burns.** Since the germinative layer of the dermis remains, such burns will usually heal from their entire surface, usually without grafts. **Third-degree burns** are called **full-thickness burns** because the dermis is damaged and the wound will heal from the edges only and will require grafting. Full-thickness burns can also include those in which the skin is destroyed, exposing muscle and even bones beneath (sometimes called **fourth-degree burns**).

When gasoline or a similar volatile liquid fuel with low viscosity and low surface tension is poured on bare skin, some will be absorbed into the epidermis, but the bulk will run off, leaving a very thin film of liquid, particularly on vertical surfaces. This thin film will burn off very quickly (less than 10 s). The skin beneath may be spared completely or reddened (first-degree burns), except where folds of the skin or clothing have retained enough fuel to sustain longer burning, which will induce severe blistering and even charring of the epidermis in extreme cases. Deep penetrations exposing the muscle or subcutaneous fat require flame exposures of several minutes, much longer than the typical gasoline "pool" fire. On horizontal skin surfaces, a deeper "pool" may be retained long enough to produce a "halo" or ring of blisters (second-degree burns) around the circumference of the pool (DeHaan 2007).

BLUNT FORCE TRAUMA

Blunt force trauma can also cause or contribute to the death of fire victims. Structural collapse or explosions can cause solid materials to strike victims. Falls or impacts with stationary surfaces (furniture or door frames) during escape attempts can induce blunt trauma that only careful examination can distinguish from an assault. Wound patterns, bloodstains, or even trace evidence can be used to interpret blunt trauma injuries and establish whether they resulted from assault or from some fire-related event. The fire investigator should consider consulting with the pathologist and the criminalist to help evaluate blunt trauma injuries, possibly linking them to features at the scene.

◆ 7.5 VISIBILITY

The optical opacity of dense smoke and its irritants impairs the vision and respiration of normal-sighted people who find their ability to travel impaired by the smoke. This optically dense smoke affects

- Exit choice and escape decisions;
- Speed of movement; and
- Wayfinding ability.

During structure fires, the occupants often depend on their ability to seek out exit signs, doors, and windows (Jin 1975). **Visibility** of an object depends on several factors such as the smoke's ability to scatter or absorb the ambient light, the wavelength of the light, whether the item viewed (e.g., an exit sign) is light emitting or light reflecting, and the individual's visual acuity (Mulholland, in SFPE 2002b, 2-265).

OPTICAL DENSITY

Calculating an estimate for visibility is based on the term D, the **optical density** per meter (OD/m). Optical density uses a collateral term known as the extinction coefficient, K, which is the product of an **extinction coefficient** per unit mass, K_m, and the mass concentration of the smoke aerosol, m, where

$$K = K_m m, \tag{7.16}$$

$$D = \frac{K}{2.3} \tag{7.17}$$

with

K = extinction coefficient (m^{-1}),

K_m = specific extinction coefficient (m^2/g),

m = mass concentration of smoke (g/m^3), and

D = optical density per meter (m^{-1}).

The values for K_m are typically 7.6 m^2/g for smoke produced during flaming combustion of wood or plastics and 4.4 m^2/g for smoke produced during pyrolysis (SFPE 2002b).

In terms of the extinction coefficient, K, one problem at a fire is determining the visibility, S, of light-emitting and light-reflecting exit signs to occupants. S is a measure of how well an individual can see through the smoke. Light-emitting signs are two to four times better than light-reflecting signs (Mulholland, in SFPE 2002b, 2-265), as the values for K reveal in equations (7.18) and (7.19),

TABLE 7.6 ◆ Mass Optical Density (D_m) for Flaming Mattresses

Type of Material	Mass Optical Density (m^2/g)
Polyurethane	0.22
Cotton	0.12
Latex	0.44
Neoprene	0.20

Source: Derived from Babrauskas 1981.

$$KS = 8 \text{ for light} - \text{emitting signs,} \tag{7.18}$$

$$KS = 3 \text{ for light} - \text{reflecting signs,} \tag{7.19}$$

where

K = extinction coefficient (m^{-1}) and
S = visibility (m).

Methods for estimating visibility based on the mass optical density are considered realistic. In studies, the optical density at which people turned back from a smoke-filled area was a visibility distance of 3 m (9.84 ft) (Bryan 1983). The study also showed a tendency for women to be more likely to turn back than men. Other factors include the ability of the persons to see exit signs directing them to safe egress from the building. Height of exit signs may be critical to their visibility.

Estimates of visibility are based on mass optical densities, D_m, derived from test data (Babrauskas 1981). Typical values of mass optical density produced by mattresses in flaming combustion are listed in Table 7.6.

The following equation is used to estimate the density of the visible smoke, D:

$$D = \frac{D_m \Delta M}{V_c}, \tag{7.20}$$

where

D = optical density per meter (m^{-1}),
D_m = mass optical density (m^2/g),
ΔM = total mass loss of sample (g), and
V_c = total volume of compartment or chamber (m^3).

EXAMPLE 7.4   Visibility

A small, 300-g (0.66-lb), polyurethane mattress cushion on a waiting bench is set afire by a juvenile arsonist and is undergoing flaming combustion. The bench is located in a 6-m-square (20-ft-square) waiting room with a ceiling height of 2.5 m (8.2 ft). Determine the closest visibility of both light-emitting and light-reflecting signs leading to the exit door. Assume that the smoke filling the waiting room is uniformly mixed.

Total mass loss of mattress	$\Delta M = 300$ g
Mass optical density	$D_m = 0.22$ m^2/g (from Table 7.1)
Volume of compartment	$V_c = (6\text{ m})(6\text{ m})(2.5\text{ m}) = 90.0\text{ m}^3$
Optical density	$D = (0.22\text{ m}^2/\text{g})(300\text{ g})/(90.0\text{ m}^3) = 0.733\text{ m}^{-1}$

Extinction coefficient	$K = 2.3\,D = (2.3)(0.733\ \mathrm{m^{-1}}) = 1.687\ \mathrm{m^{-1}}$
Visibility (light emitting)	$S = 8/K = (8)/(1.687\ \mathrm{m^{-1}}) = 4.74\ \mathrm{m}$
Visibility (light reflecting)	$S = 3/K = (3)/(1.687\ \mathrm{m^{-1}}) = 1.77\ \mathrm{m}$

The calculations indicate that a light-emitting sign can be seen in this fire at a distance of up to 4.74 m (15.5 ft), compared with 1.7 m (5.58 ft) for an unlit sign.

FRACTIONAL EQUIVALENT CONCENTRATIONS OF SMOKE

The following formulas relate to enclosed spaces. As the value of FEC_{smoke} approaches 1, the level of visual obscuration increases and the chance of escape decreases significantly.

$$FEC_{smoke} = \frac{D}{0.2} \text{ for small enclosures.} \tag{7.21}$$

$$FEC_{smoke} = \frac{D}{0.1} \text{ for large enclosures.} \tag{7.22}$$

Where D = optical density per meter of the smoke being encountered.

WALKING SPEED

As previously summarized, the optical density of the smoke affects a person's decision to choose the closest exit and ability to make proper escape decisions, as well as wayfinding ability and speed of movement. Experiments with human subjects navigating through nonirritant smoke–filled corridors indicated that speed of movement decreases with increased smoke density (Jin 1975).

Based on Jin's research, equation (7.23) provides an expression of this relationship:

$$FWS = (-1.738)(D) + 1.236 \tag{7.23}$$
$$\text{for the range } 0.13\ \mathrm{m^{-1}} \le D \le 0.30\ \mathrm{m^{-1}} \tag{7.24}$$

where
  FWS = fractional walking speed (m/s) and
  D = optical density per meter $(\mathrm{m^{-1}})$.

For this equation, the smoke optical density ranges between 0.13/m (below normal walking speed) and 0.56/m (above walking speed in darkness at 0.3 m/s). The limits on the equation do not allow for delays such as erratic walking and sensory irritation. Jin used wood smoke in his experiments; therefore, his equation should be considered valid only for nonirritant smoke.

WAYFINDING

Jin's expression does not correlate midcourse corrections in **wayfinding**, reduced visibility, and irritability of the smoke. Studies have shown that the average density at which persons turned back was a visibility distance of 3 m (9.84 ft) ($D = 0.33\ \mathrm{m^{-1}}$ and $K = 0.76$). Poor visibility and irritation of the eyes are the leading factors in reduced wayfinding, followed by irritation of the respiratory system (Jensen 1998).

SMOKE

Smoke contains water vapor, CO, CO_2, inorganic ash, toxic gases, and chemicals in aerosol form as well as soot. Soot is agglomerations of carbon from incomplete combustion large enough to produce visible particles. These particles may be very hot and are not cooled readily as they are inhaled, so they may induce edema and burns where they lodge in the mucosal tissue of the respiratory system. Soot particles are active adsorbents, so they may carry toxic chemicals and permit their ingestion or inhalation (with direct absorption by the mucosal tissues). Soot can be inhaled in quantities sufficient to block airways physically and cause mechanical asphyxiation. Soot, water vapor, ash, and aerosols in smoke can also obscure the vision of victims and prevent their escape.

◆ **7.6 TIME INTERVALS**

INTERVAL BETWEEN FIRE AND DEATH

One of the problems outlined earlier is the time interval between exposure to a fire and its fatal aftermath. Death can occur nearly instantaneously or minutes or hours later. Under these conditions it is not difficult to connect the death to its actual cause. When a person dies weeks or even months after a fire, the cause can still be the fire, but the linkage can be obscured by the extensive medical events in between.

There is a range of effects that can determine the time interval between the fire and death. With **instantaneous death,** vagal inhibition or laryngospasm occurs on inhalation of flames and hot gases, causing cessation of breathing followed very quickly by death. Explosion trauma and incineration from exposure to a fully developed fire (often as a result of structural collapse) will result in nearly instantaneous death.

Death can occur within seconds to minutes from **hyperthermia,** exposure to very hot but nonlethal gases or steam, or anoxia, the lack of oxygen. Toxic gases such as hydrogen cyanide or other pyrolysis products; asphyxia via the inhalation of carbon monoxide; or blockage of airways by soot, exposure to flames; physical trauma (with loss of blood); internal injuries; and brain injuries can cause death within minutes.

Death can occur in hours from carbon monoxide, edema from inhalation of hot gases, burns (shock), and brain or other internal injuries. Dehydration and shock from burns cause death days after the fire. Even weeks or months after the fire, deaths can occur owing to infections or organ failure triggered by fire injuries.

Note that the **cause of death** is defined as the injury or disease that triggers the sequence of events leading to death. In a fire, the cause may be inhalation of hot gases, CO, or other toxic gases, heat, burns, anoxia, asphyxia, structural collapse, or blunt trauma. The **mechanism of death** is the biological or biochemical derangement incompatible with life. Mechanisms can be respiratory failure, exsanguination, infection, organ failure, and cardiac arrest. The **manner of death** is a medicolegal assessment of the circumstances in which the cause was brought about. In the United States, these are most often homicide, suicide, accident, natural, or undetermined.

As one can appreciate, the longer the interval between the cause (the fire) and the onset of the mechanism of death (organ failure, septicemia, etc.), the more likely it is that the connection will be lost. This is especially true when the victim has been moved from trauma care hospitals to long-term care facilities, sometimes in other geographic areas. The investigator must be diligent to ensure that the cause of death is not listed on the final death certificate as some generic term such as respiratory failure, cardiac arrest, or septicemia.

SCENE INVESTIGATION

The reconstruction of the activities of a fire victim may depend on finding and documenting bloodstains (such as from impact with a wall or door jamb or handprints on a wall), mechanical (blunt trauma) injuries, or artifacts found with the body. The nature of dress (street clothes, robe, nightgown) or objects (dog leash, jewelry, flashlight, fire extinguisher, house keys, phone, keepsakes, etc.) can provide clues as to what the person was doing prior to collapse.

The position (face up or face down) is not usually significant, as people known to have died during a fire have been found in all positions. Due to pugilistic posturing, the attitude of the body bears little reliable relation to antemortem actions. As mentioned earlier, pugilistic posturing occurs as a physical reaction to heat and can occur whether the person was dead before the fire or died in it. It can cause bodies to shift position, sometimes to the point where they roll off unstable surfaces like chairs or mattresses. Young children tend to hide under beds or in closets, but finding them in other locations is not proof that they did not have the time or physical capacity to seek safety.

POSTMORTEM DESTRUCTION

A body exposed to fire can support combustion, the rate and thoroughness of which depend on the nature and condition of exposure of the body to the flames. The skin, muscles, and connective tissues will shrink as they dehydrate and char. If exposed to enough flame, they will burn and yield some heat of combustion, although quite reluctantly. The relatively waterlogged tissues of the internal organs must be dried by heat exposure before they can combust, and that dehydration step increases their fire resistance and delays their consumption.

Bones have moisture and a high fat content, especially in the marrow, so they will shrink, crack, and split and contribute fuel to an external fire. The subcutaneous fat of the human body provides the best fuel, having an effective heat of combustion of the order of 32–36 kJ/g (DeHaan, Campbell, and Nurbakhsh 1999). Like candle wax, however, it will not self-ignite or smolder and will not normally support flaming combustion unless the rendered fat is absorbed into a suitable wick. This "wick" can be provided by charred clothing, bedding, carpet, upholstery, or wood in the vicinity (as long as it forms a porous, rigid mass). The size of the fire that can be supported by such a process is controlled by the size (surface area) of the wick. Depending on the position of the body and its available wick area, fires supported by the combustion of a body will be of the order of 20–120 kW, similar to a small wastebasket fire. Such a fire will affect only items close to it, and fire damage will often be very confined.

The flames produced by combustion of body fat will be 800–900°C (1475–1650°F), and if they impinge on the body surface, they can aid the destruction of the body. The process, if unaided by an external fire, is quite slow, with a fuel consumption rate of the order of 3.6–10.8 kg/hr (7–25 lb/hr). It is possible, given a long enough time (5–10 hr), that a great deal of the body can be reduced to bone fragments (DeHaan 2001; DeHaan, Campbell, and Nurbakhsh 1999; DeHaan and Pope 2007).

If a body is exposed to a fully developed room or vehicle fire, however, the rate of destruction will be much closer to that observed in commercial crematoria. In those cases, flames of 700–900°C (1300–1650°F) and a heat intensity of 100 kW/m^2 envelop the body and can reduce it to ash and fragments of the larger bones in 1.5–3 hr (Bohnert, Rost, and Pollak 1998; DeHaan and Fisher 2003).

SUMMARY OF POSTMORTEM TESTS DESIRABLE
IN FIRE DEATH CASES

Although not all the following may be needed in all cases, it can be appreciated that once the body is released for burial or cremation, it will be too late for forensic examination (DeHaan 2007, chap. 15). The complexity of many fire death cases may necessitate finding the answers to questions that were not apparent earlier. Having comprehensive samples and data is the best route to a successful and accurate investigation.

- **Blood** (taken from a major blood vessel or chamber of the heart, not from the body cavity); tested for COHb saturation, HCN, drugs (therapeutic and abuse), alcohol, and volatile hydrocarbons.
- **Tissue (brain, kidney, liver, lung).** Tested for drugs, poisons, volatile hydrocarbons, combustion by-products, CO (as a backup for insufficient blood).
- **Tissue (skin near burns).** Tested for vital chemical or cellular response to burns.
- **Stomach contents.** Tested for establishment of activities before death and possible time of death.
- **Airways.** Full longitudinal transection of airways from mouth to lungs to examine and document extent and distribution of edema, scorching/dehydration, and soot.
- **Internal body temperature.** Should be measured at the scene. It may be elevated owing to fire exposure or antemortem hyperthermia, but unexpectedly *low* body temperatures after a recent fire should indicate that the victim was dead well before the fire.
- **X Rays.** Full-body (including associated debris in body bag) and details of teeth and any unusual features discovered (fractures, implants).
- **Clothing.** All clothing remnants and associated artifacts removed and preserved.
- **Photographs.** General (overall), with close-ups of any burns or wounds, in color, with scale.
- **Postmortem weight.** Postmortem weight of the body determined, *not* including the fire debris or body bag.

It is very useful for the fire investigator to be present when the postmortem is conducted, not only to ensure that all appropriate observations are made but also to be on hand to answer any questions that arise during the examination. Few pathologists have extensive knowledge of fire chemistry or fire dynamics, so the investigator is in a good position to advise the pathologist as to the fire conditions in the vicinity of the body.

◆ 7.7 CASE EXAMPLE 1: DEATH FOLLOWED BY FIRE—CAUSE AND DURATION

At about 9 A.M., Mr. John Doe (not his real name) discovered the severely burned bodies of his sister-in-law and his 3-year-old nephew in the dining room of their home. He had been called minutes earlier by his brother, Mr. Jim Doe, a salesman, who reported that he had left home that morning and had not been able to reach his wife (who was 8 months pregnant) by telephone. Concerned, he called his brother and asked that he stop by and check on her.

On arrival, John Doe found the house full of smoke but only small flames showing in the vicinity of the bodies that were sprawled amidst an area of burned carpet in the dining room. He reportedly batted out the small flames with a hand towel and then withdrew to call emergency 9-1-1. The local fire department responded by

FIGURE 7.5 ◆ The fire damage was limited to an oval area of carpet about 2 m² (22 ft²) in size, centered on the two bodies. The mother was sprawled roughly face-down and the child was spread-eagled on his back alongside her. *Courtesy of Elk Grove Fire Department, Elk Grove, CA.*

9:10 A.M. and, forewarned of the conditions within, limited its response to one fire-fighter with a pressurized water tank who entered the building, sprayed a very small amount of water on or near the body to extinguish the small flames still present, and then withdrew along the same path. Except for opening windows to vent the accumulated smoke, that was the extent of fire suppression.

Jim Doe returned to the house while the investigation was beginning and claimed that he had left the house at about 6:00 that morning and that everything had been in order. He had stopped to put gasoline in his car and then at a restaurant while making his sales calls for the morning, establishing his presence elsewhere after 6:00 A.M.

The fire damage was limited to an oval area of carpet about 2 m² in size, centered on the two bodies. The mother was sprawled roughly facedown, and the child was spread-eagled on his back alongside her, as shown in Figure 7.5. There were areas of scorching on the walls and baseboards of the dining room near the entry to the kitchen and two areas of damage to the vinyl floor of the kitchen immediately adjacent to the mother's lower right leg. There were areas of scorching on her legs adjacent to these floor patterns. The fire investigators quickly determined through laboratory analysis that automotive gasoline had been poured on the carpet around the bodies and in nearby areas of the living room dining room, kitchen, and hallway (Figure 7.6).

Both victims had zero COHb levels and no soot detected in the airways. Both bodies bore signs of physical (blunt force) trauma, and the child's cause of death was ascribed to severe multiple head wounds. Bloodstain patterns indicated that the

FIGURE 7.6 ◆ Fire investigators quickly determined through laboratory analysis that automotive gasoline had been poured on the carpet around the bodies and in nearby areas of the living room, kitchen, and hallway. *Courtesy of Elk Grove Fire Department, Elk Grove, CA.*

mother had been beaten or kicked in the vicinity of where the bodies were found, but fire damage to her face, neck, and upper torso precluded the establishment of an exact cause of death (Moran 2001).

The major fire-related issues pertained to the duration of the fire: What had fueled the fire? How long would a body burn in such circumstances? and What is the mass loss rate for a body when it represents the main fuel package for a fire (as opposed to being involved in a massive enveloping fire fueled by furnishings and other external fuel load)? A test protocol was designed to collect data by which some hypotheses could be tested.

Carpet and pad from the scene were recovered in large quantities to serve as comparison samples and test targets for fire testing. The carpet was 95 percent polypropylene and 5 percent nylon face yarns bonded to a polypropylene mesh backing. The pad was low-density polyurethane foam. Testing revealed that this carpet/pad combination could be ignited to support a self-sustaining combustion by the ignition of a small quantity of gasoline. The burning carpet was observed in tests to produce small flames, 5–8 cm (2–3 in.) high that advanced to consume approximately 0.5 m^2 of carpet per hour. Based on the limited amount of carpet burned very deeply (i.e., in direct contact with the gasoline pool fire), it was estimated that less than 2 L of gasoline had been used. Such a quantity would be expected to produce a fire of approximately 500 kW if ignited on carpet, with a duration of 2 to 3 min. Testing confirmed these estimates.

As can be seen in the floor plan in Figure 7.7, the living room, dining room, kitchen, and family room formed one large room, with a high vaulted ceiling extending throughout those areas, and a partial-height wall separated the living/dining portion

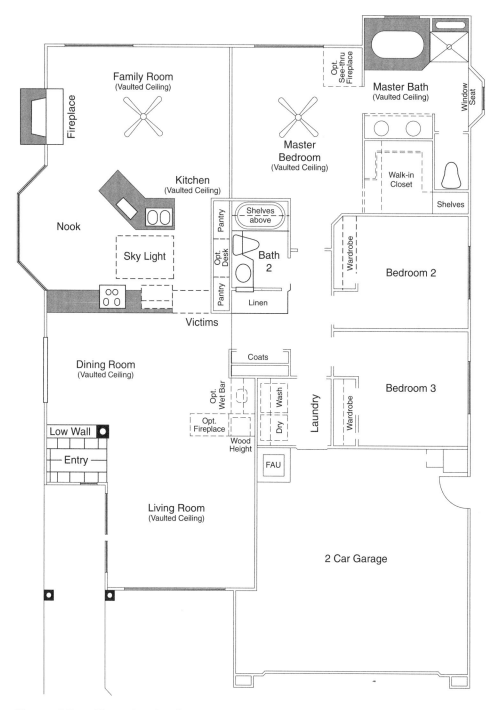

FIGURE 7.7 ◆ Floor plan showing the living room, dining room, kitchen, and family room, which formed one large area with a high vaulted ceiling. *Courtesy of Bruce Moran, Sacramento County District Attorney's Laboratory of Forensic Services, Sacramento, CA.*

from the family room/kitchen. The vaulted ceiling over the location of the bodies was an estimated 3.6 m (12 ft) from the floor and bore no heat damage. A flame plume from a 500-kW fire approximately 1 m (3.3 ft) in diameter would be (per Heskestad) about 1.8 m (6 ft) high and would not be expected to produce any ceiling damage.

The wall nearest the main plume would have been the south dining room wall and would have been more than a meter away. All walls were painted gypsum wallboard and none bore any heat damage. There was no heat damage to any of the wood furniture in the dining room. There was a narrow, arc-shaped burn mark on the carpet identified as a gasoline trail between the entry door and the bodies, but there was no damage to any furnishings except to the carpet.

Several areas of the carpet bore scorched and melted areas whose appearance could be reproduced by the burning of a splattering of small droplets of gasoline on the surface of such synthetic carpet pile. Smoke had penetrated throughout the house, and soot had lightly coated all horizontal surfaces and condensed out on vertical surfaces (showing the outlines of wall studs, ceiling joists, and other concealed structural elements).

Blood spatters on surfaces and coatings of soot on floors showed that objects such as vases, boxes, and cutlery that were strewn about to simulate a burglary had been placed in those positions after the blood was splattered but before the fire. The areas of scorching and sooting on the vinyl flooring of the kitchen were identified as being produced by gasoline or a similar ignitable liquid of relatively low boiling point by comparison with tests of such fluids on similar floor coverings. The most significant fire damage was to the bodies of the two victims. No clothing could be identified on the child's body, and the remains of a knit top and panties on Mrs. Doe would not have constituted a significant fuel load.

A reasonably small quantity of gasoline had apparently been used and so attention was focused on the combustion of the bodies as a major factor in establishing the time frame for the fire. Because Mrs. Doe had been receiving prenatal care, there was good documentation of her antemortem body weight. Compared with the postmortem weight, it was estimated that some 12 kg (25 lb) of body weight had been lost. No such data of comparable quality were available for the child victim, so it was decided that the mass lost from Mrs. Doe's body would be the most reliable measure of establishing duration of the fire.

Testing was conducted at the California Bureau of Home Furnishings (BHF) room calorimeter to establish the heat release rates and mass loss rates of body fat. Pork fat was selected for its similarity to human subcutaneous fat and availability. Tests of various small (1- to 2-kg, 2.2- to 4.4-lb) quantities of pork fat (with skin) wrapped in cotton cloth established that small fires, 20 to 70 kW, could be maintained by combustion of the rendered fat (DeHaan, Campbell, and Nurbakhsh 1999).

The size of the fire was controlled largely by the surface area of the wick (the charred cloth), and the duration of the fire was controlled by the supply of fuel. Mass loss rates appeared to be of the order of 1–2 g/s (3.6–7.2 kg/hr). Testing was also conducted in the cone calorimeter at BHF using a 10 × 10-cm (3.9 × 3.9-in.) tray, with paraffin wax as a control. It was found that at 35–50 kW/m^2 incident radiant heating, pork fat would melt and smoke but could not be ignited to support a continuous flame unless a cotton cloth wick was placed across the surface of the fat (DeHaan, Campbell, and Nurbakhsh 1999). The radiant heat would then char the cotton cloth melting the fat beneath. The melted fat wicking up via the charred cloth could be readily ignited. Cone calorimeter testing established that the effective heat of combustion for both pork and human fat was approximately 34 kJ/g and that their rates of heat release (per unit surface area of fuel) were roughly the same, 200–250 kW/m^2. The observed properties of melting behavior and ignitability for human and pork fat were also indistinguishable, confirming that test results using pork fat or pig carcasses would yield results applicable to human cadavers (DeHaan, Campbell, and Nurbakhsh 1999).

A final series of tests in which pig carcasses clothed in knit garments and ignited by means of a 1-L (0.22 gal) pour of gasoline confirmed that the charred garments and charred carpet and pad served as a wick and would support combustion of the fat being rendered out of the carcass. This combustion produced a smoky fire of approximately 50 kW, with a mass loss rate of approximately 1.5 g/s (5.4 kg/hr) for times exceeding 1 hr. Further, the gasoline pour would burn off in approximately 3 min while producing a fire of 250–400 kW. This brief fire would initiate fire in the surrounding carpet that would continue to burn for various times, consuming carpet at a rate of about 0.5 m²/hr.

Of interest in this case was the observation that the gasoline-fed fire was not of sufficient duration to cause the skin to shrink and split and allow the rendered fat to escape to support external flames. Only after the clothing and carpet had burned for 10 min or longer could the rendering process be seen to be supporting the fire.

The fire behavior supported by tests and data were considered when reaching the conclusion that the fire had been burning for 2 to 4 hr at the time of its extinguishment. Since the fire was still burning at approximately 9 A.M., it was concluded that it was ignited using gasoline sometime between 5 and 7 A.M. This time frame, combined with conflicting statements made by the suspect, led to his prosecution. He was convicted of three counts of second-degree murder and one count of arson. He is serving consecutive sentences for a total of 60 years to life.

This case is courtesy of Bruce Moran, Sacramento County District Attorney's Laboratory, Sacramento, California, and Jeff Campbell, Sacramento County Fire Department (retired).

◆ 7.8 CASE EXAMPLE 2: EXECUTION BY FIRE

The burned body of a young woman was found in a remote agricultural area. She bore burns over most of her body and was dressed in the burned remains of cotton denim pants, a long-sleeved cotton shirt, and a bra, all identified by brand name tags that survived the fire (Figure 7.8). Her ankles had been bound together with a heavy leather belt wrapped over her cotton socks, prior to the fire (Figure 7.9).

The victim was found lying on her back with her arms and legs partially flexed. There was a burned pile of clothing some 2.6 m (8 ft 8 in.) away from her head (probably in a cloth duffel bag). The clothing and other contents indicated that this bag belonged to the victim.

There were a number of small areas of scorching or charring visible on the dry soil nearby, isolated from both the clothing and the body. Although these could have been the result of isolated burning splashes of flammable liquid, they were more consistent in appearance with having been the result of contact between the burning clothing and the soil.

Examination of the scene revealed that there was a faint pattern of wrinkled fabric (as from garments like shirt and pants) impressed into the dry sandy soil between the body and the pile/bag of clothing. There was an additional pattern of wrinkled fabric in the soil to the right of the body adjacent to the torso and upper right leg.

The soil at the fire scene revealed additional information. There were deep scuff marks in the soil, of the type made by bare or stocking feet, under and beside the toes of the victim. Scorched areas of soil to the right side of the body were displaced and

FIGURE 7.8 ◆ Victim of murder by burning as found at scene, Repeated scuff marks with limited scorching in soil adjacent to torso and limbs show movement during the fire. Impressions of creased fabric of shirt on soil next to torso were produced as victim rolled on ground. Charred residue of denim jeans remains on left leg and waist. Charred cloth fragments away from body also show movement after and while burning. Shoe print in lower left corner is partially overlaid by "leg" fabric impression showing sequence.

disturbed by patterns of movement of the soil, probably by movement of the right arm and right leg of the victim. There were a number of shoe prints and tire prints in the vicinity of the body, some of which overlaid the scuff marks, but in most cases the scuff marks obscured the shoe impressions, confirming sequencing and timing.

Laboratory tests revealed the presence of gasoline in the clothing of the victim, in the burned clothing pile, and in the soil in several areas. There was no gasoline container found in the vicinity. No estimate could be made of the quantity of gasoline that had been used. No other fuels were present except for the clothing.

There was extensive scorching and blistering of the skin across nearly all exposed areas of the victim's body. The back was more burned than the front of the torso, and most of the clothing was burned away, particularly on the back and sides. There were signs of vital reaction to many burned areas of skin. The burns were only partial thickness (second degree) over nearly all the body, with the only exception being deep charring of the ankles and lower legs immediately adjacent to the cotton socks held in place by the leather belt. Nearly all the scalp hair was burned away, reduced to black masses. Charred remains of cloth were clutched in the fingers of the right hand.

Postmortem examination revealed internal heat/flame damage to the mucosa of the throat and larynx as a result of inhalation of extremely hot gases. Toxicological analysis reported a COHb level of approximately 14 percent. The fire issues here were whether gasoline had been poured on the victim while she was conscious and what the result of such an event would be.

FIGURE 7.9 ◆ Scuff marks in soil near toes and feet show repeated movement. Flexion of toes is due to heat shrinkage of tendons. Feet of victim bear only first- and second-degree burns from cotton socks burning off. Sock material trapped beneath leather belt around ankles burned longer (supported by gasoline wicking out), inducing charring of skin beneath).

Re-creations in the laboratory using a dressmaker's mannequin approximately the same height as the victim demonstrated that two full revolutions of a body that size as it rolled on the ground resulted in the same displacement as seen at the scene between the body and the clothing pile. Owing to the taper of the body from the shoulders to the feet, the path taken by a body rolling in this manner is in the form of an arc rather than a straight line.

A gasoline fire is known to produce heat transfer sufficient that direct flame contact induces blistering (partial-thickness burns) in about 5 s at 50 kW/m^2. Based on the distribution of burn damage to the body and the clothing, it was concluded that the fire damage could not have been the result of gasoline having been poured over an inert, unconscious body. The absence of protected areas and the extensive damage to the back and buttocks could have resulted only from the victim's consciously moving. The charred cloth clutched in the hand away from other fire damage indicated a conscious attempt to grasp the burning clothing during the fire.

Although flexion of the major joints as a result of fire exposure can result in some movement of an unconscious or even a deceased body, the pattern of cloth/fabric impressions in the soil was the result of a rolling movement of the body. The scuff marks in the soil to the right side of the body were most likely the result of movement of the right arm and right leg of the victim while she was lying on her right side. The body apparently fell or rolled onto its back, where it assumed its final resting position. Movement of the feet and toes would have produced the deep scuffs found there. Fire damage to the clothing and the isolated scorch marks on the soil were also consistent with conscious movement of the body.

The duration of a gasoline fire would be of the order of 2 min or less (assuming that less than a gallon was used). The duration would be shorter if the gasoline was spread on bare skin or thin cotton fabric. Such a short-duration fire would produce the limited-depth burns found here, with the only exception being the gasoline-soaked cotton socks, which acted as a wick to sustain flames long enough to induce deep burns to adjacent tissue. The presence of shoe prints overlying some of the scuff marks in the soil was an indication that the victim was set afire in the vicinity of the clothing bag and that her assailants were present during (at least) the initial stages of the fire as she rolled on the ground in an attempt to extinguish the flames. It was concluded that she was conscious for many seconds during this fire.

The three people who carried out this execution were acquaintances of the victim who decided that it would be interesting to dispose of her in this manner and transported her in a truck while they purchased the can and the gasoline. All three have pled guilty to murder charges.

◆ 7.9 CASE EXAMPLE 3: FATAL VAN FIRE

At about 5:30 on a cold November morning a farmer heard shouts of alarm coming from across the river. He looked out to see a camping-style van fully engulfed in flames. He phoned the fire department, but owing to the remote location of the site, it was 15 min or more before fire crews arrived and another 10 min before the fire was extinguished.

Inside the van were the remains of an adult female victim who had reportedly been asleep in the back of the van with her boyfriend at the time the fire was discovered. The boyfriend, after awakening and reportedly discovering the smoke and limited flames within the van, had allegedly tried to rouse the victim, who was nonresponsive, and then escaped the van by butting his head against a rear window. He attempted to rescue his friend from outside but discovered that all but one of the doors to the van were locked, and he could not reach her through the one unlocked door because a blanket hung between the cargo compartment and the driver's compartment. He claimed that they had been camping on the river just for a day. Both were recreational drug users, and he had a record for breaking-and-entering and minor drug charges.

Investigators called to the scene discovered that all the windows to the "Sportsman-style" van had been broken but could not tell if any had been broken by mechanical force (the side and rear windows were tempered glass, which shatters into similar small pieces whether broken by thermal shock or mechanical impact), but all appeared to have been soot covered. The van had been gutted, with most damage to the cargo and driver's compartments (Figure 7.10), and there were signs that the fire had penetrated into the engine from inside the van. The sheet metal of the roof had been badly distorted by the fire. There were no signs that the vehicle had been in operation at the time of the fire. The vehicle's fuel tank was still half full and the fuel system was normal.

Fire damage outside the vehicle was limited to the grass immediately around and under it; there were no indications of an external grass fire growing to involve the vehicle. The van was heavily loaded with bags and boxes of clothing, food, camping supplies, tools, and blankets. There were a propane camping lantern and camp stove with two 1-lb propane bottles, but they were not in use. All exposed fuels in the van were fire damaged.

A friend who had visited the man and woman the night before the fire said that they were eating snacks (with no cooking) and were using small votive candles for

FIGURE 7.10 ◆ Exterior damage to the gutted van, showing most damage to the cargo and driver's compartments. *Courtesy of Solano County Sheriff's Dept., Fairfield, CA.*

light. Luckily the fire chief in this jurisdiction always kept a point-and-shoot camera in his vehicle and was able to get several photos of the still-smoldering interior of the vehicle before the contents were removed to permit overhaul. The upholstery of the seats and dash and paneling of the cargo compartment were completely consumed (Figure 7.11). The fire damage clearly demonstrated that the fire within this compartment had gone to flashover, with ventilation supplied by the windows (closed at the start of the fire but failing as the fire grew).

The decedent was found in a partial sitting position on the right side of the cargo compartment (Figure 7.12) facing the rear of the van. The body had suffered massive fire damage; nearly the entire right half of the body had burned away (exposing the spinal column), with major bones reduced to fragments. The victim had a 41 percent COHb level, with no alcohol and very low levels of amphetamine metabolites. She was young (23 years old), healthy, and physically unimpaired. Extensive fire exposure had destroyed the soft tissues of the head and neck and had caused the exposed skull to calcine and fracture into a number of pieces. There were no underlying hemorrhages indicating antemortem trauma. The fire had compromised the lungs and airways such that the presence of combustion products could not be determined.

The COHb level left no doubt that the victim had been alive and breathing for some time during the fire and had died as a result of exposure to smoke and flames. The extensive damage to the body was ascribed by the pathologist to incineration by a flammable liquid fire, despite later laboratory findings that revealed no petroleum distillates in the small segment of carpet recovered from beneath her.

No flammable liquid residues were detected on the exterior of the van or beneath it. The extensive damage to the van, the absence of accidental ignition sources,

Figure 7.11 ◆ Interior of the vehicle immediately after extinguishment and before overhaul. The victim is sitting facing the camera, with her back resting against the back of the passenger seat. *Courtesy of Solano County Sheriff's Dept., Fairfield, CA.*

and the nonoperational status of the van led investigators to conclude that the fire had been deliberately ignited with a flammable liquid on or around the decedent, which prevented her escape. The boyfriend was charged with murder.

The surviving victim (boyfriend) was found outside the van dressed in T-shirt and undershorts (he was the person shouting for help whom the farmer heard). He had a coating of soot on his face, arms, and hands but no burns. His facial hair (including mustache) was unburned, but hair at the top of his head was singed. He had fresh abrasions on his knuckles, one on the top of his head, and scratches and abrasions on his lower legs. Tiny fragments of glass were found in his scalp hair. His blood was not sampled immediately but was taken later and was found to bear low CO concentrations, low levels of amphetamine metabolites, and no alcohol. He claimed that he and the decedent had gone to sleep side by side wrapped in blankets on the floor of the cargo compartment and awakened to discover the fire. He denied any conflict with the decedent.

A review of the evidence (by one of the authors) at the request of the local district attorney's investigator resulted in the conclusion that the physical, medical, pathological, and toxicological evidence was more consistent with an accidental fire occurring as described by the accused than with a murder with flammable liquid. A fire-related human behavior analysis revealed that if the survivor had been sleeping on his back in a smoke-filled van, his face and arms would be expected to be coated with soot but not exposed to the hot gas layer until he sat up and attempted escape. At that time, his scalp hair would be exposed to the highest temperatures. Beating at the window glass or fumbling in the dark for the door locks could have produced the

FIGURE 7.12 ◆ Fire victim as found in the vehicle. Note the extensive destruction by flames, with exposed ribs, spinal column, and internal organs. The skull has calcination and fracture damage resulting from prolonged fire exposure. *Courtesy of Solano County Sheriff's Dept., Fairfield, CA.*

abrasions on his right hand. Butting his head against the rear window until it broke could have produced the abrasion on the top of his head and the glass fragments in his hair. Sliding out the high rear window would be expected to induce cuts and scratches to the lower legs.

The 41 percent COHb level in the decedent is similar to the levels encountered in victims of accidental house fires. In the opposing hypothesis, throwing gasoline on someone in sufficient quantities to prevent his/her escape would be likely to leave un-burned residues after the fire in protected areas under the body and would also be likely to produce a flash fire of sufficient size to induce singed facial hair on someone close by (i.e., the thrower). The ignition of a quantity of gasoline vapor in the vicinity of a victim's face can induce rapid cessation of breathing so that CO levels in the blood would be very low. The first few seconds of a gasoline vapor/air fire are also very low in CO, so the gases that are inhaled are unlikely to produce an elevated COHb level. Such a scenario would not produce the glass fragments, singed scalp hair, or leg injuries. A gasoline-fueled fire would also be unlikely to induce such extreme damage to the decedent's body unless it involved massive quantities of fuel.

The fire issues involved were the following.

- ◆ What were the contributions of the fuel load in the van?
- ◆ Did the size of the van (compartment) and the breakage of windows play a role?
- ◆ Could a fire in a closed van produce a fire that would be escapable and produce the physical and clinical evidence seen on the survivor?
- ◆ Could such a fire progress to flashover?

FIGURE 7.13 ◆ The test van obtained had nearly the same arrangement of windows but had one solid side panel instead of a large window, so a vent panel was cut through the metal and left attached by a hinge so it could be opened during the fire test to simulate failure of the window by thermal shock. *Courtesy of Solano County Sheriff's Dept., Fairfield, CA.*

- If the fire did progress to flashover, could postflashover fire produce the damage to the body and to the van without the presence of any ignitable liquids?

A long-wheelbase Sportsman van similar to the one in the incident was obtained by the district attorney's office. Based on calculations of the window area and height, it was determined that when all the windows failed, enough air would be admitted to the van to support a fire of 3.2 MW. The internal volume of the van was calculated to be 9.8 m^3, so there was considerably more air inside even a closed van than there would be in the average passenger vehicle.

The van obtained (Figure 7.13) had nearly the same arrangement of windows but had one solid side panel instead of a large window, so a vent panel was cut through the metal and left attached by a hinge so it could be opened during the fire test to simulate failure of the window by thermal shock. The interior finish was duplicated, with paneling, carpeting, and seats similar to the original.

Wooden tool boxes and a wool blanket curtain were fitted to simulate the original, and some 120 kg (265 lb) of clothing, blankets, plastic crates, cardboard, and miscellaneous combustibles was added, estimated to be about half the added fuel mass in the original van, as shown in Figure 7.14.

Temperatures were monitored via three thermocouples mounted on a tree in the center of the van near the position of the decedent and one thermocouple near the rear door where the survivor allegedly made his escape. Ignition was produced with an open flame in ordinary combustibles and the van closed up. A misplaced votive candle was probably the ignition source for the actual fire. Given the nature of the fire and the fuels available, it was very unlikely to have been a smoldering source.

In the test, within 3 min heavy smoke had formed throughout the van and temperatures in the smoke layer ranged from 80 to 100°C (176 to 212°F), as shown in Figure 7.15. Soot was observed condensing on all windows. The rear window was broken from outside by mechanical impact at 3 min 45 s. Gas layer temperatures

FIGURE 7.14 ◆ Wooden toolboxes and a wool blanket curtain were fitted to simulate the original, and some 120 kg of clothing, blankets, plastic crates, cardboard, and miscellaneous combustibles was added, estimated to be about half of the added fuel mass in the original van. *Courtesy of Solano County Sheriff's Dept., Fairfield, CA.*

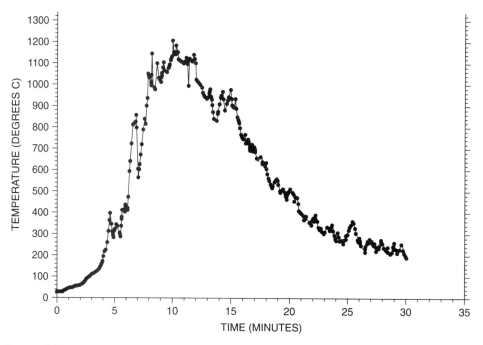

FIGURE 7.15 ◆ Test results for fire in a test vehicle documenting the interior temperature versus time at a thermocouple located near the victim's location. *Courtesy of Fred Fisher, Fisher Research & Development, Inc., Vacaville, CA.*

FIGURE 7.16 ◆ Thermal damage to exterior of test van. *Courtesy of Solano County Sheriff's Dept., Fairfield, CA.*

exceeded 600°C (1112°F) by 7 min, and flashover occurred as windows began to fail from thermal shock. The side vent was opened as all the windows failed between 6 and 8 min into the fire. By 8 min, temperatures in the hot gas layer exceeded 1000°C (1832°F). Temperatures of over 600°C (1112°F) were maintained for over 12 min as the fire consumed the fuel load and went into decay. Total burn time was 30 min. Damage to the exterior and interior of the van duplicated that in the original, as shown in Figures 7.16 and 7.17.

Temperatures and heat fluxes in excess of those found in commercial crematoria were measured in the center of the van. The victim's body would have been exposed to these intense fire exposures from her right side, near the center of the compartment, where the postflashover burning would have been most intense. The fuel load available in this test limited the duration of the fire, but not its intensity. The intensity was determined by the surface area of fuel involved in the postflashover burning and the ventilation available to support the combustion.

Although no heat release data were captured during this test, the test team, with their extensive experience, estimated that the heat release rate was in excess of 3 MW during the peak of the fire (at 10 to 12 min). Because the ventilation inside the test van was the same as in the fatal van, the size and intensity of the fires would have been equivalent. With the more substantial fuel load in the scene van, it would have been capable of supporting a full postflashover fire for much longer than the test van could.

Witness observation by the farmer confirmed that the van was fully engulfed when he first saw it and it burned that way until extinguished some 25–30 min later. Based on the results of this test, the district attorney agreed that an accidental fire

FIGURE 7.17 ◆ Thermal damage to interior of test van. *Courtesy of Solano County Sheriff's Dept., Fairfield, CA.*

could not be ruled out and, in fact, was more consistent with the physical evidence. All charges were dropped.

Test data in this case are courtesy of Fred Fisher, P.E., of Vacaville, California. The authors thank Professor R. Brady Williamson of University of California, Berkeley, Dr. Charles Fleischman of University of Canterbury, Christchurch, New Zealand, and members of the Solano County Fire Investigation Team for their contributions and for making this test possible.

◆ 7.10 CASE EXAMPLE 4: MURDER BY FIRE, OR ACCIDENT?

At approximately 4 P.M. on a weekday, fire and ambulance services were alerted to attend a female that had been burned and to a fire in a kitchen in a suburban apartment block. On arrival ambulance and fire personnel found an adult female deceased, with obvious burn injuries, at the street frontage gate. Information from a male indicating he was the partner of the deceased provided the location of a suspected fire in an apartment kitchen. The apartment was located and inspected, to find that a small fire had occurred within the kitchen, with slight fire damage caused and the fire extinguished. Firefighters requested an investigator to attend owing to the fatality.

The initial theory provided by firefighters at the scene was that the male was working on a model car engine and the fuel being used had ignited, with the female involved by the ensuing fire. A further theory was developed by police at the scene, that the male and female had been quarrelling, fuel had been accidentally spilled on

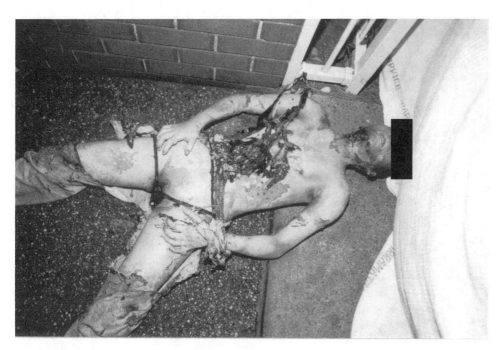

FIGURE 7.18 ◆ Upper body of fire victim, found outside residence. Note fire damage to clothing, upper body, and face. *Courtesy of Ross Brogan, NSW Fire Brigades, Greenacre, NSW, Australia.*

the kitchen floor, and an accidental ignition had involved the clothing of the female, with her running out of the apartment, downstairs into the yard, where she was found on fire and subsequently carried 50 meters to the front gate, where she succumbed to her injuries from the fire, where she was found by attending emergency services.

The body of the deceased was inspected in situ, and it was noted that she had been wearing a cotton-type material tracksuit, with long legs and long sleeves; what appeared to be nylon type socks were on her feet, and she had also been wearing underwear. The clothing was severely fire damaged, principally from the hips/thighs up and involved the upper body and head (Figure 7.18). The upper body was severely burned and the head and chin area suffered extreme burn injuries; both hands and forearms had suffered severe burn injuries as well. The legs of the pants, however, were intact and not consumed by the fire (Figure 7.19).

In a trail from the body at the front gate, along the path leading to the front door, and up the internal stairs to the second level were found burned remnants of clothing and burned pieces of skin. Several pieces of burned skin adhered to the stair railing outside the fire-affected apartment.

Inside the apartment it was found that a small fire had damaged sections of the kitchen. There were some small scorch marks on the tiled floor and some slight scorch marks on the bottom of the timber cupboards, but more important, several areas of the top surface of the cupboards had been fire damaged (Figure 7.20). A paper towel roll had a V pattern from the benchtop upward on its front surface; a microwave oven had a V pattern of scorching on the front window surface, and a kitchen appliance on the benchtop had plastic feet, one of which was slightly melted. An odor of methylated spirits was evident within the kitchen and was emitted from beneath the

FIGURE 7.19 ◆ Legs of fire victim. Note pants legs are intact. Nylon socks are melted. *Courtesy of Ross Brogan, NSW Fire Brigades, Greenacre, NSW, Australia.*

FIGURE 7.20 ◆ The kitchen where the incident occurred. Note small fire damage at center bottom of cupboards and on paper towel roll at countertop near microwave. *Courtesy of Ross Brogan, NSW Fire Brigades, Greenacre, NSW, Australia.*

FIGURE 7.21 ◆ Fire damage on kitchen countertop at scene of fire. Note V pattern on paper towel roll and burn mark on microwave front window. *Courtesy of Ross Brogan, NSW Fire Brigades, Greenacre, NSW, Australia.*

microwave oven (Figure 7.21). A partially melted plastic container was located on the floor adjacent to the bottom of the kitchen cupboards (approx. 4-L capacity, 1 gal).

A one-fifth scale model car engine was located within an adjoining room, and details were recorded of the model and serial numbers. Subsequent research through a model car supplier revealed that this engine was fueled by methylated spirits (denatured alcohol) and not petrol (gasoline), as suggested by the male occupant as the fuel for the fire.

After inspection of the injuries suffered by the deceased and of the fire damage within the kitchen, a discussion with police was held. The notion that the deceased had been accidentally caught in the fire, which had been ignited by spilled fuel on the kitchen floor, was dismissed as a fiction. The reasoning used to test this theory was that the deceased was still wearing nylon socks, and the lower section of her cotton clothing was intact and not fire damaged; if the fire had been on the floor of the kitchen, then the socks and lower body clothing would have been the first to catch alight, based on normal fire behavior. The theory advanced was that the deceased had had fuel splashed/poured/thrown over the upper half of the body and ignited, to have caused the injuries and damage found. Subsequent scientific testing showed that the clothing and areas of the kitchen had been contaminated with methylated spirits.

Police interviewed the suspect male and he subsequently provided information that he had quarrelled with his female partner, had been in possession of a 4-L plastic container of methylated spirits (approx. half full) and had been so enraged that he had thrown some of the liquid onto her, thrown the container onto the floor, and had kicked it around in his rage. He had then held a disposable lighter up and away from her and ignited it in an effort to scare her. The liquid on the floor had accidentally ignited, and she became involved and had run down the stairs and out of the apartment.

FIGURE 7.22 ◆ Testing of theory with clothing and flammable liquid in a fatal fire death case. *Courtesy of Ross Brogan, NSW Fire Brigades, Greenacre, NSW, Australia.*

He had followed, smothered the fire with a bedsheet, and carried her to the front gate.

A further search of the apartment revealed some of the male's clothing, on the bathroom floor, burned and wet. This evidence, along with burns that he had suffered showed he had caught alight as well during the incident and had gone to the shower and extinguished himself prior to following the female downstairs. The male was subsequently charged by police with murder.

In an effort to test the investigator's theory, the clothing that had been worn by the deceased was scientifically analyzed, and similar sets of clothing purchased. Two store mannequins were also obtained and a test conducted to fire test the clothing. The Fire Service Training College Fire Training Tower was used, and all tests were videotaped and photographed, to be used as evidence (Figure 7.22). The mannequins were clothed in the identical clothing and two tests were conducted using methylated spirits, one with liquid poured only on the floor around the base of the mannequin and another with liquid poured over the upper body and onto the floor.

The liquid was ignited in both cases and the resulting fire recorded. Even though the concrete room was bare of furnishings, fittings, and cupboards it was considered a basic test to determine whether the resulting fire produced similar injuries to those of the deceased. In both cases with the liquid on the floor, the lower body clothing and socks were consumed early in the fire (Figures 7.23, 7.24, 7.25, and 7.26). With the liquid applied to the upper body of the mannequin, the resulting burn damage to the face and upper body of the mannequin proved to be identical with the actual injuries suffered by the deceased; both sets of photographs could be placed side by side and the injuries were seen to be identical (Figures 7.27 and 7.28). The injuries suffered by

FIGURE 7.23 ◆ Test underway with flammable liquid at floor level only. *Courtesy of Ross Brogan, NSW Fire Brigades, Greenacre, NSW, Australia.*

FIGURE 7.24 ◆ Fire test with fire being extinguished. *Courtesy of Ross Brogan, NSW Fire Brigades, Greenacre, NSW, Australia.*

FIGURE 7.25 ◆ Fire damage to mannequin after fire test from floor-level pour. *Courtesy of Ross Brogan, NSW Fire Brigades, Greenacre, NSW, Australia.*

FIGURE 7.26 ◆ Fire test underway with flammable liquid on floor and on upper body and clothing. Note that dark coloring on clothing is liquid on upper body. *Courtesy of Ross Brogan, NSW Fire Brigades, Greenacre, NSW, Australia.*

FIGURE 7.27 ◆ Mannequin fully involved in test with liquid at floor and on upper body. (Judge considered that this graphic photo may upset jurors.) *Courtesy of Ross Brogan, NSW Fire Brigades, Greenacre, NSW, Australia.*

FIGURE 7.28 ◆ Results of test showing fire damage/injury to mannequin that was compared with photographs of the actual victim. *Courtesy of Ross Brogan, NSW Fire Brigades, Greenacre, NSW, Australia.*

the deceased showed the fire had risen from beneath the face and affected the underchin area, lower portion of the upper lip, the nostrils and underside of the eyelids, and the lower lobes of the ears of the victim.

With these data and evidence the case was prepared for trial on the murder charges. After a conference with the Crown Prosecutor it was decided that further tests would be conducted in a kitchen-type configuration to test the validity of the actual scene configuration to validate the veracity of the testing and the theories involved. The public housing authority was approached and a vacant townhouse obtained with a kitchen configuration similar to that of the actual incident scene. This configuration was thought to lend authenticity to the test results, rather than the open concrete room used in the initial tests.

First, a 4-L (1-gal) container similar to the one found in the apartment was obtained and marked with graduations on the outside to show liquid levels, in approximately 250-mL increments. The container was filled with colored water and "throw" tests were conducted to try to give an indication of how much liquid was expelled with each throw, at differing liquid levels. These results were to then be used to show how much liquid may have been expelled onto the victim during the actual incident. The tests were videotaped and photographed again, for evidence purposes. These results were used in the court trial.

The fire tests were conducted using clothing identical to that that the victim had been wearing, this time in the kitchen similar in configuration to the fire scene. Videotaped and photographed, these tests were again available for the trial, as evidence. Again, the tests where the liquid was applied to the upper body of the mannequin resulted in burn damage and injuries identical with those of the victim, to the mind of the investigator, and the prosecutor, that the theory proposed was valid and practical, having been tested by scientific methodology.

At the trial, after viewing the video evidence of the tests the trial judge refused to allow the videos to be shown to the jury, stating that they appeared to be prejudicial to the defendant and a possible appeal point after the trial. However, the judge allowed the theory testing and results to be discussed during presentation of evidence. In all, the investigator spent five and a half days in the witness box presenting evidence-in-chief and under cross examination, and presenting evidence to establish his own credibility, the credibility of the hypothesis, and the validity of the tests that proved the story of the defendant to be false. The male was subsequently convicted of manslaughter.

It was interesting that during the cross-examination the defense attorney proposed a theory aligned with the fact that it was (the investigator had explained the scientific basis of ignition of flammable liquid vapor and not the actual liquid), in his words, "*a spiderweb of vapor hanging in the air*" that ignited and ignited the upper clothing of the deceased. This theory was vigorously debated, with the scientific basis of the theory explained during the cross-examination, and ultimately dismissed, partly because a question from the jury indicated that a member of the jury had some scientific training and complained that the defense were using "junk science theories" to deceive the jury on something that was quite plainly a valid scientific theory proposed by the prosecution. The other factor was that if the "spider web of vapor" was hanging in the air, then the lower clothing was just as susceptible to ignition as the upper clothing. The jury found the defendant guilty.

This case provided courtesy of Inspector Ross Brogan AFSM, CFI—NSW Fire Brigades, Sydney, Australia.

In a fatal fire, the cause of death and the cause of the fire are independent but linked by circumstances. Each must be established, and only then can the link between them be determined.

Accidental fires can accompany deaths by accident, suicide, homicide, or even natural causes. Incendiary fires can be associated with homicide (as a direct cause of death or simply as part of the crime event) but also with accidental or natural-cause deaths. For a fire death investigation to be successful (i.e., accurate and defensible), the following guidelines must be observed.

- **Treat as a crime scene.** Every fire with a death or major injury should be treated as a potential crime scene and not prejudged as accidental. The scene should be secured, preserved, documented, and searched by qualified personnel acting as a cooperative team.
- **Documentation is essential.** This includes accurate floor plans, with dimensions and major fuel packages included, and comprehensive photographic coverage. Photos must include presearch survey photos, photos during search and layering, and photos of all views of the body prior to removal, during removal, and during postmortem examination. This documentation is essential to proper reconstruction (as described in Chapter 4).
- **Avoid moving the body.** The body must not be moved until it has been properly examined by the fire investigator and the pathologist or coroner's representative and thoroughly documented by photos and diagrams. The debris under and within 0.9 m (3 ft) of the body should be carefully layered and sifted. All clothing or fragments should be preserved. Artifacts (jewelry, weapons, etc.) must be documented and collected.
- **Assess the fuels.** The fire investigator has to assess the fuels already at the scene (structure as well as furnishings), the role fuels may or may not have played in ignition, flame spread, heat release rates, time of development, and creation of flashover conditions.
- **Perform full forensic exams.** Every fire death deserves a full forensic postmortem including toxicology and X-rays. Toxicology samples should be tested for alcohol and drugs, as well as CO, and should include both blood and tissue. The clothing should remain with the body and be documented and evaluated in situ before removal, if possible, then properly preserved. The internal (liver) body temperature should be taken as soon as possible (preferably at the scene).
- **Examine pets.** Deceased pets should be X-rayed and necropsied. Injuries to living pets should be noted and documented. Blood from deceased pets should also be tested for COHb saturation.
- **Examine living victims.** "Nonfatal" burn victims should be photographed and blood samples taken for analysis later if needed. External clothing (pants, shoes, shirt) should be saved and properly preserved.
- **Fully appreciate the fire environment.** Pathologists and homicide detectives must appreciate the fire environment—temperatures, heat and its transfer, flames, and smoke—and the distribution of fire products and the variables in human response to those conditions. In best practice, the pathologist visits the scene and sees the body in situ to appreciate its conditions of exposure (to flame, heat, and smoke), the nature of debris, and its location and position.
- **Carry out forensic reconstructions.** A full reconstruction may involve criminalistics evidence such as blood spatter or transfers, fingerprints, tool marks, shoe prints, and trace evidence. The criminalist should be part of the scene team along with the homicide detective, fire investigator, and pathologist.

As we have seen, a death involving fire is not a simple exposure to a static set of conditions at a single moment in time. Fire is a complex event, and a fire death investigation is even more complex and challenging. A coalition of talents and knowledge working together as a team is the only way to get the right answers to the big questions: What killed the victim? Was the fire accidental or deliberate? and How did those two events interact?

▪▪▪

Problems

7.1. Discuss the common problems and pitfalls associated with fatal fire investigations. Compare them with a recent ongoing case in the national media. What parallels can you find?

7.2. Referring to Example 7.1, determine the visibility of the same fire in an enclosed hallway measuring 3 × 15 m (9.84 × 49.2 ft) with a ceiling height of 2.5 m (8.2 ft).

7.3. Referring to Example 7.2, calculate the fractional incapacitating dose for a male, a child, and an infant in the same scenario.

7.4. Review the newspaper coverage of a recent fatal fire in your community. Determine who investigated it, what the conclusions were, and whether criminal charges resulted.

▪▪▪

Suggested Reading

Adair, T. W., L. DeLong, M. J. Dobersen, S. Sanamo, R. Young, B. Oliver, and T. Rotter. 2003. Suicide by fire in a car trunk: a case with potential pitfalls. *Journal of Forensic Sciences* 48 (3): 1113–17.

Brogan, R. 2000. The Speed Street fire—Was it murder by—or prior to—the fire? *Fire & Arson Investigator* 50 (July): 17–21.

DeHaan, J. D. 2007. *Kirk's fire investigation,* 6th ed., chap. 15. Upper Saddle River, NJ: Prentice Hall.

DeHaan J. D., and F. L. Fisher. 2003. Reconstruction of a fatal fire in a parked motor vehicle. *Fire & Arson Investigator* 53 (2): 42–46.

Levin, B. C., P. R. Rechani, F. Landron, J. R. Rodriguez, L. Droz, F. M. deCabrera, S. Kaye, J. L. Gurman, H. M. Clark, and M. F. Yoklavich. 1990. Analysis of carboxyhemoglobin and hydrogen cyanide in blood from victims of the Dupont Plaza Hotel fire in Puerto Rico. *Journal of Forensic Science* 35 (1): 151–68.

Pope, E. J., O. C. Smith, and T. G. Huff. 2004. Exploding skulls and other myths about how the human body burns. *Fire & Arson Investigator* (April): 23–28.

Purser, D. A. 1995. Toxicity assessment of combustion products. In *SFPE handbook of fire protection engineering,* 2nd ed., chaps. 2–8. Quincy, MA: SFPE/NFPA.

Raj, P. K. 2006. "Hazardous heat." *NFPA Journal* (September/October): 75–79.

Society of Fire Protection Engineers. 2000. *Engineering guide to predicting 1st and 2nd degree skin burns.* Bethesda, MD: SFPE.

Fire Testing

8 CHAPTER

I have devised seven separate explanations, each of which would cover the facts as far as we know them. But which of these is correct can only be determined by the fresh information which we shall no doubt find waiting for us.

—Sir Arthur Conan Doyle,
"The Adventure of the Copper Beeches"

Testing for fire reconstruction purposes can span the range from simple field tests requiring no equipment to benchtop flame tests to extensively instrumented, full-scale test burns in real buildings (as in Figure 8.1). Fire tests can help confirm or reject hypotheses about the fire's ignition or spread, test and validate the predictions of computational models or those of experienced investigators, or demonstrate the roles of various factors in the fire and its effects on occupants. This chapter explores, in summary form, many of the tests useful in fire investigation.

◆ 8.1 ASTM AND CFR FLAMMABILITY TESTS

A number of test standards and methods applicable to fire investigation were listed in Table 1.3 (see Chapter 1). Several of the ASTM tests can be used to assess the ignitability of materials or the nature and speed of flame propagation, and they are discussed briefly here. However, as you will see, there are limitations to the application of results of these tests to an actual case. Many limitations are mostly due to the geometry or size of the sample.

A flame will spread much more quickly upward than it will downward or outward in the same fuel. A fire started at the top edge of a sofa back will spread much more slowly than if the same fabric is ignited at the base of the sofa. A fire ignited at the top of hanging draperies will be much more likely to cause the hooks and drapery rod to fail and collapse with the draperies, possibly resulting in drop-down ignition of materials beneath, whereas draperies ignited at the bottom are often nearly completely consumed before they cause failure of their supports.

Fire resistance properties today are sometimes built into a combination of layers of fabrics and backings that result in acceptable ignition resistance. Separating and testing the layers individually may yield very misleading results. The residual moisture in test samples also can affect their ignitability, and so many specimens are often conditioned at 24–48 hr under specific criteria prior to testing.

Figure 8.1 ◆ Testing for fire reconstruction purposes can span the range from simple field tests requiring no equipment to extensively instrumented full-scale test burns on real buildings. *Courtesy of Michael Dalton, Knox County Sheriff's Office.*

Most ASTM standard methods include an advisory comment regarding use of the method to compare materials or their performance under controlled conditions to assess the possible contributions a material *might* make in a real fire and not to predict what a product *will* do when exposed to a real fire. Some of the tests reference other entities such as the National Fire Protection Association (NFPA) and the U.S. Consumer Product Safety Commission (CPSC).

- **Flammability of Clothing Textiles (16 CFR 1610-U.S.).**—This test requires that a sample of fabric 50 × 150 mm (2 × 6 in.) placed in a holder at a 45° angle and exposed to a flame for 1 s not ignite and spread flame up the length of the sample in less than 3.5 s for smooth fabrics or 4.0 s for napped fabrics.
- *ASTM D 1230*, **Standard Test Method for Flammability of Apparel Textiles (not identical with 16 CFR 1610).**—CPSC requires fabrics introduced into commerce to meet requirements of 16 CFR 1610. This test is suitable for textile fabrics as they reach the consumer or for apparel other than children's sleepwear or special protective garments. A sample of fabric 50 × 150 mm (2 × 6 in.) is held at a 45° angle in a metal holder and a controlled flame is applied to the bottom end for 1 s. The time required for the flame to proceed up the fabric a distance of 127 mm (5 in.) is recorded. This test is reportedly less expensive and time-consuming than 16 CFR 1610.
- **Flammability of Vinyl Plastic Film (16 CFR 1611-U.S.).**— This test requires that vinyl plastic film (for wearing apparel) placed in a holder at a 45° angle and ignited not burn at a rate exceeding 1.2 in. per second.

◆ **Flammability of Carpets and Rugs, 16 CFR 1630 (Large Carpets) and 16 CFR 1631 (Small Carpets).**—A specimen of carpet is placed under a steel plate with a 20.32-cm (8-in.)-diameter circular hole. A methenamine tablet is placed in the center of the hole and ignited. The duration of burning and heat release rate of the tablet duplicate those of a typical dropped match. If the specimen chars more than 3 in., in any direction, it is considered a "fail." The ambient radiant heat flux is minimal because the sample is tested at room temperature. The ignition source is a modest 50- to 80-W flame of brief duration. Testing has shown that some carpets that pass this test will be readily ignited and spread flames if the radiant heat flux is larger as a result of a larger and more prolonged ignition source. The carpet is tested in the flat, horizontal position, so the same carpet mounted vertically may behave very differently. Note that *ASTM E 648* is the standard test (*Critical Heat Flux of Floor Coverings Using a Radiant Panel*).

◆ ***ASTM D 2859*, Standard Test Method for Ignition Characteristics of Finished Textile Floor Covering Materials.**—This method uses a steel plate 22.9 cm (9 in.) square and 0.64 cm (0.25 in.) thick with a 20.32 cm (8-in.)-diameter circular hole. A methenamine tablet is placed in the center and ignited in a draft-free enclosure. Samples are thoroughly oven-dried and cooled in a desiccator before being tested. The product fails if the charred portion reaches within 25 mm (1 in.) of the edge of the hole in the steel plate. Eight specimens are tested. The Flammable Fabrics Act (FFA) regulations require that at least seven of the eight specimens pass this test.

◆ **Flammability of Mattresses and Pads (16 CFR 1632-U.S.).**—A minimum of nine regular tobacco cigarettes are burned at various locations on the bare mattress—quilted and smooth portions, tape edge, tufted pockets, and so forth. The char length of the mattress surface must not be more than 50 mm (2 in.) from any cigarette in any direction. The test is repeated with the cigarettes placed between two sheets covering the mattress.

◆ **Flammability of Children's Sleepwear (16 CFR 1615 and 1616-U.S.).**—Each of five 88.9 × 25.4-cm (35 × 10-in.) specimens is suspended vertically in a holder in a cabinet and exposed to a small gas flame along its bottom edge for 3 s. The specimens cannot have an average char length of more than 18 cm (7 in.), no single specimen can have a char length of 25.4 cm (10 in.) (full burn), and no single sample can have flaming material on the bottom of the cabinet 10 s after the ignition source is removed. These requirements are for finished items (as produced or after one washing and drying) and after the items have been washed and dried 50 times.

◆ ***ASTM E 1352*, Standard Test Method for Cigarette Ignition Resistance of Mock-Up Upholstered Furniture Assemblies.**—This test uses reduced-scale plywood mock-ups, 47 × 56 cm (18 × 22 in.), upholstered with simulations of seat, backrest, and armrests to test ignitability of furniture to dropped smoldering cigarettes. The test is used for furniture to be used in public occupancies, nursing homes, hospitals, and the like. The cigarettes are positioned on each of the various surfaces and along crevices between seats, armrest, and back. The distance of char extension from each cigarette or ignition by open flame is recorded for comparison.

◆ ***ASTM E 1353*, Standard Test Method for Cigarette Ignition Resistance of Components of Upholstered Furniture.**—This test uses mock-ups to test individual components—cover fabrics, welt cords, interior fabrics, and filling or batting materials in the form, geometry, and combination in which they are used in real furniture. This test uses single cigarettes, but each is covered with a single layer of cotton sheeting that retains more heat and makes it a more severe test than one in open air.

◆ ***ASTM E 648*, Standard Test Method for Critical Radiant Flux of Floor-Covering Systems Using a Radiant Heat Energy Source.**—This test uses a horizontally mounted floor covering specimen 20 × 99 cm (8 × 39 in.) with a gas-fired radiant heater (with a gas pilot burner) mounted at a 30° angle above it that produces a radiant heat flux of 1–10 kW/m². The distance the flame propagates along the sample indicates the minimum radiant heat flux for ignition and propagation.

◆ 8.2 ASTM TEST METHODS FOR OTHER MATERIALS

The ASTM's (2004a) *Fire Test Standards,* 6th ed., lists methods for testing other materials per ASTM Committee E5 on Fire Standards.

- *ASTM D 1929*, **Standard Test Method for Determining Ignition Temperature of Plastics (ISO 871).**—This test uses a cylindrical hot-air furnace to heat the test sample in a pan. A thermocouple monitors the temperature of the sample while the temperature is adjusted so that pyrolysis gases venting through the top of the chamber can be ignited by a small pilot flame held near it, which establishes the flash ignition temperature (FIT). The same apparatus can be used to establish the spontaneous ignition temperature (SIT) by eliminating the pilot flame and observing the sample visually for flaming or glowing combustion (or a rapid rise in the sample temperature). SITs for common plastics are 20°C–50°C (68°F–122°F) higher than FITs by this technique. The test requires 3 g of material for each test in the form of pellets, powder, or cut-up solids or films.

- *ASTM E 119*, **Standard Test Methods for Fire Tests of Building Construction and Materials.**—A specimen (wall assembly, floor, door, etc.) is exposed to a standard fire exposure to evaluate the duration for which those assemblies will contain a fire or retain their integrity. A large furnace (horizontal or vertical) is used to create a standard time–temperature environment. The assembly is exposed to the prescribed conditions and its surface temperature is measured, its structural integrity is monitored, and the time for penetration of flame through it is determined. These are the performance criteria (see Figure 8.2).

- *ASTM E 659*, **Standard Test Method for Autoignition Temperature of Liquid Chemicals.**—A small sample (100 μL) of liquid is injected into the mouth of a glass flask heated to a predetermined temperature, and the flask is observed for the presence of a flash flame inside the flask and a sudden rise in internal temperature (as monitored by an internal thermocouple). If no ignition is observed, the temperature is increased and the test repeated until the material ignites reliably. This method establishes the "hot flame" autoignition temperature (AIT) in air. Ignition delay times may also be recorded. The temperature at which small sharp rises occur in the internal temperature alone is the "cool flame" autoignition temperature.

    The method can also be used for solid fuels that melt and vaporize or sublime completely at the test temperatures, leaving no solid residues. The test conditions are controlled by the heat transfer between the glass flask and the fuel introduced and the confined geometry of the spherical flask. In real-world ignitions, the nature of the surface and the manner of contact will determine the convective heat transfer coefficient and may modify times or temperatures. Any geometry (tube or enclosure) that keeps the fuel in contact with the hot surface is going to produce ignitions at lower temperatures than an open, flat surface where buoyancy of the heated vapors can transport the fuel away from the heated surface.

FIGURE 8.2 ◆ Surface rate of flame spread test (ASTM E 1321) uses a radiant heat panel at an angle to the surface being tested. *Courtesy of NIST.*

- ◆ *ASTM D 3675*, **Standard Test Method for Surface Flammability of Flexible Cellular Materials Using a Radiant Heat Energy Source.**—This method uses a gas-fired radiant heat panel 30 × 46 cm (12 × 18 in.) in front of an inclined specimen of material 15 × 46 cm (6 × 18 in.), oriented so that ignition (with a pilot flame) first occurs at the upper edge of the sample and the flame front moves downward. The rate at which the flame moves downward is observed and a flame spread index is calculated. The test is suitable for any material that may be exposed to fire. At least four specimens must be tested.
- ◆ *ASTM D 3659*, **Standard Test Method for Flammability of Apparel Fabrics by Semi-restraint Method.**—This test simulates the burning characteristics of a garment hanging vertically from the shoulders of a wearer. Test specimens are 15 × 38 cm (6 × 15 in.) (five specimens required) and are oven-dried and weighed prior to testing. The sample is hung vertically from a cross bar, and the flame of a small gas burner is positioned against the bottom edge of the fabric for 3 s and then removed. The weight (percentage area) destroyed by the flame and the time required are the criteria for comparison. (Cross-reference to FF 3-71, Flammability of Children's Sleepwear, Sizes 0–6X.)
- ◆ *ASTM E 84*, **Standard Test Method for Surface Burning Characteristics of Building Materials.**—This is the "Steiner Tunnel" test, which mounts specimens on the underside of the top of an insulated tunnel 7.6 m (25 ft) long, 30 cm (12 in.) high, and 45 cm (17.75 in.) wide. A gas burner at one end ignites the sample and the rate of flame spread along the length of the specimen is observed and recorded. The test specimen must measure at

least 51 cm × 7.32 m (20 in. × 24 ft). An optical density measurement system (source and photocell) is mounted in the vent pipe of the apparatus. Samples of products of combustion can be taken downstream of the photometer. The "upside-down" configuration of this test apparatus is not often found in real-world situations except with combustible ceiling coverings.

- *ASTM E 1321*, **Standard Test Method for Determining Material Ignition and Flame Spread Properties.**—This method tests for the ignition and flame spread properties of a vertically oriented fuel surface when exposed to an external radiant heat flux. The results can be used to calculate the minimum radiant heat flux and temperature needed for ignition and flame spread. The test configuration is a large, vertically mounted specimen and a gas-fired radiant panel heater mounted at an angle to it. A gas pilot flame is used, and the rate and distance of flame spread along the length of the specimen are recorded. (See Figure 4.28.)

- *ASTM E 800*, **Standard Guide for Measurement of Gases Present or Generated during Fires.**—This guide describes methods for properly collecting and preserving combustion gas samples during fire tests and analytical methods for O_2, CO, CO_2, N_2, HCl, HCN, NO_x, and SO_x, using gas chromatography, infrared, or wet chemical methods. As we have seen, some fuels produce high concentrations of toxic or irritant gases under various fire conditions. Testing materials known to be present in a fatal fire may yield clues as to what role their combustion gases played in the fatality.

- **California TB 603 Standard for Mattresses, Box Springs, and Futons.**—After January 1, 2005, the state of California required that bedding materials comply with new testing procedures. The test is designed to simulate real-world flaming ignition sources and measure the energy released from burning bedding materials. The top and side of the mattress are exposed to gas burners for a period of time (as in Figure 8.3). Failure of the test can be either (1) a peak heat release rate of 200 kW or greater within 30 min after ignition or (2) a total heat release of 25 MJ or greater within 10 min of ignition. As of July 1, 2007, all mattresses sold for residential use in the United States must have passed the 16 CFR 1633 test, which is similar, except the total heat released should not exceed 15 MJ.

◆ 8.3 FLASH AND FIRE POINTS OF LIQUIDS

Tests for flash and fire (flame) points of ignitable liquids can be as simple as placing a few drops of the fuel in a petri dish or watch glass at room temperature and waving a lighted match near the surface of the liquid. Ignition will confirm that the material is a Class I A or B Flammable Liquid with a flash point below room temperature [23°C (75°F)]. More reliable testing requires more controlled evaporation and confinement of the vapors with a more reproducible ignition source. These tests are briefly described next. Flash point testing in general involves taking a quantity of liquid fuel and placing it in the cup of a water bath whose temperature is gradually raised. The cup may be open to the atmosphere or closed with a small shutter. A very small flame is introduced periodically, and the cup is observed for the presence of a brief flash of flame. The fire point of many liquids may be established by the open-cup method by increasing the temperature until a self-sustaining flame is established rather than just a "flash" of flame.

Closed-cup testers retain some of the vapor produced and make it easier to ignite. Closed-cup flash points are typically a few degrees lower than open-cup determinations of the same fuel.

FIGURE 8.3 ◆ The TB 603 test in use on a mattress. All mattresses sold in the United State. for residential use after July 1, 2007, have to pass the 16CFR1633 test, which has a total heat release of 15 MJ, compared with the 25-MJ limit of this test. *Courtesy of Bureau of Home Furnishings.*

- *ASTM D 56*, **Standard Test Method for Flash Point by Tag Closed Cup Tester.**— Applicable to low-viscosity liquids with flash points below 93°C (200°F). Requires 50 mL of liquid for each test. (Sometimes abbreviated TCC.)
- *ASTM D 92*, **Standard Test Method for Flash and Fire Points by Cleveland Open Cup.**—Applicable to all petroleum products with flash points above 79°C (175°F) and below 400°C (752°F) except fuel oils. Requires at least 70 mL for each test. (Sometimes abbreviated COC.)
- *ASTM D 93*, **Standard Test Methods for Flash Point by Pensky–Martens Closed-Cup Tester.**—Applicable for petroleum products with a flash point in the range of 40°C–360°C (104°F–680°F) including fuel oils, lubricating oils, suspensions, and higher-viscosity liquids. Requires at least 75 mL of fuel for each test.
- *ASTM D 1310*, **Standard Test Method for Flash Point and Fire Point of Liquids by Tag Open-Cup (TOC) Apparatus.**—Applicable for liquids with flash points between −18°C and 165°C (0°F and 325°F) and fire points up to 165°C (325°F) (*note:* Will work at

subambient temperatures). Fire point criterion: when fuel ignites and burns for at least 5 s. Requires 75 mL of sample for each test.

- *ASTM D 3278*, **Standard Test Method for Flash Points of Liquids by Small-Scale, Closed-Cup Apparatus.**—Suitable for flash point determination of fuels with a flash point between 0°C and 110°C (32°F to 230°F) in small quantities (2 mL required for each test). Employs a microscale apparatus sold as the Setaflash Tester. Test methods similar to those for ISO 3679 and 3680 (*note:* Will work on subambient determinations).
- *ASTM D 3828*, **Standard Test Method for Flash Point by Small-Scale Closed Tester.**— Similar to ASTM D 3278. Used to establish whether a product will flash at a given temperature, using 2–4 mL of sample for each test.

Owing to the multiplicity of flash point test methods, the ASTM (2000) offers a *Standard Test Method for Selection and Use of ASTM Standards for the Determination of Flash Points of Chemicals by Closed-Cup Methods* (ASTM E 502-84).

◆ 8.4 CALORIMETRY

Classical **bomb calorimetry** is used to measure the total heat of combustion of fuels (as in ASTM D 2382). The term *bomb* refers to a sealed vessel that can withstand internal pressures. The procedure involves burning a weighed specimen in a sealed container with an excess of oxygen until the specimen is completely oxidized. The heat generated by the combustion is measured by measuring the increase in temperature of the sealed "bomb" and using its specific heat and mass to calculate the heat released in the combustion. This test is unsuitable for fuels comprising multiple substances, and it yields the *total* heat of combustion, not the *effective* heat of combustion (which allows for losses due to incomplete combustion).

Because almost all common fuels yield nearly the same amount of heat per mass of oxygen consumed (13 kJ/g), one could easily calculate the heat generated by measuring the amount of oxygen consumed in the combustion process. This procedure is called **oxygen consumption** (or depletion) **calorimetry**. If a specimen is burned in air and the waste products are all drawn into a system of ducts, the O_2, CO, and CO_2 concentrations can be measured. If the ventilation rate into the test chamber and through the exhaust duct is measured, the amount of oxygen consumption can be measured. Optical sensors in the ductwork can also monitor the optical density and obscuration of smoke products generated. Note that both the oxygen consumption and the heat release rate are calculated from data collected.

This approach has been refined into four general categories of size and application: cone, furniture, room, and industrial calorimeters. In the **cone calorimeter,** developed by Babrauskas at NIST and described in *ASTM E 1354*, a 10×10=cm (4×4 =in.) specimen in a metal tray is exposed to a uniform incident radiant heat flux from a cone-shaped electric heater mounted above it (as in Figure 8.4).

The radiant flux striking the sample can be controlled by adjusting the temperature of the heater element. A small electric arc source is introduced to ignite the plume of smoke generated by the heating and then removed. The flame and smoke vent upward through the center of the cone heater, and the gases are ducted to where the flow rate, O_2, CO, and CO_2 concentrations and smoke obscuration are measured. The entire specimen tray is mounted on a sensitive electronic balance, so the mass loss rate is continually calculated. The analysis calculates the heat release rate (expressed per unit surface area of fuel), mass loss rate, and effective heat of combustion. Even though the samples

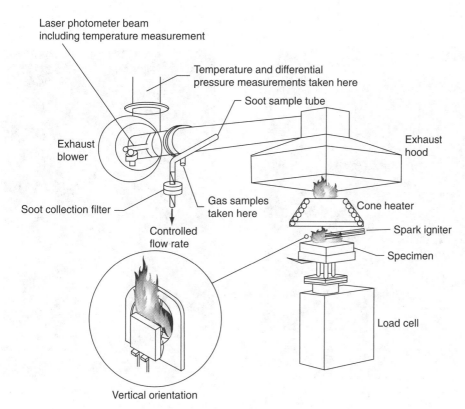

Laser photometer beam
including temperature measurement

Temperature and differential
pressure measurements taken here

Soot sample tube

Exhaust
blower

Exhaust
hood

Soot collection filter

Gas samples
taken here

Cone heater

Controlled
flow rate

Spark igniter

Specimen

Load cell

Vertical orientation

FIGURE 8.4 ◆ In the cone calorimeter, developed by Babrauskas at NIST and described in *ASTM E* (1354, a) 10 × 10-cm 4 × 4-in. specimen in a metal tray is exposed to a uniform incident radiant heat flux from a cone-shaped electric heater mounted above it. *Courtesy of Dr. Vytenis Babrauskas, Fire Science & Technology, Issaquah, Washington.*

tested are reduced scale, testing has shown that the results have a high degree of correlation to real-world fire performance (Babrauskas 1997). The system will work with liquid or solid fuels (even low-melting-point thermoplastic materials) in the horizontal configuration. Wood and other rigid fuels can also be tested in the vertical configuration.

Furniture calorimeters were developed to test the heat release rate of single items of full-scale real furniture by the same method. Here, the furniture is mounted on a load cell and burned in the open, with all combustion products drawn into a vent hood. The flow rate, O_2, CO, and CO_2 concentrations are measured in the ductwork. The calculated heat release rate can then be combined with the mass loss data from the load cell to calculate the effective heat of combustion. Examples are shown in Figure 8.5. A variation, called a LIFT apparatus, tests wall and ceiling coverings by igniting them as vertical wall or corner configurations.

When multiple pieces of furniture, along with carpet, wall linings, and other fuels, are to be evaluated as a fuel package, the **room calorimeter** is used. In this case a room is built (often to a standard size of 2.4 × 3.6 × 2.4 m (8 × 12 × 8 ft), and the entire room becomes the "collector" for the exhaust vent. The combustion products vent through the door opening and are drawn up into the exhaust hood and measured. This arrangement allows heat release rate to be measured constantly as fire spreads from one item to another, even culminating in near-flashover conditions.

The largest-scale devices are often called "industrial" calorimeters. These basically comprise a giant exhaust hood and appropriate ducting and instrumentation and are located in a high-bay building. Typical measuring capacities are 12 kW (cone calorimeter), 1.5–5 MW (furniture calorimeter), 3–6 MW (room calorimeter), and 10–50 MW

(a)　　　　　　　　　　　　　　　(b)

FIGURE 8.5 ◆ (a) The furniture calorimeter at ATF's Fire Research Lab is capable of measuring fires up to 1 MW. *Courtesy of ATF/FRL.* (b) The 4-MW calorimeter at ATF's Fire Research Lab (being calibrated here) is one of the largest in the world. *Courtesy of ATF/FRL.*

(industrial calorimeter). The latter would be most useful for performing fire scene reconstructions; unfortunately, most of these devices are owned by organizations (e.g., UL, ATF) that do not offer commercial testing services for fire reconstruction. A new unit is being constructed at Southwest Research Institute Research Institute (San Antonio, Texas) that will be available for commercial fire reconstruction projects.

◆ 8.5 FURNISHINGS

Tests for ignitability of upholstered furniture vary widely, from cigarette ignition to small flame to larger gas-fired burners of known heat output (typically 17–40 kW). They may be applicable to standardized, small-scale mock-ups of furnishings to the actual items taken from the production line. Pass/fail criteria may be based on observable flame spread, penetration, percentage mass loss, peak heat release rate, quantity of smoke produced, or toxicity of fire gases.

TEST METHODS

Various test methods in use today include those described under the U.S. Code of Federal Regulations (CFR), British Standards (BS), International Standards Organization (ISO), ASTM, and California Bureau of Home Furnishings Technical Bulletins (TB). Krasny, Parker, and Babrauskas (2001) have assembled a comprehensive description and discussion of the current techniques, summarized in Table 8.1. See also *Kirk's Fire Investigation,* 6th ed., for additional information (DeHaan 2007).

TABLE 8.1 ◆ Summary of Furniture Item Flammability Tests					
Test	*Cigarette Ignition*	*Flame Ignition*	*HRR*	*Smoke*	*Toxicity*
ISO 8191	X	X			
BS 5852	X	X			
ATSM E 1352	X				
ASTM E 1353	X				
NFPA 261	X				
NFPA 260	X				
California TB116, 117	X				
Upholstered Furniture Act Council	X				
BIFMA	X	X			
CFR 1632	X				
BS 6807	X	X			
Calif. TB 133		X	X	X	X
Calif. TB 129		X	X	X	X
Calif. TB 121		X		X	X
ASTM E 1537		X		X	X
ASTM E 1590		X		X	X
ASTM E 162		X			
ASTM D 3675		X			
ISO 5660			X	X	X
ASTM E 1474			X	X	X
ASTM E 1354			X	X	X
NFPA 264A			X	X	X
ASTM F 1550M				X	
ISO TR 5924				X	
ASTM E 662				X	
ASTM E 906			X	X	
ISO TR 9122					X
ISO TR 6543					X
ASTM E 1678					X

Source: Derived from Krasny, Parker, and Babrauskas 2001.

As with the bench-scale tests described earlier, there are limitations and cautions regarding applying the results of these tests to the reconstruction of "real" fires. Cautions include requirements for conditioning samples for 24–48 hr at a particular temperature and humidity. The investigator should be aware of what influence such variables might have on the final result. The geometry of the test compartment is also important. The same piece of furniture may yield different heat outputs or fire patterns if tested in a corner or against a wall versus in the center of a large room that minimizes radiant feedback or ventilation effects.

The location, manner, means, and duration of ignition can play a significant role in the fire performance of furnishings. A vinyl-covered chair seat, for instance, can provide substantial resistance to small ignition sources if an ignition source is placed on the seat (PVC upholstery tends to char, intumesce, and shield the substrate). But even a small flaming source placed under the seat where the flame can contact the polyurethane padding will readily ignite the chair.

CIGARETTE IGNITION OF UPHOLSTERED FURNITURE

Cigarette ignition of upholstered furniture was a major fire cause for decades owing to the predominant use of cotton or linen upholstery fabrics over cellulosic padding of cotton, coir, kapok, or similar materials, all of which were easily ignited by a glowing cigarette.

Holleyhead (1999) published an extensive review of the literature regarding cigarette ignition of furniture. A great deal of misinformation once existed, however, about the time required for such ignition to result in flaming combustion. Investigators erroneously concluded that such ignitions always required 1–2 hr, inferring that any flaming ignition in a shorter time must have been the result of deliberate ignition by open flame.

Extensive testing by the California Bureau of Home Furnishings and NIST (then NBS), first reported in the early 1980s, demonstrated that ignition time was variable. Transition to flaming combustion was reported in as little as 22 min from the placement of a glowing cigarette between the cushions or between the cushions and an arm or backrest. Others required 1 to 3 hr, and some never transitioned to open flames and smoldered for hours before self-extinguishing. Even identically built mock-ups demonstrated a wide variety of behaviors. A statistical analysis of these results was published by Babrauskas and Krasny (1997)

Tests by the authors have demonstrated similar time and ignition results. If the fabric is torn and the cigarette drops into direct contact with a non–flame retardant–treated cotton padding, transition to flame has been observed in as little as 18 min in still air. The presence of drafts can influence the progress and increase the heat release rate of smoldering cellulosics. One informal test by the author with a cigarette wrapped loosely in a cotton towel and left outside in a light, variable wind resulted in flames about 15 min after placement. The mechanism for transition from smoldering to flame has so far defied modeling or accurate predictions. Hand-rolled cigarettes present a lower ignition risk because they tend to self-extinguish much more than do commercially made tobacco cigarettes. Several states have recently passed legislation mandating "fire-safe" or self-extinguishing cigarettes. These will reduce chances for cigarette ignition of upholstered materials in the future. Fire-safe cigarettes typically use a "banded" paper with alternately high and low porosity in the bands, causing the smoldering to cease.

Latex (natural) rubber foam is also easily ignited by contact with a glowing cigarette. It should be noted that the cigarette itself will be consumed in a maximum of 20–25 min after being lit (somewhat longer if tucked between seat cushions). Ignition times longer than that obviously result from a self-sustaining smolder in the surrounding fuel initiated by the cigarette, as confirmed by the significant quantity of white smoke that is produced for some minutes prior to flame (far more than can arise from a single cigarette).

◆ 8.6 ADDITIONAL PHYSICAL TESTS

SCALE MODELS

A great deal of useful information can be obtained from building scaled-down replicas of rooms or furnishings for testing, at reduced cost and complexity compared with full-scale re-creations. These models are particularly valuable for testing ignition and incipient fire hypotheses (in which the interaction of room surfaces or conditions does not play a major role).

When radiant heat or ventilation factors are involved, the test must take into account that all factors in a fire do not scale down in the same linear fashion as the room dimensions. For instance, ventilation through an opening is controlled by the area of the opening and the square root of its height. It is possible to calculate corrections for such factors, but it is not as simple as building a doll house, with all doors and windows in the same proportion as in full scale. Radiant heat flux falls off with the inverse square ($1/r^2$) of the separation distance, so scale models must keep those relationships in mind as well.

Floor or wall materials reduced in thickness to scale proportions may respond to thermal insult as "thermally thin" targets if they are less than a few millimeters thick and may respond differently than "thick" sections of the same material. Speed of smoke movement and time to flashover will also not be "to scale." Special materials may be selected for models of walls and ceiling so their thermal behavior will duplicate, in scale, that of gypsum or plaster walls.

FLUID TANKS

A great deal of the information used to create and test the models of smoke and hot gas movement discussed in previous chapters was developed in **fluid tanks** with no fire or smoke at all. Because movement of smoke in a building is driven by the buoyancy of the hot gases versus the density of normal room air, the same process can be simulated by using two liquids of different densities (such as fresh and salt water). If one fluid is dyed, its movement and mixing can easily be observed and recorded.

Often, a scale (one-eighth to one-fourth) model of the room(s) is built of clear Plexiglas™ and immersed and inverted in a large tank of the lighter (less dense) liquid. The heavier liquid is then introduced into the model in a manner to simulate the generation of gases from a fire in the room. Scaling factors are important, and the relative densities and viscosities of the liquids have to be calibrated to simulate the mixing of gases in a buoyant flow, but the technique can provide answers to some investigative problems and confirm or reject computational models and hypotheses (Fleischmann, Pagni, and Williamson 1994).

FIELD TESTS

Identification of the first fuel ignited is critical to the reconstruction process. If the first fuel is more easily susceptible to open flame ignition than to glowing or hot surface ignition (such as natural gas, gasoline vapors, or polyethylene plastic), this helps focus the scene search on possible ignition sources of the appropriate type. Cellulosic fuels or other natural materials such as wool are much more easily ignited by hot surface sources like discarded cigarettes or glowing electrical connections. Most such materials can be reliably identified by visual inspection, but information as to their behavior when ignited is much more "on point" for the fire investigator.

A simple **flame** or **ignition susceptibility test** using a match or lighter applied to one corner while the small test sample is held in still air will reveal how readily it will ignite and whether it will support flaming combustion when the ignition source is removed, as described in *NFPA 705* (NFPA 1997); however *NFPA 701* (NFPA 2004a) is now considered the preferred practice. Under these conditions, cellulosic materials will support a yellow flame with a gray smoke. When the flame is blown out, these materials tend to support glowing combustion. The ash left behind will be gray to black in color and powdery or crumbly in texture. There will be no hard-melted droplet of residue. Thermoplastic synthetic materials will melt as they burn, so melted, burning droplets will result. They also tend to shrink and shrivel as they melt and often burn with a blue-based flame, with smoke ranging from negligible to white (polyethylene) to heavy and inky black (polystyrene). These materials do not support smoldering combustion.

Thermosetting resins are usually more difficult to ignite than other fuels. They tend to smolder when flame is removed, produce an aggressive smoke, and leave a hard, semiporous residue. Elastomers (rubbers) can be either natural (latex) or synthetic and can behave like either. Some will burn very readily; others, like silicone rubber, burn much more reluctantly. Silicone rubber leaves a brilliant white, powdery ash, whereas other elastomers leave a hard, dark porous mass. Samples of any unknown material suspected of being the first fuel ignited should be taken if there is any doubt of their contribution to the ignition or growth of the fire.

Caution should be exercised in both conducting and interpreting such informal ignition tests. Note that the *NFPA 705* test has been criticized as being subjective and variable owing to the procedures and personnel involved. There is also a risk of injury to personnel and a reported cause of serious accidental fires. A strong suggestion is to use *NFPA 701, Methods of Fire Tests for Flame Propagation of Textiles and Films*, when conducting flame or ignition susceptibility tests (as in Figure 8.6).

The presence of fire retardants may significantly affect ignition and flame spread. It should also be noted that materials like carpet are much more easily ignited from a corner or edge than if the same ignition source is applied to the center of the specimen. Ignition and flame spread are enhanced by vertical sample orientation and may be retarded by draft or moisture in the sample, so the results should be noted but not taken as proof of identity or of actual fire behavior (NFPA 1997).

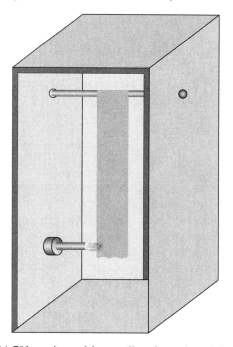

FIGURE 8.6 ◆ The *NFPA 701* test is used for small-scale testing of the flammability of fabrics. The *NFPA 701* apparatus uses a free-hanging vertical strip [150 × 400 mm/ (6 × 16 in.)] of the fabric to be tested in an enclosed test chamber. The flame from a laboratory burner is applied at the bottom of the strip for 45 s and then removed. The flame spread behavior of the fabric is observed, and after extinguishment, the residue is weighed.

FULL-SCALE FIRE TESTS

Considerable knowledge has been gathered from fire tests conducted in real buildings slated for demolition. It is rare that the dimensions, ventilation, and construction materials of a test building will match those conditions of a specific fire that is being studied, but a great deal of reliable knowledge about fire behavior has been gleaned from these tests. Aside from physical suitability, there are often issues of fire exposure to surrounding properties, environmental concerns, logistics of access, and limitation of fire protection services. If such buildings are used, provisions should be made for multiple video camera viewports and the inclusion of thermocouple and radiometers as described next.

One author found a house nearly identical to one in which six fire fatalities occurred. The test house had been scheduled for demolition. It was repaired and refurnished to duplicate the initial fire scene. Thermocouples, gas analyzers, and video cameras were used to capture fire tests with accelerated and nonaccelerated ignitions. The test data revealed that a nonaccelerated fire could have been responsible (DeHaan 1992).

Because of a number of training-fire deaths in the United States, fire departments have been considerably more reluctant to expose their personnel to the risks of large fires in original buildings. The majority of fire personnel recognize the value of training firefighters in such structures and will plan "training burns" in accordance with the safety provisions outlined in *NFPA 1403—Standard on Live Fire Training Evolutions* (NFPA 2002a). The investigator should be aware that those provisions include prefire venting of ceilings and roofs, removal of most doors and all glass, and limited fuel loads (no ignitable liquids or gases). The resulting fire behavior and, most important, fire patterns therefore do not generally conform to those features of fires in buildings with normal furnishings, doors, glass windows, and intact ceilings and roofs. It is suggested that fire investigators not attempt to use such firefighting exercises to gather behavior/pattern data.

Full-scale rooms that duplicate particular conditions and control other variables can be purposely built for fire tests. Guidance for room fire experiments is given in *ASTM E 603, Standard Guide for Room Fire Experiments* (ASTM 2003a). This guide addresses assembling tests that will be used to evaluate the fire response of materials, assemblies, or room contents in real fire situations that cannot be evaluated in small-scale tests. Provisions for measuring the optical density of smoke, temperatures, and heat fluxes in the compartment are described. The documentation and controls necessary are also described.

◆ 8.7 CASE EXAMPLE 1: CUBICLE CONSTRUCTION

Effective room mock-up cubicles can be assembled at low cost for full-scale tests. Based on a design by Mark Wallace, these are basically four 2.43 × 2.43-m (8 × 8-ft) wood-framed panels (typically wood studs on 61-cm (24-in.) centers with gypsum wallboard with a similar panel resting on top to form a ceiling (see Figure 8.7).

A larger unit can be constructed easily, with the plans shown in Figure 8.8. The design provides for a "knockdown" wall that can be removed easily after the test to facilitate documentation.

Doors and windows can be cut as needed. A header must be left above each door 25–45 cm (10–18 in. deep). Cubicles are easily erected on four wood pallets covered with 13-mm (0.5-in.) sheet plywood or on (2 × 4- or 4 × 4-in.) joists. Since these are

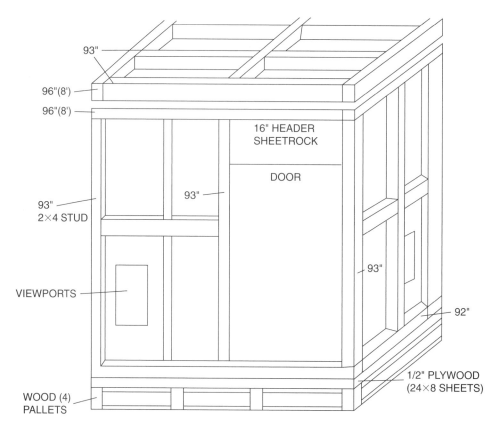

FIGURE 8.7 ◆ Effective room mock-up cubicles can be assembled at a low cost for full-scale tests. *Courtesy of J. D. DeHaan.*

intended for short-duration (<30 min) tests, lack of insulation will not usually affect the results of interior fires. Electrical outlets can easily be added.

Viewports are easily added by cutting holes of approximately 25 × 30 cm (10 × 12 in.) in one or more walls near floor level. An unframed piece of ordinary window glass is then glued to the *interior* surface of the wall using silicone sealant (and allowed 24 hr to harden before the fire). The absence of a frame means that the entire sheet of glass is exposed to the same heat flux and will not develop the same stresses that cause failure before flashover. Experience reveals that such windows will usually last until flashover. Heat-resistant glass such as that used in fireplace screens or over doors can ensure integrity postflashover if that is desired.

Thermocouples can easily be added through small holes drilled through the gypsum where desired. A minimum of three thermocouples on one wall is suggested: one near the ceiling, one midlevel, and one approximately 15 cm (6 in.) above the floor at a point well away from the door or other vents. A second set can be installed on the opposite side of the room, or additional thermocouples can be placed inside the door header (to record flameover) or above target fuel packages. The smallest-diameter thermocouples that are practicable should be used, since larger wire gauges lead to systematic depression in the measured temperatures. A size of 0.005 in. is often suitable. The finer the diameter, however, the more prone the thermocouples are to breakage. Data are best captured on a datalogger (such as an Omega 550), although they can also be captured on a video camera monitoring digital outputs of several individual pyrometers at once for later manual transcription.

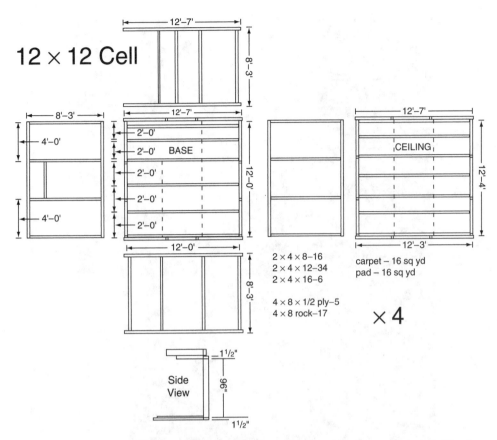

12 × 12 Cell

BASE

CEILING

2 × 4 × 8–16
2 × 4 × 12–34
2 × 4 × 16–6

4 × 8 × 1/2 ply–5
4 × 8 rock–17

carpet – 16 sq yd
pad – 16 sq yd

× 4

Side View

FIGURE 8.8 ◆ Example plans for a larger test cubicle unit that can be constructed easily. The design provides for a knockdown wall that can be removed easily after the test to facilitate documentation. *Courtesy of Mike O'Brien, Chief, Suisun City Fire Department, Suisun City, California.*

Such cubicles allow creation of environments that closely simulate real rooms [additional 1 × 2.5-m (4 × 8-ft) segments can be added for larger rooms] while keeping safety risks to a minimum (no risks of toxics, asbestos, structural collapse, etc.). See Figure 8.9 for an example of a postflashover effect created in such a test cubicle. One entry wall can be framed separately so it can be easily opened and laid flat on the ground after the fire for ease of examination and photography. Video recordings should be made of all tests using multiple cameras and synchronized time clocks where possible. A visual and audio cue should accompany ignition so that it marks $t =$ 0 on all recordings simultaneously. It should be remembered that the small size 2.5 × 2.5 m (8 × 8 ft) will produce flashover in less time (approximately 30 percent less) than a typical 3 × 4-m (10 × 13-ft) room given the same fuel load.

Scientifically valid small-scale tests can be conducted to test flame spread ignitability without special equipment as long as possible roles of the test variables are taken into account. For instance, carpet or furniture can be tested in the open, with the realization that it would burn more quickly (producing greater heat release rates) if tested in a compartment because of the radiant heat reflected back from walls and ceiling.

Carpets need to be secured to the floor in the same manner as in the actual structure. Synthetic carpets shrink and curl as they burn. If the edges thus lift, the combustion of the carpet will be seriously enhanced and distort the heat release rate and fire patterns. Carpets must be backed with the same pad as during the fire being duplicated,

FIGURE 8.9 ◆ Results of postflashover fire in a furnished 2.4 × 2.4-m (8 × 8-ft) office cubicle. Note extensive destruction of synthetic carpet and combustion of drywall paper on right. Glass viewport in right-rear corner failed prior to flashover. Note clean-burn ventilation pattern extending up the rear wall and extensive damage to carpet beneath the opening. There was only carpet and pad in the right-rear corner. *Courtesy of J. D. DeHaan. Tests courtesy of California Association of Criminalists and Huntington Beach Fire Department.*

since interaction between the pad and the carpet can cause self-sustaining combustion although neither material might sustain flame spread by itself.

The ignition source used should duplicate the one known to be involved, or various kinds should be tested. For instance, polypropylene carpet will resist ignition by a single dropped match at room temperature, as it will pass the required methenamine tablet test; however, a slightly larger ignition source such as a burning crumpled sheet of newspaper or any additional external heat flux and the carpet will self-sustain combustion. Once ignited, such carpet will sustain combustion at rates up to approximately 0.5–1 m²/hr (5–11 ft²/hr) with small flames about 5–7 cm (2–3 in.) tall.

If a fire test is conducted in a compartment, care must be taken that the major fuel packages are located in the same position in each test (or in the same position as in the scene being replicated). Packages in corners will burn differently than packages against the wall or in the middle of the room. A large fire located away from an open door will burn differently than the same fire set near the door (Figure 8.10).

(a)

FIGURE 8.10 ◆ (a) Interior of 2.4 × 2.4-m (8 × 8-ft) cubicle after a 20-min fire. Bed on right was ignited by open flame to clothing on top and required nearly 16 min to be completely involved. The room went to flashover less than 1 min later, igniting carpet and chair in left-rear corner (with pig haunch and legs). Note the extensive combustion of the corner of the bed nearest the door (ventilation effect). Glass viewport in rear corner failed prior to flashover. *Courtesy of Special Agent Mike Marquardt, CFI, ATF, Grand Rapids, MI.*

(b)

FIGURE 8.10 ◆ (b) View from doorway of same compartment after removal of bed, chair, and remaining carpet from left side of compartment. Note exposed areas of carpet and 13-min (0.5-in.) plywood floor consumed in less than 4 min of postflashover fire and top-down burning of 5 × 10-cm (2 × 4 = in.) wood floor joists. No accelerants were used. *Courtesy of Special Agent Mike Marquardt, CFI, ATF, Grand Rapids, MI. Tests courtesy of the Iowa Chapter of the IAAI.*

◆ 8.8 CASE EXAMPLE 2: IOWA CUBICLE TESTS

Examples of the valuable data that can be obtained from cubicle tests are those obtained from tests conducted September 25, 2002, in Waterloo, Iowa. In these tests, the Iowa Chapter of the International Association of Arson Investigators (IAAI) and the Bureau of Alcohol, Tobacco and Firearms (ATF) built two cubicles of nearly identical fuel loads.

FIGURE 8.11 ◆ Prefire view of the furnished 2.4 × 2.4-m (8 × 8-ft) cubicle with drywall walls and ceilings. The fire was ignited on the skirt of the chair on the left. *Photo courtesy of Special Agent Mike Marquardt, CFI, ATF, Grand Rapids, MI. Test courtesy of the Iowa Chapter of the IAAI.*

Each cubicle contained a synthetic upholstered chair, a wood dresser, curtains, a kitchen chair, and synthetic carpet with polyurethane padding. The first cubicle was ignited by an open flame to the skirting of one chair. This fire required 10.5 min to grow to flashover. The second cubicle was ignited with gasoline across the middle of the floor, with flashover occurring in approximately 70 s. Both cubicles experienced the same maximum temperatures and same damage to the carpet and wood floor. The tests also served as a good example of physical reconstruction and data gathering.

Figures 8.11–8.21 detail the Iowa cubicle tests.

◆ 8.9 CASE EXAMPLE 3: LARGE-SCALE TESTS

The most sophisticated (and most expensive) form of full-scale testing is the re-creation of an entire compartment in a laboratory with calorimetry, radiometry, continuous gas sampling, and video monitoring. An excellent example of this was the re-creation of the entire seating bay of the Stardust Disco, conducted at the Fire Research Station, Gaston, UK (FRS 1982).

The single-component or furniture mock-up tests described previously are useful for gauging ignitability and fuel properties, but only a full-scale reconstruction (such as the Stardust test) can reveal the complex interactions between the growing fire and the various fuels that, in that case, resulted in rapid growth to flashover of a small initial fire.

There are very few facilities in the world that can provide the resources necessary. NIST, Factory Mutual Engineering, Underwriters Laboratories, California Bureau of Home Furnishings (Department of Consumer Affairs), Aberdeen Proving Grounds,

FIGURE 8.12 ◆ Postfire view after 13 min, which was postflashover for 5 min. *Photo courtesy of Special Agent Mike Marquardt, CFI, ATF, Grand Rapids, MI. Test courtesy of the Iowa Chapter of the IAAI.*

FIGURE 8.13 ◆ Pattern on floor showing intense charring of plywood and irregular destruction of carpet and pad, most extensive at the door. *Photo courtesy of Special Agent Mike Marquardt, CFI, ATF, Grand Rapids, MI. Test courtesy of the Iowa Chapter of the IAAI.*

FIGURE 8.14 ◆ Same cubicle with furniture replaced in its prefire positions. Note the floor-to-ceiling combustion direction patterns on the second chair and dresser. *Photo courtesy of Special Agent Mike Marquardt, CFI, ATF, Grand Rapids, MI. Test courtesy of the Iowa Chapter of the IAAI.*

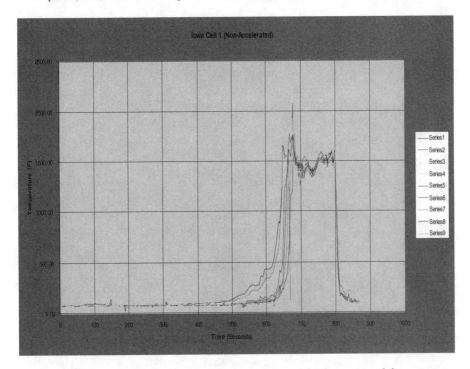

FIGURE 8.15 ◆ Temperature plot for the thermocouple tree in the center of the compartment at 0.3-m (1-ft) intervals. Note postflashover temperatures of approximately 816°C (1500°F). *Photo courtesy of Special Agent Mike Marquardt, CFI, ATF, Grand Rapids, MI. Test courtesy of the Iowa Chapter of the IAAI. Data courtesy of David Sheppard, Fire Protection Engineer, ATF Research Laboratory, Ammendale, MD.*

FIGURE 8.16 ◆ Prefire views of cubicle furnished in the same fashion as previously. The thermocouple tree is in the center of the cubicle. *Photo courtesy of Special Agent Mike Marquardt, CFI, ATF, Grand Rapids, MI. Test courtesy of the Iowa Chapter of the IAAI.*

FIGURE 8.17 ◆ Accelerated cubicle fire approaches flashover temperature at 60 s. *Photo courtesy of Special Agent Mike Marquardt, CFI, ATF, Grand Rapids, MI. Test courtesy of the Iowa Chapter of the IAAI.*

FIGURE 8.18 ◆ Postfire view after 4 min of gasoline-accelerated fire. The ceiling and front drywall collapsed from fire-fighting action. Note protected areas on walls behind both large chairs. *Photo courtesy of Special Agent Mike Marquardt, CFI, ATF, Grand Rapids, MI. Test courtesy of the Iowa Chapter of the IAAI.*

FIGURE 8.19 ◆ Floor patterns indistinguishable from that of nonaccelerated fire. Note destruction of carpet and padding and charring of floor in areas toward door where no accelerant was poured. *Photo courtesy of Special Agent Mike Marquardt, CFI, ATF, Grand Rapids, MI. Test courtesy of the Iowa Chapter of the IAAI.*

FIGURE 8.20 ◆ Reconstruction of furniture placement. Note fire patterns on chairs outward from the rear center of the floor, where the accelerant was poured. *Photo courtesy of Special Agent Mike Marquardt, CFI, ATF, Grand Rapids, MI. Test courtesy of the Iowa Chapter of the IAAI.*

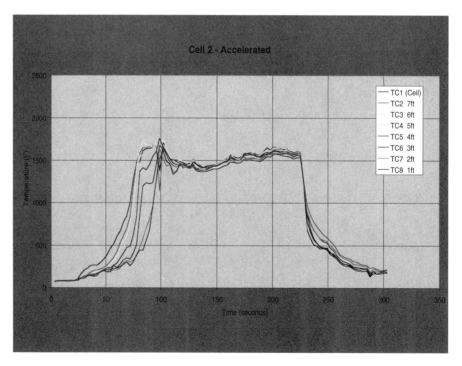

FIGURE 8.21 ◆ Temperature plot for thermocouple tree. Note ultrafast fire growth to flashover at 70–90 s. Postflashover period is approximately 130 s. *Photo courtesy of Special Agent Mike Marquardt, CFI, ATF, Grand Rapids, MI. Test courtesy of the Iowa Chapter of the IAAI. Data courtesy of David Sheppard, Fire Protection Engineer, ATF Research Laboratory, Ammendale, MD.*

Building Research Establishment (Gaston, UK) and Southwest Research Institute all have large burn laboratory and testing facilities. These agencies have provided invaluable testing for many fire investigators over the years, often at no cost to public agencies.

The ATF Fire Research Laboratory (FRL), Ammendale, Maryland, is a scientific fire research laboratory dedicated to supporting many research needs of the fire investigation community. It also houses ATF's National Forensic Laboratory. The FRL was designed by a team of fire scientists, engineers, and fire investigation specialists from NIST, Factory Mutual Engineering, Underwriters Laboratories, Hughes Associates, and the University of Maryland.

An example of a large-scale test, simplified for the purposes of a single demonstration, is shown in Figures 8.22, 8.23, 8.24, and 8.25, corresponding to 3, 8, 11, and 14 min after initiation of a room fire, respectively. The Fire Research Station staff assembled a $2.5 \times 3.75 \times 2.4$-m ($8.2 \times 12.4 \times 8$-ft) re-creation of a living room (lounge) under the 9×9-m (30×30-ft) calorimeter in the Large Burn Hall at the FRS.

The structure was wood frame with ceramic fire insulation board liner (to permit reuse of the basic structure). The interior was lined with gypsum plasterboard, and the floor was covered with vinyl-backed carpet tiles (over a layer of sand to protect the concrete floor of the burn hall). There was a shallow 25-cm (10-in.) header across one side, which was otherwise left entirely open to simulate a "patio room" enclosure. This very large opening would provide enough air to ensure that the fire would not be ventilation limited even if it went to flashover. From the relationship for maximum mass flow of air through an opening of area A_0 and height h_0,

$$\dot{m}_{air} = 0.5A_0\sqrt{h_0} = (0.5)(8.06)(1.46) = 5.88 \text{ kg/s}, \tag{8.1}$$

and since 1 kg of air will support 3 MJ of heat release,

$$\dot{Q}_{max} = (3000 \text{ kJ/kg}^3)(\dot{m}) = 17,640 \text{ kW}. \tag{8.2}$$

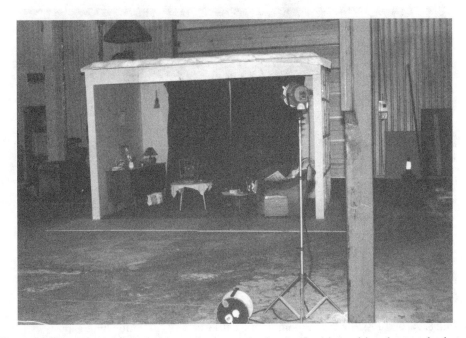

FIGURE 8.22 ◆ Full-scale living room fire test at 3 min after ignition, with only wastebasket and newspapers alight (incipient fire stage). *Courtesy of J. D. DeHaan. Test courtesy of FRS, Building Research Establishment, Garston, UK.*

FIGURE 8.23 ◆ Full-scale living room fire test at 8 min after ignition. Draperies have burned almost completely; drop-down ignites chair and table in far-left corner (growth phase). *Courtesy of J. D. DeHaan. Test courtesy of FRS, Building Research Establishment, Garston, UK.*

FIGURE 8.24 ◆ Full-scale living room fire test at 11 min after ignition. Fire is post-flashover, with maximum heat release rate of 5.2 MW observed at 10.7 min. Carpet fully involved. *Courtesy of J. D. DeHaan. Test courtesy of FRS, Building Research Establishment, Garston, UK.*

FIGURE 8.25 ◆ Full-scale living room fire test at 14 min after ignition. Wood cabinet at left is alight. Filled with the remaining draperies, it will sustain a large fire that grows to 2 MW, then decays. Sofa and chairs are almost completely consumed (decay phase). *Courtesy of J. D. DeHaan. Test courtesy of FRS, Building Research Establishment, Garston, U.K.*

The ventilation limit for this room would be of the order of 15–17 MW. This calculation assumes 100 percent efficiency. At 50 percent, Q_{max} would be 8825 kW (8.8 MW).

The furnishings included non-flame-retardant draperies (no window); wood tables, chairs, and cabinets; a polyvinyl chloride beanbag chair; miscellaneous newspapers and books; and a three-seat sofa of recent production. This sofa was made using flame-retardant fabrics that delayed the growth of the fire substantially.

Three Chromel/Alumel thermocouples were placed in the room to monitor gas temperatures at ceiling and nose height. Data from those thermocouples were logged every second during the test. The fire gases were collected in the calorimeter hood above (not visible in Figure 8.23).

Heat release rate and CO, CO_2, and O_2 gas concentrations were monitored continuously in the calorimeter. Continuous video recordings were made and still photos were taken by observers at 30-s to 1-min intervals. The newspapers in the wastebasket at the far end of the sofa (near the beanbag chair) were ignited by open flame. The heat release rate of this test is shown in Figure 8.26, and the temperatures are shown in Figure 8.27. They can be compared with the development stages of the fire shown in Figures 8.22–8.25.

This fire was allowed to decay to 25 min and the remaining flames were then extinguished by water spray. Owing to the sustained postflashover burning in the room, there were no reliable indicators of duration or direction of propagation remaining on visual inspection.

This demonstration showed that when a single major fuel package such as a living room sofa is flame retardant, the growth to flashover can be delayed considerably (a 1990-vintage sofa would have promoted flashover in a similar room in less than 3 min if ignited by open flame in a similar fashion). Data from such tests can be used to confirm the accuracy of time/condition predictions of models, corroborate witness statements, demonstrate what stage of a fire might have incapacitated or killed a victim trapped in it, or demonstrate the role furnishings or fire protection systems play in preventing future fires.

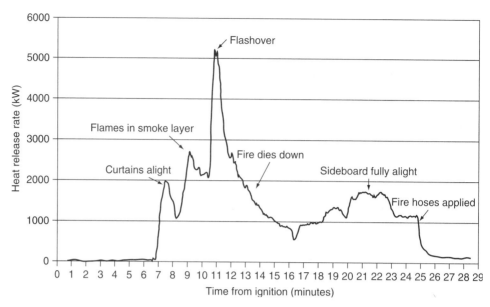

FIGURE 8.26 ◆ Heat release rate recorded in full-scale fire test. *Courtesy of FRS, Building Research Establishment, Garston, UK.*

A series of full-scale fire tests were carried out under the auspices of the U.S. Fire Administration, National Institute of Justice (NIJ), National Association of Fire Investigators (NAFI), and Eastern Kentucky University by Ron Hopkins, Patrick Kennedy, and their coinvestigators. The most recent series concerned the survivability of fire patterns in postflashover tests (Hopkins, Gorbett, and Kennedy 2007). Some of their results are shown in Figure 8.28.

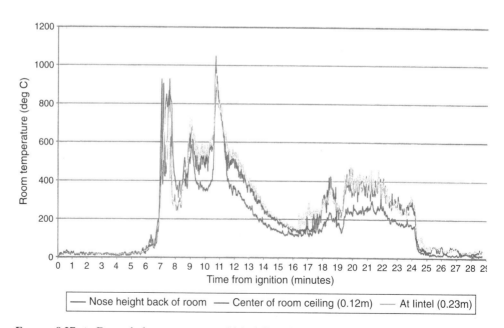

FIGURE 8.27 ◆ Recorded temperatures within full-scale room fire test. *Courtesy of FRS, Building Research Establishment, Garston, UK.*

(a)

FIGURE 8.28 ◆ Room fire tests conducted at Eastern Kentucky University revealed that post-flashover burning of short duration (2–4 min) does not obliterate V patterns and other useful fire patterns. (a) Fire pattern on wall near origin in postflashover bedroom test shows V pattern extending out from origin. Fire started on bed, flashover at 790 s, extinguishment at 1005 s. Postflashover fire duration: 215 s (3.6 min). *Courtesy of Ron Hopkins, TRACE Fire Protection and Safety, Richmond, KY.*

FIGURE 8.28 ◆ (b) Fire pattern on wall near origin in postflashover living room test shows V pattern extending out from origin, and effect of major thermal plume from sofa. Fire started on sofa, flashover at 640 s, extinguishment at 836 s. Postflashover fire duration: 196 s (3.2 min). *Courtesy of Ron Hopkins, TRACE Fire Protection and Safety, Richmond, KY.*

(b)

◆ 8.10 SUMMARY AND CONCLUSIONS

Fire testing includes a tremendous range of tests, from simple field flammability tests such as *NFPA 705,* to bench-scale fabric tests such as ASTM and 16 CFR tests, to full-scale tests in buildings. Useful data from such tests can range from single observations to temperatures and radiant heat fluxes and (thanks to oxygen depletion calorimetry) even heat release rates of large fuel packages. Tests are often carried out to collect more information about basic fire processes. This information is then published in peer-reviewed publications and authoritative treatises, or even on websites. It is then available for use in predicting ignition and fire events. As we will see in the next chapter's Case Studies, such tests can fill a critical role in testing hypotheses about the ignition, spread, and effects of fire.

The criteria for using fire test information in forensic fire scene reconstructions are whether the test is correctly done, whether the data are accurately collected and tabulated, and whether the test is appropriate and applicable to the situation under consideration. Fire tests are designed to collect certain data in a reproducible manner and the investigator must ask if the fuels, conditions, and ignition mechanism reproduce the fire in question. Does the test comport to be under a published test protocol such as one from ASTM or NFPA? If so, does it actually follow that protocol? If not, what factors were considered in its design? What variables were considered and how were they controlled? Fuel moisture, fuel mass and quantity, physical state, ambient temperature, heat flux, and oxygen content all play important roles in ignition, flame spread, and heat release. If it is a "custom-designed" test, what data can be fairly and accurately collected and analyzed? Was there a planned series of tests to examine the sensitivity and reproducibility of the data? Because of the importance fire test data can have in both criminal and civil fire investigations, it is imperative that tests are conducted and data are interpreted without misrepresentation in a balanced, impartial and reproducible manner.

■ ■

Problems

8.1. Find an example of a legitimate fire testing facility in your state. Visit or call this organization and determine what type of testing it performs.

8.2. Obtain a collection of at least 10 tests involving the use of a calorimeter from published research.

8.3. Research the methods of conducting field fire testing.

8.4. Obtain small samples of different clothing or upholstery fabrics from a fabric store (with identified content) and conduct *NFPA 705* ignition tests on each (in a safe location). Collect data and compare observations to descriptions included here.

■ ■

Suggested Reading

DeHaan, J. D. 2001. Full-scale compartment fire tests. *CAC News* (second quarter): 14–21.

——— 2002. Our changing world—Part 2; Ignitable liquids: Petroleum distillates, petroleum products, and other stuff. *Fire & Arson Investigator* 52, no. 3 (April).

——— 2007. *Kirk's fire investigation,* 6th ed., chap. 10. Upper Saddle River, NJ: Prentice Hall.

Holleyhead, R. 1999. Ignition of solid materials and furniture by lighted cigarettes: A review. *Science and Justice* 39 (2): 75–102.

Krasny, J. F., W. J. Parker, and V. Babrauskas. 2001. *Fire behavior of upholstered furniture and mattresses.* New York: William Andrews.

National Fire Protection Association. 1997. *NFPA 705—Recommended practice for a field flame test for textiles and films.* Quincy, MA: NFPA.

National Fire Protection Association. 2000. *NFPA 555—Guide on methods for evaluating potential for room flashover.* Quincy, MA: NFPA.

Urbas, J. 1997. Use of modern test methods in fire engineering. *Fire and Arson Investigator* 47 (December): 12–15.

CHAPTER 9

Case Studies

Mr. Mac, the most practical thing that you ever did in your life would be to shut yourself up for three months and read twelve hours a day at the annals of crime. Everything comes in circles.... The old wheel turns, and the same spoke comes up. It's all been done before, and will be done again.

Sir Arthur Conan Doyle, "The Valley of Fear"

The purpose of this chapter is to introduce the concept of detailed case histories as a learning tool. As with any science, particularly that of forensic fire scene reconstruction, documented case histories play a large role in introducing underlying concepts, expanding the knowledge of fire science, and suggesting new areas of research to explain the phenomena documented. Examples are offered to demonstrate how data collected in controlled tests can be of value in formulating or testing hypotheses.

The cases presented here reflect many of the concepts raised previously in this textbook. Although many principles are discussed, the case analyses are not all conclusive, leaving the reader to apply additional concepts in reviewing, discussing, or expanding on the case histories.

◆ 9.1 CASE EXAMPLE 1: UNUSUAL FIRE DEATH

At approximately 2:30 P.M., the fire department responded to a "smoke showing" call from an apartment complex just 200 yd from their station. Discovering smoke coming from the closed door of one apartment and there being no response to a knock, the firefighters forced the door open. The movement of the door was limited by an object lying on the tile entry area behind it. A limited flaming fire was observed on this object.

The fire was quickly extinguished with a mist hose stream. No further fire damage was found to the rooms or their furnishings, as shown in Figure 9.1. The burning object turned out to be the body of the apartment's occupant, a middle-aged female (Figure 9.2).

The victim's body had suffered considerable fire damage to the upper-left torso, shoulder, and upper arm, with penetration through skin and subcutaneous body fat to underlying connective tissue in the affected areas. The body had apparently been lodged against the base of the wood entry door, and an inverted V pattern of char

FIGURE 9.1 ◆ This fire was quickly extinguished with a mist hose stream. Soot deposits are visible on the walls and ceiling. No further fire damage was found to the rooms or their furnishings, except for limited charring to the inside of the door. *Courtesy of Santa Ana Fire Department, Santa Ana, California.*

FIGURE 9.2 ◆ The victim's body suffered considerable fire damage and was found lodged against the base of the wood entry door. An inverted V pattern of char extended from near the floor to some 3.3 ft (1 m) up the door. Smoke staining is noticeable surrounding the charred area. *Courtesy of Santa Ana Fire Department, Santa Ana, California.*

extended from near the floor to some 1 m (3.3 ft) up the door. Smoke staining surrounded the charred area. Smoke had permeated the entire one-bedroom apartment, and soot had settled on all horizontal surfaces and condensed on walls, ceilings, and cabinets, thus revealing their internal structures by differential condensation. This suggested a small smoky fire of considerable duration.

At postmortem exam, the victim was found to have been a nonsmoker in good health, with a COHb saturation of less than 3 percent but a blood alcohol concentration of 0.31 percent weight per volume. She was known to be an alcoholic and had been seen at 10:30 A.M. under the influence. She was wearing the remains of slacks and a loose top. Fragments of a cloth placemat were found under the body.

Owing to the minimal firefighting efforts and absence of fire damage, a trail of burned cloth fragments and melted plastic was easily tracked from her location (the only entrance/exit) back to the kitchen's electric range. One element of the range was still energized and was partially covered with a heat-damaged sheet of aluminum foil (Figure 9.3).

FIGURE 9.3 ◆ One element of the range was still energized and was partially covered with a heat-damaged sheet of aluminum foil. Note the frozen-food dish and drink glass to the left of the range controls. *Courtesy of Santa Ana Fire Department, Santa Ana, California.*

Wrappers found in the kitchen showed that microwave meals in plastic dishes were commonly prepared, but there was no microwave oven. Apparently, she was using the electric range as an impromptu hotplate when the plastic dish ignited. She apparently used a placemat (one missing from the table) and rushed for the front door with the flaming dish. The flames ignited her loose top garment and she was overcome by the inhalation of flames before any substantial CO was produced or inhaled. She collapsed behind the door, and the fire, fueled by rendering body fat, with the placemat and her clothing acting as a wick, burned until she was discovered.

For analysis of the plume against the door using the *NFPA 921* flame height calculation from Chapter 3 and a wall effect factor of $k = 2$,

$$H_f = 0.174(kQ)^{2/5}, \tag{9.1}$$

$$\dot{Q} = \frac{79.18\,H_f^{5/2}}{k}, \tag{9.2}$$

where

H_f = flame height (m),

$\dot{Q}$ = heat release rate (kW),

k = wall effect factor, with

1 = no nearby walls,

2 = fuel package at wall, and

4 = fuel package in corner, and

$$\dot{Q} = \frac{(79.18)\,(1.0)^{5/2}}{2} = 39.6 \text{ kW; or} \approx 40 \text{ kW} \tag{9.3}$$

This heat release rate is in the range of values observed during previously published tests using pig carcasses (DeHaan, Campbell, and Nurbakhsh 1999).

Calculations of the heat release rate needed to produce plume damage on the rear of the door and calculations of the approximate body mass loss (5–10 kg, 11–22 lb) all indicated a fire of some 30–50 kW had burned for about 2 hr (based on DeHaan, Campbell, and Nurbakhsh 1999). In the behavioral analysis, note the cocktail glass to the left of the range controls and the elevated blood alcohol level, which also suggest alcohol impairment of the victim's judgment in this case.

This case is courtesy of Robert Eggleston, Santa Ana (California) Fire Department (retired).

◆ 9.2 CASE EXAMPLE 2: HEAT RELEASE RATE HAZARD OF CHRISTMAS TREES

Each year, Christmas trees are involved in a number of household and commercial fires. The National Fire Protection Association (NFPA) estimated in a study from 1999 to 2002 that an annual average of 310 Christmas trees (both real and artificial) were the first item ignited in residential fires in the United States. These Christmas tree fires result in an average of 14 deaths, 40 civilian injuries, and $16.2 million in property losses annually. The staggering statistic is that a civilian death occurred in one in every 22 reported Christmas tree fires (Rohr 2005).

This NFPA research also pointed out that, on average, electrical problems or malfunctions were the cause of approximately 40 percent of the fires. Heat sources placed too close to the tree were the cause in 24 percent of the cases, and juveniles playing with fire caused fires in 7 percent of the cases. Lastly, in 8 percent of the cases, candles were the direct heat source for ignition.

To assess this hazard and address some of the misinformation about ignition, heat release rates, tree varieties, and flame retardants, Dr. Vytenis (Vyto) Babrauskas, Dr. Gary Chastagner, and Eileen Stauss conducted an extensive study of Christmas trees and their contributions to fires. The results of one major part of this study are presented in this section (Babrauskas, Chastagner, and Stauss 2001).

INTRODUCTION

Burning items that produce a high heat release rate are the ones most likely to lead to fire injuries or fatalities. This is true not only because of possible thermal injuries but also because the production of toxic gases is highly correlated with heat release rates (Babrauskas 1997, 1998; Babrauskas and Peacock 1992). In most households, upholstered furniture and mattresses are the combustible items likely to produce the highest heat release rate values. A mattress may produce 500 to 2500 kW, and an upholstered chair about the same, whereas a large sofa may produce 2000 to 3000 kW (Babrauskas and Krasny 1985; Babrauskas et al. 1997). Other occupant goods, such as bookcases, television sets, and dressers, normally show heat release rate values only a fraction above these values. Exceptions are wardrobes and Christmas trees.

Several studies examining the heat release rates of Christmas trees have appeared in the literature. Ahonen and coworkers in Finland tested a fresh-cut tree and two that represented a week's indoor storage. When ignited with a small source, the fresh tree had a heat release rate of about 80 kW, and the others had a rate of about 600 kW (Ahonen, Kokkala, and Weckman 1984).

The California Bureau of Home Furnishings tested nine Christmas trees ignited with a match (Damant and Nurbakhsh 1994). Their results also showed that moisture was important, with very dry trees producing 900–1700 kW. A fresh-cut tree, however, could not be ignited by small ignition sources, such as matches.

Stroup and others (1999) at NIST conducted fire tests of Scotch pine Christmas trees using a large calorimeter. Radiative and total fluxes were measured at a location 2 m from the tree centerline. Seven of the trees were allowed to dry for approximately 3 weeks and burned easily when ignited with an electric match. An eighth tree, which could not be ignited, was freshly cut and kept in water. These trees ranged in height from 2.3 to 3.1 m (7.5 to 10 ft) and in weight from 9.5 to 20 kg (21 to 44 lb). Heat release rates were measured from 1620 to 5170 kW.

TEST PROCEDURES

Babrauskas, Chastagner, and Stauss (2001) tested Christmas trees in a heat release rate calorimeter capable of measuring over 10,000 kW. The basic procedure for doing such calorimetry tests are described in *ASTM E 1354-03, Standard Test Method for Heat and Visible Smoke Release Rates for Materials and Products Using an Oxygen Consumption Calorimeter* (ASTM 2003b). The tests were run in a large laboratory hall where air supply was effectively unrestricted. Christmas trees are commonly placed in a corner of a room, so to simulate the small amount of radiant heat feedback due to walls, an "open room" was built. Using a floor plan of 2.4 × 3.6 m (8 × 12 ft), two walls 3.0 m (10 ft) high were erected, as was a ceiling. The remaining two walls were absent, apart from short skirts below the ceiling used to channel the exhaust gases to the collection hood. The walls were made of noncombustible fiber-reinforced cement board. The center of each Christmas tree was located 0.9 m (3 ft) from each wall, which left a slight clearance in front of the walls. The test Christmas tree was placed on a load platform, which allowed a continuous record of mass loss to be obtained. In the heat release rate calorimeter, smoke production and CO and CO_2 production were measured along with the heat release rate. In addition, four radiometers and a large number of thermocouples were installed at various locations.

The trees tested were all Douglas fir (*Pseudotsuga menziesii*) and were approximately 2 m (6.5 ft) tall. Various watering programs were used, ranging from displaying trees in stands that were always filled with water to the other extreme of never watering the tree. To examine systematically the effect of watering programs, a majority of the trees were burned following a 10-day display period. These data were supplemented with data from tests with additional trees in which the length of the display period and care were varied.

Before each test, the trees were weighed. The weight of the trees ranged from 6.4 to 22.4 kg (14.1 to 49.3 lb), largely owing to differences in the moisture content of the trees. Apart from the 22.4-kg (49.3-lb) specimen, which had a very high moisture content and did not show a high heat release rate, the next-heaviest specimen was 16 kg (35.3 lb). The average weight was 9 kg (19.9 lb).

In the formulation of the test program, it was hypothesized that the moisture content at the time of burning would be the main controlling variable (in contrast, for example, with the length of time the tree was displayed). Thus, in the study, the moisture content of the foliage was accurately determined. In previous research programs, not enough trees were tested, nor were accurate moisture measurements available for estimating the role of moisture. Moisture content was determined by sampling branches, weighing them, then oven-drying them at slightly above 100°C (212°F) and weighing them again. Moisture content (MC) is defined as

$$ MC = \frac{Mass_{wet} - Mass_{dry}}{Mass_{dry}} \tag{9.4} $$

Note that since the definition uses the dry mass in the denominator, the moisture content can easily exceed 100 percent.

In most U.S. households, there will be a significant amount of additional combustibles associated with Christmas trees—decorations of various sorts, wrappings at the base, and gift packages. In the tests described here, no tree decorations were used. Located at the base, however, was a representative collection of gift boxes. The boxes were cardboard and had customary Christmas wrappings and ribbons. Most of the boxes contained a small amount of tissue paper and no other combustibles. Two toys, a plastic truck and a stuffed animal (peak heat release rate = 6 kW) were the only nonpackaging fuels included. The packages were arranged in an identical manner around the base for each test.

Three tests were run without packages and two other tests were run without the simulated walls to determine the effect, if any, these items had on the fire development. To start the fire, the procedure specified by the California State Fire Marshal for fire testing of Christmas trees [a Bunsen burner with a 38-mm (1.5-in.)-high flame; time of application = 12 s] was first applied to a branch about halfway up the tree. The burner simulates a small flame source such as a cigarette lighter. If the tree would not ignite from this source, then the same flame was applied to a gift package below the tree or to a stuffed toy located in the tree.

RESULTS

Photographs of the tests are shown in Figures 9.4 through 9.10. The first quantitative aspect that was determined in these tests was the moisture content needed for the California test burner, applied directly to the tree, to cause ignition of the tree. The results indicated that MC = 50 percent was a suitable demarcation between trees that ignited from this source and trees that did not. Of 48 trees tested, only 3 failed to obey this rule, and one test was inconclusive. All the trees that did not ignite from the small flame ignited from the burning gift-package assembly or stuffed toy.

The gift-package assembly, when tested by itself, showed peak heat release rate values of 98–118 kW, with four replicates tested. Thus, the package assembly's own heat release rate was insignificant compared with that of a tree. Visual observations indicated the process of ignition in the tests where the gift packages ignited the tree. The packages burn at a moderate but steady rate, and the fire effectively circles around the base of the tree. The hot combustion products rise, which has the effect of progressively drying out the tree branches. By the time the packages have stopped burning, the branches are dry enough that they catch fire and burn.

FIGURE 9.4 ◆ An example tree ready for testing. All test trees included gift packages under the tree and a single stuffed toy animal in the middle of the tree. Ignition was by a small flame to a branch, by a small flame to the stuffed toy, or by a small flame to the gift packages. No tree failed to burn from at least one of these ignition locations. *Courtesy of Dr. Vytenis Babrauskas, Dr. Gary Chastagner, and Eileen Stauss.*

FIGURE 9.5 ◆ Fierce burning from a dry (moisture content [MC] = 11 percent) tree. Peak heat release rate = 2700 kW. *Courtesy of Dr. Vytenis Babrauskas, Dr. Gary Chastagner, and Eileen Stauss.*

FIGURE 9.6 ◆ Postfire condition of tree shown burning in Figure 9.5. Only the main trunk remains. *Courtesy of Dr. Vytenis Babrauskas, Dr. Gary Chastagner, and Eileen Stauss.*

FIGURE 9.7 ◆ A rapidly burning tree will ignite combustibles nearby. Here nearby fabric targets are shown being ignited. *Courtesy of Dr. Vytenis Babrauskas, Dr. Gary Chastagner, and Eileen Stauss.*

Figure 9.11 shows the heat release rate curves for three tests. The MC = 20 percent and MC = 38 percent specimens were ignited by the small flame burner placed next to a branch. It can be seen that about 50 s elapses before any detectable heat release rate occurs, but after that point the burning is exceedingly fierce. In about 1 min the tree is essentially burned up, with only minor burning going on after that point.

The trunks and the large branches never burned in these tests, although in a room fire other items would likely contribute to the fire, which would generally cause even these members to be consumed. In the tests reported here, about one-fourth to one-third of the mass of each tree remained unburned.

The MC = 70 percent specimen showed a distinctly different behavior. It took about 80 s before there was significant heat release rate. Again, about two-thirds to three-fourths of the mass was consumed for the package-ignited trees, but their time course was much slower. They typically showed a curve with two or more "humps" and the time required for the bulk of the tree to be burned was about 300–500 s, instead of 80 s.

The moistest package-ignited tree tested, at MC = 154 percent, showed a peak heat release rate of 444 kW, which is still a sizable fire. Even this tree lost 60 percent

FIGURE 9.8 ◆ This Christmas tree would not burn when a small flame was applied to a tree branch. Applying the same flame to a small stuffed animal hung on the tree ignited the toy and the tree. Peak heat release rate = 525 kW. *Courtesy of Dr. Vytenis Babrauskas, Dr. Gary Chastagner, and Eileen Stauss.*

FIGURE 9.9 ◆ A fire-retardant (FR) tree burning. Some FR specimens generated a great deal of smoke and mist. *Courtesy of Dr. Vytenis Babrauskas, Dr. Gary Chastagner, and Eileen Stauss.*

FIGURE 9.10 ◆ This FR-treated tree burned vigorously and essentially burned up completely. *Courtesy of Dr. Vytenis Babrauskas, Dr. Gary Chastagner, and Eileen Stauss.*

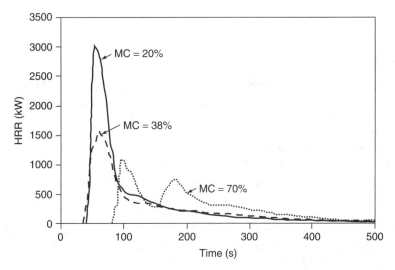

FIGURE 9.11 ◆ Example heat release rate curves at several moisture levels. *Courtesy of Dr. Vytenis Babrauskas, Dr. Gary Chastagner, and Eileen Stauss.*

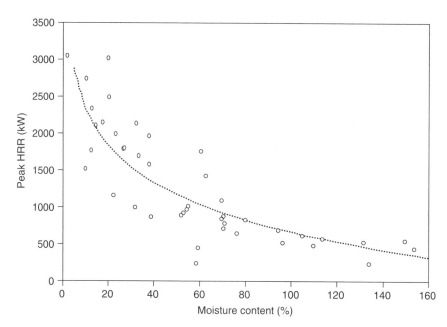

FIGURE 9.12 ◆ The relation between MC and HRR for non-FR trees. All trees Douglas fir, ~2 m (6 ft) high. *Courtesy of Dr. Vytenis Babrauskas, Dr. Gary Chastagner, and Eileen Stauss.*

of its mass during the test. Because moisture content data were available for each tree, a relation between moisture content and peak heat release rate could be found (Figure 9.12).

Two clear trends emerged from these tests: (1) an increasing moisture content decreases the peak heat release rate; and (2) none of the trees are "nonburning," or even of negligible fire hazard, if ignited with a large ignition source. The equation derived from the dotted curve in Figure 9.12, which can be used to predict heat release rate based on the moisture content of average-size non–fire retardant trees, is

$$\dot{Q} = 4050 - 734 \ln(\text{MC}), \tag{9.5}$$

where *ln* denotes natural logarithm.

CONCLUSIONS

Christmas tree fires are well documented in fire-incident investigations and have the potential for highly catastrophic results. For example, there is a detailed FEMA report (David 1991) concerning a Christmas tree fire that killed seven people. As well as in homes, Christmas trees have been involved in catastrophic fires in public buildings. A 10.5-m (35-ft) tree ignited in the lobby of one luxury hotel in the predawn, activating an automatic sprinkler, and caused the evacuation of 2300 guests (McDonald 2005).

Some experimental data are available from previous laboratory studies, but these were limited and did not explore a wide range of treatments, nor did they quantify the effect of moisture content. The present study shows that trees with a MC <15 percent will produce heat release rates in the range of 2000 to 3000 kW. These values pertain to a common tree height of 2 m (6.6 ft). Taller or shorter trees can be expected to give proportionately different results.

The test results confirmed that moisture content (apart from tree species and tree height, which were not tested) was the main variable governing the burning characteristics. The length of the display period and the quality of the watering regimen both serve to determine the moisture content. Additional studies by Babrauskas et al. (unpublished) indicated that trees treated with flame retardants did not show an improved fire behavior. Even more regrettably, in some cases fire retardant treatments caused the trees to dry out much sooner.

Douglas firs with a moisture content >50 percent were found to be unlikely to be directly ignited by small flames. However, such small ignition sources can readily ignite gift-wrapped packages, and this secondary ignition can then result in a serious tree fire. Defective lights or electrical cords are the most commonly reported ignition sources associated with Christmas tree fires. The arcs that occur in the failure of such devices are physically quite small, but an electrical fault may preheat combustibles in the vicinity of the fault prior to overt arcing. This ignition mechanism has not been studied but deserves to be investigated in some detail.

It is normally estimated that a small-to moderate-sized room can be driven to flashover by a fire of about 1500 kW if ventilation is through a single open doorway (Babrauskas 1984). This type of calculation is easier to perform for a bedroom than for a living room, where complex, interconnected spaces are often involved. Nonetheless, heat release rate values of the kind documented here represent a highly significant hazard, since in actual houses, Christmas trees are unlikely to be adjacent to cement board surfaces. Instead, combustible curtains, chairs, sofas, and other items will likely be close by, and these can be expected to ignite readily from the kinds of fires that have been documented here.

Good and prudent advice is frequently given to consumers to keep their Christmas trees well watered. This is essential, since if tree basins are allowed to run dry, moisture levels drop. None of the trees that were continually displayed in water during these tests burned when exposed to the small flame ignition source. However, it is not uncommon for trees to be displayed for much longer periods of time than were used in these tests. Trees are commonly bought a few days after Thanksgiving and may not be taken down until Epiphany (January 6), a time span of about 6 weeks. Based on previous postharvest studies, Douglas firs would not be expected to be able to maintain a high moisture level for this length of time even if they were displayed in water. Other species of Christmas trees such as Noble fir or Fraser fir that are able to maintain high moisture levels for extended periods of time would be much better suited for use under these conditions. This probably explains why these species are now the most commonly purchased Christmas trees on the market.

This case and testing are courtesy of Dr. Vytenis Babrauskas, Dr. Gary Chastagner, and Eileen Stauss, Washington.

◆ 9.3 CASE EXAMPLE 3: VEHICLE FIRES AND EXPLODING GASOLINE CANS

Several cases have been documented in which a person is burned to death as a result of a vehicle fire, and a witness claims that a plastic container of gasoline in the passenger compartment had leaked or spilled and that a fire had (somehow) been ignited in the vapors created. The fire then "must have" traveled back to the container and ignited its vapors, causing it to explode in a massive fireball that prevented rescue of the occupant. Such scenarios raise several issues that are testable by the means described in this text.

IGNITABILITY OF VAPORS

Assuming that there is some portion of the passenger compartment that contains vapors within the flammable range (1.4–7.5 percent in air), what is required to ignite them? In his extensive review, Holleyhead (1996) demonstrated that ignition of gasoline vapors by a hot surface, in particular, glowing cigarettes, is nearly impossible.

According to the literature, the autoignition temperature (AIT) of automotive gasoline is of the order of 400°C–500°C (752°F–932°F), and the temperature of the glowing coal of a cigarette can be as high as 800°C–900°C (1472°F–1652°F). Before concluding that the higher source temperature means that there would be ignition, these two numbers have to be evaluated in the context of the contact and heat transfer between the source and the fuel.

If a cigarette is not being puffed, the hot coal has a layer of ash surrounding it. This ash has a very low thermal conductivity that minimizes heat transfer. Gasoline has a minimum ignition energy (MIE) of 0.24 mJ at its ideal mixture, so at least this amount of energy has to be transferred. The laminar flow around the coal also minimizes contact between the hot surface and the fuel molecules. As a result of these effects, there is an insufficient quantity of heat transferred to cause ignition.

Recent tests by the ATF Fire Research Laboratory involved contact of burning tobacco cigarettes of different brands with gasoline vapors from a pool at room temperature. A total of 137 attempts were made using both smoldering and actively drawn cigarettes with no ignitions observed. Further testing with the same results involved a gasoline-saturated cotton blouse (Malooly and Steckler 2005).

If the cigarette is being puffed, the ash layer is largely absent and the fuel might be drawn into intimate contact with the porous glowing coal. In this situation, however, the gases are being drawn through the coal so quickly that there is insufficient contact time for enough heat to be transferred, and the oxygen content of the smoke stream is so low that a flame will not be able to propagate. The one exception is if there is an impurity in the tobacco or the paper that permits a brief flicker of flame to occur as the cigarette is being puffed. In that case, the flame, if in contact with an ignitable mixture, may ignite it. For this reason, hand-rolled cigarettes pose a slight risk, whereas tests of commercially made tobacco cigarettes have never demonstrated an ignition of gasoline vapors (Holleyhead 1996).

Gases with a lower minimum ignition energy such as hydrogen and acetylene (MIEs of 0.01–0.02 mJ) and carbon disulfide vapor (with its very low AIT) have been shown to be ignitable by a glowing cigarette. Hot surface ignition, such as from the electric element of the vehicle's cigarette lighter, is not possible owing to its relatively low temperature and the flat surface contact area. An electric arc caused by switching off electrical components of the vehicle or the flame of a lighter being used will cause ignition of the accumulated vapors if they are present at the correct concentrations and in contact with the ignition source. Because most such connections are well concealed in the instrument panel of a modern vehicle, such contact would be very difficult to achieve.

VAPOR CONCENTRATIONS AND HUMAN TENABILITY

The second issue is vapor concentrations and human tenability. The concentration of gasoline vapor in air must exceed 1.4 percent (14,000 ppm) to be ignitable under normal ambient conditions. The normal human nose is sensitive to gasoline vapors at concentrations below 20 ppm (and sometimes reported to be < 1 ppm). Gasoline vapors are an irritant to the eyes and to the mucosal membranes of the nose and throat at concentrations above 200 ppm and will be intolerable at concentrations above 2000 ppm, even in the absence of a sense of smell. This means that gasoline vapors will be detectable and a gasoline-vapor environment will be untenable for a conscious person at concentrations far below their lower flammable limit (LFL).

Because gasoline vapors are about three times heavier than air, they will tend to sink in still air. The movement of air in a vehicle caused by vehicle movement, fan, passenger movement, or even thermal currents from warm bodies will circulate, mix, and dilute the vapors. It would be very difficult, if not impossible, for an ignitable concentration of over 14,000 ppm in one part of the passenger compartment to be tolerated by conscious passengers.

SOURCE OF VAPORS

The third issue is the source of the vapors. Liquid fuel must evaporate before it can be ignited. As we have seen previously, the rate of evaporation is controlled by the surface area of the liquid pool, temperature, and nature of the surface. An open fuel container will produce vapors only from the exposed fuel surface; that is, a container with a cap loosely in place will leak vapors, but at a very slow rate.

If the liquid is spilled on the seat or floor of the vehicle during transport, it will evaporate much faster. If the surface area of the supposed spill can be estimated, one can estimate the maximum evaporation rate based on published data (DeHaan 1995). Only the fuel that is in the vapor state at the time of ignition will support that ignition (with ongoing flame dependent on a continuous source of vapors).

The problem, then, is putting a suitable ignition source (flame or electrical arc) into contact with a suitable accumulation of vapor. If a floor-level spill is hypothesized, the ignitable vapors may be found only at or near floor level (owing to the dilution and mixing in air). Laboratory testing of the floor mats should reveal the residues of significant amounts of liquid. Careful examination of the vehicle may reveal the presence of a suitable ignition source (dropped matchstick, broken wire to seat motor, etc.). Except under the most unusual vehicle conditions, the ignition source will be found in the same "compartment" as the fuel vapor accumulation (i.e., trunk, engine, passenger area). Interviewing a witness may reveal the circumstances of ignition.

THE CONTAINER

Assuming that there was a spill (ignored by passengers), enough vapor, a suitable source, and ignition of the accumulated vapors, one would expect a deflagrating combustion of the vapors in the vehicle, but would there be an explosion of the vapors in the can?

The answer is in the vapor pressure of the gasoline. At normal ambient temperatures above 0°C (32°F), the combined vapor pressure of the most volatile components in automotive gasoline (butane, isobutane, pentane, hexane) is sufficient (>200 mmHg) that (if there is any significant quantity of liquid fuel in the container) the vapor concentration at equilibrium conditions in the container is well above the upper 7.5 percent (57 mmHg) flammability limit of the vapors. This means that only the vapors that have escaped from the container and mixed adequately with the surrounding air to be within their flammable range will support combustion.

The result is not an explosion in the gasoline container but a modest flame supported at the mouth of the container by the vapors coming from within. As shown in the test in Figure 9.13, a small flame will be supported at the loose cap. If it is a plastic container, this flame will eventually melt (in 1–10 min) the neck and top of the container, exposing a larger surface area of vent. In most cases the container will burn from the top down, melting at the fuel level inside as it burns. If there is enough boilover of excess fuel, the sides will collapse and release a large liquid pool. This process nearly always leaves liquid fuel trapped in the melted folds of the base of the container that is readily detected by careful

FIGURE 9.13 ◆ Fire test sequence (top to bottom) for evaluating a 20-L plastic can filled with gasoline, with a loose cap. After 8 min, a small fire at the cap is supported by vapors leaking from the cap. After 19 min, the top has melted away, producing an equivalent pool fire of 400 kW, and after 24 min, spillover of boiling gasoline occurs. *Courtesy of New Zealand Police, Christchurch, New Zealand.*

visual inspection of the remnant in the lab. If the container is a metal can, the soldered seams may eventually melt and the heat conducted through the metal will cause increasingly faster evaporation of the remaining fuel, but no explosion. Only when the container is nearly empty (or if it was so at the start of the fire), may there be a stage where there is not enough fuel vapor to sustain a saturation (equilibrium) vapor pressure inside, and the mixture inside the can may sustain a deflagration as it boils dry.

The same parameters hold true for light alcohols whose lower vapor pressure and wide flammability range can produce an explosive atmosphere inside the container under some conditions. This combination of factors has apparently been responsible for injury-producing explosions as people have squirted or poured ethanol or methanol onto fires. It has been demonstrated in a recently published paper by Hasselbring (2007) that if a gasoline container is being emptied of its liquid content, the turbulence caused by air rushing back through the orifice can create nonequilibrium conditions. If the vapors are ignited as the contents are being dumped, the flame can propagate back into the container and cause a deflagration that ruptures a plastic gas container. Very low ambient temperatures (< 6°C, 43°F) and partial evaporation of the gasoline can also create a combustible-range vapor concentration inside the container.

This was the mechanism thought to be responsible for the fuel tank explosion on TWA Flight 800, where a small quantity of low-volatility Jet A (aviation kerosene) fuel in a large center wing tank reached ignition conditions because it was heated externally prior to takeoff (Dornheim 1997). In that scenario, with the fuel's vapor concentration increased and the external atmospheric pressure decreased as the plane gained altitude, the vapor–air mixture became susceptible to ignition from an electric arc.

THE EXTERNAL FIRE

When there is a spill onto the floor mat of the vehicle, or the entire top of the container is melted away, the heat release rate of the resulting fire can be estimated from existing data on pool fires. What is needed to determine this heat release rate is the estimated surface area of the pool.

When melted away, the exposed top of the plastic container in Figure 9.3, for instance, is approximately 0.4×0.4 m (16×16 in.), which produces an effective "pool" 0.16 m^2 (1.72 ft^2) in area.

To estimate the heat release rate of fire emitted from the exposed top of the burning container, we refer to Chapter 2, equation (2.5), when the horizontal burning area of a material is known along with the burning rate per unit area (mass flux):

$$\dot{Q} = \dot{m}'' \Delta h_{\text{eff}} A, \tag{9.6}$$

where

$\dot{Q} = $ total heat release rate (kW),

$\dot{m}'' = $ effective mass burning rate per unit area (g/m^2-s),

$\Delta h_{\text{eff}} = $ heat of combustion (kJ/g), and

$A = $ burning area (m^2).

Table 2.3 lists the typical range of values of mass flux and heat of combustion for gasoline (see Chapter 2). Substituting these values into the equation for heat release rate, we find the maximum value for $\dot{Q}$ to be

$$\dot{Q} = (53 \text{ g/m}^2\text{-s}) \, (43.7 \text{kJ/g}) \, (0.16 \text{ m}^2) = 370 \text{ kW}. \tag{9.7}$$

However, for smaller-diameter fires, $\dot{m}''$, the effective mass burning rate per unit area (g/m^2-s) is typically lower in value, thus producing a smaller heat release rate. For example, if $\dot{m}''$ was 36 g/m^2-s, the maximum value for $\dot{Q}$ would be estimated to be 250 kW.

Similar calculations can be carried out for the area of the footwell or floor mat assuming maximum area and saturation. This fire will be sustained if there is adequate ventilation as the fuel level regresses at 1–4 mm/min until the container fails or all the fuel is consumed.

◆ 9.4 CASE EXAMPLE 4: ASSESSING CONSTRUCTION AND WALL COVERINGS

This case study emphasizes the importance of assessing building features such as construction and wall coverings to evaluate their possible contributions to fire spread and production of unusual burn patterns, as well as the importance of developing a valid hypothesis by comparing evidence from the scene to witness-reported information and features of the uninvolved portion of the structure.

A fire in a three-story wood-frame university dormitory building claimed five lives and injured eight, all residents of the building. The 14.3 × 37.8-m (47 × 124-ft) building contained 46 studio apartments, and each floor had a central hallway served by enclosed stairwells at each end. The walls and ceilings of the halls and stairwells were plasterboard. The stairs themselves were wood, and the hallway floors were plywood covered with vinyl floor tiles. The walls of the first (ground) and third-floor halls and the stairwells were covered with thin [approximately 3- to 4-mm ($\frac{3}{16}$-in.)] plywood paneling nailed to the walls and held in place with wooden baseboards and wainscoting.

The stairwell doors were fire-rated gypsum core doors, fitted with spring closers. However, they also had flip-up doorstops, and interviews with residents indicated that they were frequently blocked open. The doors to individual rooms were hollow-core wood interior doors. There were heat detectors in the stairwells and halls directly connected to the fire department's headquarters alarm panel, but no fire suppression system.

The fire extensively damaged the east end of the second floor and the stairwell to the third floor. There was no fire damage to the first floor, but damage did extend about halfway down the east stairs toward the first floor. There was less damage on the third floor, and significant portions of the wood paneling survived. It was the burn patterns on this paneling that initially suggested that multiple fires had been set in the third-floor hallway using flammable liquids.

There were 20 or more areas where localized fire damage to the flooring had occurred and extended up the walls in burned areas 0.61–0.9 m (2–3 ft) high, with most being 0.15–0.3 m (6–12 in.) wide. The burned areas were on both sides of the third-floor hallway at somewhat regular intervals (Figure 9.14) and ranged from the panel-

FIGURE 9.14 ◆ Areas of localized fire damage to the third-floor flooring and walls at regular intervals along the length of the corridor. *Courtesy of Sgt. Paul Echols, Carbondale [Illinois] Police Department.*

ing being scorched or lightly charred to its being completely destroyed (along with the baseboard), exposing the plasterboard beneath. All these areas were found to be located where the paneling was not nailed tightly to the wall studs [i.e., centered between the nailings at 41- or 81-cm (16- or 32-in.) centers] (Figure 9.14), allowing the paneling to bulge slightly away, thus allowing an air space to develop and producing a thermally thin fuel.

This thermally thin material was exposed to radiant heat from the hot gas layer that filled the corridor from the stairwells and from the burning paneling that had ignited in the hot gas layer, peeled away, and dropped to the floor. The wainscoting secured most of the lower half of the paneling in the third-floor hall, leaving these isolated V patterns coming from floor level.

Investigation revealed that ceiling penetrations in the stairwells allowed fire gases to penetrate the floor joist space between the second and the third floors. The joists were arranged longitudinally (along the length of the building), providing a channel for air movement into the third-floor hallway. The plywood flooring had been nailed to these longitudinal joists with no provision for cross support where the plywood sheets butted against one another.

Floor tiles that overlapped these butt joints were seen to be cracked and broken. Floor tiles along the walls were loose or displaced from the flexing across the joints, and a few had been replaced. This allowed air from the joist space to escape through these joints and permit more intense burning of the adjacent baseboard and paneling as they were heated by radiant heat from the fire gases trapped in the hall. Where a butt joint coincided with a thermally thin area of paneling, a localized area of damage was created (Figure 9.15) by the radiant heat from the intense/deep hot gas layer (as

FIGURE 9.15 ◆ Localized areas of damage at ventilation points. *Courtesy of Sgt. Paul Echols, Carbondale [Illinois] Police Department.*

evidenced by the burned areas of walls above the wainscoting). The "burn-throughs" of flooring/wall structure occurred at 1.22- or 2.44-m (4- or 8-ft) intervals where the plywood flooring joined.

One large area of such damage was originally thought to be the result of a large burning pool of ignitable liquid against a wall. However, interviews with survivors (supported by physical evidence) revealed that one rescuer had crawled through this very area dragging a victim during the fire and had not been exposed to an ignitable liquid or localized fire at that time. This situation illustrates the value of thorough interviews and comparison of information from the scene with that from witnesses to validate or exclude a hypothesis. There were also areas of localized scorching on floors and baseboards that were clearly consistent with the brief-duration fires fueled by portions of the upper paneling that peeled off the wall and ignited in the hot gas layer. These would be classified as drop-down fires. Examination of the floor on the second floor and undamaged paneling on the first floor confirmed the features described.

This fire was ultimately identified as an arson, with two origins identified in piles of clothing and trash outside the door of one apartment immediately next to the stairs and in the stairwell immediately adjacent. No flammable liquids were involved. The staircases were affected by fall-down, buoyant flow, and radiation processes. Figure 9.16 shows the extension of fire from the second floor down the stairs by radiant heat and fall-down of the paneling, with a firefighter standing in the open door in the second-floor hallway.

Figure 9.16 is of the stairs between the second and the third floors immediately above one of the origins of the fire. Much more extensive damage to risers (including penetrations) and to wall coverings and handrails was present in this stairwell than in the one below it. Of note, all the fatalities and most of the eight injuries occurred in rooms on the third floor (not on the floor of origin).

Thin materials are usually thought to be draperies, loose garments, calendar pages, and the like. They can also be multilayered materials that delaminate when exposed to heat. These include plywoods, veneers, and some wallpapers or paints that blister because moisture or solvent is trapped beneath an impermeable outer skin, and wall or floor coverings that can separate from a solid underlayment owing to adhesive or fastener weaknesses. Skin can also display this effect under some conditions as a blister forms, and the distended epidermal layer loses contact with the dermal layer beneath.

The generally accepted equations for thermally thin and thermally thick materials from Chapter 2 [equations (2.11) and (2.12)] are repeated here. In the case of thermally thin conditions, for $\delta \leq 1$ mm,

$$t_{ig} = \frac{T_{ig} - T_\infty}{\dot{q}''/\rho c_p \delta}. \tag{9.8}$$

Under thermally thick conditions, $\delta > 1$ mm,

$$t_{ig} = \frac{\pi}{4} k \rho c_p \left(\frac{T_{ig} - T_\infty}{\dot{q}''} \right)^2, \tag{9.9}$$

where

t_{ig} = time to ignition (s),

k = thermal conductivity (W/m-K),

ρ = density (kg/m³),

c_p = specific heat capacity (kJ/kg-K),

δ = thickness of material,

FIGURE 9.16 ◆ Extension of the fire down the stairs by radiant heat and fall-down of the paneling. *Courtesy of Sgt. Paul Echols, Carbondale [Illinois] Police Department.*

T_{ig} = piloted fuel ignition temperature,

T_∞ = initial temperature, and

$\dot{q}''$ = radiant heat flux (kW/m²).

From Table 2.5 (Chapter 2), the characteristics of yellow pine are a thermal conductivity, k, of 0.17 W/m-K, a density, ρ, of 640 kg/m³, a specific heat capacity, c_p, of 2.85 kJ/kg-K, and a thermal inertia, $k\rho c_p$ of 0.255 kW²-s/m⁴-K². With a piloted ignition temperature, T_{ig}, of 450°C (842°F), an ambient initial temperature, T_∞, of 20°C (68°F), and a paneling surface thickness, δ, measuring 3–4 mm (0.12–0.16 in.), the thermally thick time-to-ignition formula is used:

$$t_{ig} = \left(\frac{\pi}{4}\right)(0.255)\left(\frac{450 - 20}{\dot{q}''}\right)^2 \tag{9.10}$$

or, simplified,

$$t_{ig} = \frac{37012.0}{(\dot{q}'')^2}. \tag{9.11}$$

In evaluating a range of possible heat fluxes, 20, 30, 40, and 50 kW/m², the calculated times to ignition for the pinewood paneling are 93, 41, 23, and 15 s, respectively. Where this paneling was beginning to delaminate, $\delta = 1$ mm, and thin-material ignition would be expected in approximately 26 s at 30 kW/m².

Therefore, for low heat fluxes (<50 kW/m²) and short durations (<30 s), thin materials are much more likely to be ignited (or even melted) than are thick objects made of the same materials. These considerations have to be kept in mind when documenting indicators of heat damage.

◆ 9.5 CASE EXAMPLE 5: PLUME HEIGHT AND TIME OF BURNING BODY

Early one morning, at 8:20 A.M., employees noted a plume of smoke coming from brush at the edge of an industrial park. Responding to what was thought to be an unattended campfire of the homeless, the fire department found the fire to be consuming the body of a young woman.

The small flames were quickly doused and the scene preserved by 8:28 A.M. The body was found lying on its back, severely burned (approximately 50 percent of the body weight remaining, with the lower legs burned off). Remnants of a heavy cotton blanket or drapery were found with the body. Next to the body was a burned, carpet-covered board of the type found in camper shells and trailers, braced on some low-hanging branches, as shown in Figure 9.17.

FIGURE 9.17 ◆ Responding fire department personnel found the fire consuming the body of a young woman next to a burned board. *Courtesy of Sgt. Bruce M. Wiley, San Jose Police Department, San Jose, California.*

FIGURE 9.18 ◆ Photo of the body location shows an inverted-V pattern of thermal damage to foliage of shrubs overhanging the position of the body. Scorching and shriveling extended to a peak about 2.1 m (6.9 ft) above the ground where the body burned. Charring extended some 0.6–0.9 m (2–3 ft) from the ground. *Courtesy of Sgt. Bruce M. Wiley, San Jose Police Department, San Jose, California.*

The brush above the body bore an inverted-V pattern of damage, extending from ground level, involving charred branches 0.6–0.9 m (2–3 ft) from the ground, scorched leaves and branches at 1.5 m (5 ft), and withered and scorched leaves at 2.1 m (7 ft). As shown in Figure 9.18, there was no sustained combustion of the brush, since the body was located in a natural gap somewhat between the shrubs.

The body was not readily visible from the street owing to a gentle slope downward. The victim was identified as a young local woman. Her COHb was less than 10 percent (negligible, due to her history as a smoker). She had a low blood alcohol concentration and tested negative for drugs. There were no signs of a struggle at the scene, no weapons, and no blood. The dry hardpan soil had been walked over repeatedly by fire personnel, so no shoe prints were recovered. Based on postmortem weight and an estimated antemortem weight obtained from the victim's mother, the mass loss of the body was estimated as 30 kg (66 lb).

The main issues were time of death, cause of death, and duration of burning. Owing to massive destruction of the soft tissue, no wounds could be identified (an informant later said that she had been shot in the side of the neck around midnight), and time of death was not established (although there were no signs of decomposition).

As far as establishing the time of burning, the McCaffrey (1979) method was used for estimating the centerline maximum temperature rise, velocity, and mass flow rate of a plume, as described in Chapter 3, equations (3.15), (3.16), and (3.17). The temperature profiles for the smoke plume above various size fires can be estimated as

$$T_0 - T_\infty = 21.6\, \dot{Q}^{2/3} Z^{-5/2}, \qquad (9.12)$$

where

T_0 = maximum ceiling temperature (°C),

T_∞ = ambient air temperature (°C),

$\dot{Q}$ = total heat release rate (kW), and

Z = measurement along the centerline (m).

These relationships correspond with the plot in Figure 9.19 (Quintiere 1998, 142).

A fire of about 100 kW at ground level would be calculated to have gas temperatures just about right ($T = 100°C–150°C$ at 2 m, 450°C at 1 m). This maximum probably occurred as the cotton blanket/curtain wrapped around the victim burned, but assuming that it was sustained during the burning of the body, the thermal plume damage observed sets the maximum heat release rate of the body at 100 kW.

Based on data from DeHaan, Campbell, and Nurbakhsh (1999) and DeHaan (2001), a burning body by itself via the wick mechanism will demonstrate a mass loss

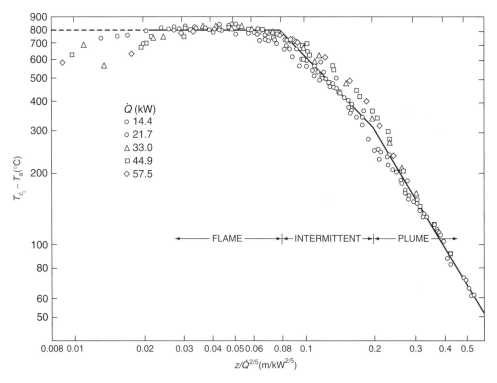

FIGURE 9.19 ◆ Centerline fire plume temperatures. *From* Principles of Fire Behavior, *1st ed., by J. Quintiere © 1998. Reprinted with permission of Delmar Learning, a division of Thomson Learning; www.thomsonrights.com. Fax 800-730-2215.*

rate of about 3 g/s (10.8 kg/hr). Allowing for fluid loss, time to ignition, and a total mass loss of about 30 kg yields a minimum time of about 4 hr. Anything less than 1 g/s (30 kW) would not sustain itself, and this fire was still burning, with flames showing in the crotch area of the body when it was found and extinguished at 8:28 A.M. That minimum rate is 3.6 kg/hr (7.9 lb/hr), which sets a maximum time of about 6 hr. The body had been placed face down on a wide board with carpet, and as the limbs flexed, the body apparently shifted and rolled off the board and onto its back. The leaf litter under the body was much less badly burned than the board (where nearly all the ashes and debris were found), and there was no protected area on the body. Nearly all the skin was burned away from all surfaces of the torso. Based on this calculation of 4 to 6 hours and physical evidence (burned blanket, absence of flammable liquid residues, death before fire exposure), which corroborated the witness statements, the individuals responsible were tried and convicted of murder.

This case study is courtesy of Officer Robert Froese, San Jose Police Department, San Jose, California.

◆ 9.6 CASE EXAMPLE 6: ARSON-HOMICIDE

In 1981, a man died as the result of third-degree burns over most of his body. His boarding-house roommates reported that he had returned to the house a short time earlier, well under the influence of alcohol, and was last seen asleep on the living room sofa. Sometime later, the roommates, who were watching television in another room, heard shouts, saw a flash of flame, and were horrified to see the victim stagger out of the living room with his clothes on fire. He walked to an adjacent bedroom, where an attempt was made to smother the flames with a blanket, and then to a bathroom, where his clothes were stripped off and dropped in the bathtub.

A neighbor had seen the flames and notified the fire department. Fire crews arrived to find the victim in the bathroom with a trail of burned clothing and burned blankets in the adjacent bedroom. The fire in the living room was limited to the sofa base and its cushions and was extinguished with a minimum of water. Scene investigation revealed a glass jar with residues of gasoline on the floor of the dining room, burns to the cushions of the sofa (Figure 9.20), and numerous isolated surface burns of the carpet in front of the sofa. Residues of gasoline were identified in the carpet, the victim's clothing, and the jar. The victim died some 18 hr later from external burns (98 percent second and third degree) and inhalation of hot gases.

A roommate confessed to having thrown gasoline on the victim as he allegedly attempted to rise from the sofa and assault him. The accused claimed that the victim had repeatedly assaulted him physically and sexually in recent days and that he had made plans to stop him if he tried it again. Roommates had seen him in the house with a glass jar of clear liquid similar to gasoline minutes before the fire.

Authorities reasoned that the victim had been asleep on the sofa and that the accused had poured the gasoline on him as he lay on his side. The front of the sofa had been consumed, and there were limited burns to the loose seat and backrest cushions of the sofa; the scattering of burns across the carpet in front of the sofa was noted but not interpreted at the time. The accused was convicted of murder (with a torture enhancement!) and sentenced to life.

During the appeal process it was requested that the evidence of the burn patterns be reexamined. Unfortunately, most of the scene photos are too poor in quality to

FIGURE 9.20 ◆ Scene investigation revealed burns to the seat cushions of the sofa and a glass jar with residues of gasoline on the floor of the dining room. *Courtesy of Norman Fehle, LAPD Arson (Retired).*

permit reprinting, but the sketch in Figure 9.21 demonstrates the critical features. The front face of the sofa (which appeared to be cotton padded) was burned nearly to floor level, and fire had penetrated under the cushions.

The seat cushions appeared to be cotton covering over foam rubber and had been thrown on the floor during extinguishment. Of the surfaces visible in the photos, only one bore visible damage, which was limited to longitudinal burns parallel to the edges. The two backrest cushions still in place bore only limited, isolated burn damage. There was a scattering of lightly burned areas on the carpet in front of the sofa extending some 0.91 m (3 ft) away from it. These appeared to be the scorched or melted areas created when a small quantity of flammable liquid burns off the face of the carpet without penetrating the pile of the carpet.

Because the victim had walked about the house while on fire, all his clothing had been burned, leaving only charred remnants in various rooms. The postmortem photos and report revealed generalized third-degree burns uniformly distributed over most of his body. As a result, no interpretation could be offered as to the area of clothing first ignited.

Because synthetic fabrics can burn with an intensity similar to that of gasoline, the damage they can produce cannot always be distinguished from that produced by gasoline. An attempt was made to reconstruct the apparent burn damage to the carpet and

FIGURE 9.21 ◆ Plan view sketch of the couch, pillows, and remains of burned clothing. *Courtesy of J. D. DeHaan.*

sofa. It was thought unusual that a fire ignited by pouring gasoline on a reclining body would damage the front of the sofa cushions, but not the tops or the adjacent backrest cushions. Extensive spillage or splatters on the carpet some distance from the sofa would also not be an expected consequence of a pour onto a body on top of the cushions.

Using a sofa of approximately the same proportions with loose seat and backrest cushions, tests were conducted to establish the nature of splash and splatter patterns created by throwing a liquid toward the sofa. The sofa was partially covered with brown wrapping paper, a glass jar of the same capacity as that recovered at the scene was used, and isopropyl alcohol was used as the liquid. Isopropyl alcohol has viscosity and surface tension properties very similar to those of gasoline, so it will darken paper surfaces where it contacts them and yet will evaporate completely after a short period of time and is nontoxic.

A separate cushion propped against the front of the sofa simulated the presence of a body sitting on the edge of the sofa or standing immediately in front of it. The "thrower" was standing approximately 2 m (6 ft) from the front of the vertical cushion. Figure 9.22 shows the splatter pattern created when the alcohol (about 300 mL) was thrown low against the target (about knee height). A very heavy concentration is seen in a single "pool" at the base of the cushion. Although it cannot be seen on the fabric, a quantity of liquid was detected on the front of the sofa cushions on both sides of the target. It was decided to cover these cushions with paper to detect this side splatter.

FIGURE 9.22 ◆ The splatter pattern created when alcohol (about 300 mL) was thrown low against the target, at about knee height. *Courtesy of J. D. DeHaan.*

Figure 9.23 shows the fluid still in flight when thrown higher up on the cushion, and Figure 9.24 shows the wetted areas about 30 s later. Note the wetted areas on the front of the cushions and the more limited distribution of concentrated (pooled) liquid but more discrete splatters on the floor in front of it.

Figure 9.25 shows the effect of throwing the liquid against a textured fabric surface rather than a smooth paper one. Note the distribution of discrete spots on the floor and limited splatter on the front edge of the seat cushions. Figure 9.26 shows the distribution on a horizontal surface. The main splatter reaches about 1.8 m (70 in.) from the near edge (where the forward motion of the jar was checked). The farthest

FIGURE 9.23 ◆ The fluid still in flight when thrown higher up on the cushion. *Courtesy of J. D. DeHaan.*

FIGURE 9.24 ◆ The wetted areas about 30 s later. Note the wetted areas on the front of the cushions and the more limited distribution of concentrated (pooled) liquid but more discrete splatters on the floor in front of it. *Courtesy of J. D. DeHaan.*

splatter is about 2.5 m (100 in.) from the near edge. This large concentrated pool does not match the distribution on the carpet at the scene.

Of the limited series of tests conducted it appeared that the photos in Figures 9.24 and 9.25 represented the pattern of distribution of gasoline noted at the scene. These tests showed that it was more likely that the gasoline was not poured on the recumbent victim as he slept but thrown at him from some distance as he sat on the edge of the sofa or attempted to rise from it. These tests supported the accused's contention that he acted in self-defense, and he was released on parole having served 18 years for the crime.

FIGURE 9.25 ◆ The effect of throwing the liquid against a textured fabric surface rather than a smooth paper one. *Courtesy of J. D. DeHaan.*

FIGURE 9.26 ◆ The effect of throwing the liquid against a horizontal smooth paper surface. *Courtesy of J. D. DeHaan.*

◆ 9.7 CASE EXAMPLE 7: ACCIDENTAL FIRE DEATH

A woman was found dead on the landing outside her bedroom door. She was covered in soot and her body was outlined in soot on the carpet around her. She was partially undressed, as shown in Figure 9.27.

There was a large amount of soot and mucus in the victim's nostrils. Postmortem blood analysis revealed that she had a high COHb concentration, a significant blood alcohol level, and high levels of antidepressants. There was no evidence of violence. A very localized fire had taken place in/under a nightstand in her bedroom, as shown in Figure 9.28.

FIGURE 9.27 ◆ A partially undressed woman found dead on the landing outside her bedroom door, covered in soot. *Courtesy of Ross Brogan, New South Wales Fire Brigades, Fire Investigation & Research Unit, Greenacre, NSW, Australia.*

FIGURE 9.28 ◆ A localized fire in and under the bedroom nightstand. Note the plume extension from beneath the nightstand, radiant heat–driven extension into the carpet, heater (plugged in), shoes and throw rug in disarray, prescription medicines on the table and desk, and glass and bottle of alcohol. *Courtesy of Ross Brogan, New South Wales Fire Brigades, Fire Investigation & Research Unit, Greenacre, NSW, Australia.*

Because there was very limited fire, little suppression activity was necessary, and care was taken to preserve the scene for the investigator. Note the minimal plume extension from beneath the nightstand, and radiant heat–driven extension into the sisal carpet. Under the nightstand was a small plastic-cased heater (plugged in). The remains of clothing were found charred into the remains of the heater, having apparently fallen from the chair. Note the shoes and throw rug in disarray. The bedclothes were disturbed, there was a TV remote control on the bed, and there were prescription medicines on the table and desk, along with a large glass of alcohol and a partial bottle of alcohol.

The scenario and physical evidence were entirely consistent with the victim's partly undressing while consuming alcohol and medications, then reclining on the bed to watch television. She apparently fell asleep after partially undressing and dropping clothing haphazardly onto the chair. The clothing fell off the chair, onto the heater, and ignited into a small, but very smoky, fire. The deceased then awoke and attempted to escape but was overcome once she stood and started breathing in the accumulated smoke layer more deeply, reaching the landing before collapsing from the combined effects of CO, smoke, alcohol, and drugs.

This case is courtesy of Ross Brogan, New South Wales Fire Brigades, Greenacre, NSW, Australia.

◆ 9.8 CASE EXAMPLE 8: PIZZA SHOP EXPLOSION AND FIRE

An early-morning explosion and fire did considerable damage to a pizza shop. A considerable quantity of gasoline had been poured inside the premises (Figure 9.29). The remains of a plastic oil can were found melted on the floor, with residues of gasoline still inside the container (Figure 9.30).

FIGURE 9.29 ◆ Interior of pizza shop showing fire damage from a large gasoline pour on the floor. *Courtesy of Ross Brogan, New South Wales Fire Brigades, Fire Investigation & Research Unit, Greenacre, NSW, Australia.*

FIGURE 9.30 ◆ Melted and scorched plastic oil container (center) still contains gasoline. Note the direction indicators of melted plastic and scorched items on the shelf, which act as vectors pointing toward heat. *Courtesy of Ross Brogan, New South Wales Fire Brigades, Fire Investigation & Research Unit, Greenacre, NSW, Australia.*

The gas stove in the kitchen (with pilot light) was identified as the ignition source. The rear door appeared to have been broken by the explosion and then exposed to a sooty fire. The remains of a toilet paper roll trailer were found adjacent to the rear door (the roll bearing gasoline residues) (see Figure 9.31). Near the rear stairs was a badly scorched athletic shoe with surface scorching and melting consistent with brief exposure to flames, as shown in Figure 9.32.

FIGURE 9.31 ◆ Broken glass forced outward from the explosion within the shop. Note the toilet paper roll inside the door with paper extending up the stair to the right. *Courtesy of Ross Brogan, New South Wales Fire Brigades, Fire Investigation & Research Unit, Greenacre, NSW, Australia.*

FIGURE 9.32 ◆ A scorched and burned athletic shoe indicates the path of flight up the steps. *Courtesy of Ross Brogan, New South Wales Fire Brigades, Fire Investigation & Research Unit, Greenacre, NSW, Australia.*

Farther away were blood droplets and the remains of burned clothing in the alleyway leading to the apparent location of the "getaway car" (Figure 9.33). Two individuals arrived at a local hospital at about the same time as the fire report, each reporting a different address as the location where he suffered his burns and other injuries. The fire service responded to both addresses and found no fires at either. One suspect was apparently inside the structure pouring the gasoline while the other was outside preparing the trailer. Bloodstains from the step and parking area were matched to one of the suspects.

The individual inside succumbed to his burn injuries some 3 months after the fire. The surviving suspect eventually implicated the pizza shop owner in a "hired torch" scheme.

This case is courtesy of Ross Brogan, New South Wales Fire Brigades, Greenacre, NSW, Australia.

◆ 9.9 CASE EXAMPLE 9: STAIRWELL SMOKE FILLING

This fire occurred at approximately 7:00 P.M. in a two-story two-bedroom apartment. The fire was reported by an occupant, who then left, closed the door to the apartment, and awaited fire crews outside. No mention was made for some time that another occupant, an elderly adult male, was in his bedroom at the end of the upstairs hallway, presumably asleep, as was his habit at that time.

FIGURE 9.33 ◆ Burned clothing remnants between rear door and vehicle. Bloodstains on pavement from suspect injured by explosion-propelled glass. *Courtesy of Ross Brogan, New South Wales Fire Brigades, Fire Investigation & Research Unit, Greenacre, NSW, Australia.*

The fire had involved a large quantity of newspapers, magazines, and food containers that had been strewn on the stairs from the midflight landing down to the foyer on the main floor, as shown in Figures 9.34, 9.35, and 9.36. The main structure was reinforced concrete and masonry. The interior walls were painted plasterboard.

There was a vinyl wallpaper (woodgrain) on one wall of the stairwell. The hall floors and steps were carpeted. The entry foyer floor was vinyl tile. The flame plume had extended primarily along one wall, with areas of consumption of wall covering right to the gypsum at the base of the steps, with reduced blistering and charring of the paint up to a height of about 2.7 m (9 ft). The ceiling over the stairs was about 5.2 m (17 ft) above the floor of the foyer; the stairs were 0.8 m (34 in.) wide. The hot smoke layer was recorded by the discoloration and blistering of the paint on the walls of the hall and stairwell and was at approximately 2.2 m (7 ft) from the foyer floor level. The upper hallway was approximately 6.1 m (20 ft) long, 2.4 m (8 ft) high, and 0.9 m (35 in) wide. A hallway door to a second bedroom was tightly closed (as was a furnace room door) at the time of the fire, as minimal smoke/heat penetration was noted in these rooms. The door to a 2.2 × 2.4 × 2.4-m (7 × 8 × 8-ft) bathroom at the top of the stairs was open, and there was extensive heat and smoke damage to the room and its contents.

The responding firefighters found the closed entry door to be burning from within and fire extending out into the common hallway. The door was forced open, and a substantial fire was discovered in the stairwell and in the entry foyer. The fire was extinguished with fog/water application. There was no immediate search for

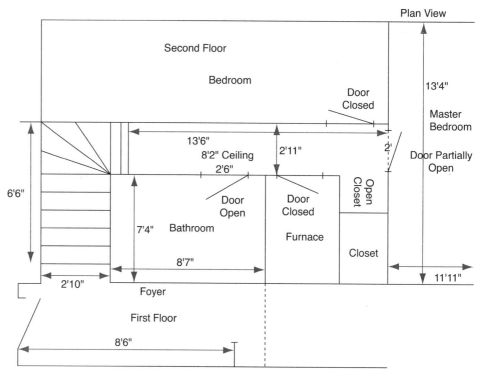

FIGURE 9.34 ◆ Floor plan of apartment. Fire on stairs. Elderly victim was in master bedroom at end of upstairs hallway. *Courtesy of J. D. DeHaan.*

victims, as no one was reportedly present. The victim was discovered when firefighters opened windows in the upstairs bedrooms to vent smoke and steam.

The door at the end of the hallway was reportedly nearly closed when firefighters searching the apartment entered (Figure 9.37). The bedroom [approximately 4×3.7 m (13×12 ft)] was heavily charged with smoke. The windows were closed. The victim was found on his bed by the sound of his labored breathing. He was removed from the room and resuscitated with oxygen but succumbed to inhalation injuries and

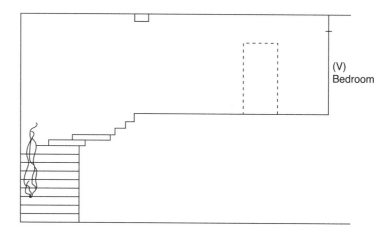

FIGURE 9.35 ◆ Elevation drawing of apartment. *Courtesy of J. D. DeHaan.*

FIGURE 9.36 ◆ View from foyer to midstair landing, with hall and the victim's bedroom door to the right. Main fire was on stairs to left. *Courtesy of Halton Regional Police Service, Oakville, Ontario, Canada.*

extensive first- and second-degree burns on the front half of his torso and face the next day. The level of the hot smoke layer in the room was 1.5–1.8 m (5–6 ft) from the floor, as evidenced by discoloration of the walls and the softening of a plastic clock face on one wall. The collapse of a plastic lamp shade on a table and light smoke deposits visible on various surfaces throughout the room as well as the observations of the firefighter indicated that smoke had largely filled the room. On admission to hospital the victim was found to have a COHb saturation of 23.3 percent (after some 40 min on oxygen). He was seen to have soot in the nostrils.

The fire damage was limited to the stairwell, and the issue became, What were the conditions in the bedroom such that the victim could have suffered those injuries? He was observed to have no singed hair and he appeared to be dressed in a cotton T-shirt and underpants that did not appear to be thermally damaged (his outer clothing was found on a sofa in the room). It was conjectured that the fire gases produced by the stairwell fire had penetrated around the partially open door during the course of the fire and its suppression. The radiant and convected heat of the accumulated hot gases (supplemented in part by the steam generated by the suppression activity) induced first- and second-degree burns on the victim.

The heat release rate of a fire at the foot of the stairs can be estimated by applying the flame height calculation method from *NFPA 921* (NFPA 2004a), described in Chapter 3, on fire pattern analysis:

$$H_f = 0.174 \, (k\dot{Q})^{2/5}, \tag{9.13}$$

$$\dot{Q} = \frac{79.18 \, H_f^{5/2}}{k}, \tag{9.14}$$

where

H_f = flame height (m),
$\dot{Q}$ = heat release rate (kW), and
k = wall effect factor, with

    1 = no nearby walls,
    2 = fuel package at wall, and
    4 = fuel package in corner.

The area of direct flame contact (both intermittent and continuous) is determined by establishing the area of most damage to the wall covering, which extended to a height, H_f, of approximately 2.7 m (8.85 ft) from the foyer floor, as shown in Figure 9.36. Calculation of the heat release rate, $\dot{Q}$, with a wall effect factor of $k = 2$ (fuel package at wall) gives

$$\dot{Q} = \frac{(79.18) \, (2.7)^{5/2}}{2} = 474 \text{ kW (or} \sim 500 \text{ kW)}. \tag{9.15}$$

From to the position of the fire, we can assume that the smoke production of a 500-kW fire accumulated in the stairwell and hallway above, as shown in Figures 9.37 and 9.38. According to the Zukoski (1978) method for smoke production rates, the equation for estimating the mass flow rate of a plume above the flame at 20°C (68°F) is

$$\dot{m}_p = 0.065 \, \dot{Q}^{1/3} \, Y^{5/3}, \tag{9.16}$$

where

$\dot{m}_p$ = rate of smoke-filled gas production (kg/s),
$\dot{Q}$ = total heat release rate (kW), and
Y = distance from virtual point source to bottom of smoke layer (m).

For a fire of $\dot{Q} = 500$ kW and $Y = 5$ m (16.4 ft) at a time close to its start, the estimated smoke production rate along the centerline is

$$\dot{m}_p = (0.065) \, (500)^{1/3} \, (5)^{5/3} = 7.51 \text{ kg/s}. \tag{9.17}$$

As the corridor fills, the height becomes $Y = 2.5$ m (8.2 ft), and

$$\dot{m}_p = (0.065) \, (500)^{1/3} \, (2.5)^{5/3} = 2.37 \text{ kg/s}. \tag{9.18}$$

With a maximum density of 1.2 kg/m³ these fill rates, at a minimum, are 6.1 and 1.94 m³/s. With an average T of 200°C (392°F), the density would be reduced by a

FIGURE 9.37 ◆ View of the hallway from the landing. The victim's bedroom was at the end of the hallway. The smoke pattern confirmed that the door to the bedroom was partially open during the fire. Note the lines of demarcation of damaged and undamaged wall surfaces. *Courtesy of Halton Regional Police Service, Oakville, Ontario, Canada.*

factor of 273/473 (T_a/T_g), or approximately 0.7 kg/m³, and the fill rates would be, correspondingly, 10.6 and 3.3 m³/s.

Given the dimensions of the upper hall (including the staircase), 6.1 m long × 2.4 m high × 0.9 m wide (20 × 7.87 × 2.95 ft), its volume was only 13.2 m³. Even allowing for the filling of the bathroom, which measured 2.2 × 2.4 × 2.4 m (7.2 × 7.9 × 7.9 ft), with a volume of 13.7 m³ (484 ft³), the hall and bathroom [26.9 m³ (950 ft³)] would have been filled with smoke in 5 s or less after the fire reached 500 kW. The buoyancy of the gases accumulating at the hall door would have caused them to push around the partly opened door from floor to ceiling and begin to fill the bedroom beyond. The bedroom was 35.5 m³ (1254 ft³) in volume, and the more buoyant hot gases coming around the door would have produced a 0.6- to 0.9-m (2–3-ft) deep layer (from the ceiling) within a few minutes.

FIGURE 9.38 ◆ View down the corridor from the bedroom toward the stairwell. Note the demarcation surface effect of smoke and heat patterns on the far wall. *Courtesy of Halton Regional Police Service, Oakville, Ontario, Canada.*

The radiant heat from a gas layer in a compartment was established by Quintiere and McCaffrey (Quintiere 1998, 62; Quintiere and McCaffrey 1980), where smoke layers with temperatures between 200°C and 300°C (392°F and 572°F) were found to produce radiant heat fluxes of 3–5 kW/m² on the floor of the compartment. Because the radiant heat threshold for second-degree burns is of the order of 4 kW/m² (Quintiere 1998, 60), it can be seen that radiant heat from a hot gas layer alone can induce second-degree burns (given 30 s) or more at temperatures of approximately 250°C (482°F) (far below flame temperatures, which range from 500°C to 1200°C).

In addition, convective heat transfer is occurring as the room fills with hot fire gases and steam and adds to the thermal insult to which the body is subjected, increasing the likelihood of burns to victims in rooms filling with combustion products. This fire is a good illustration of the application of (1) indicators, (2) flame height versus heat output relationship, and (3) Zukoski's smoke-filling estimations.

This case is courtesy of Halton Regional Police Service, Oakville, Ontario, Canada.

◆ **9.10 CASE EXAMPLE 10: THE BURNING BED**

A fire was reported at 0830 hours in a ground-level apartment of a residence by a neighbor who saw smoke coming from a window. The fire department made entry a few minutes later to find no smoke or heat on the main floor but considerable smoke in the ground-floor garage and utility area. Entry into the adjacent apartment revealed very heavy smoke and only moderate heat. Firefighters entered through a double-doored corridor to find a flaming fire on a bed, which was quickly knocked down. A badly burned body was found on the burned double bed (Figure 9.39). Fire damage in the room was limited to the body, the bed, the adjacent wall (Figure 9.40), and a scorch mark on the ceiling directly above the bed. Smoke levels throughout the studio-style apartment were down to ~0.5 m (1.6 ft) of the floor (based on postfire photos). The apartment was well sealed from the utility area, and all doors and windows were closed (except for a 1-cm (0.4-in.) gap where one window did not close completely. As shown in the floor plan (Figure 9.41), the main living area was 6.7 × 3.9 m (22 × 13 ft) with a flat ceiling 2.3 m (7.5 ft) above the floor. Exterior weather conditions were cool and drizzly with no strong winds.

The bed was only recently purchased (receipts still lay nearby), and film in the decedent's camera was processed to reveal the arrangement of furniture, bedding, and decoration just a day prior to the fire. The mattress was nearly completely consumed (including under the torso) and much of the bedding was consumed. Fortunately, the remains of the bed and bedclothes were well documented and preserved as evidence. The decedent had a low COHb level (commensurate with her smoking habit) and no indications of inhalation of hot gases or soot. The body was burned over 90 percent of its surface, and some 7 kg (15.5 lb) of body mass had

FIGURE 9.39 ◆ Bedroom with badly burned body and bed. Note smoke level in room. *Courtesy of San Francisco Arson Task Force, San Francisco, California.*

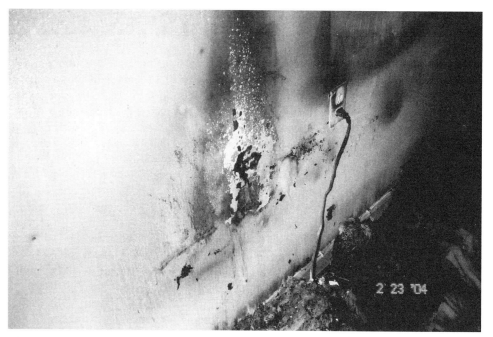

(a)

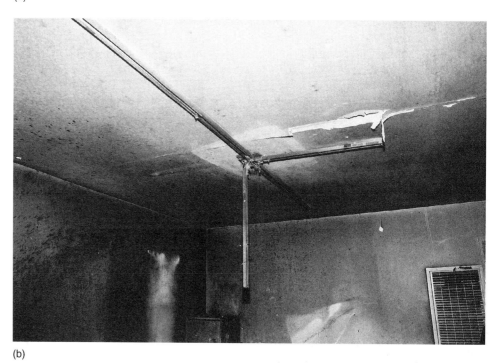

(b)

FIGURE 9.40 ◆ (a) Limited damage to wall at head of mattress was the only damage to structure. (b) Minimal fire damage to ceiling above bed. *Courtesy of San Francisco Arson Task Force, San Francisco, California.*

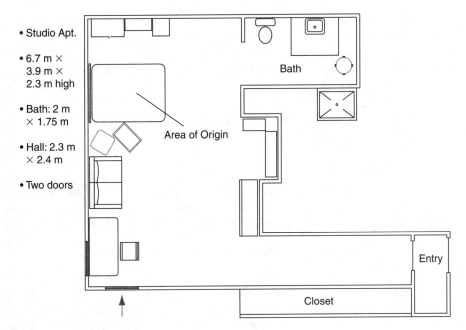

- Studio Apt.

- 6.7 m × 3.9 m × 2.3 m high

- Bath: 2 m × 1.75 m

- Hall: 2.3 m × 2.4 m

- Two doors

Bath

Area of Origin

Entry

Closet

FIGURE 9.41 ◆ Floor plan of apartment. Note extended entry hallway with double doors. *Courtesy of San Francisco Arson Task Force, San Francisco, California.*

been consumed (based on prefire information from the decedent's mother) versus her postmortem weight.

The origin of the fire was the bedclothes or bed itself, and the synthetic nature of those materials precluded a smoldering cigarette and falling incandescent lamp as ignition sources. The nearby gas-fired wall heater was in good repair and showed no physical evidence of having ignited nearby combustibles (no smoke deposits, scorch marks, or melted adhering residues). The apartment was in good repair with no reported malfunctions. An empty whiskey bottle under the bed (not the decedent's) indicated the possible use of alcohol as an accelerant, but chemical tests on the remaining residues of bedclothes were negative.

The central issue was the duration of the fire. It is well known (from tests by NIST and others) that modern foam-padded mattresses like this one will readily produce a fire of 1 MW or greater once ignited by an open flame as a "fast t^2" fire. Such a fire [with a calculated flame plume for a 1.2-MW fire of 2.9 m (9.5 ft)] would be expected to reach the ceiling and induce major thermal damage to it. Ventilation, or lack thereof, was the key. As we saw previously per *NFPA 555*, the duration of burning can be calculated for a t^2 fire in a tightly closed compartment via

$$t = [(3V_0/\sigma)(\Delta h_c \rho_0)]^{1/3} \qquad (9.19)$$

In this case, the fire was started on the bed, some 0.6 m (2 ft) above the floor. This meant that $\dot{Q}$ never exceeded 400–500 kW (flame plume height <1.7 m (5.6 ft)). For a fast t^2 mattress fire (with $\sigma = 0.047$), $\dot{Q}$ would have reached 2 MW in unlimited oxygen, and clearly that did not occur.

The solution for t (above) can be calculated:

$$\Delta h_c = 13 \, \text{kJ/g}$$

$$\rho_0 = 1.2 \, \text{kg/m}^3$$

If the whole apartment provides air to the fire, total volume is $\sim 80\ m^3$ (860 ft²), but for a fire on the bed, only the volume of air above or level with the fire (top of the bed) can contribute, so that reduces the volume of air to 59 m³ (635 ft²). At 21 percent oxygen, this means that $V_{O_2} = 12.4\ m^3$ (133 ft²) at a maximum. With that value,

$$t = [(3V_0/\sigma)(\Delta h_c \rho_0)]^{1/3}$$
$$= [((3)(12.4)/(0.047))((13\ KJ/g)(1200\ g/m^3))]^{1/3}$$
$$= [(791)(13)(1200)]^{1/3}$$
$$= 230\ s$$

Because synthetics must burn as a flame rather than smolder, the effective V_{O_2} is reduced by 50 percent of actual volume (per *NFPA 555*) to 6.2 m³ (67 ft²), because flaming combustion will not be sustained if the O_2 content is less than 50% of normal. Then

$$t = [(396)(13)(1200)]^{1/3}$$
$$= 182\ s$$

That would mean the duration of the initial flaming fire was on the order of 3–4 min.

Clearly, some leakage was occurring at the tops of the walls (allowing smoke into the adjacent garage/utility area), around entry doors (double—found closed but not locked), and windows (see Figure 9.42). The combustion of the bed, bedding, and body all required open flame, which would not be supported in a vitiated layer. The combustion of 7 kg (15.5 lb) of body mass requires a flaming fire of some 1–2 hr when the body is the main fuel being consumed. The entering firefighter saw a flaming fire on the bed, but that was the result of a flare-up when fresh air was admitted as he entered the hall doors. A neighbor had seen smoke coming from a rooftop vent on the house (the water heater vent from the adjacent garage) an hour before the fire was called in, so there was some considerable smoke in the garage as early as 0730 hours. It was ultimately concluded that the fire was started deliberately on the bed some 2–5 hr prior to extinguishment and that it was quickly vitiated by the descending smoke layer and cycled between brief spells of open-flame combustion and smoldering of cotton bed linens for most of that time (as illustrated in Figure 9.43). This time factor was critical in discrediting an alibi for the decedent's boyfriend.

In September 2006, an experiment was conducted to test that mechanism. A tightly sealed room $3 \times 2.85 \times 2$ m high ($9.8 \times 9.4 \times 6.6$ ft) was fitted with a urethane mattress pad [$0.8 \times 1.7 \times 0.1$ m thick ($2.6 \times 5.6 \times 0.3$ ft)] with a cotton fabric cover on a bed mock-up 0.6 m (2 ft) high. A 0.6×0.6-m (2×2-ft) vent was fitted to a rear wall, at a sill height of 1 m (3.3 ft). Fixed glass windows from floor level to 0.5 m (1.6 ft) allowed visual monitoring of the fire. The bed was ignited by open flame at one corner, and temperatures of the ceiling and the center of the mattress were monitored by thermocouples. The door was closed within 30 s of ignition. Within 5 min of ignition, the ceiling temperature had risen from 20°C to 161°C (68°F to 322°F) and the temperature at the center of the mattress had risen to 80°C (176°F). At 10 min smoke deposits were visible on the tops of the view windows [down 0.04 m (0.13 ft) on the glass]. The glass became warm to the touch at the top but cool nearer the floor. At this time the ceiling temperature had fallen quickly to 89°C (192°F), and by 11:25 min the ceiling and bed temperatures had both dropped further [to 84°C and 72°C (183°F and 162°F)], so the vent was opened at that time. At 20:20 min the ceiling temperature had returned to 89°C (192°F) and rose steadily to 189°C (372°F) at 24:05 min. At 21:00 min flames were again visible on the bed, and the mattress temperature hit 220°C

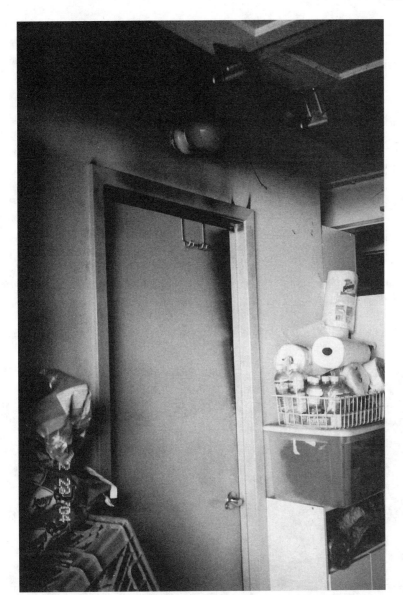

FIGURE 9.42 ◆ Entry from garage showing limited smoke penetration. *Courtesy of San Francisco Arson Task Force, San Francisco, California.*

(428°F); 30 s later, it hit 720°C (1328°F) as the mattress burned quickly [mattress temperature dropping to 102°C (216°F) by 24:05 min]. The remaining fire was extinguished at 26:20 min. The results are seen in Figures 9.44 and 9.45. The smoke horizon is at 0.4 m (1.3 ft) throughout the room (below the level of the bed), and the mattress is entirely consumed, but fire damage to the room is limited to an area of the wall adjacent to the pillow. This was a very close duplication of the conditions seen in the fatal fire, confirming the hypothesis of a cyclical fire behavior controlled by the smoke level in the room.

Case study courtesy of John D. DeHaan.

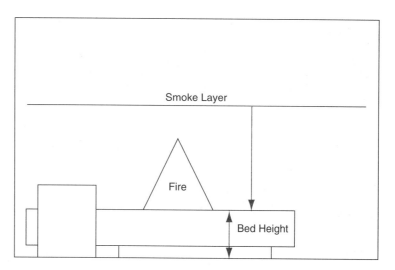

FIGURE 9.43 ◆ Simplified room cross section showing descent of smoke layer. *Courtesy of J. D. DeHaan.*

FIGURE 9.44 ◆ Smoke layer descending in test room. *Courtesy of J. D. DeHaan.*

(a)

(b)

FIGURE 9.45 ◆ (a) Postfire: complete consumption of foam mattress. (b) Postfire: smoke level at bed level. *Courtesy of J. D. DeHaan.*

◆ 9.11 CASE EXAMPLE 11: THE CHILDERS BACKPACKERS FIRE

A June 2000 fire in a hostel in Australia resulted in the deaths of 15 guests and injuries to many more of the 70 guests and three staff who escaped. Owing to the construction of the building, suppression was very difficult in a central area, resulting in considerable destruction and collapse. As shown in Figures 9.46, 9.47, and 9.48, the scene consisted of what were once three separate buildings (built in stages between 1902 and 1940).

The original two-story accommodation block was timber framed with masonry walls (ordinary construction), and the coverings of walls, floors, and some ceilings were all finished wood. Some rooms retained their original stamped-tin ceilings. The TV lounge (dining hall) and the kitchen block were of masonry construction—cement coating on brick. Second-story areas over the kitchen and TV/dining lounge contained sleeping rooms. The areas between the original buildings were roofed over to form an atrium. Open wood staircases, "bridges," and balconies opened onto this atrium. A sloped sheet-iron roof covered the building, and a fibro-cement panel roof with skylights covered the atrium. A modern service block had been added to the rear of the building with toilets and shower rooms. The weather at the time was cool and very humid with some fog.

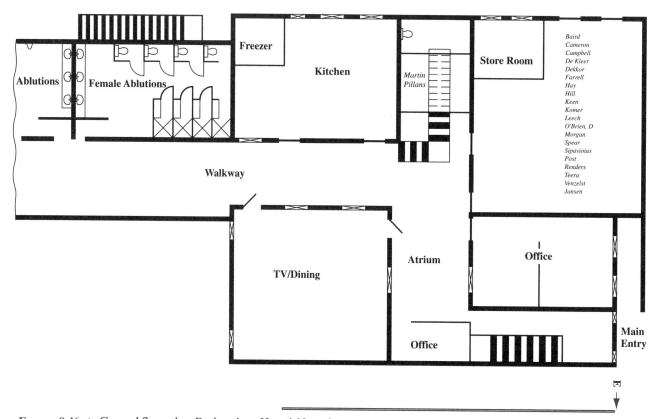

FIGURE 9.46 ◆ Ground floor plan: Backpackers Hostel. Note three separate buildings (kitchen, TV/dining, and original accommodation block (on right)). *Courtesy of Queensland State Coroner and Queensland Police Service, Queensland, Australia.*

FIGURE 9.47 ◆ Top floor plan. *Courtesy of Queensland State Coroner and Queensland Police Service, Queensland, Australia.*

Guests were awakened by the sounds of breaking glass, shouts of alarm, and footsteps on the wood stairs and balconies between 12:20 and 12:30 A.M. All guests who fled via the atrium saw fire erupting from the doors or windows of the TV lounge and none elsewhere. The fire was reported at 0031 hours (12:31 A.M.) by a guest who saw fire in the TV lounge. Fire crews were on scene by 0037 hours to find the center of the complex well involved and numerous guests needing rescue from adjoining rooftops. Extensive interviews with all the guests revealed that several people had been in or passing by the TV lounge between 11:50 P.M. (2350 hours) and midnight (0000 hours) and had noticed nothing out of the ordinary. Smoke detectors were fitted, but the fire alarm system had been disconnected about a month before the fire pending repairs. One guest had seen an unidentified man in the room "around midnight" observing a fire in paper towels in a plastic wastebin in the southwest corner of the room. He cautioned the stranger and watched as the man carried the plastic can out the south door of the main hall. (Remnants of the now-burned can were seen during the evacuation of the building and were visible in postfire photos.)

The scene was thoroughly documented by the Queensland Police Accident Investigation Squad using laser-based Total Stations and digital still and video cameras. Apple QuickTime VR interactive imaging was used to produce an interactive "tour" that allowed the viewer to move from room to room and get a 360° look at each

FIGURE 9.48 ◆ Front of accommodations block facing street (panoramic photo). *Courtesy of Queensland State Coroner and Queensland Police Service, Queensland, Australia.*

room. Because the building was a heritage listed building, there had been a survey of the building in 1993 (before its conversion to a hostel). In addition, guests provided photos and videos taken during previous stays, and promotional videos and flyers provided (nearly current) views of the interior of the TV lounge.

Portions of the wood-framed rooms on the floor above the TV lounge had collapsed into the lounge, but portions of the room's ceiling were still intact (see Figure 9.49). The debris in the lounge was layered out to reveal the remains of wood framing of the room's chairs, sofas, and tables. These remains were compared with the prefire documentation to establish the location of major furnishings. In addition, staff members and guests were asked to draw diagrams of the furniture layout the night of the fire.

Based on this documentation, the TV lounge was reconstructed accurately despite the extensive destruction. This room was 7.6 × 8.0 m (25 × 26 ft) in size with a 3.9-m (12.8-ft)-high ceiling and was described as having painted masonry (cement rendering over brick) walls with fibro-cement ceiling panels. It had two doors, one in the west wall (with no door) in the southwest corner and one on the north wall in the northwest corner (with a wood door in place but in the full-open position). Each door was 0.92 m (3 ft) wide and (6.9 ft) 2.1 m with a 0.55-m (1.8-ft)-high glass transom above each. There were two large windows in the south (exterior) wall [each 1.0 m (3.3 ft) wide × 1.75 m (5.7 ft) high], two large windows in the west wall (opening onto the west hallway) [each 1.25 m (4.1 ft) wide × 1.75 m (5.7 ft) high], and one large window in the north wall [1.27 m (4.2 ft) wide × 1.95 m (6.4 ft) high] (opening onto the atrium). Sills for all windows were 1.4–1.7 m (4.6–5.6 ft) above the floor. As shown in Figure 9.50a, the room was furnished with a number of upholstered armchairs, two- and three-seat

Figure 9.49 ◆ Interior of TV lounge. *Courtesy of Queensland State Coroner and Queensland Police Service, Queensland, Australia.*

(a)

Figure 9.50 ◆ (a) Prefire: still from promotional video showing furniture. *Courtesy of Queensland State Coroner and Queensland Police Service, Queensland, Australia.*

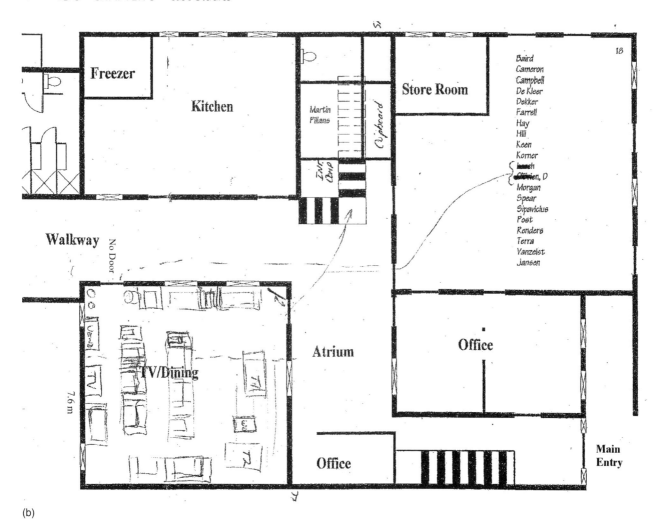

Baird
Cameron
Campbell
De Kleer
Dekker
Farrell
Hay
Hill
Keen
Korner
~~Leash~~
~~Obrien~~, D
Morgan
Spear
Sipavicius
Post
Renders
Terra
Vanzeist
Jansen

FIGURE 9.50 ◆ (*Continued*) (b) Sketch of room showing reconstructed furniture at the time of the fire. *Courtesy of J. D. DeHaan.*

sofas, two large wooden picnic tables, miscellaneous wood tables, a large TV, a stereo system, snack vending machine, and two plastic wastebins in the southwest corner. There were no draperies or other wall coverings, but there was carpeting of undetermined type on the floor. The upholstered furniture constituted the major fuel load in the room, but neither the owners nor the staff could identify the components or the age (except for an estimate that it was used but in good condition, dating from the late 1970s to late 1980s). This means that the pieces were most likely synthetic fabric (or synthetic/cotton blend) over either latex foam or polyurethane foam padding. Examination of the remains by on-scene examiners failed to reveal any identifiable residues owing to extensive and prolonged burning. The location of furniture items was re-created based on witness descriptions and prefire photos, as shown in Figure 9.50b.

Wood trim on the door frame and wood frames of the tables and sofas were burned all the way to floor level, and areas of the cement wall coating were spalled down to floor level. No identifiable remnants of the floor covering were recovered, and nearly all combustibles within the TV lounge had been consumed. These observations are all indicators that the fire in the TV lounge had proceeded to flashover

FIGURE 9.51 ◆ Interior of north-south hall. Entry to TV lounge on left. Fire hose not in service. *Courtesy of Queensland State Coroner and Queensland Police Service, Queensland, Australia.*

and had burned as a postflashover fire for some time. Extensive venting of flames from the window and door openings (as evidenced by burn patterns on the exterior walls) would also have been produced by postflashover burning.

As illustrated in Figure 9.51 and 9.52, patterns of fire damage to the other portions of the building (actually two buildings, north and south, joined by the roofed atrium) were traceable to flame plumes from the windows and doors of the TV lounge impinging on the combustible wood structure of the atrium balcony, stairway, and "bridge"; radiant heat igniting exposed combustibles or causing the failure of glass windows in the rooms above the TV lounge (south building); or to immersion in or contact with the hot gas layer in the atrium. The hot gases venting from the TV lounge and those produced by the combustion of the wood balcony and stairs in the atrium caused a hot gas layer to form just under the ceiling. This hot gas layer then caused failure of glass door panels and transom panels in the south-facing portion of the north building and also penetrated into the roof void of the north building. Hot gases penetrating the north building from the roof void and from the hallway (balcony) door ignited the combustible painted wood walls and ceilings and caused the upper-floor rooms to burn from the top down. There was also penetration of the ground-floor rooms of the north building resulting from failure of the glass transoms above the entry doors and failure of the ceiling joists, which caused the metal ceilings in the ground-floor rooms to collapse (late in the fire).

The wood balcony in the atrium was almost completely destroyed, as was the wood "bridge" that connected that balcony (and the north-south hallway in the north building) to the first-floor north-south hallway in the south building. There was less damage to the ground-floor north-south hallway (to the west of the TV lounge), and it appeared that there was less extension out the west-side windows and

FIGURE 9.52 ◆ Main entry hall—looking south. Note collapsing landing of floor above.
Courtesy of Queensland State Coroner and Queensland Police Service, Queensland, Australia.

door and out the south windows than there was out the window and door on the north side of the TV lounge. This indicates that the largest concentration of fire was probably located toward the northwest or north-central portion of the TV lounge. These observations and the condition of remaining wood furniture frames within the lounge strongly suggest that the fire was first established in that northwest quadrant of the room, most probably in the sofas and chairs located there. Because of the extensive damage to the contents of the room, no more specific area of origin could be established.

The ceiling of the TV lounge was largely collapsed at its center with extensive damage to the wood floor joists and decorative wood beam coverings above the center of the room, as well as penetration upward through the floor into the accommodation rooms above. This ceiling damage indicated that a flame plume at least 3.5 m (11.5 ft) high was sustained in the center of the room (equivalent to a fire with a minimum heat release rate of 2.5–3 MW.

The analysis of this fire used numerous relationships described in this text. Based on the dimensions of the room and its ventilation openings, it was possible to calculate the size of the fire necessary to cause flashover in the room. Based on the patterns of fire damage visible postfire and especially on the descriptions of witnesses (who reported flames exiting the doors and windows of the TV lounge when the fire was first seen), it was certain that flashover had occurred prior to penetration of the ceiling and floor above and prior to any downward collapse of the upper-floor rooms.

At the time of the original case analysis, a version of the Thomas correlation from *FPETool* was used to estimate the minimum heat release rate to cause flashover. Based on the size of the room and interior finish, it was estimated that a fire with a minimum heat release rate of 4–5 MW would be needed to produce flashover. The correlations in the *NRC—Fire Dynamics Tools* spreadsheets were used to calculate the effects of ventilation variables. [For purposes of these calculations h_0 was assumed to be 2.6 m (8.5 ft) for all openings.] If only the two doors to the TV lounge were open, A_0 would be 3.68 m² (39.7 ft²), and the minimum Q_{f0} (per MQH formula) would be 2 MW, Babrauskas: 4.1, and Thomas: 3.9. If the transoms were open (as they probably were) or failed from fire exposure, A_0 would be 4.78 m² (51.5 ft²), and the results would be MQH: 2.3 MW, Babrauskas: 5.6 MW. When all windows and doors were open, the total A_0 would be 15.12 m² (162.8 ft²), and MQH: 2.8, Babrauskas: 8.5, and Thomas: 6.1 MW.

The ventilation factor would be calculated from the equation

$$\dot{Q}_{max} = 1260 A_0 \sqrt{h_0}. \tag{9.20}$$

Until the ceiling failed, the maximum fire supportable by the two open doors alone would be 7.5 MW. With doors and transoms open, the maximum would be 9.7 MW. With all doors and windows open, the fire could reach 30.6 MW (if there were enough fuel available!).

A traditional cotton-upholstered, cotton-padded sofa cannot achieve a heat release rate anywhere near that level (being more typically 300–500 kW). Given that a modern (polyurethane foam/synthetic fabric) three-seat sofa can achieve maximum heat release rates of 2–3 MW in isolation (i.e., large rooms), it can be concluded that a single such sofa would not be capable of bringing such a room to flashover. To do so would require at least two such sofas burning at or near maximum heat release rates at about the same time. Because a sofa ignited by open flame requires about 4–5 min to reach its maximum heat release rate (in a large room) and then begins to decay to a smaller fire—and a total of about 8–10 min to burn from one end to the other—a fire ignited at one location by a single flaming source would have to ignite at least two other pieces of similar furniture to achieve flashover potential. To do so would require at least 15 min for flames to spread from one item to the others by direct flame contact given the end-to-end arrangement of sofas and chairs in the lounge before the fire. (Based on the apparent location of the remaining furniture framework, there did not appear to be a rearrangement or stacking of the furniture prior to the fire that could have reduced the time for full involvement.) It was concluded that multiple ignition from multiple ignition sources or a single flame source applied at several locations would be required to produce the fire observed in the time available. Flame spread from a single item ignited in the southwest corner (i.e., near the wastebin seen afire earlier) would not be expected to produce the required end-to-end fire progression in the time available and would be expected to produce more flame extension out the west windows and door than out the north. Modern synthetic upholstery materials are generally not susceptible to ignition by contact with a small smoldering source such as a dropped cigarette or glowing electrical connection. Latex foam rubber and cellulosic materials (cotton upholstery or cotton, sisal, or jute padding) on the other hand can be ignited by such sources or by open flame. Because the materials used in this furniture could not be identified, either type of ignition source has to be considered theoretically competent. However, ignition by glowing/smoldering source is much slower than ignition by open flame, requiring many minutes before open flame is observed. For instance, a fire started by placing a burning cigarette on chair or sofa upholstery even under the best conditions of fuel and ventilation will require 22–90 min or more before flaming combustion is established. During that time (prior to open flame), large quantities of smoke are produced that would be detectable to any conscious person in the vicinity. Witnesses in and around the room less than 20 min before the fire was observed as a fully developed fire detected no smoke. Once a smoldering fire progresses to open flame, the fire grows much like a flame-ignited fire for some minutes before the furniture item is fully involved.

As the fire in the lounge approached flashover, the glass windows in the room and the glass transoms above the doors failed, allowing more venting of hot gases and smoke and flame into adjoining areas. It was estimated that with all the doors and windows fully open, a fire of more than 30 MW could be supported in this room.

From a human tenability and fire safety viewpoint, several reconstructions were assessed. As the fire gases escaped from the room of origin, they rose to the low "ceiling" created by the atrium roof, where they quickly accumulated at transom level. [The "header" height between the atrium ceiling and the tops of the transoms was estimated to be 0.1 m (0.3 ft) or less.] Here the hot gases caused failure of the glass transoms above the guest room doors in the north building and penetrated the rooms and the ceiling and roof void. As the wooden wall, ceiling, and roof components of the north building ignited, the fire spread through the ceiling space and burned the north accommodation rooms from the top down. Doors on the upper-floor level of the south building facing the atrium "bridge" appeared to be simple glass-panel wood doors. Doors (if any) from the north building to the balcony could not be identified, but photos show no evidence of any substantial or fire-rated door. Such non-fire-rated doors would have failed quickly and allowed combustion gases from the atrium to enter the upper-floor halls of both north and south buildings and quickly ignite the exposed painted wood wall and ceiling components of the hallways, aiding rapid fire spread.

Fire gases in the atrium were trapped under the ceiling and also banked down against the balcony exit doors of the upper-floor rooms that faced the atrium. These doors once opened onto the open rear balcony of the original building. Some of these rooms had exit doors facing the central corridor but some did not, and one "dormitory" room had wooden bunk beds built against the corridor doors, rendering them useless. The one exterior window of this room was fitted with security bars. All 10 occupants of this room died there, most of them found in a pile against the blocked window.

The person identified as having set the fire was convicted of arson and murder in 2002. It was set as a revenge fire, as he had been evicted from the hostel previously and had argued with some guests.

Case study courtesy of Queensland Police Service and John D. DeHaan.

◆ 9.12 CASE EXAMPLE 12: WOLF HOUSE

On August 22, 1913, fire gutted the nearly completed dream house of author Jack London, located on a hilltop overlooking Glen Ellen, California. Its massive stone outer walls on the lower two floors and heavy (rough log) timber upper structure were considered fire resistant. The house, as depicted in the architect's rendering (Figure 9.53), was basically a three-story (masonry with wood floor) structure with an open central pool and courtyard, opening at one end through a stone pergola. The

FIGURE 9.53 ◆ Architect's rendering of Jack London's house, from the south. *Courtesy of California State Parks.*

1394 m² (15,000 ft²) house was nearly completed after more than 2 years of construction, and the Londons planned on moving into it just a few days later. After the fire, the ashes and charred timbers were cleared and dumped down an adjacent hillside, in preparation for rebuilding. Replacement timbers were cut and aged, but the house was never rebuilt. The Londons lived in "The Cottage" on the grounds until Jack's death in 1916. In 1960 the property was donated to the state of California as a state park. The building ruins were stabilized but not modified, as seen in Figure 9.54. Various hypotheses were offered about the cause of the fire, including arson (by radicals or by London himself). These were eventually discounted for lack of evidence. Wolf House was the culmination of one of London's dreams and he was crushed not by the loss of a house but by the destruction of so much beauty. Accidental causes were minimal because electrical service had not yet been connected, and it was a typical clear, still August night in the Valley of the Moon, although very hot (40°C, 100°F) (after a very hot day). London had visited the house about 5:00–6:00 P.M. to inspect the day's progress and detected nothing amiss. An insurance investigator toured the fire scene but did no excavation and found no obvious causes. An insurance policy for $6000 was paid off (against a house worth over $70,000 in 1913 dollars). The cause was always considered undetermined.

In May 1995 a multidisciplinary team of engineers and fire experts under the direction of Professor Robert Anderson offered their donated services to the State Parks Department to see whether the fire's origin and cause could be established. The Parks Department provided a trove of newspaper stories, interviews, photos, letters, and building plans for our use, and unlimited access to the site for 4 days. The facts of the event were established from those sources. The fire was reported shortly after midnight by residents of Glen Ellen, who responded to the scene (about a mile from town) with what little equipment was available. They reported the house to be nearly completely involved on all levels on their arrival. One of the first people to spot the fire was a young girl who could see the already large fire from her room. When asked

FIGURE 9.54 ◆ Remains of building after the fire. View of west wall (looking east). The windows at the center of the lower floor open into the dining room. The windows above are of the library above. The large roof-level concrete structure to center left was a water cistern. *Courtesy of California State Parks.*

why she was not in bed asleep at such a late hour, she told reporters she couldn't sleep because "it was the hottest night she could remember." That bit of information became crucial later.

It was clear from the notes and letters that London was modifying and customizing the dream home as it was being built. Plans dated as late as June 1913 (see Figures 9.55, 9.56, and 9.57) did not reflect some of the features such as door and stairway placement found in the "as built" remains. A careful survey was carried out to verify what openings existed to provide a path for the spread of flames and hot gases throughout so much of the structure. Although the burned timbers had been removed, and much of the heat and smoke patterns erased by decades of exposure to the elements, traces remained. It was evident that a staircase (which was not on the plans) had been built from the dining room (along its east wall) to the library above. London's notes and letters talked of a spiral staircase extending from the library to his writing room on the third level (also not on the plans). Both the staircase and the spiral stairs would provide a chimney for fire gases to spread quickly from a ground-floor origin to roof line. Doorways between the ground-level "party room" and the expanded kitchen were also different from the plans.

London also often spoke of his dislike of "artificial" or manmade coverings for walls, floors, and ceilings (even for plaster and paint) and his insistence on rough "natural finishes." The heavy supporting timbers were redwood trunks with their bark intact. Although the floors were usually stone (or wood), the walls were rough-cut

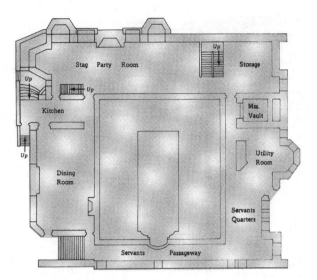

FIRST LEVEL

FIGURE 9.55 ◆ Floor plan (proposed) of ground level. The staircase at the end of the party room was apparently replaced by a stairway built against the east wall of the dining room (opening into the library above). The kitchen was expanded to the north, and the end of the party room was closed off with a north-south wall with two doors. *Courtesy of California State Parks.*

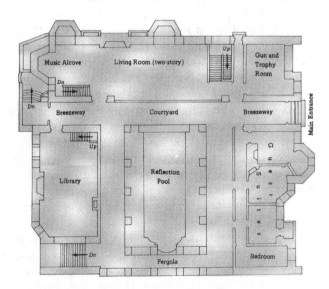

SECOND LEVEL

FIGURE 9.56 ◆ Floor plan (proposed) of main floor with entrance at the northwest corner and a two-story living room (main hall) and the library. No evidence was found of interior stairway near the "music alcove." *Courtesy of California State Parks.*

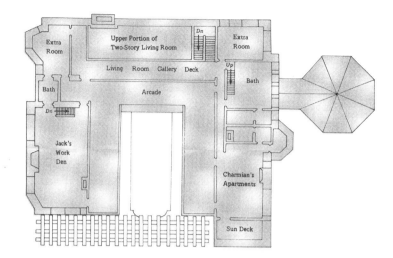

THIRD LEVEL

FIGURE 9.57 ◆ Floor plan (proposed) of second floor showing Jack's work den. The staircase at its north end was replaced with an open spiral staircase near the center of the room (described by London). *Courtesy of California State Parks.*

wood planks or concrete covered with bamboolike mats (small fragments of this matting were found in hidden crevices during the 1995 exam). Beam pockets in the "party room" and living room walls were found to contain the ends of the ceiling joists. The burn patterns on these revealed that the fire had burned from ground level upward in those rooms. Large open doorways and stairwells aided in the spread of fire gases from room to room. Open-air pergolas, balconies, and incomplete window installation guaranteed adequate ventilation for any fire once started. The all-wood upper floors provided ignition of other portions by radiant heat and falling embers. There was no furniture present to support fire spread.

Once the interior structure could be reconstructed, fire engineering analysis was carried out (assisted by *Hazard I* computer modeling) to test various hypotheses about the fire's origin. Observations by first-arriving witnesses indicated spread involving at least three levels, so ignition somewhere on the ground level was most likely. Owing to numerous openings, stone load-bearing walls, and still air conditions outside, a fire started on an upper floor would be unable to ignite the lower floors until structural collapse occurred. The one area that would account for the most rapid vertical and horizontal spread was the kitchen/dining room area because of the large doorways and staircases in this area communicating to the adjacent wings of the building (see Figures 9.58 and 9.59).

With a working hypothesis as to area of origin in mind, consideration focused on causes and possible ignition sources. The kitchen appliances were not yet in service, and with the very hot weather there had been no fireplace or heating system use. The main electrical service had not yet been connected. Some electrical switches, receptacles, and conduits were recovered from the dump site adjacent to the house. None of these showed any electrical activity. The house was only days from completion, so all hot work for plumbing had been completed long ago. One note, however, mentioned that the cabinets and woodwork in the kitchen and dining room were being "finished" on the day of the fire. With London's dislike for paint and insistence on natural

FIGURE 9.58 ◆ North wall of ground-level party room. Photo taken during 1995 investigation. *Courtesy of Robert N. Anderson.*

FIGURE 9.59 ◆ Photo taken during 1995 investigation of west end of party room. Door on left opened into kitchen (not shown in plans). Remains of raw wood covering were found on adjacent concrete walls. *Courtesy of Robert N. Anderson.*

products, it is almost certain that wood finishing entailed hand-rubbed linseed oil finishes. The timeline for the hours preceding the fire was reconstructed from notes and interviews in the archives. Work was finished for the day (being a Friday) at about 4:00 P.M. London toured the house about 5:00–6:00 P.M. He was the last person known to be in the structure. The fire was seen, already quite large, just after midnight. It is well known that linseed oil–type finishes on rags have a propensity to self-heat to ignition, particularly at elevated ambient temperatures. Such self-heating can vary with conditions, but a time frame of 6–8 hours is quite common. In the early stages any odors generated may be dismissed as a normal paint smell, and any smoke may be dissipated by drafts through open windows and doors, thereby avoiding detection.

We can never know, of course, what started this historic fire, but virtually all reasonable hypotheses about origin and cause can be examined and ruled out. The remaining hypotheses about both origin and cause fit all available data and, thus, are readily defensible. The investigation did offer cautions about establishing structural conditions from a variety of sources, since building plans may not reflect the "as built" condition.

Case study courtesy of Prof. Robert Anderson, Los Altos Hills, California, and John D. DeHaan.

◆ 9.13 SUMMARY AND CONCLUSIONS

This chapter introduced the concept of detailed case histories as a learning tool in applying and understanding the principles of fire engineering analysis. With case histories, many underlying concepts, fire science principles, and new areas of research are uncovered. Controlled tests can be of value in formulating or testing a hypothesis for the origin of a fire or its growth, the behavior of the occupants, or an act or omission that contributed to the fire loss.

Although many principles are discussed, the case analyses are not all-conclusive; the reader is free to apply additional concepts in reviewing, discussing, or expanding on the case histories.

Problems

9.1. Find a recent fire death case in which "smoking materials" was the identified cause. What data would you collect and what tests would you use to confirm this conclusion?

9.2. Examine the photographs in one of the examples. Try to determine additional information from this review that did not appear in the case study.

9.4. Using the equations provided regarding plume height against a wall, determine the maximum-size fire that would produce plume damage as seen in Case Example 1 (p. 386). Assuming a fuel mass loss rate of 4 kg/hr and a Δh_{eff} of 32 kJ/g, calculate the duration of the fire.

9.5. Review three of the historic cases covered in the *NFPA Journal* and evaluate how today's knowledge could have helped solve these cases more quickly through testing of alternative hypotheses.

Suggested Reading

DeHaan, J. D., S. J. Campbell, and S. Nurbakhsh. 1999. The combustion of animal fat and its implications for the consumption of human bodies in fires. *Science and Justice* 39 (1): 27–38.

DeHaan, J. D., and S. Nurbakhsh. 2001. Sustained combustion of an animal carcass and its implications for the consumption of human bodies in fires. *Journal of Forensic Science* 46 (5): 1076–81.

Holleyhead, R. 1996. Ignition of flammable gases and vapors by cigarettes: A review. *Science and Justice* 36 (4): 257–66.

Lentini, J. J., D. M. Smith, and R. W. Henderson, 1993 Unconventional wisdom: The lessons of Oakland. *Fire and Arson Investigator*. 43 (June): 18–20.

Zeigler, D. L. 1993. The One Meridian Plaza fire—A team response. *Fire and Arson Investigator* 43 (September): 38–39.

CHAPTER 10

Future Tools for the Fire Investigator

*I had come to an entirely erroneous conclusion which shows, my dear Watson, how danger-
ous it always is to reason from insufficient data.*

—Sir Arthur Conan Doyle,
"Adventures of the Speckled Band"

◆ 10.1 INTRODUCTION

There are many new tools available for the fire investigator that can improve the
documentation, analysis, interpretation, and reconstruction of fire events, and many
more in development. As with most new technologies, many of these are costly at
present but will decrease in price as development continues. A number of these have
been briefly described in the relevant sections of this book (digital scanning cameras, photogrammetry, Total Station surveying, etc.). This chapter examines such devices in more depth and suggests concepts that investigators can use today and into
the future.

◆ 10.2 PRESENT AND EMERGING TECHNOLOGIES

Several present and emerging technologies have shown promise. Investigators often
seek out their own solutions, particularly when approached by universities, government laboratories, and vendors who think they may have an insight into the overall
problems.

Reference in this book to any product, process, or service by trade name, trademark, manufacturer, or otherwise does not necessarily constitute its endorsement or
recommendation by the authors, their agencies, or the U.S. Government.

DIGITAL CAMERAS

Computer technology has made dramatic improvements in the capabilities of even relatively low-cost digital cameras. It is common now for cameras to offer image capture quality of 5MB or higher, with 512MB flash memory readily available. Digital cameras are now available with SLR lens-and-shutter assemblies, so a wide variety of tasks can be addressed. One of the most useful innovations is the incorporation of a microphone and sound recording chip into the camera. With each photo taken, the photographer can dictate a description of the photo up to 20 s in length. The chip can then be downloaded and the narration converted to a printed photo log. This eliminates the need to set the camera down and make a log entry with each photo (or to try to recreate a log from memory after viewing the photos). Such features are available on the Nikon D100, Sony 727 and 828, and some Olympus digital cameras. For those who prefer to use a film camera, small voice-activated digital voice recorders are available at low cost from retailers such as Radio Shack. These can record up to 1 hr on a single chip in a device the size of a half-pack of cigarettes without the use of fragile tape cassettes.

DIGITAL SCANNING CAMERAS

With the advent of easy-to-use computer image software such as Roxio PhotoStitch, QuickTime VR, and PTGui for creating panoramic photos from a series of still photos, investigators have realized what an advantage such panoramas can be, especially for briefing counsel and making courtroom presentations. One of the first systems was an optical technology combined with digital processing called the iPix system. It allowed the investigator to capture a 360° view of a scene and produce a fully immersive, navigable (virtual reality) image. It used a 185° fish-eye lens on a digital camera mounted on a level tripod. The camera was rotated horizontally 180° between two scenes and iPix software stitched the two images together. The result allowed the viewer to "virtually" enter the scene and look up, down, to either side, or even directly behind. The iPix technology has been used successfully at fire scenes under the Arson Information Management System (AIMS 2000 project, discussed later) and by a variety of police and fire departments.

A new addition to panoramic photography has been introduced by Hewlett-Packard. This company now produces cameras with "in-camera" panorama preview and stitch features on certain digital cameras (www.hp.com).

A newer technology that offers greater flexibility and improved images is the digital scanning camera. In such products, the camera, mounted on a suitable tripod, rotates and scans a 360° view of the room (or outdoor area) in a single pass, as described in Chapter 4. The resulting image requires no stitching and is ready to view. The Panoscan MK-3 (see Figure 4.22a) uses a laptop PC to capture high-dynamic-range, high-resolution images in a scan time of less than 60 s (panoscan.com). It requires no special lighting equipment, and its image format is available as a flat panorama (see Figure 4.22b) or as a navigable virtual reality "movie" using Apple QuickTime VR, Java iPix, and other compatible players. The camera uses a Mamiya 645 wide-angle (160° vertically) lens (focal length adjustable from 2.4-mm fish-eye to 300 mm) but can also scan in IR and UV. Its high-speed shutter, triple A-to-D converters, and special processors result in high-quality scans even at 8-s scan times. The Spheron VR system is an alternative that was also described in Chapter 4 (spheronvr.com). Such systems currently range in price from $30,000 to $60,000 including software and control system. Also available is the Crime Scene Virtual Tour (crime-scene-vr.com). The latter product uses the Java Virtual Machine (Sun Java) software to produce a fully integrated, virtual scanned image with links to floor plans, close-up still photos, and investigative notes. It is presently available for under $10,000.

Photogrammetry (measurement) capability is available on both the Panoscan and the Spheron VR systems. In each case, after the area is captured in a single scan, the camera is elevated some (18 in.) on a vertically extendable mast, and the area is scanned again (see Figure 4.39). Combining the images digitally (the Panoscan system uses proprietary software called PanoMetric™) then allows an investigator to re-create a three-dimensional VR image where measurements between any two points can be taken with considerable accuracy by clicking on them in the 3D image, as shown in Figure 10.1. The resulting data are then saved in DXF format for export to AutoCAD, Maya, and other programs. The data from the Spheron VR scan is visualized through proprietary software called Scene/works (contact: rsmck@csimapping.com).

Name	AngU	AngL	DistU	DistL
Point 1	9.6118	17.161	185.91	191.84
Point 2	9.3562	16.926	185.62	191.45
Point 3	12.427	25.524	101.79	110.16
Point 4	12.321	25.903	97.918	106.34
Point 5	-22.17	-14.90	195.19	187.05
Point 6	-22.38	-15.99	221.05	212.64
Point 7	-35.67	-24.96	124.75	111.78
Point 8	-37.06	-25.48	114.97	101.64
Point 9	5.4285	19.412	99.770	105.30
Point 10	4.7060	17.486	110.21	115.16
Point 11	-33.72	-23.47	131.82	119.53
Point 12	-35.99	-25.33	124.82	111.73
Point 14	-21.93	-14.67	195.48	187.44
Point 15	1.8509	9.7890	182.40	184.99
Point 16	1.9105	10.552	167.24	170.03
Point 17	-23.96	-16.17	181.21	172.42

Distance	Name	Start	End
190.4201	Point 1–Point 2	Point 1	Point
172.0330	Point 1–Point 3	Point 1	Point
173.8684	Point 2–Point 4	Point 2	Point
104.7400	Point 1–Point 5	Point 1	Point
116.3098	Point 2–Point 6	Point 2	Point
203.2748	Point 5–Point 6	Point 5	Point
94.6839	Point 3–Point 7	Point 3	Point
90.5487	Point 4–Point 8	Point 4	Point
187.4401	Point 7–Point 8	Point 7	Point
168.4865	Point 5–Point 7	Point 5	Point
198.5529	Point 6–Point 8	Point 6	Point
35.3531	Point 9–Point 10	Point 9	Point
82.8236	Point 9–Point 12	Point 9	Point
82.2440	Point 10–Point 11	Point 1	Point
79.1906	Point 16–Point 17	Point 1	Point
35.5251	Point 15–Point 16	Point 1	Point
78.9438	Point 14–Point 15	Point 1	Point
35.5487	Point 14–Point 17	Point 1	Point

Figure 10.1 ◆ An example of two stereoscopic Panoscan images from the same location but at two different camera heights (note different view angles). The digital output on the right shows the distances (in centimeters) between any two selected points. *Courtesy of Panoscan Inc.*

TOTAL STATION SURVEYING SYSTEMS

The Total Station is a resource that has been widely used by accident investigation teams in large police departments (as well as by commercial surveying companies) for many years in the United States, the United Kingdom, and Australia, yet it is rarely used by fire investigators, probably because of its high initial cost.

As described in *Kirk's Fire Investigation* (DeHaan 2007) and in Chapter 4 of this text, this system uses a laser beam from a surveyor's theodolite to measure the distance from the reference point (which in turn is located by GPS data) to any point in view as seen in Figure 10.2. The data system captures the distance, direction, and elevation of every point at which the laser is aimed and codes it as an edge, corner, point, or feature, as instructed by the operator.

The computer data are then downloaded to a PC for translation into a finished diagram with scale dimensions. The newest generation of systems will operate remotely at distances of up to 100 m (330 ft) without a reflector or marker and at much greater distances when a reflector or marker can be positioned to reflect the laser beam back to the system with a higher efficiency. Plotting software is then used to produce accurate scaled diagrams of the scene (in both plan and elevation views). Today, using a wireless link called fieldPro™, a Total Station can be connected directly to an AutoCAD system to create 2D and 3D diagrams at the scene. See Figure 10.3 for an example.

The newest versions (such as the Leica FMS307) have a small video screen to display a preview version of the plot to ensure that all important data points are measured. The system can be used indoors or out and can capture critical measurements of a large scene far more accurately than can be achieved with a tape measure or visible-light optical range finder. Such systems deserve more evaluation and use by fire and explosion investigators. They are currently offered by Leica, Topcon, and Sokkia.

GPS MAPPING

One innovative approach to mapping large scenes using handheld GPS units has recently been improved by Magellan. Using a GPS device (the Mobile Mapper CE™) the size of a cell phone, with GPS Differential™ software extension, the examiner walks to various points at the scene and keys in a code (ArcPad®) denoting a particular set of features. The data are then downloaded and processed by a specialized software (Mobile Mapper™ Office) to produce a plot of the scene accurate to ±0.5 m (depending on the number of satellites in "view" and the stabilization time at each location). Such devices allow the mapping of large and complex scenes where viewlines for Total Station surveying would be obscured by terrain or features (trees, equipment, etc.). The handheld device contains all the usual features of GIS/GPS location systems (contact: mobilemapping@magellangps.com).

LASER SCANNING SYSTEMS

An emerging technology that combines Total Station measurement accuracy with the convenience of digital scanning cameras is laser scanning. Just as flat-bed laser scanners have become commonplace in offices for capturing electronic images of photos and documents, laser scanners to capture external images have become more widely used and lower in cost. Several manufacturers now offer PC-driven scanners that are placed in a room or at an outdoor scene and scan a laser beam

FIGURE 10.2 ◆ A Leica TPS400 Total Station in use surveying a building. *Courtesy of Leica Geosystems, Inc.*

vertically and then rotate horizontally (using mirrors) taking distance and angular deflection measurements at the rate of several thousand per second. Distance may be measured by time of flight (TOF) (where the delay time between a pulsed output and its returned reflection determines distance) or phase shift (where the beam is continuous and the interference between outgoing and returned light waves determines distance).

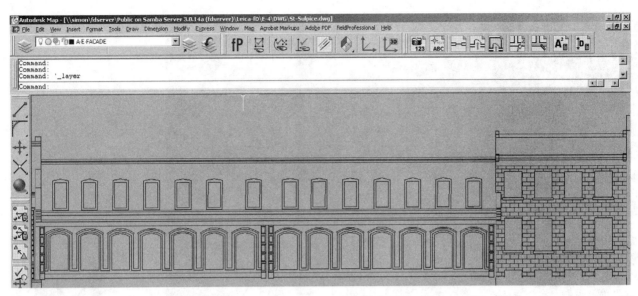

FIGURE 10.3 ◆ Using a fieldPro™ device, a Leica Total Station can be linked directly to an AutoCAD system to produce 2D drawings or 3D models at the site. *Courtesy of Leica Geosystems, Inc.*

All this information is then analyzed by specialized software on a laptop computer to produce navigable 3D images that allow for highly accurate measurements and re-creations. Two examples of a modern laser scanner are shown in Figure 10.4, and an example of a room scan image is shown in Figure 10.5. Some scanners are integrated with a high-resolution digital camera (or are coupled with an external digital camera) that captures colors and textures of surfaces, information that is then merged with the laser measurements to produce a fully navigable 3D virtual reality (VR) image of any interior or exterior scene. A scene can be scanned from two or more locations and then the images "registered" or stitched together to produce a very realistic rotatable VR image. Scans of separate areas can also be linked together.

A technology firm in North Carolina (3rd Tech Inc., 2500 Meridian Parkway, Suite 150, Durham, NC 27713, 3rdtech.com/) has developed a laser-scanning device that uses a time-of-flight, 5-MW laser range finder that, when mounted on a photographer's or surveyor's tripod, will scan an entire room with the laser beam, taking 25,000 measurements per second.

Leica Geosystems (5051 Peachtree Corners Circle, #250, Norcross, GA 30092; Forensic Account Manager, tony@lgshds.com) offers a ScanStation scanner that uses an "eye-safe" green laser that will operate successfully in rooms or outdoors from full daylight to total darkness. Their TruView data systems are fully compatible with most third-party software and allow simple translation to CD or FTP format for remote analysis.

The data from such scans can be analyzed and displayed in many ways. The 3D VR image can be rotated and viewed from any angle or any "position" within the

(a)

(b)

FIGURE 10.4 ◆ (a) A Leica ScanStation laser scanner in use at a wildland fire scene. *Courtesy of Leica Geosystems, Inc.* (b) A DeltaSphere laser scanner with laptop controller. *Courtesy of 3rdTech, Inc.*

image. Outlines, wireframes, and diagrams can be created. "Hot spot" links to close-up photos, lab reports, notes, or chain-of-custody information can be added. Additional distance and angle measurements can be taken and recorded at any time. See Figure 10.6 for some examples.

There are currently three major manufacturers who offer laser scanners with different capabilities and limitations (Leica Geosystems, 3rd Tech, and Reigl). Some

FIGURE 10.5 ◆ DeltaSphere panoramic (360°) view of a mock crime scene. The operator specified the vertical field of view (65°) and the horizontal field of view (360°). He specified the point density (13.33 points per degree), and the rest was automatic. Each vertical line of these data contains almost 1000 measurements. It takes the DeltaSphere 12 min to make more than 9 million measurements in a complete scan. (In this scene the operator chose not to scan the ceiling but could have captured all the data overhead as well, in the same amount of time.) The accuracy is better than 6 mm (0.25 in.). The data that appear curved, like the edge of the floor or the top of the closet, result from displaying these 3D data as a 2D projection. These are not distortions in the data. When displayed in 3D these edges display correctly. *Courtesy of 3rdTech, Inc.*

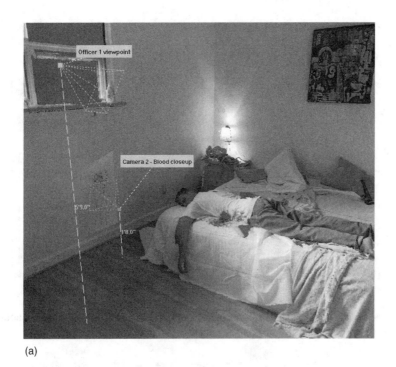

(a)

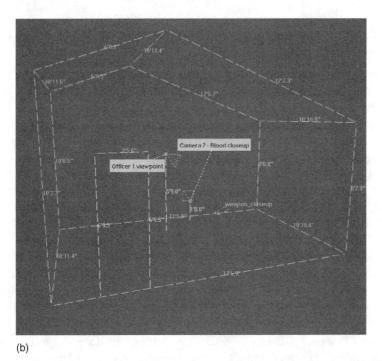

(b)

FIGURE 10.6 ◆ (a) A laser scan of a mock crime scene using the DeltaSphere with high-resolution digital images blended in using the Scene Vision-3D software to create a realistic VR image. The "viewpoint" of an observer and the location of the close-up camera are shown in the image. (b) A simple, accurate line diagram (outline) created from a Delta-Sphere image of a scan using Scene Vision-3D functions. Even data not visible in the scanned image such as corners hidden by the furniture or parts of the ceiling can be re-created.

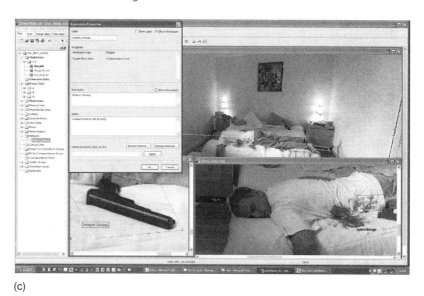

(c)

FIGURE 10.6 ◆ *Continued* (c) Close-up photos of evidence and scene notes have been added linked to "hot spots" on the scanned image. *Courtesy of 3rdTech Inc.*

have already found their way into police and security agency use. Selection considerations include the following:

Accuracy (resolution): Point (scan spot) size and point spacing (equivalent to the dpi rating of digital images). How has accuracy been measured, by whom, and under what conditions?

Distance range: Some scanners have a minimum range of 2 m (6.6 ft), so they would not be useful for interior scans. Others have a maximum range of 16 m (52 ft), so they would not be useful for large indoor scenes or exterior scenes. (Distance upper limits are dependent on the reflectivity of the target [e.g., 300 m (980 ft) at 90 percent reflectivity versus 134 m (440 ft) at 18 percent reflectivity], so the conditions of the test and nature of the target are crucial. Ranges are offered up to 280 m (900 ft), but the size of the laser spot at that distance will limit the resolution and accuracy.)

Wavelength: Some wavelengths perform better with dark (charred) surfaces, which will be a concern to fire investigators. Some wavelengths encounter interference with ambient light and suffer from "day blindness" and require the scanned area to be darkened. Others (e.g., Leica) will work in daylight or full darkness.

Accuracy (distance): Accuracy of distance measurements depends on the type (TOF or phase shift) and range. Accuracy is typically on the order of 4–7 mm at 50 m (160 ft).

Field of view: Some scanners will scan from vertical (90° up to ~60° down, others can scan only from 45° up to 45° down and require acquisition (a rescan) to capture information from directly overhead.

Scan rate: Most systems will complete a 360° scan of a room in 12–30 min, taking 4000–25,000 measurements per second.

Safety: All systems use a Class I laser, so eye safety is a concern. Some systems are designed (by scan rate) so that duration of exposure to an unprotected eye will not exceed the federal safety limits. Others require eye protection for users or others in the room.

Court acceptability: Have the results of such scanning systems been found reliable and acceptable to courts?

Photography: Some systems (Leica) have an integrated high-resolution camera that scans with the laser. Others (3rd Tech) have a supplemental camera that attaches to the scanner and captures a photographic image simultaneously with the laser scan. Either technology captures photographic color images overlaid on precise measurement grids.

Ease of "registration" or stitching: Most scenes require scanning from at least two positions. Systems (software) vary in the ease with which that can be done. Some systems produce a data format that can be entered directly in AutoCAD and similar drafting programs.

Such scanners are not inexpensive, but they permit rapid, extremely accurate measurement of all critical features in a fire scene, such as room dimensions, ventilation openings, and structural features (beams, sloping ceilings, ductwork). The captured images can be revisited at any time to conduct additional measurements. Data from some software systems can be imported into CAD systems for rendering to fixed images or creating a physical model. Readers are encouraged to contact vendors or their manufacturer websites (at the end of this chapter) for detailed information.

Depending on the wavelength of their lasers, laser scanners can capture variations in surface condition (reflectivity) that may not be apparent in standard color photographs. Compare the "false-color" laser scan of a burned test cubicle (Figure 10.7a) with a color photo of the same scene (Figure 10.7b). Note the thermal plume damage variations to the back wall that are not apparent in the color photo. Large, complex scenes can be captured in less than 1 hr with great precision. See Figure 10.8 and 10.9 for examples.

DIGITAL MICROSCOPES

Many investigators carry an eye loupe or linen tester magnifier to scenes to enable them to examine small objects, but capturing those images is a problem (which may be critical if the object is likely to change with time owing to evaporation, handling, or storage). A handheld digital microscope (MiScope®) is available that combines built-in LED lighting, precision optics, and a digital movie camera. It offers 40–140X magnification (field of view 6×8 mm to 1.5×2 mm), and the image is fed directly to a laptop computer (via a USB port) for examination and capture and labeling (as still, movie, or time-lapse images). It is available for less than $300. A high-resolution version (MiScope MP®) offering 12–140X magnification with 2.3-μm resolution is available for under $700 (zarbeco.com) (Zarbeco LLC, 12 Ridge Road, Randolph, NJ 07869, 973-989-2673) (see Figure 10.10).

PORTABLE X-RAY SYSTEMS

Laboratory X-ray systems (such as Faxitron) have been used by crime labs and engineering labs for many years to evaluate evidence of all types after recovery. These are similar to medical X-ray systems using variable voltages and exposure times to suit a wide variety of target materials (Faxitron X-Ray Corp., 225 Larkin Dr., Suite 1, Wheeling, IL 60090, faxitron.com).

(a)

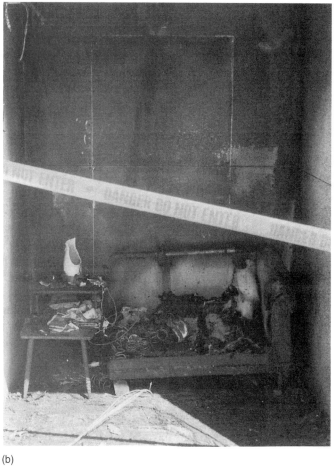

(b)

Figure 10.7 ◆ (a) False-color laser scan of a burned room reveals burn pattern features (changes in reflectivity caused by heat). *Courtesy of Leica Geosystems Inc.* (b) Normal photograph of the same scene does not capture the gradations in damage. *Courtesy of J. D. DeHaan.*

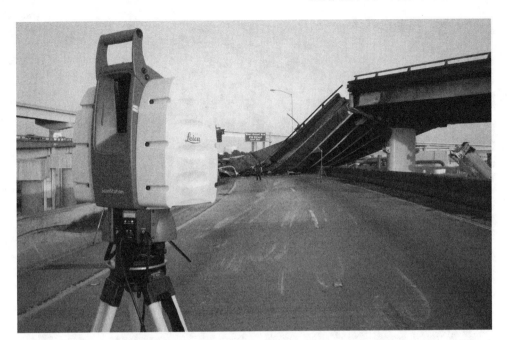

FIGURE 10.8 ◆ A massive overpass collapse triggered by a burning gasoline tanker truck was fully documented by the California Highway Patrol using a Leica ScanStation, allowing salvage and repair work to begin within a few hours of the crash. *Courtesy of Leica Geosystems Inc.*

FIGURE 10.9 ◆ The "point cloud" of data representing the scanned measurements of the overpass collapse forms the basis for any measurements needed in the reconstruction. *Courtesy of Leica Geosystems Inc.*

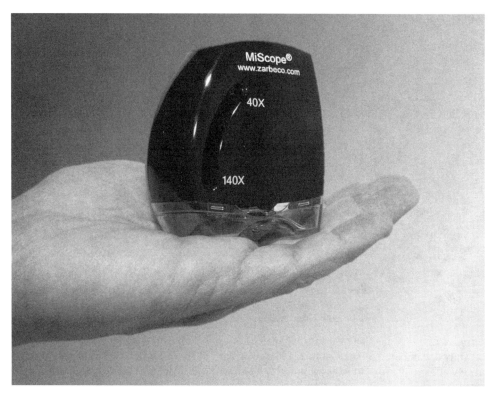

FIGURE 10.10 ◆ A handheld digital microscope (magnification 40X–140X) has built-in LED lighting. Its images are displayed on a laptop computer and captured as movie or still images. *Courtesy of Zarbeco LLC.*

Portable X-ray systems have been used for some years by bomb disposal squads to evaluate suspected devices in situ, but the recent developments in spoliation issues have made on-scene X-ray examination much more desirable for the fire investigator. One such system is a battery-powered X-ray system weighing less than 2.5 kg (5 lb). The XR150 system by Golden Engineering uses electrically generated pulses of X-rays of very short duration (50 ns) but very high power (150 kV) (see Figure 10.11). The images are accumulated based on the type and thickness of the material (1–2 pulses for paper envelopes, 100 for steel up to 1 cm (0.4 in.) thick. Images can be produced on a "green screen" fluoroscope, a "live" digital image system, or on a conventional 20 × 25-cm (8 × 10-in.) Polaroid film sheet. Such devices would allow a trained and qualified investigator to assess the internal condition of electrical devices, appliances or mechanical systems before they are disturbed or removed from the scene (thus minimizing risk of damage or loss). Such systems are currently available for under $5000 (Golden Engineering, P.O. Box 185, Centerville, IN 47330, goldenengineering.com).

PORTABLE X-RAY FLUORESCENCE

Another valuable technique, X-ray fluorescence elemental analysis, long used in the laboratory, can now be performed on a portable handheld unit the size of a power drill (as in Figure 10.12). It uses low-energy X-ray (10–40 kV) beams focused on a nearby surface to identify all but the lightest elements in one scan of a few seconds. It can be used to help detect and identify chemical incendiaries, low explosives, bromine-based flame retardants, chemical treatments, coatings, and fragments of

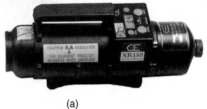

(a)

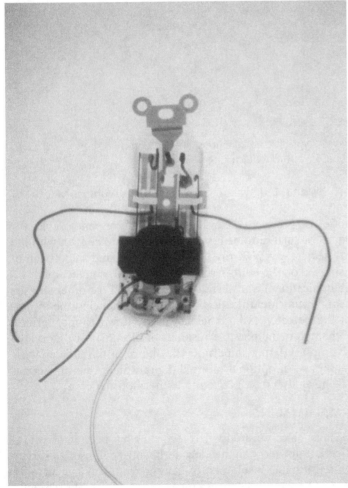

(b)

FIGURE 10.11 ◆ (a) A portable X-ray unit can take images of electrical equipment and appliances at a scene using a 150-kV X-ray source. (b) Its image can be captured on film, fluoroscope screen, or digital devices. *Courtesy of Golden Engineering Inc.*

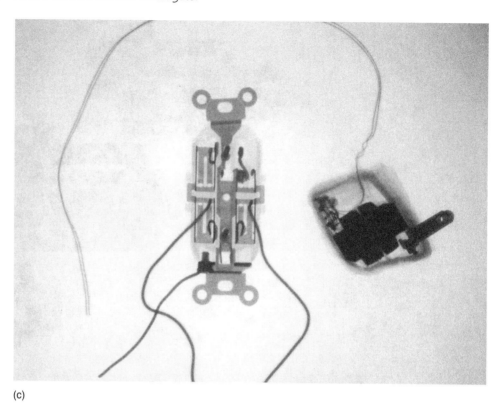

(c)

FIGURE 10.11 ◆ *Continued* (c) A typical X-ray of a household receptacle is shown on the left, plug-in power supply on right. *Courtesy of Golden Engineering Inc.*

solids of all types (Innov-X Systems, 100 Sylvan Street, Suite 100, Woburn, MA 01801 781-938-5005, innovxsys.com).

Thermo Fisher Scientific also produces a handheld X-ray fluorescence device for elemental analysis. The object or surface to be tested is brought into contact with the window of the Niton XL© device. A 2-W X-ray tube is triggered, and a spectrum of the elements present is displayed on the built-in screen or (via USB connection or integrated Bluetooth™ communications) transferred directly to a PC or other storage device. The standard system permits identification of elements from chlorine to uranium. A helium purge of the detector extends its sensitivity down to magnesium, making it very useful for characterizing possible residues of incendiary or inorganic explosive materials. The Niton© system also permits the choice of full-area analysis or a 3-mm "small-spot" analysis. An optional internal digital camera permits documentation of the sample being analyzed (contact: niton@thermofisher.com).

INFRARED VIDEO THERMAL IMAGING

Pioneered by Inframetrics, false-color infrared video recording has been in use in industrial and military applications for more than a decade, having been used to survey remotely and quickly high-tension line insulators for heating from leakage currents, transformers for overheating, and chemical delivery pipes and storage tanks for leaks, and to measure operating temperatures of surfaces of aircraft or vehicles.

Its applications to fire research would seem self-evident, providing the capability of remotely and accurately measuring the surface temperatures of fuel surfaces or

FIGURE 10.12 ◆ A handheld energy-dispersive X-ray fluorescence (XRF) analyzer can simultaneously identify up to 25 elements in a sample at a scene, helping identify chemical incendiaries, metal alloys, and other materials. *Courtesy of Innov-X Systems Inc.*

masses of hot gases or of the external surfaces of walls, windows, or other structural elements as they are exposed to fires.

A very early version was used by Helmut Brosz in 1991 in monitoring fire development in compartments (Brosz, Posey, and DeHaan, unpublished). Although De-Haan (1995, 1999) used Inframetrics equipment to measure the surface temperatures of carpet during the evaporation of volatile fuels, very few papers on the topic have been published in the fire research literature. Infrared video imaging can distinguish areas of a surface whose temperatures differ by only 0.25°C (1.8°F) in static images. Its applicability to capturing transient events such as the chaotic movement of the high-temperature flames in postflashover fires has been demonstrated (DeHaan 2001). See Figure 10.13 for an example.

The high cost of the necessary equipment has been reduced significantly in recent years with the dramatically increased power of low-cost computer components. It is to be hoped that more researchers will use this type of technology to study surface temperatures that presently are impractical to measure accurately using traditional hard-wire thermocouple devices. Several versions are made today by FLIR Systems (www.FLIRthermography.com).

MRI AND CT IMAGING

Progress in magnetic resonance imaging (MRI) and computerized tomography (CT, or multiple-position X-ray imaging) has been very rapid. Recently, MRI and multiple-slice CT were demonstrated, in the examination of a very badly charred body, to establish the presence or absence of fractured bones and reconstruction of impact injuries as well as to document the presence of surgical implants and the like (Thali et al. 2002). Such techniques have also been suggested for use in examination of melted or charred artifacts from fire scenes where nondestructive testing for pockets of

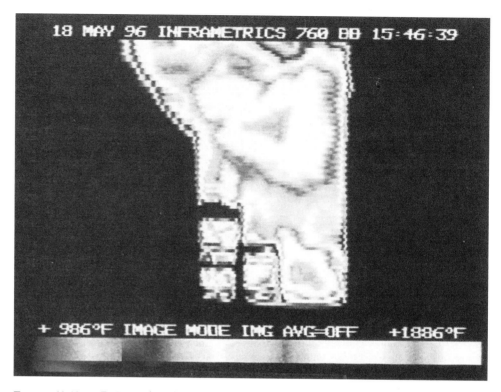

FIGURE 10.13 ◆ DeHaan (2001) used Inframetrics infrared (FLIR) video to study the dynamics of chaotic postflashover fires. *Courtesy of J. D. DeHaan.*

liquid or similar nonradiopaque inclusions would be desirable. Although expensive, such techniques may provide answers where X-rays cannot.

POLYNOMIAL TEXTURE MAPPING

A new method for increasing the photorealism of maps and complex surfaces is known as polynomial texture mapping (PTM). This technique could be used to enhance and visualize fire patterns on various surfaces such as wood paneling, gypsum, and concrete. PTM was originally used to enhance the surfaces of hard-to-read ancient clay tablets for interpretation by archaeologists, and a pilot system is available through a Hewlett-Packard/Compaq website (hpl.hp.com/ptm/).

LIGHTING

A major problem for interior fire scenes or exterior scenes at night is the provision of suitable lighting. Quartz-halogen worklights are widely used, but they have limited coverage and produce glare and blindspots. One new innovation uses a halogen/metal halide lamp inside a translucent balloon mounted on an extendable mast. An "Airstar" balloon creates uniform white glare-free light over a large area. These balloons range in size from 200 W to 4000 W and with masts from 1.8 to 9 m (6 to 36 ft) in height and can cover areas from 200 m^2 (2100 ft^2) to 10,000 m^2 (108,000 ft^2) (contact: airstar-light.us).

THE ATF FIRE RESEARCH LABORATORY

The ATF Fire Research Laboratory in Ammendale, Maryland, has been completed. It has the physical facilities to observe and measure fires of all sizes, from bench-scale tests to very large fires involving entire rooms, vehicles, and even multiple furnished rooms. It is anticipated that its unique blend of fire engineers, fire and forensic scientists, and public-sector fire investigators will create a wealth of knowledge about the properties of fires involving commonly encountered fuels such as cardboard boxes, liquid fuel spills, furniture, and even vehicles in "real-world" events.

The lab has been functional since 2004 and has been used to test many scenarios, validate or refine many of the existing models, and also refute a few of them (or at least show the limitations of their applicability) with scientifically valid studies with demonstrable uncertainties or error rates. This is exactly the kind of knowledge that the *Daubert/Kumho* decisions have begun to require of all fire investigators. The lab holds exciting promise for the future of fire investigation as it aims toward meeting the goals set out in 1997 at the planning conference described in the preface to this book (Dillon, Hill, and Sheppard 2003; Nelson and Tontarski 1998). See Figure 10.14 for examples of the large-scale tests that are possible at this facility.

(b)

(a)

FIGURE 10.14 ◆ Full-scale fire tests such as these at the ATF Fire Research Laboratory, Ammendale, Maryland, yield important data to test hypotheses about fuels, fire growth, ventilation effects, and effects on structural materials. (a) Postflashover fire test in the large burn room. (b) Even well-ventilated room fires can be studied in the medium burn room as they approach flashover. *Courtesy of ATF/FRL.*

Fire investigators, particularly those who operate in rural areas, often transform their vehicles into mobile offices. These vehicles carry all the necessary tools, forms, maps, photographic, and evidence collection equipment needed to document fully a fire scene and compose a draft preliminary investigative case report. Operating in a reactive mode in high-visibility cases, fire investigators are expected to deliver their investigative reports and photographs on a timely basis but are hindered by both technical and administrative hurdles.

Based on lessons learned from the Major Fire Investigations Program, the Federal Emergency Management Agency's (FEMA) U.S. Fire Administration (USFA) molded new information technologies to improve the quality and productivity of investigations conducted at fire scenes, explosions, disasters, and other major events. The source of many of these lessons learned by the USFA was insights into the on-site field investigations of the World Trade Center (1993) and Oklahoma City (1995) bombing cases.

The Arson Intervention and Mitigation Strategy (AIMS 2000) project, under the direction of the USFA, supports local, state, and federal arson investigations through the introduction and dissemination of innovative fire investigation technologies. The purpose of this section is to report on the productivity successes of the AIMS 2000 strategy.

Facing the concerns of increasing case loads and shrinking budgets, investigators often inquire as to the availability of U.S. military–developed technologies that could increase their efficiency on the fire scene. Military technology transfers, however, are sometimes difficult to sustain, particularly in the areas of training, hardware support, and procurements.

For example, U.S. military special operations personnel during Operation Just Cause and Desert Storm carried portable mobile field offices (the size of commercial aircraft–sized luggage) onto foreign battlegrounds. These transportable units, which typically contained a laptop computer, digital camera, color scanner, and printer gave soldiers the ability to investigate, photograph, and record battlefield targets accurately. This information was quickly assembled in the field and transmitted to field commanders for review and dissemination.

The operational requirements for conducting and documenting investigations at scenes of major fires and explosions are similar to those of the U.S. military and include a factor requiring a unique balance of quality and productivity. The AIMS 2000 strategy exploited the military solution and resulted in a novel effort to produce a field-deployable "virtual office" dubbed the Transportable Rapid Information Package (TRIP) Kit. Common practices, challenges, and proposed solutions form the functional foundation for the TRIP Kit design, as shown in Figure 10.15 and Table 10.1.

DESIGN SPECIFICATION

The TRIP Kit design was first based on a military model, which was later expanded to include the wider functionality needed by fire and arson investigators, as shown in Table 10.2. The TRIP Kit is the product of a national strategy to

- Improve the productivity and quality of criminal fire and explosion investigations by standardizing the documentation process,
- Use case management to enhance the investigative efficiency and effectiveness of fire investigation unit personnel,
- Demonstrate commercial off-the-shelf (COTS) software and hardware,

FIGURE 10.15 ◆ The AIMS 2000 strategy exploited the military solution and resulted in a novel effort to produce a field-deployable "virtual office" dubbed the Transportable Rapid Information Package (TRIP) codeveloped by the USFA and U.S. TVA Police. *Courtesy of FEMA/USFA;* www.usfa.fema.gov/dhtml/fire-service/aims2.cfm.

TABLE 10.1 ◆ Technological Assessment Matrix

Major Goal	*Current Practice*	*Current Challenge*	*TRIP Kit Solution(s)*
Case management	Nonstandard approach to scene	Reliance on *NFPA 921* and *1033*	Template-based scene exams
Incident reporting systems	Complete separate NFIRS and NIBRS forms	Reduce reporting duplication	Collect data, print NFIRS and NIBRS reports
Fire investigation unit management	Cases assigned and closed as needed	Develop case solvability scoring	Incorporate case solvability factors
Litigation and prosecutive support	Cases based on written files	Automate case files	Case-based data search and retrieval
Investigative efficiency and productivity	Case assignments by callout	Measure efficiency	Time–motion studies

Source: Kenneth J. Kuntz, National Fire Programs Division, USFA.

TABLE 10.2 ◆ TRIP Functional Capabilities

Feature	Description
Basic virtual office unit	Laptop configured with color scanner, printer, and office software tools
Fire reports	Field data collection of information needed for NFIRS, NIBRS, and fire modeling
Interviews	Written field witness interviews using prompted questions
Evidence collection	Photography by digital and video cameras, printing of evidence labels
Fire scene diagramming	Computer-assisted drafting of building floor plans and fire scene
Mapping and GPS	Plotting the fire scene location on a digital map using GPS
CD-ROM library	Retrieval of fire reports, NFPA fire codes, NFIRS data profiles, and NIST models
Network conferencing	Wireless network with other TRIP units with video desktop conferencing
Telecommunications	Cellular and pager communications, faxing of reports from scene

Source: Kenneth J. Kuntz, National Fire Programs Division, USFA.

- ◆ Increase the communications and data transfer capability at the scene, and
- ◆ Provide electronic references and computer fire modeling capability.

It is also intended that the system will provide arson data through the states to the USFA's National Fire Incident Reporting System (NFIRS) and the FBI's National Incident Based Reporting System (NIBRS). In addition, AIMS 2000 system users would be able to query a wide variety of federal electronic public access information retrieval services currently available via websites, e-mail, or a CD-ROM library carried with the TRIP Kit. Through partnerships agreements, the USFA recently completed its pilot tests with TRIP Kits deployed at the Tennessee State Fire Marshal's Office.

PRODUCTIVITY IMPROVEMENT

It has been documented in several authoritative studies that computer technologies can significantly

- ◆ Enhance the quality and productivity of criminal investigations;
- ◆ Improve arrest rates, clear unsolved cases, provide linkages with other similar cases, and share time-sensitive information among multiple agencies;
- ◆ Reduce an agency's operational costs, and
- ◆ Meet demands from citizenry for increased services.

The collaborative efforts of sharing information and resources in major case investigations are often a measure of success. For example, in a multiagency criminal investigation in the southeastern United States, 150 local and federal investigators worked on a case containing 100,000 pages of information and more than 7000 leads.

Using computer technology for document storage, retrieval, and search strategies was an essential tool for successful management of this case (Shill 1998).

A management study explored the positive impact of the use and reliance of computer databases. Investigators reported informational as well as individual performance benefits (Danziger and Kraemer 1985). The study also reported that investigators who make extensive use of computer technology are rewarded with substantially higher productivity gains compared with investigators who avoid new technology. Investigators who use computers reported assistance and arrests, clearances, and linkages between persons in custody and unsolved cases.

FUNCTIONAL LESSONS LEARNED

Many lessons were learned during the initial testing phases of TRIP Kit technologies.

GPS and Computer Mapping. Providing the ability to look for maps using a global positioning system (GPS), the deployable computer has been helpful in determining exact jurisdictional boundaries in relation to the crime scene. Also, this use provides the ability to integrate digital computer maps into formal reports.

Cases in which lightning strikes are suspected or must be ruled out are becoming more common. GPS data collection allows the investigator to determine if a lightning strike was the source of fire ignition by comparing it with information sources from commercial lightning strike databases (Vaisala Inc. 2002).

Video and Teleconferencing. A unique function of the TRIP Kit is to capture imagery digitally and transfer it electronically by Internet connection. This enables investigators to provide good on-site capture of photos for preliminary reports and provides the ability for remote "photoconferencing." This capability is used by the state of Tennessee to relay photographs of major fires and explosions for use in timely briefings of upper management, who may need the information to allocate resources to continue to support the case.

Fire Modeling. The use of fire modeling is becoming more prevalent, and tests show that accurate collection of building measurements is needed for scene diagrams and models. The fire model later adds interpretive value to investigations.

Computer capabilities include access to fire modeling heat release data and other information stored in CD-ROM formats. These sources include the NFPA *Fire Protection Handbook* (NFPA 2000b), NFPA codes and guides such as *NFPA 921* (NFPA 2004a), and *The SFPE Handbook of Fire Protection Engineering* (SFPE 2002b).

Records Management. A records management department, when facing automation, must address a critical "build versus buy" decision on the acquisition of software to handle a comprehensive department-wide records management system. Suggested requirements indicated that printed case reports contain color text, department logos shown on reports, scanned photographs added as attachments, and custom field layouts. Statistical information must be collected at the same time as other case data to avoid replication of work.

EXPLORING WIRELESS TECHNOLOGIES

The TVA Police experimented with low-cost wireless technologies to extend the limits of TRIP Kits, particularly at the scenes of major cases. The concept is that the use of wireless technologies will improve productivity and real-time information availability.

For an example of an application in industry, Microsoft Corporation during its implementation of wireless technologies has deployed about 4000 access points at its Redmond, Washington, campus and satellite branch offices. About 35,000 employees internationally are using wireless LAN (WLAN) cards in their laptops. Other lessons learned included increased security of their network. Wireless networks are appearing today in hotels, commercial businesses, government facilities, and even entire cities. Microsoft believes that the use of wireless technologies yields a 1- to 1½-hr productivity gain per employee per day (Intel 2003).

INFORMATION TECHNOLOGY TRAINING IS KEY

The Metropolitan Toronto Police Department reported that information technology has saved the force $21.3 million in cost avoidance and significantly increased productivity (Wolchak et al. 1995). Toronto, like other jurisdictions receiving increasing demands from their citizenry for expanded services, is using information technology to meet public expectations. These demands also stress the already committed austere budget environment.

Internationally, policing strategies have become more receptive to new technology (Icove, Wherry, and Schroeder 1998). The New Zealand police strategic plan focuses on people, business redesign, and technology. Their strategy for technology includes gains in productivity and performance via operational information management systems, particularly those that have effective interaction between the police and the community they protect (Policing New Zealand 2000; crime.co.nz).

The TRIP Kit is a truly mobile office, as shown in Figure 10.16. Investigators in the field have at their disposal the tools necessary to document and store time-sensitive information electronically. Experience has shown that capturing digital images of crime scene photographs, video, and audio increases police productivity (Wise 1995). The TRIP Kit combines such institutional experiences from these and other studies that report to increase investigator productivity and efficiency using low-cost computer information technologies.

COMPUTER SKILL ENHANCEMENT FOR INVESTIGATORS

In this initial project, TRIP Kits were issued to federal and state law enforcement agents whose responsibility included criminal investigations of fires and explosions. In the initial phase of the project, the only knowledge about operating laptop computers came from "home use," as reported by the participants.

The project managers quickly realized that these investigators needed to obtain baseline knowledge in applying computer skills to operate and maintain their TRIP Kits successfully. The final course consisted of 4 days of instruction in using operating system and major office applications. These applications included the electronic mail, word processing, spreadsheet, and presentation software modules. These applications are the core tools used in preparing criminal investigative reports and courtroom presentations.

The investigators also received training in several of the peripherals used in the TRIP Kits, such as digital cameras and a screen-capturing program. They were shown and had the opportunity to integrate digital photos into documents and presentation graphics.

As with the introduction of any innovative technology, the USFA and TVA Police quickly learned that education in information technology is an essential element of success in the testing, evaluation, and deployment of TRIP Kits. With as little as 4

FIGURE 10.16 ◆ The Special Operations Response Team (SORT) operated by the Tennessee State Fire Marshal's Office uses mobile technologies at scenes of major fires and explosions to assimilate information, document the scenes, and transfer imagery to their headquarters using cell phone and satellite telephone communications. *Courtesy of D. J. Icove.*

days of instruction, criminal investigators self-reported a 29 to 48 percent improvement in skill levels after they completed a modular seminar concentrating on word processing, spreadsheets, and presentation graphics.

A second level of training came with using TRIP Kits integrated into the State of Tennessee and TVA's respective police information management systems. The skills obtained during the first phase produced a "level playing field" during the second critical training phase. Additional training is anticipated as these investigators become more accustomed to the technology.

FUTURE CAPABILITIES

The TRIP Kit concept addresses the major contemporary issues of case management, incident reporting systems, unit management, litigation support, and investigator productivity. It will truly be a model investigative tool for the next decade (Madrzykowski 2000).

In summary, there was significant improvement in the overall capability of investigators when exposed to modular training seminars consisting of customized training on pertinent applications. Investigators had more confidence and improved productivity and efficiency, consistent with results of other research studies. Clearly, an organized instructional design is needed to undertake comprehensive training.

As the literature demonstrates, computer and information technologies can significantly improve the quality and productivity of criminal investigations, reduce an agency's operational costs when using integrated information management systems, and meet demands for increased services. Finally, some minimal computer skill information technology training should take place along with delivery of TRIP Kits or similar technologies to fire and arson investigators to ensure maximum benefit.

◆ 10.4 SUMMARY AND CONCLUSIONS

The future progress of fire investigation, then, is really dependent not just on advances in technology but on the ever-increasing cooperation and communication among the disparate elements of the fire community. The daring training project that came to fruition as InterFire brought together the resources of the International Association of Arson Investigators (IAAI); U.S. Fire Administration (USFA); National Fire Protection Association (NFPA); Bureau of Alcohol, Tobacco, and Firearms (ATF); and American Re-Insurance (as well as individual contributors from all over the United States) to produce a revolutionary, interactive CD training guide and "instant" resource library for every fire investigator. This tremendously successful effort demonstrated what could be accomplished by interdisciplinary cooperation. Today the CFItrainer.net operated by the IAAI extends high-quality training opportunities to investigators everywhere who have Internet access.

The contact among fire investigators, engineers, and scientists has become more routine through shared sources of information. The IAAI, ATF, SFPE, and NFPA all host conferences and special training sessions that are intended not for investigators only or researchers only, but for both. Interscience Communications of London hosts major biennial conferences such as InterFlam and Fire and Materials at which fire investigation is a major topic, along with state-of-the-art academic or industrial research. The Technical Working Group for Fire and Explosions (TWGFEX) also holds informative biennial interdisciplinary conferences at the National Center for Forensic Sciences in Orlando, FL.

Every year the IAAI and its chapters hosts national and regional training conferences that bring investigators and researchers together in a variety of formats. The International Association of Fire Safety Sciences hosts a triennial international conference that usually includes papers of special interest to the investigative community.

For those unable or unwilling to travel, there are a host of peer-reviewed publications that offer very useful information. These include the *NFPA Journal, Fire Technology*, the IAAI's *Fire and Arson Investigator*, *Fire and Materials*, *Fire Safety Journal*, and the *Journal of Forensic Sciences*. Although it is not practical for an individual to belong to more than a few professional organizations, these journals and the published proceedings from the conferences mentioned here make it possible to keep up with the ever-increasing wealth of information. It is this information that will make the greatest difference to the quality of fire investigation in the future.

Suggested Reading

DeHaan, J. D. 2001. Full-scale compartment fire tests. *CAC News* (Second Quarter): 14–21.

DeHaan, J. D. 2004. Advanced tools for use in forensic fire scene investigation, reconstruction and documentation. Interflam 2004, 10th International Fire Science and Engineering Conference, Edinburgh, July 3–7.

FLIR Systems: www.flirthermography.com
IPIX: www.ipix.com
3rd Tech, Inc.: www.3rdtech.com
Vaisala: www.lightningstorm.com

◆◆◆ **Appendix A**

References

Accu-Line. 2003. Accu-Line Drawing Products, Buildersight CO. www.Accu-line.com

ADL. 2003. Anti-government extremist convicted in Colorado IRS arson. Anti-Defamation League. http://www.adl.org/learn/news/IRS_Arson.asp, Posted August 7, 2003.

Ahonen, A., M. Kokkala, and H. Weckman. 1984. Burning characteristics of potential ignition sources of room fires. Research Report 285. Espoo, Finland: Technical Research Centre.

Albers, J. C. 2002. Pour pattern or product of combustion? *Fire & Arson Investigator,* California Conference of Arson Investigators, July, p. 5.

Alpert, R. L. 1972. Calculation of response time of ceiling-mounted fire detectors. *Fire Technology* 9:181–95.

Alpert, R. L., and E. J. Ward. 1984. Evaluation of unsprinklered fire hazards, *Fire Safety Journal* 7:127–143.

APA. 1994. *Diagnostic and statistical manual of mental disorders,* 4th ed. (DSM-IV). Washington, DC: American Psychiatric Association.

Armor Forensics Inc. 2003. 13386 International Parkway, Jacksonville FL, 32218. www.armorholdings.com.

Associated Press. 2001. Shadowy eco-militants stepping up arson campaign. Terrence Petty, AP, June 1.

ASTM. 1989. *E 1138-89: Terminology of technical aspects of products liability litigation.* West Conshohocken, PA: American Society for Testing and Materials Committee E30.40 on Technical Aspects of Products Liability Litigation (withdrawn 1995).

ASTM. 1999. *E 860: Standard practice for examining and testing items that are or may become involved in products liability litigation.* West Conshohocken, PA: American Society for Testing and Materials Committee E30.05 on Forensic Sciences.

ASTM. 2000. ASTM E 502-84: *Standard test method for selection and use of ASTM standards for the determination of flash point of chemicals by closed-cup methods.* West Conshohocken, PA: American Society for Testing and Materials.

ASTM. 2001. *Forensic science book of standards,* Vol. 14.02. West Conshohocken, PA: American Society for Testing and Materials Committee E30 on Forensic Sciences.

ASTM. 2002. *E 1321–97a: Standard test method for determining material ignition and flame spread properties.* West Conshohocken, PA: American Society for Testing and Materials Subcommittee E05.22.

ASTM. 2003a. *E 603-01: Standard guide for room fire experiments.* West Conshohocken, PA: American Society for Testing and Materials Subcommittee E05.13.

———. 2003b. *E 620: Standard practice for reporting opinions of technical reports.* West Conshohocken, PA: American Society for Testing and Materials Subcommittee E30.11.

———. 2003c. *E 1188: Standard practice for collection and preservation of information and physical items by a technical investigator.* West Conshohocken, Pa: American Society for Testing and Materials.

———. 2003d. *E 1354-03: Standard test method for heat and visible smoke release rates for materials and products using an oxygen consumption calorimeter.* West Conshohocken, PA: American Society for Testing and Materials Subcommittee E05.21.

―――. 2004. *Fire test standards*, 6th ed. West Conshohocken, PA: American Society for Testing and Materials Committee E5 on Fire Standards.

Ayala, F. J., and B. Black. 1993. Science and the courts. *American Scientist* 81:230–39.

Babrauskas, V. 1975. COMPF: A program for calculating post-flashover fire temperatures (UCB FRG 75-2). University of California, Berkeley: Fire Research Group.

―――. 1980a. Estimating room flashover potential. *Fire Technology* 16:94–103, 112.

―――. 1980b. Flame lengths under ceilings. *Fire and Materials* 4 (no. 3):119–26.

―――. 1981. Applications of predictive smoke measurements. *Journal of Fire and Flammability* 12:51–64.

―――. 1982. Will the Second Item Ignite? National Bureau of Standards, NBSIR, 81–2271.

―――. 1984. Upholstered furniture room fires—Measurements, comparison with furniture calorimeter data, and flashover predictions. *Journal of Fire Sciences* 2(1):5–19.

―――. 1988. Upholstered furniture room fires—Measurements, comparison with furniture calorimeter data, and flashover predictions. *Journal of Fire Sciences* 2:5–19.

―――. 1996. Fire modeling tools for FSE: Are they good enough? *Journal of Fire Protection Engineering* 8(2): 87–96.

―――. 1997. The role of heat release rate in describing fires. *Fire & Arson Investigator* 47 (June): 54–57.

―――. 1998. Fire safety improvements in the combustion toxicity area: Is there a role for LC_{50} tests? In *Flame Retardants' 98*, 213–24. London: Interscience Communications.

―――. 2002. The heat release rate hazard of Christmas trees. Personal communication.

―――. 2003. *Ignition handbook*. Issaquah, WA: Fire Science Publishers, and Bethesda, MD: the Society of Fire Protection Engineers. ISBN 0-9728111-3-3.

―――. 2004a. *Truck Insurance v. MagneTek*: Lessons to be learned concerning presentation of scientific information. *Fire & Arson Investigator*. International Association of Arson Investigators, October.

―――. 2004b. Glass breakage in fires. Issaquah, WA: Fire Science and Technology Inc. www.doctorfire.com/glass.html.

―――. 2005. Charring rate of wood as a tool for fire investigators. *Fire Safety Journal* 40:528–54.

―――. 2006. Mechanisms and modes for ignition of low-voltage, PVC-insulated electrotechnical products. *Fire and Materials* 30:151–74.

Babrauskas, V., D. Baroudi, J. Myllymaki, and M. Kokkala. 1997. The cone calorimeter used for predictions of the full-scale burning behavior of upholstered furniture. *Fire and Materials* 21:95–105.

Babrauskas, V., G. Chastagner, and E. Stauss. 2001. Flammability of cut Christmas trees. International Assocation of Arson Investigators, Annual Training Conference, Atlantic City, NJ.

Babrauskas, V., B. F. Gray, and M. L. Janssens. 2007. Prudent practices for the design and installation of heat-producing devices near wood materials. *Fire and Materials* 31:125–35.

Babrauskas, V., and S. J. Grayson, eds. 1992. *Heat release in fires*. Basingstoke, UK: Taylor and Francis, ISBN 0419161007.

Babrauskas, V., and J. F. Krasny. 1985. *Fire behavior of upholstered furniture*. NBS Monograph 173. Gaithersburg, MD: U.S. National Bureau of Standards.

―――. 1997. Upholstered furniture transition from smoldering to flaming. *Journal of Forensic Sciences* 42:1029–31.

Babrauskas, V., J. R. Lawson, W. D. Walton, and W. H. Twilley. 1982. Upholstered furniture heat release rates measured with a furniture calorimeter. NBSIR 82-2604. Gaithersburg, MD: National Bureau of Standards.

Babrauskas, V., and R. D. Peacock. 1992. Heat release rate: The single most important variable in fire hazard. *Fire Safety Journal* 18:255–72.

Babrauskas, V., R. D. Peacock, and P. A. Reneke. 2003. Defining flashover for fire hazard calculations. Part II. *Fire Safety Journal* 38:613–22.

Babrauskas, V., and W. D. Walton. 1986. A simplified characterization of upholstered furniture heat release rates. *Fire Safety Journal* 11:181–92.

Bailey, C. 2006. One stop shop in structural fire engineering. http://www.mace.manchester.ac.uk/project/research/structures/strucfire/.

Bailey, J. W. 1983. Archaeology: Help for arson investigation from an unexpected source. *Fire Chief* (March): 24–27.

Beering, P. S. 1996. *Verdict: guilty of burning—What prosecutors should know about arson.* Indianapolis, IN: Insurance Committee for Arson Control, May.

Beller, D., and J. Sapochetti. 2000. Searching for answers to the Cocoanut Grove fire of 1942. *NFPA Journal* (May/June): 85–92.

Berrin, E. R. 1977. Investigative photography. Technology Report No. 77-1. Society of Fire Protection Engineers, Bethesda, MD.

Bessey, G. E. 1950. Investigations on building fires. Part II: The visible changes in concrete or mortar exposed to high temperatures. National Building Studies, Technical Paper No. 4, Department of Scientific and Industrial Research, Building Research Station, Garston, England, 6–18.

Bleay, S., G. Bradshaw, and J. E. Moore. 2006. Fingerprint development and imaging newsletter: Special edition. Publication No. 26/06. Home Office Scientific Development Branch, UK, April.

Blocker, C. M., and D. J. Icove. 2001. Improved quality and productivity of criminal investigators through information technology education. TVA University Conference Paper Series. Knoxville: U.S. Tennessee Valley Authority, February, 30–37.

Bohnert, M., T. Rost, and S. Pollak. 1998. The degree of destruction of human bodies in relation to the duration of the fire. *Forensic Science International* 95:11–21.

Brannigan, F. L., R. G. Bright, and N. H. Jason. 1980. *Fire investigation handbook.* NBS Handbook 123. Washington, DC: National Bureau of Standards, U.S. Department of Commerce.

Bostrom, L. 2005. Methodology for measurement of spalling of concrete. *Fire & Materials 2005 Conference.* London: Interscience.

Bryan, J. L. 1983. An examination and analysis of the dynamics of the human behavior in the Westchase Hilton Hotel Fire, Houston, Texas, on March 6, 1982. Report prepared for the National Fire Protection Association, March.

Bryan, J. L., and D. J. Icove. 1977. Recent advances in computer-assisted arson investigation. *Fire Journal* 17: 1.

Buchanan, A. H. 2001. *Structural design for fire safety.* New York: Wiley.

Budnick, E. K., D. D. Evans, and H. E. Nelson. 1997. In *Fire Protection Handbook,* 18th ed., sec. 11, chap. 10, 11-97–11-107. Ed. A. E. Cote, J. L. Linville, M. K. Appy, and R. P. Beredetti. Quincy, MA: National Fire Protection Association,

Bukowski, R. W. 1991. Fire models: The future is now! *Fire Journal* 85(2):60–69.

———. 1992. Analysis of the Happyland Social Club fire with HAZARD I. *Fire & Arson Investigator* 42(3):36–47.

Bukowski, R. W. 1995a. How to evaluate alternative designs based on fire modeling. *Fire Journal* 89(2):68–74.

———. 1995b. Predicting the fire performance of buildings: Establishing appropriate calculation methods for regulatory applications. In *Proceedings of ASIAFLAM95 International Conference on Fire Science and Engineering,* 9–18, Kowloon, Hong Kong, March 15–16.

———. 1996. Modelling a backdraft incident: The 62 Watts Street (New York) fire. *Fire Engineers Journal* 56(185): 14–17; *NFPA Journal,* (Nov/Dec 1995): 85–89.

Burnette, G. E. 2000. Spoliation of evidence: A fire scene dilemma. *InterFIRE*. www.interfire.org.

———. 2003. Fire scene investigation: The *Daubert* challenge. Personal communication. (See also www.interfire.org).

———. 2005. Client bulletin: New case applying the *Daubert* standard. Personal communication.

Butler, C. P. 1971. Notes on charring rates in wood. Fire Research Note No. 896. London: Department of the Environment and Fire Officers Committee, Joint Fire Research Organization.

Campagnolo, T. 1999. The Fourth Amendment at fire scenes. *Arizona Law Review* 41 (Fall): 601.

Canfield, D. 1984. Causes of spalling of concrete at elevated temperatures. *Fire & Arson Investigator* 34(4):22–23.

Canter, D. 2003. *Mapping murder*. London: Virgin Books.

Canter, D., and K. Fritzon. 1998. Differentiating arsonists: A model of firesetting actions and characteristics. *Legal and Criminological Psychology* 3:73–96.

Carey, N. 2002. Powerful techniques, arc fault mapping. *Fire Prevention* (UK) (March): 47–50.

Cavanagh, K., E. Du Pasquier, and C. Lennard. 2002. Background interference from car carpets—The evidential value of petrol residues in cases of suspected vehicle arson. *Forensic Science International* 125: 22–36.

Cavanagh-Steer, K., E. Du Pasquier, C. Roux, and C. Lennard. 2005. The transfer and persistence of petrol on car carpets. *Forensic Science International* 147(2005): 71–79.

Chakrabarti, B., T. Yates, and A. Lewry. 1996. Effects of fire damage on natural stonework in buildings. *Construction and Building Materials* 10(7):539–44.

Chesbro, K. J. 1994. Taking *Daubert*'s focus seriously: The methodology conclusion distinction. *Cardozo Law Review* (Yeshiva University) 15(6–7):1745–53.

Christensen, A. M. 2002. Experiments in the combustibility of the human body. *Journal of Forensic Science* 47(3):466–70.

Christensen, A. M., and D. J. Icove. 2004. The application of NIST's fire dynamics simulator to the investigation of carbon monoxide exposure in the deaths of three Pittsburgh fire fighters. *Journal of Forensic Science* 49(1).

Chu Nguong, Ngu. 2004. Calcination of gypsum plasterboard under fire exposure. Fire Engineering Research Report 04/6. Christchurch, NZ: University of Canterbury, Dept of Civil Engineering.

Clifford, R. C. 2000. *Qualifying and attacking expert witnesses.* Costa Mesa, CA: James, 3–64.

Collier, P. C. R. 1996. Fire in a residential building: Comparisons between experimental data and a fire zone model. *Fire Technology* 32(3):195–217.

Coulson, S. A., and R. K. Morgan-Smith. 2000. The transfer of petrol onto clothing and shoes while pouring petrol around a room. *Forensic Science International* 112:135–41.

Curtin, D. P. 1999. Curtin's short courses in digital photography. www.shortcourses.com-chapter11.htm.

Damant, G., and S. Nurbakhsh. 1994. Christmas trees—What happens when they ignite? *Fire and Materials* 18:9–16.

Danziger, J. N., and K. L. Kraemer. 1985. Computerized data-based systems and productivity among professional workers: The case of detectives. *Public Administration Review* (January–February): 196–209.

David, J. 1991. Seven-fatality Christmas tree fire, Canton, Michigan (December 22, 1990). Emmittsburg, MD: Federal Emergency Management Agency.

Deans, J. 2006. Recovery of fingerprints from fire scenes and associated evidence. *Science and Justice* 40(3):153–68.

Decker, J. F., and B. L. Ottley. 1999. *The investigation and prosecution of arson.* Charlottesville, VA: Lexis Law.

DeHaan, J. D. 1987. Are localized burns proof of flammable liquid accelerants? *Fire & Arson Investigator* 38(1):45–49.

———. 1992. Fire—Fatal intensity. *Fire & Arson Investigator* 43(1).

———. 1995. The reconstruction of fires involving highly flammable hydrocarbon liquid. PhD diss., University of Strathclyde, Glasgow, Scotland. (Published by University Microfilms International, 1996.)

———. 1999. The challenge of fire investigations. Interflam '99 Fire Science and Engineering Conference, Edinburgh, Scotland.

———. 2000. Why is it important to recognize flashover in a room fire? California Institute of Criminalistics. www.interfire.org./res_file/flashovr.htm.

———. 2001. Full-scale compartment fire tests. *CAC News* (Second Quarter): 14–21.

———. 2002 Our changing world Part 2: Ignitable liquids: petroleum distillates, petroleum products, and other stuff. *Fire & Arson Investigator* 52, no. 3 (April).

———. 2002. Our changing world Part 3: Detection limits: Is more sensitive necessarily more better? *Fire & Arson Investigator* 52, no. 4 (July).

———. 2002. Our changing world. Part 2: Ignitable liquids: Petroleum distillates, petroleum products and other stuff, Fire and Arson Investigators. International Association of Arson Investigators (July) 20–23.

———. 2004. Advanced tools for use in forensic fire scene investigation, reconstruction and documentation. Interflam 2004, 10th International Fire Science and Engineering Conference, Edinburgh, July 3–7.

———. 2005. Reliability of computer modeling in fire scene reconstruction. *Fire & Arson Investigator* (January): 40–47.

———. 2007. *Kirk's fire investigation*, 6th ed. Upper Saddle River, NJ: Prentice Hall.

DeHaan, J. D., S. J. Campbell, and S. Nurbakhsh. 1999. The combustion of body fat and its implications for fires involving bodies. *Science and Justice* 39:27–38.

DeHaan, J. D., and F. L. Fisher. 2003. Reconstruction of a fatal fire in a parked motor vehicle. *Fire & Arson Investigator* 53(2):42–46.

DeHaan, J. D., and S. Nurbakhsh. 2001 Sustained combustion of an animal carcass and its implications for the consumption of human sciences 46, no. 5 (September).

DeHaan, J. D., and E. Pope. 2007. Combustion properties of human and large animal remains. Proceedings of Interflam, Fire Research Conference, Interscience, London, September 3–5.

DeWitt, W. E., and D. W. Goff. 2000. Forensic engineering assessment of FAST and FASTLite fire modeling software. National Academy of Forensic Engineers, December, 9–19.

Dillon, S. E., S. Hill, and D. Sheppard. 2003. Role of the ATF Fire Research Laboratory in fatal fire investigations. Presentation lecture, American Academy of Forensic Sciences (AAFS) 55th Annual Meeting, Chicago, February 17–22.

Dornheim, M. A. 1997. Fuel ignites differently in aircraft, lab environments. *Aviation Week and Space Technology* (July 7): 61–63.

Douglas, John E., Ann W. Burgess, Allen G. Burgess, and Robert K. Ressler, eds. 1992 and 1997 (2nd ed.). The crime classification manual: A standard system for classifying crimes. San Francisco: Jossey-Brass.

Drysdale, D. 1999. *An introduction to fire dynamics*, 2nd ed. New York: Wiley.

DSM-IV. 1994. *Diagnostic and statistical manual of mental disorders,* 4th ed. Washington, DC: American Psychiatric Association.

Eastman Kodak. 1968. Basic police photography. Publication No. M77. Rochester, NY: Eastman Kodak Co.

———. 1971. Fire and arson photography. Publication No. M67. Rochester, N.Y.: Eastman Kodak Co.

Ebdon, D. 1983. *Statistics in geography*. Oxford: Blackwell, 109.

Eos Systems Inc. 2003. 101-1847 West Broadway, Vancouver, British Columbia, Canada V6J 1Y6. www.eossystems.com.

Esposito, J. C. 2005. Fire in the Grove: The Cocoanut Grove tragedy and its aftermath. Cambridge, MA: Da Capo Press.

Evans, D. D., and D. W. Stroup. 1986. Methods to calculate the response time of heat and smoke detectors installed below large unobstructed ceilings. *Fire Technology* 22(1):54–65.

Factory Mutual. 1996. Fire protection for offices: From office to ashes in 7 minutes—Can you afford it? Publication P8802 (Rev 4-97). Norwood, MA: Factory Mutual Engineering Corp.

Fang, J., and N. Breese. 1980. Fire development in basement rooms. NBSIR 80-2120. Washington, DC: National Bureau of Standards, October.

FBI. 1996. *Crime in the United States for 1995*. Washington, DC: U.S. Department of Justice.

———. 1999. Definitions and guidelines for the use of imaging technologies in the criminal justice system. *Forensic Science Communications* 1 no. 3 (October).

———. 2004. Scientific Working Group on Imaging Technology (SWGIT) references/resources. *Forensic Science Communications* (March).

Feld, J. M. 2002. The physiology and biochemistry of combustion toxicology. In *Proceedings of Fire Risk and Hazard Assessment Research Applications Symposium*. FP Research Foundation, July 24–26.

FEMA. 1997a. USFA fire burn pattern tests (FA 178). Emmitsburg, MD: Federal Emergency Management Agency, U.S. Fire Administration, July 16.

———. 1997b. Arson in the United States. Report FA-174. Emmitsburg, MD: Federal Emergency Management Agency, U.S. Fire Administration, Tri-Data Corp., August.

Fleischmann, D. M., P. J. Pagni, and R. B. Williamson. 1994. Salt water modeling of fire compartment gravity currents. In *Proceedings of the 4th International Symposium, International Association for Fire Safety Science*, ed. T. Kashiwagi, 253–64. Ottawa, Ontario, Canada, July 13–17. Boston: IAFSS.

FLIR Systems, 16 Esquire Rd., N. Billerica, MA 01862. www.flirthermography.com.

Forney, G. P., and W. F. Moss. 1992. Analyzing and exploiting numerical characteristics of zone fire models. Report NISTIR 4763. Gaithersburg, MD: National Institute of Standards and Technology, March.

Forney, G. P., and K. B. McGrattan. 2006. User's guide for Smokeview version 4—A tool for visualizing fire dynamics simulation data. NIST Special Publication 1017. Gaithersburg, MD: National Institute of Standards and Technology, March.

FPRF. 2002. Recommendations of the Research Advisory Council on Post-fire Analysis: A white paper. Quincy, MA: Fire Protection Research Foundation, February.

FRCP. 2000. Orders of the Supreme Court of the United States adopting and amending rules: Order of April 17, 2000. Federal Rules of Civil Procedure as amended effective on December 1, 2000. www.gamb.uscourts.gov.

Friedman, R. 1992. An international survey of computer models for fire and smoke. *SFPE Journal of Fire Protection Engineering* 4(3):81–92.

Fritzon, K. 2001. An examination of the relationship between distance travelled and motivational aspects of firesetting behaviour. *Journal of Environmental Psychology* 21:45–60.

FRS. 1982. Anatomy of a fire (video). Fire Research Station, Garston, Watford, UK.

FRS. 2002. Fire Research Station, the Research-Based Consultancy and Testing Company of BRE, Garston, Watford, UK. www.bre.co.uk/frs.

Fulghum, D. 1997. ANG pilot: Jet hit by object. *Aviation Week and Space Technology*, March 10.

Geller, J. L. 1992. Arson in review: From profit to pathology. *Journal of Clinical Forensic Psychiatry* 15:623–45.

Geller, J. L., J. Erlen, and R. L. Pinkas. 1986. A historical appraisal of America's experience with pyromania—A diagnosis in search of a disorder. *International Journal of Law and Psychiatry* 9(2):201–29.

Georges, D. E. 1967. The ecology of urban unrest in the city of Newark, New Jersey, during the July 1967 riots. *Journal of Environmental Systems* 5(3):203–28.

———. 1978. The geography of crime and violence: A spatial and ecological perspective. Resource paper for college geography No. 78-1. Washington, DC: Association of American Geographers.

Gianelli, P., and E. Imwinkelried. 2007. *Scientific evidence*, 4th ed. Charlottesville, VA: Lexis (Michie).

Goldbaum, L. R., T. Orellano, and E. Dergari. 1976. *Annals of Clinical and Laboratory Science* 6:372.

Graham, M. 1994. *Cleary and Graham's handbook of Illinois evidence*, 6th ed., para. 702.4 at 565. Quoted in J. F. Decker and B. L. Ottley, *The investigation and prosecution of arson*, Lexis Law Publ., Matthew Bender, Charlottesville, VA, 1999, 20.

Gratkowski, M. T., N. A. Dembsey, and C. L. Beyler. 2006. Radiant smoldering ignition of plywood. *Fire Safety Journal* 41(6):427–43.

Grosselin, S. D. 1998. The application of fire dynamics to fire forensics. Master's thesis, Worcester Polytechnic Institute, Fire Protection Engineering, Worcester, MA.

Grosshandler, W. L., N. P. Bryner, D. Madrzykowski, and K. Kuntz. 2005. Report of the technical investigation of the Station Nightclub fire. NIST NCSTAR 2, vol. 1. Gaithersburg, MD: National Institute of Standards and Technology, June.

Hajpal, M. 2002. Changes in sandstones of historical monuments exposed to fire or high temperatures. *Fire Technology* 38:373–82.

Harmon, R. B., R. Sosner, and M. Wiederight. 1985. Women and arson: A demographic study. *Journal of Forensic Sciences* 30(2):467–77.

Harris, E. C. 1975. The stratigraphy sequence: A question of time. *World Archaeology* 7:109–21.

Harris, R. J., M. R. Marshall, and D. J. Moppett. 1977. The response of glass windows to explosion pressures. The Institution of Chemical Engineers, Symposium Series No. 49, Papers of the Symposium on Chemical Process Hazards, University of Manchester Institute of Science and Technology, April 5–7, 87–101.

Harris, R. J. 1983. The investigation and control of gas explosions in buildings and heating plants. London: British Gas Corp., E&FN Spon, 96–100.

Hasemi, Y., and M. Nishihata. 1989. Fuel shape effects on the deterministic properties of turbulent diffusion flames, 275–84. In *Proceedings of the Second International Symposium, International Association of Fire Safety Science*. Washington, D.C.: Hemisphere.

Hasemi, Y., and T. Tokunaga. 1984. *Combustion Science Technology* 40: 1–18.

Hasselbring, L. C. 2007. Case studies: Exploding portable gasoline containers. *Fire and Materials 2007*. San Francisco, January 29–31.

Hayasaka, H. 1997. Unsteady burning rates of small pool fires. In *Proceedings of 5th Symposium on Fire Safety Science*, ed. Y. Hasemi. Tsukuba, Japan.

Health Canada. 1995. *Investigating human exposure to contaminants in the environment: A handbook for exposure calculations*. Ottawa, Ontario, Canada: Health Protection Branch, Health Canada, Ministry of National Health and Welfare. www.hc-sc.gc.ca

Hertz, K. D., and L. S. Sorensen. 2005. Test method for spalling of fire-exposed concrete. *Fire Safety Journal* 40:466–76.

Heskestad, G. H. 1982. Engineering relations for fire plumes. SFPE Technology Report 82-8. Boston: Society of Fire Protection Engineers, 6.

————. 1988. Fire plumes. In *The SFPE handbook of fire protection engineering*, ed. J. DiNenno, sec. 1, chap. 6, 1-107–1-115. Quincy, MA: National Fire Protection Association.

Heskestad, G. H., and M. A. Delichatsios. 1977. Environments of fire detectors—Phase 1: Effects of fire size, ceiling height, and material. NBS-GCR-77-86 and NBS-GCR-77-95. Gaithersburg, MD: National Bureau of Standards.

Hewitt, Terry-Dawn. 1997. A primer on the law of spoliation of evidence in Canada. *Fire & Arson Investigator* 48, no. 1 (September):17–21.

Hietaniemi, Jukka. 2005. Probabilistic simulation of glass fracture and fallout in fire. VTT Working Papers 41, ESPOO 2005. Finland: VTT Building and Transport.

Holleyhead, R. 1996. Ignition of flammable gases and liquids by cigarettes: A review. *Science and Justice* 36(4):257–66.

————. 1999. Ignition of solid materials and furniture by lighted cigarettes—A review. *Science and Justice* 39(2):75–102.

Hopkins, R. L., G. Gorbett, and P. M. Kennedy. 2007. Fire pattern persistence and predictability on interior finish and construction materials during pre- and post-flashover compartment fires. *Fire and Materials 2007*, San Francisco, January 29–31.

Huff, T. G. 1994. Fire-setting fire fighters: Identifying and preventing arsonists in fire departments. Focus on: Arson. *IAFC On Scene* 8, no. 15 (August 15): 6–7.

————. 1997. *Killing children by fire. Filicide: A preliminary analysis.* Quantico, VA: Federal Bureau of Investigation.

Hunt, S. 2000. Computer fire models. *Section News NFPA* 1(2):7–9.

Hurd, R. M. 1903. *Principles of city values.* New York: Records and Guide.

Hurley and Monahan. 1969. Arson: The criminal and the crime. *British Journal of Criminology* 9(1):4–21.

Icove, D. J. 1979. Principles of incendiary fire analysis. Unpublished PhD dissertation, Knoxville: College of Engineering, University of Tennessee.

————. 1995. Fire scene reconstruction. First International Symposium on the Forensic Aspects of Arson Investigations, Federal Bureau of Investigation, Fairfax, VA, July 31.

Icove, D. J., and J. D. DeHaan. 2004. *NFPA 921*'s impact on fire scene reconstruction. *Fire Protection Engineering* No. 21 (Winter): 10–16.

————. 2006. "Hourglass" burn patterns: A scientific explanation for their formation. International Symposium on Fire Investigation Science and Technology (ISFI 2006), June 26–28, Cincinnati, OH.

Icove, D. J., J. E. Douglas, G. Gary, T. G. Huff, and P. A. Smerick. 1992. Arson. In *Crime classification manual*, ed. J. E. Douglas, A. W. Burgess, A. G. Burgess, and R. K. Ressler, 165–66. New York: Macmillan.

Icove, D. J., E. C. Escowitz, and T. G. Huff. 1983. The geography of violent crime: Serial arsonists, 46. In *Agenda and Abstracts, 8th Annual Geographic Resources Analysis Support System (GRASS) Users Conference*, Reston, VA, March 14–19.

Icove, D. J., and M. H. Estepp. 1987. Motive-based offender profiles of arson and fire-related crimes. *FBI Law Enforcement Bulletin* (April).

Icove, D. J., and M. M. Gohar. 1980. Fire investigation photography. In *Fire investigation handbook*. Gaithersburg, MD: National Bureau of Standards.

Icove, D. J., and G. Haynes. 2007. Guidelines for conducting peer reviews of complex fire investigations. *Fire and Materials 2007*, San Francisco. January 29–31.

Icove, D. J., and P. R. Horbert. 1990. Serial arsonists: An introduction. *Police Chief* (Arlington, VA) (December): 46–48.

Icove, D. J., P. E. Keith, and H. L. Shipley. 1981. An analysis of fire bombings in Knoxville, Tennessee. U.S. Fire Administration Grant EMW-R-0599. Knoxville: Knoxville Police Department Arson Task Force, Department of Public Safety, December 17.

Icove, D. J., V. B. Wherry, and J. D. Schroeder. 1998. *Combating arson-for-profit: Advanced techniques for investigators*. Columbus, OH: Battelle Press. ISBN 1-57477-023-3.

Inberg, S. H. 1927. Fire tests of office occupancies. *NFPA Quarterly* 20:243.0.

Inciardi, J. A. 1970. The adult firesetter: A typology. *Criminology* 8 (August).

Intel. 2003. Microsoft Corporation: 35,000 employees go wireless with WLAN. www.intel.com/cal/business/casestudies/pdf/microsoft.pdf.

iPIX. 2003. Interactive Pictures Corp., 1009 Commerce Park Drive, Suite 400, Oak Ridge, TN 37830. www.ipix.com

Iqbal, N., and M. H. Salley. 2002. Development of a quantitative fire scenario estimating tool for the U.S. Nuclear Regulatory Commission Fire Protection Inspection Program. Washington, DC: Fire Protection Engineering and Special Projects Section, Nuclear Regulatory Commission.

———. 2004. Fire dynamics tools (FDTs): Quantitative fire hazard analysis methods for the U.S. Nuclear Regulatory Commission Fire Protection Inspection Program. Washington, DC.

Jia, F., E. R. Galea, and M. K. Patel. 1999. Numerical simulation of the mass loss process in pyrolyzing char materials. *Fire and Materials* 23:71–78.

Janssens, M. 2000. *Introduction to mathematical fire modeling*, 2nd ed. Lancaster, PA: Technomic. ISBN 1566769205.

Jansson, R. 2006. Thermal stresses cause spalling. Brand Posten, SP Swedish National Testing and Research Institute. (No. 33): 24–25.

Jensen, G. 1998. Wayfinding in heavy smoke: Decisive factors and safety products: Findings related to full-scale tests. IGPAS, InterConsult Group ICG. www.interconsult.com (May 1).

Jin, T. 1975. Visibility through fire smoke. Part 5: Allowable smoke density for escape from fire. Report of Fire Research Institute of Japan. (No. 42): 12.

Karlsson, B., and J. G. Quintiere. 1999. *Enclosure fire dynamics*. Boca Raton, FL: CRC Press, 64.

Karter, M. J., Jr. 2005. Fire loss in the United States during 2004. Quincy, MA: National Fire Protection Association, Fire Analysis and Research Division.

———. 2006. Fire loss in the United States during 2005. NFPA Journal. Quincy, MA: National Fire Protection Association. September.

Kawagoe, 1958. Fire behavior in rooms. Report no. 27. Tokyo: Building Research Institute.

Kennedy, J., and P. Kennedy. 1985. Fires and explosions—Determining the cause and origin. Chicago, IL: Investigations Institute.

Kennedy, P. M. 2004. Fire pattern analysis in origin determination. In proceedings of International Symposium on Fire Investigation (ISFI), Cincinnati, OH, June.

Kennedy, P. M., J. H. Shanley, W. C. Alletto, R. Cory, J. Herndon, and J. Ward. 1997. Report of the USFA program for the study of fire patterns. Federal Emergency Management Agency, U.S. Fire Administration. FA 178. July.

Keski-Rahkonen, Olavi. 1988 Breaking of window glass close to fire. *Fire and Materials* 12(2): 61–69.

———. 1991. Breaking of window glass close to fire, II: Circular panes. *Fire and Materials* 15(1):11–16.

Khoury, G. A. 2000. Effect of fire on concrete and concrete structures. *Progress in Structural Engineering Materials* 2:429–42.

King, C. G., and J. I. Ebert. 2002. Integrating archaeology and photogrammetry with fire investigation. *Fire Engineering*. (February): 79.

Kolczynski, P. J. 2000. *Preparing for trial in federal court*, 2nd ed. Costa Mesa, CA: James.

König, J., and L. Walleij. 2000. Timber frame assemblies exposed to standard and parametric fires. Part 2: A design model for standard fire exposure. Report No. 100010001. Tratek. Stockholm: Swedish Institute for Wood Technology Research.

Krasny, J. F., W. J. Parker, and V. Babrauskas. 2001. *Fire behavior of upholstered furniture and mattresses*. New York: William Andrews.

Lau, P. W. C., R. White, and I. Van Zeeland. 1999. Modelling the charring behavior of structural lumber. *Fire and Materials* 23:209–16.

Law, M. 1978. Fire safety of external building elements—The design approach. *AISC Engineering Journal* (Second Quarter).

Lawson, J. 1977. An evaluation of fire properties of generic gypsum board products. NBSIR 77-1265. Washington, DC: Center for Fire Research, NIST.

Lawson, J. R., and J. G. Quintiere. 1985. Slide-rule estimates of fire growth. NBSIR 85-3196. Gaithersburg, MD: National Bureau of Standards.

LeBeau, J. L. 1987. The methods and measures of centrography and the spatial dynamics of rape. *Journal of Quantitative Criminology* 3: 125–41.

LeMay, J. 2003. Using "scales photography." *Law Enforcement Technology Magazine* (October 2002) and *Forensic Scientist* 1, no. 1 (April):87–89.

Lentini, J. J. 1992. Behavior of glass at elevated temperatures. *Journal of Forensic Sciences* 37(5):1358–62.

———. 2001. Standards impact: The forensic sciences. *ASTM Standardization News* (February): 17–19.

Lentini, J. J., J. A. Dolan, and C. Cherry. 2000. The petroleum-laced background. *Journal of Forensic Sciences* 45(5):968–89.

Lentini, J. J. 2006. *Scientific protocols for fire investigation.* Boca Raton, FL: CRC Press.

Levin, B. 1976. Psychological characteristics of firesetters. *Fire Journal* (March): 36–41.

Lewis, N. D. C., and H. Yarnell. 1951. *Pathological firesetting (pyromania).* Nervous and Mental Disease Monographs, No. 82. New York: Coolidge Foundation.

Lieber, A. L. 1978. *The lunar effect: Biological tides and human emotions.* New York: Dell.

Lilley, D. G. 1995. Fire dynamics. American Institute of Aeronautics and Astronautics, AIAA-95-0894, Meeting and Exhibits, Reno, NV, January 9–12.

Lipson, A. S. 2000. *Is it admissible?* Costa Mesa, CA: James, 44-2–44-8.

Ma, T. S. M. Olenick, M. S. Klassen, R. J. Roby, and J. L. Torero. 2004. Burning rate of liquid fuel on carpet (porous media). *Fire Technology* 40, no. 3 (July): 227–46.

MacQueen, J. 1967. Some methods for classification and analysis of multivariate data. In *Proceedings of the 5th Berkeley Symposium of Probability and Statistics*. Berkeley: University of California Press.

Madrzykowski, D. 2000. The future of fire investigation. *Fire Chief* (October): 44–50.

Madrzykowski, D., and R. L. Vettori. 2000. Simulation of the dynamics of the fire at 3146 Cherry Road NE, Washington, DC, May 30, 1999. NISTIR 6510. Gaithersburg, MD: National Institute of Standards and Technology, Center for Fire Research, April.

Madrzykowski, D., and W. D. Walton. 2004. Cook County Administration Building Fire, October 17, 2003. NIST SP 1021. Gaithersburg, MD: National Institute of Standards and Technology, July.

Magnusson, S. E., and S. Thelandersson. 1970. Temperature–time curves of the complete process of fire development. Theoretical study of wood fuel fires in enclosed spaces. Civil and Building Construction Series No. 65. Stockholm: Acta Polytechnica Scandinavia.

Malooly, J. E., and K. Steckler. 2005. Ignition of gasoline by cigarette. Washington, DC: Bureau of Alcohol, Tobacco, Firearms and Explosives, May.

Mann, J., N. Nic Daeid, A. Linacre. 2003. An investigation into the persistence of DNA on petrol bombs. *Proceedings of the European Academy of Forensic Sciences*, Istanbul.

McCaffrey, B. J. 1979. Purely buoyant diffusion flames: Some experimental results. NBSIR 79-1910. Gaithersburg, MD: National Bureau of Standards.

McCaffrey, B. J., J. G. Quintiere, and M. F. Harkleroad. 1981. Estimating room fire temperatures and the likelihood of flashover using fire test data correlations. *Fire Technology* 17(2): 98–119.

McDonald, J. 2005. Christmas tree ignites at Disney's Grand Californian. *Orange County Register*, December 28.

McGill, D. 2003. Fire Dynamics Simulator, FDS 683, participants' handbook. School of Fire Protection, Seneca College, Toronto, ON, January.

McGrattan, K. B., ed. 2006. Fire Dynamics Simulator (version 4) technical reference guide. NIST Special Publication 1018. Gaithersburg, MD: National Institute of Standards and Technology, March.

McGrattan, K. B., and G. P. Forney. 2006. Fire Dynamics Simulator (version 4) user's guide. NIST Special Publication 1019. Gaithersburg, MD: National Institute of Standards and Technology, March.

McGrattan, K., B. Klein, S. Hostikka, and J. Floyd. 2007. Fire Dynamics Simulator (version 5) user's guide. NIST Special Publication 1019-5. Gaithersburg, MD: National Institute of Standards and Technology.

McGraw, J., and F. Mowrer. 1999. Flammability of painted gypsum wallboard subjected to fire heat fluxes. *Interflam 99 Proceedings.* London: Interscience.

Merck index, 11th ed. 1989. Rahway, NJ: Merck Co.

Microsoft. 2002. Visio Crime Scene Template. www.microsoft.com

Milke, J. A. 2000. Smoke management for covered malls and atria. *Fire Technology* 26(3): 223–43.

Milke, J. A., and F. W. Mowrer, 2001. Application of fire behavior and compartment fire models seminar. Tennessee Valley Society of Fire Protection Engineers (TVSFPE), Oak Ridge, September 27–28.

Mitler, H. E. 1991. Mathematical modeling of enclosure fires. NISTIR 90-4294. Gaithersburg, MD: National Institute of Standards and Technology, May.

Mitler, H. E., and J. A. Rockett. 1987. Users' guide to FIRST, a comprehensive single-room fire model. Report CIB W14/88/22 (USA); NBSIR 87-3595; 138 pp. Gaithersburg, MD: National Bureau of Standards, September.

Moodie, M., and S. E. Jagger. 1991. The technical investigation of the fire at London's King's Cross Underground Station. *Journal of Fire Protection Engineering* 3(2):49–63.

Moore, D. T. 2006. Critical thinking and intelligence analysis. Washington, DC: Joint Military Intelligence College, Center for Strategic Intelligence Research.

Moran, Bruce. 2001. Personal communication. March 29.

Morgan-Smith, R. 2000. Persistence of petrol on clothing. Paper presented at ANZFSS Symposium of Forensic Sciences, Gold Coast, Queensland, Australia, March.

Mowrer, F. W. 1990. Lag time associated with fire detection and suppression. *Fire Technology* 26, no. 3 (August): 244–65.

———. 1992. Methods of quantitative fire hazard analysis. TR-100443. Research Project 3000-37. Prepared for Electric Power Research Institute (EPRI) by the Society of Fire Protection Engineers. Boston.

Mowrer, F. W. 1998. Window breakage induced by exterior fires. NIST-GCR-98-751. Washington, DC: US Dept. of Commerce.

———. 2001 Calcination of gypsum wallboard in fire. Paper presented at NFPA World Fire Safety Congress, Anaheim, CA, May 13–17.

———. 2003. Spreadsheet templates for fire dynamics calculations. University of Maryland, Department of Fire Protection Engineering, September. Spreadsheets and documentation are posted on the Fire Risk Forum website, www.fireriskforum.com.

Mudan, K. S., and P. A. Croce. 1995. Fire hazard calculations for large open hydrocarbon fires, 3–203. In *The SFPE handbook of fire protection engineering,* 2nd ed. Bethesda, MD: Society of Fire Protection Engineers.

Myers, R. A. M., and R. A. Cowley. 1979. CO poisoning. *Journal of Combustion Toxicology* 6:86.

Nelson, G. L. 1998. Carbon monoxide and fire toxicity: A review and analysis of recent work. *Fire Technology* 34 (1).

Nelson, H. E. 1987. An engineering analysis of the early stages of fire development—The fire at the Dupont Plaza Hotel and Casino on December 31, 1986, 50. Gaithersburg, MD: National Institute of Standards and Technology, Center for Fire Research, April.

———. 1989. An engineering view of the fire of May 4, 1988, in the First Interstate Bank Building, Los Angeles, California. NISTIR 89-4061. Gaithersburg, MD: National Institute of Standards and Technology, Center for Fire Research, March.

———. 1990. Fire growth analysis of the fire of March 20, 1990, Pulaski Building, 20 Massachusetts Avenue, NW, Washington, D.C. NISTIR 4489. Gaithersburg, MD: National Institute of Standards and Technology, Center for Fire Research, December.

Nelson, H. E. 1991. Engineering analysis of the fire development in the Hillhaven Nursing Home Fire, October 5, 1989. NISTIR 4665. Gaithersburg, MD: National Institute of Standards and Technology, Center for Fire Research, September.

Nelson, H. E. 2002. From phlogiston to computational fluid dynamics. *Fire Protection Engineering* (Winter): 9–17. www.sfpe.org

Nelson, H. E., and R. E. Tontarski. 1998. Fire research for fire investigation. HAI Report 98-5157-001. Proceedings of the International Conference. ATF Inaugural Conference, Baltimore, MD, November 11–14.

Netherwood, R. E. 1966. The relationship between lunar phenomena and the crime of arson. Thesis, Master of Criminology, University of California, Berkeley.

Newman, J. S. 1993. Integrated approach to flammability evaluation of polyurethane wall-ceiling materials. Polyurethane World Congress, Washington, DC: Society of the Plastics Industry, October 10–13.

NFPA. 1943. The Cocoanut Grove Night Club fire, Boston, November 28, 1942 (preliminary report). Quincy, MA: National Fire Protection Association, January 11.

———. 1997. *NFPA 705—Field flame test for textiles and films, recommended practice.* Quincy, MA: National Fire Protection Association.

———. 1998. *NFPA 906—Guide for fire incident field notes.* Quincy, MA: National Fire Protection Association.

———. 2000a. *NFPA 555—Guide on methods for evaluating potential for room flashover.* Quincy, MA: National Fire Protection Association.

———. 2000b. *Fire protection handbook,* 18th ed. Quincy, MA: National Fire Protection Association.

———. 2000c. *NFPA 92B—Guide for smoke management systems in malls, atria, and large areas.* Quincy, MA: National Fire Protection Association.

———. 2002a. *NFPA 1403—Standard on live fire training evolutions.* Quincy, MA: National Fire Protection Association.

———. 2002b. *NFPA 72—National fire alarm code,* Appendix B. Quincy, MA: National Fire Protection Association.

———. 2003a. *Fire protection handbook*, 19th ed. 2 vols. Quincy, MA: National Fire Protection Association.

———. 2003b. *NFPA 1033—Standard for professional qualifications for fire investigator.* Quincy, MA: National Fire Protection Association.

———. 2003c. U.S. fire problem overview report. National Fire Protection Association. Quincy, MA, June.

———. 2004a. *NFPA 921—Guide for fire and explosion investigations.* Quincy, MA: National Fire Protection Association.

———. 2004b. *NFPA 701—Standard methods of fire tests for flame propagation of textiles and films.* Quincy, MA: National Fire Protection Association.

NIJ. 2000. *Fire and arson scene evidence: A guide for public safety personnel.* NCJ 181584. Washington, DC: National Institute of Justice, June.

NIST. 1991. Users' guide to BREAK1, the Berkeley algorithm for breaking window glass in a compartment fire. NIST-GCR-91-596. Gaithersburg, MD: National Institute of Standards and Technology.

———. 1997. *Full-scale room burn pattern study.* NIJ Report 601-97. Washington, DC: National Institute of Justice, December.

NRC. 2006. Verification and validation of selected fire models for nuclear power plant applications. Vol. 4: Consolidated fire growth and smoke transport (CFAST), preliminary report, NUREG-1824. Rockville, MD: Nucleon Regulatory Commission.

———. 2000. The need for scientific fire investigations. *Fire Protection Engineering* 8 (Fall): 4–8.

Ogle, R. A., and J. L. Schumacher. 1998. Fire patterns on upholstered furniture: Smoldering versus flaming combustion. *Fire Technology* 34(3):247–65.

Orloff, L., A. T. Modak, and R. L. Alpert. 1977. Burning of large-scale vertical surfaces, 1345. In *16th International Symposium on Combustion*, Combustion Institute, Pittsburgh, PA.

Orville, R. E., and G. R. Huffines. 1999. Annual summary: Lighting ground flash measurements over the contiguous United States: 1995–97. *Monthly Weather Review* 127:2693–2703.

Park, R. E., and E. W. Burgess. 1921. *An introduction to the science of sociology.* Chicago: University of Chicago Press.

Parker, T. W., and R. W. Nurse. 1950. Investigations on building fires, Part I. The estimation of the maximum temperature attained in building fires from examination of the debris. National Building Studies, Technical Paper No. 4, Department of Scientific and Industrial Research, Building Research Station, Garston, England, 1–5.

Paulsen, T. 1994. The effect of escape route information on mobility and wayfinding under smoke-logged conditions, 693–704. In *Fire Safety Science—Proceedings of the Fourth International Symposium*, International Association of Fire Safety Science, Ottawa.

Peacock, R. D., G. P. Forney, P. Reneke, R. Porter, and W. W. Jones. 1993. CFAST, the consolidated model of fire growth and smoke transport. Gaithersburg, MD: National Institute of Standards and Technology, NIST Technical note 1299. February.

Peacock, R. D., W. W. Jones, P. A. Reneke, and G. P. Forney. 2005. CFAST—Consolidated model of fire growth and smoke transport (version 6). NIST Special Publication 1041. Gaithersburg, MD: National Institute of Standards and Technology, December.

Peacock, R. D., P. A. Reneke, R. W. Bukowski, and V. Babrauskas. 1999. Defining flashover for fire hazard calculations. *Fire Safety Journal* 32(4):331–45.

Peige, J. D., and C. E. Williams. 1977. *Photography for the fire service.* Oklahoma City: International Fire Service Training Association, Oklahoma State University.

Penney, D. G., ed. 2000. *Carbon monoxide toxicity.* Boca Raton, Fla.: CRC Press.

Perry, R. H., and D. W. Green, eds. 1984. *Perry's chemical engineers' handbook,* 6th ed. New York: McGraw-Hill.

Peterson, J. E., and R. D. Stewart. 1975. Predicting the carboxyhemoglobin levels resulting from carbon monoxide exposure. *Journal of Applied Physiology* 39:633–38.

Policing New Zealand. 2000. *New Zealand police today—Policing 2000.* Video. www.crime.co.nz.

Pope, E. J., and O. C. Smith. 2003. Features of preexisting trauma and burned cranial bone. Presentation lecture, American Academy of Forensic Sciences (AAFS) 55th Annual Meeting, Chicago, February 17–22.

———. 2004. Identification of traumatic injury in burned cranial bone: An experimental approach. *Journal of Forensic Sciences* 49(3):431–40.

Purser, D. 2001. Human tenability. The technical basis for performance-based fire regulations. United Engineering Foundation Conference, San Diego, CA, January 7–11.

———. 2002. Toxicity assessment of combustion products, 2-83–2-171. In *The SFPE handbook of fire protection engineering,* 3rd ed. Quincy, MA: National Fire Protection Association.

Putorti, A. D. 2001. Flammable and combustible liquid spill/burn patterns. NIJ Report 604-00, NCJ 186634. Washington, DC: National Institute of Justice, March.

Pyrosim. 2007. Pyrosim user manual (version 2006.1). Manhattan, KS: Thunderhead Engineering. http://www.thunderheadeng.com/pyrosim/PyrosimManual.pdf.

Quinsey, V. L., T. C. Chaplin, and D. Upfold. 1989. Arsonists and sexual arousal to fire setting: Correlation unsupported. *Journal of Behavioral and Experimental Psychiatry* 20 (no. 3): 203–208.

Quintiere, J. G. 1994. A perspective on compartment fire growth. *Combustion Science and Technology* 39:11–54.

———. 1998. *Principles of fire behavior.* Albany, NY: Delmar. ISBN 0827377320.

———. 2006. *Fundamentals of fire phenomena.* West Sussex, England: Wiley. ISBN 0470091134.

Quintiere, J. G., and B. S. Grove. 1998. Correlations for fire plumes. NIST-GCR-98-744. Gaithersburg, MD: National Institute of Standards and Technology.

Quintiere, J. G., and M. Harkleroad. 1984. New concepts for measuring flame spread properties. NBSIR 84-2943. Gaithersburg, MD: National Bureau of Standards.

Quintiere, J. G., and B. J. McCaffrey. 1980. The burning of wood and plastic cribs in an enclosure. Vol. 1, NBSIR 80-254. Gaithersburg, MD.

Raj, P. K. 2006. Hazardous heat, *NFAA Journal* (September/October): 75–79

Reardon, J. J. 2002. The warning signs of a faked death: Life insurance beneficiaries can't recover without providing "due proof" of death. *Connecticut Law Tribune* June 10, p. 5.

Rengert, G., and J. Wasilchick. 1990. Space, time, and crime: Ethnographic insights into residential burglary. Grant 88-IJ-CX-0013. Final Report to the U.S. Department of Justice. Philadelphia: Department of Criminal Justice, Temple University.

Rich, L. 2007. Rooms to go. Personal correspondence.

Richardson, J. K., L. R. Richardson, J. R. Mehaffey, and C. A. Richardson. 2000. What users want fire model developers to address. *Fire Protection Engineering* (Spring): 22–25.

Rider, A. O. 1980. The firesetter: A psychological profile. Part 1. *FBI Law Enforcement Bulletin* 49(6):6–13.

Robbins, E. S., and L. Robbins. 1967. Arson with a special reference to pyromania. *New York State Journal of Medicine* (March).

Rogerson, P., and Y. Sun. 2001. Spatial monitoring of geographic patterns: An application to crime analysis. *Computers, Environment and Urban Systems* 25:539–56.

Rohr, K. D. 2001. An update to what's burning in home fires. *Fire and Materials* 25:43–48.

———. 2005. Products first ignited in U.S. home fires. Quincy, MA: National Fire Protection Association, Fire Analysis and Research Division, April.

Rossmo, D. K. 1998. Expert system method of performing crime site analysis. U.S. Patent 5781704.

———. 1999. *Geographic profiling.* Boca Raton, FL: CRC Press.

Routley, J. G. 1995. Three firefighters die in Pittsburgh house fire, Pittsburgh, Pennsylvania. Report 078. Emmitsburg, PA: U.S. Fire Administration, Major Fires Investigation Project.

Rule, A. 1999. *Bitter harvest*. New York: Pocket Books.

Salley, M. H., J. Dreisbach, K. Hill, R. Kassawara, B. Najafi, F. Joglar, A. Hammins, K. McGrattan, R. Peacock, and B. Gautier. 2007. Verification and validation—How to determine the accuracy of fire models. *Fire Protection Engineering* 34 (Spring).

Sanderson, J. L. 1995. Tests and further doubt to concrete spalling the theories. *Fire Findings* 3(4):1–3.

———. 1997. Horizontal surface testing yields clues to fire's origin. *Fire Findings* 5, no. 2 (Spring): 1–3

———. 2002. Depth of char: Consider elevation measurements for greater precision. *Fire Findings* 10, no. 2 (Spring): 6.

Sapp, A. D., G. P. Gary, T. G. Huff, and S. James. 1993. Characteristics of arsons aboard naval ships. Monograph. Quantico, VA: FBI National Center for the Analysis of Violent Crime, U.S. Department of Justice.

Sapp, A. D., G. P. Gary, T. G. Huff, D. J. Icove, and P. R. Horbert. 1994. Motives of serial arsonists: Investigative implications. Monograph. Quantico, VA: FBI National Center for the Analysis of Violent Crime, U.S. Department of Justice.

Sapp, A. D., T. G. Huff, G. P. Gary, and D. J. Icove. 1995. A motive-based offender analysis of serial arsonists. Monograph. Quantico, VA: FBI National Center for the Analysis of Violent Crime, U.S. Department of Justice.

Schorow, S. 2005. *The Cocoanut Grove fire.* Beverly, MA: Commonwealth Editions.

Schroeder, R. A. 2004. Fire investigation and the fire engineer. *Fire Protection Engineering Magazine* no. 21 (Winter): 4–9.

Schroeder, R., and R. Williamson. 2001. Application of materials science to fire investigation. *Fire and Materials*. London: Interscience.

———. 2003. Post-fire analysis of construction materials—Gypsum wallboard. In *Proceedings Fire and Materials 2003*, San Francisco, January 28–29. (See also http://www.schroederfire.com/) and *Fire and Materials 2000.* New York: Wiley.

SFPE. 1995a. *The SFPE handbook of fire protection engineering*, 2nd ed. Quincy, MA: National Fire Protection Association.

———. 1995b. *Fires-T3: A guide for practicing engineers*. Bethesda, MD: SFPE Task Group on Documentation of Computer Models, Society of Fire Protection Engineers.

———. 1999. *Engineering guide for assessing flame radiation to external targets from pool fires.* Bethesda, MD: SFPE Task Group on Engineering Practices, Society of Fire Protection Engineers.

———. 2000a. *Engineering guide to predicting 1st and 2nd degree skin burns*. Bethesda, MD: SFPE Task Group on Engineering Practices, Society of Fire Protection Engineers.

———. 2002b. *The SFPE handbook of fire protection engineering*, 3rd ed. Quincy, MA: National Fire Protection Association.

———. 2002c. *SFPE engineering guide to piloted ignition of solid materials under radiant exposure.* Bethesda, MD: SFPE Task Group on Engineering Practices, Society of Fire Protection Engineers.

———. 2002d. *SFPE engineering guide—The evaluation of the computer model DETECT-QS.* Bethesda, MD: Society of Fire Protection Engineers.

———. 2002e. *Guidelines for peer review in the fire protection design process.* Society of Fire Protection Engineers, Bethesda, MD, October 8.

———. 2003. *Engineering guide: Human behavior in fire.* Society of Fire Protection Engineers, Bethesda, MD, June.

Shields, T. J., G. W. H. Silcock, and M. F. Flood. 2001. Performance of a single glazing assembly exposed to enclosure corner fires of increasing severity. *Fire and Materials* 22:123–52.

Shill, R. 1998. Beyond search and retrieval: Five benefits of document management. *New York Law Journal*. New York Law Publishing Co., March 31, 1998.

Shipley, H. L., D. J. Icove, and P. E. Keith. 1981. An analysis of fire bombings in Knoxville, Tennessee. U.S. Fire Administration Grant No. EMW-R-0599. Knoxville Police Arson Task Force, Department of Public Safety, December 17.

Short, N. R., S. E. Guise, and J. A. Purkiss. 1996. Assessment of fire-damaged concrete using color analysis. *InterFlam '96 Proceedings*. London: Interscience.

Silcock, G. W. H., and T. J. Shields. 2001. Relating char death to fire severity conditions. *Fire and Materials* 25:9–11.

Skelly, Michael J., Richard J. Roby, and Craig L. Beyler. 1991. Experimental investigation of glass breakage in compartment fires. *Journal of Fire Protection Engineering* 3(1):25–34.

Smith, D. W. 2006. The pitfalls, perils, and reasoning fallacies of determining the fire cause in the absence of proof: The negative corpus methodology. International Symposium on Fire Investigation Science and Technology (ISFI 2006), Cincinnati, OH, June 26–28.

Smith, F. P. 1991. Concrete spalling: Controlled fire test and review. *Journal of Forensic Science* 31(1):67–75.

Smith, O. C., and E. J. Pope. 2003. Burning extremities: Patterns of arms, legs, and preexisting trauma. Presentation lecture, American Academy of Forensic Sciences (AAFS) 55th Annual Meeting, Chicago, February 17–22.

Snook, B., P. J. Taylor, and C. Bennell. 2004. Geographic profiling: The fast, frugal, and accurate Way. *Applied Cognitive Psychology* 18:105–21.

Spearpoint, M. J., and J. G. Quintiere. 2001. Predicting the ignition of wood in the cone calorimeter. *Fire Safety Journal* 36(4):391–415.

Spitz, W. U., and D. J. Spitz, eds. 2006. *Spitz and Fisher's medicolegal investigation of death: Guidelines for the application of pathology to crime investigation*, 4th ed. Springfield, IL: Charles C. Thomas.

Stanley. 2005. *Stanley introduces reliable, accurate true laser measuring for everyone.* Press release, Stanley Works, www.stanleytools.com, July 21.

Stanton, J., and A. Simpson. 2002. Filicide: A review. *International Journal of Law and Psychiatry* 25:1–14.

Steinmetz, R. C. 1966. Current arson problems. *Fire Journal* (September).

Stickevers, J. 1986. Factors to consider when attempting to determine point of origin. *Fire Engineering* (January): 36–46.

Stoll, A. M., and L. C. Greene. 1959. Relationship between pain and tissue damage due to thermal radiation. *Journal of Applied Physiology* 14(3):373–82.

Stroup, D. W., L. A. DeLauter, J. H. Lee, and G. L. Roadarmel. 1999. Scotch pine Christmas tree fire tests. Report of Test. FR 4010. Gaithersburg, MD: National Institute of Standards and Technology (December).

Svare, R. J. 1988. Determining fire point-of-origin and progression by examination of damage in the single phase, alternating current electrical system. Invited paper—Ministry of Public Security, Beijing, People's Republic of China. Presented in Shanghai, PRC, April. Published in *Journal of People to People, International Arson Investigation Delegation to the People's Republic of China and Hong Kong.*

Sweeney, G. O., and P. R. Perdew. 2005. Spoliation of evidence: Responding to fire scene destruction. *Illinois Bar Journal* 93 (July): 358–67.

Takeda, H., and J. R. Mehaffey. 1998. "WALL2D: A model for predicting heat transfer through wood-stud walls exposed to fire. *Fire and Materials* 22:133–40.

Takeuchi, S., N. Saitoh, K. Kuroki, et al. 2005. Visualization of petroleum accelerants in arson cases. Paper presented at International Association of Forensic Sciences, Hong Kong, August 21–26.

Tassios, T. P. 2002. Monumental fires. *Fire Technology* 38:311–17.

Tewarson, A. 1995. In *The SFPE handbook of fire protection engineering*, 2nd ed., chap. 3-4. Quincy, MA: National Fire Protection Association.

Thali, M. J., K. Yen, T. Plattner, W. Schweitzer, P. Vock, C. Ozdoba, and R. Dirnhofer. 2002. Charred body: Virtual autopsy with multi-slice computed tomography and magnetic resonance imaging. *Journal of Forensic Sciences* 47, no. 6 (November).

Thomas, G. 2002. Thermal properties of gypsum plasterboard at high temperatures. *Fire and Materials* 26(1):37–45.

Thomas, P. H. 1971. Rates of spread of some wind-driven fires. *Forestry* 44:155–75.

Thomas, P.H. 1974. Fire in model rooms. Conseil Internationale du Batiment (CIB) Research Program, Building Research Establishment, Borehamwood, Hertfordshire, England.

Tobin, W. A., and K. L. Monson. 1989. Collapsed spring observation in arson investigation: A critical metallurgical evaluation. *Fire Technology* 25(4):317–35.

Tou, J. T., and R. C. Gonzalez. 1974. *Pattern recognition principles* Reading, MA: Addison-Wesley, 94–97.

Vaisala Inc. (formerly Global Atmospherics). 2003. 2705 E. Medina Rd., Tucson, AZ. 85706. www.lightningstorm.com

Vandersall, T. A., and J. M. Wiener. 1970. Children who set fires. *Archives of General Psychiatry* 22 (January).

Vasudevan, R. 2004. Forensic engineering analysis of fires using Fire Dynamics Simulator (FDS) modeling program. *Journal of the National Academy of Forensic Engineers* 21, no. 2 (December): 79–86.

Vreeland, R. G., and M. B. Waller. 1978. The psychology of fire setting: A review and appraisal. Grant no. 7-9021. Gaithersburg, MD: National Bureau of Standards, December.

Wheaton, S. 2001. *Personal accounts: Memoirs of a compulsive firesetter*. Washington, DC: American Psychiatric Association. psjournal@psych.org.

Wilkinson, P. 2001. Archaeological survey site grids. *Practical Archaeology* 5 (Winter): 21–27.

Wise, B. 1995. Catching crooks with computers. *American City and County* 110(6):54.

Wolchak et al. 1995. Public sector stars, Framingham, MA: Computerworld. 22.

Wolford, M. R. 1972. Some attitudinal, psychological, and sociological characteristics of incarcerated arsonists. *Fire & Arson Investigator* 22.

Wu, Y., and D. D. Drysdale. 1996. Study of upward flame spread on inclined surfaces. HSE Contract Research Report No. 122/1996.

You, H-Z. 1984. An investigation of fire plume impingement on a horizontal ceiling. 1—Plume region. *Fire and Materials* 8(1):28–39.

Zicherman, J., and P. A. Lynch. 2006. Is pyrolysis dead?—Scientific processes vs. court testimony: The recent 10th circuit court and associated appeals court decisions. *Fire & Arson Investigator* (January).

Zukoski, E. E. 1978. Development of a stratified ceiling layer in the early stages of a closed-room fire. *Fire and Materials* 2(2).

LEGAL REFERENCES

103 Investors I, L.P. v. Square D Company, Case No. 01-2504-KHV, 372 F.3d 1213 (U.S. App. 2004). LEXIS 12439, U.S. Dist. LEXIS 8796, 10th Cir. Kan., 2004. Decided May 10, 2005.

Abon Ltd. v. Transcontinental Insurance Company, Docket No. 2004-CA-0029, 2005 Ohio 302 (Ohio App. 2005). LEXIS 2847. Decided June 16, 2005.

Eid Abu-Hashish and Sheam Abu-Hashish v. Scottsdale Insurance Company, Case No. 98 C 4019, 88 F. Supp. 2d 906 (U.S. Dist. 2000). LEXIS 3663. Decided March 16, 2000.

Allstate Insurance Company, as Subrogee of Russell Davis v. Hugh Cole Builder Inc. Hugh Cole individually and dba Hugh Cole Builder Inc., Civil Action No. 98-A-1432-N, 137 F. Supp. 2d 1283 (U.S. Dist. 2001). LEXIS 5016. Decided April 12, 2001.

American Family Insurance Group v. JVC American Corp., Civil Action No. 00-27(DSD-JMM) (U.S. Dist. 2001). LEXIS 8001. Decided April 30, 2001.

Booth, Jacob J., and Kathleen Booth v. Black and Decker Inc., Civil Action No. 98-6352, 166 F. Supp. 2d 215, (U.S. Dist. 2001). LEXIS 4495, CCH Prod. Liab. Rep. P16, 184. Decided April 12, 2001.

Chester Valley Coach Works et al. v. Fisher-Price Inc., Civil Action No. 99-CV-4197 (U.S. Dist. 2001). LEXIS 15902, CCH Prod. Liab. Rep. P16,18. Decided August 29, 2001.

Commonwealth of Pennsylvania v. Paul S. Camiolo, Montgomery County, No. 1233 of 1999.

Cunningham v. Gans, 501 F. 2d 496, 500 (2nd Cir. 1974).

David Bryte v. American Household Inc., Case No. 04-1051, CA-00-93-2, 429 F.3d 469 (4th Cir. 2005; 2005 U.S. App.). LEXIS 25052. Decided November 21, 2005.

Daubert v. Merrell Dow Pharmaceuticals Inc. 509 U.S. 579 (1993); 113 S. Ct. 2756, 215 L.Ed.2d 469.

Daubert v. Merrell Dow Pharmaceuticals Inc. (Daubert II), 43 F.3d 1311, 1317 (9th Cir. 1995).

Farner v. Paccar Inc., 562 F. 2d 518, 528–29 (8th Cir. 1977).

Fireman's Fund Insurance Company v. Canon U.S.A. Inc., Case No. 03-3836, 394 F.3d 1054 (U.S. App. 2005). LEXIS 471; 66 Fed. R. Evid. Serv. (Callaghan) 258; CCH Prod Liab. Rep. P17,274. Filed January 12, 2005.

Frye v. United States, 293 F.1013 (DC Cir. 1923).

General Electric Company v. Joiner, 66 U.S.L.W. 4036 (1997).

Kumho Tire Co. Ltd. v. Carmichael, 119 S. Ct. 1167 (1999). U.S. LEXIS 2199 (March 23, 1999).

LaSalle National Bank et al. v. Massachusetts Bay Insurance Company et al., Case No. 90 C 2005 (U.S. Dist. 1997). LEXIS 5253. Decided April 11, 1997. Docketed April 18, 1997.

Michigan Miller's Mutual Insurance Company v. Benfield, 140 F.3d 915 (11th Cir. 1998).

Marilyn McCarver v. Farm Bureau General Insurance Company, Case Number 2004-3315-CK-M, State of Michigan, County of Berrien, Decided February 1, 2006.

James B. McCoy et al. v. Whirlpool Corp. et al., Civil Action No. 02-2064-KHV; 214 F.R.D. 646 (U.S. Dist. 2003). LEXIS 6901; 55 Fed. R. Serv. 3d (Callaghan) 740.

In Re: Paoli Railroad Yard PCB Litigation, 35F.3d 717 (U.S. App. 1994). LEXIS 23722; 30 Fed. R. Serv.3d (Callaghan) 644; 25 ELR 20989; 35 F.3d at 743. Filed August 31, 1994.

Pappas, Andronic, et al. v. Sony Electronics Inc. et al., Civil Action No. 96-339J, 136 F. Supp. 2d 413 (U.S. Dist. 2000). LEXIS 19531, CCH Prod. Liab. Rep. P15,993. Decided December 27, 2000.

Polizzi Meats Inc. v. Aetna Life and Casualty, 931 F. Supp. 328 (D.N.J. 1996).

Bethie Pride v. BIC Corporation, Société BIC, S.A., Civil Case No. 98-6422; 218 F.3d 566 (U.S. App. 2000). LEXIS 15652; 2000 FED App. 0222P (6th Cir.); 54 Fed. R. Serv. 3d (Callaghan) 1428; CCH Prod. Liab. Rep. P15,844. Decided July 7, 2000.

Royal Insurance Company of America as Subrogee of Patrick and Linda Magee v. Joseph Daniel Construction Inc., Civil Action No. 00-Civ.-8706 (CM); 208 F. Supp. 2d 423 (U.S. Dist. 2002). LEXIS 12397. Decided July 10, 2002.

Snodgrass, Teri, Robert L. Baker, Kendall Ellis, Jill P. Fletcher, Judith Shemnitz, Frank Sherron, and Tamaz Tal v. Ford Motor Company and United Technologies Automotive Inc., Civil Action No. 96-1814 (JBS) (U.S. Dist. 2002). LEXIS 13421. Decided March 28, 2002.

State Farm and Casualty Company, as subrogee of Rocky Mountain and Suzanne Mountain, v. Holmes Products; J.C. Penney Company, Case No. 04-4532 (3rd Cir. 2006). LEXIS 2370. Argued January 17, 2006. Filed January 31, 2006. (Unpublished).

Ronald Taepke v. Lake States Insurance Company, Fire No. 98-1946-18-CK, Circuit Court for the County of Charlevoix, State of Michigan. Entered December 8, 1999

TNT Road Company et al. v. Sterling Truck Corporation, Sterling Truck Corporation, Third-party Plaintiff v. Lear Corporation, Third-Party Defendant. Civil No. 03-37-B-K (U.S. Dist. 2004). LEXIS 13463; CCH Prod. Liab. Rep. P17,063 (U.S. Dist. 2004). LEXIS 13462 (D. Me., July 19, 2004). Decided July 19, 2004.

Truck Insurance Exchange v. MagneTek Inc. (U.S. App. 2004). LEXIS 3557 (February 25, 2004).

John Witten Tunnell v. Ford Motor Company. Civil Action No. 4:03CV74; 330 F. Supp. 2d 731 (U.S. Dist. 2004). LEXIS 24598 (W.D. Va., July 3, 2004). Decided July 2, 2004.

Travelers Property and Casualty Corporation. v. General Electric Company, Civil Action No. 3; 98-CV-50(SRU); 150 F. Supp. 2d 360 (U.S. Dist. 2001). LEXIS 14395; 57 Fed. R. Evid. Serv. (Callaghan) 695; CCH Prod. Liab. Rep. P16,181. Decided July 26, 2001.

United States of America v. John W. Downing, Crim. No. 82-00223-01 (U.S. Dist. 1985). LEXIS 18723; 753 F.2d 1224, 1237 (3d Cir. 1985). Decided June 20, 1985.

United States of America v. Ruby Gardner, No. 99-2193, 211 F.3d 1049 (U.S. App. 2000). LEXIS 8649, 54 Fed. R. Evid. Serv. (Callaghan) 788.

United States v. Markum, 4 F. 3d 891 (10th Cir. 1993).

United States v. Ortiz, 804 F.2d 1161 (10th Cir. 1986).

United States of America v. Lawrence R. Black Wolf Jr., CR 99-30095 (U.S. Dist. 1999). LEXIS 20736. Decided December 6, 1999.

Vigilant Insurance v. Sunbeam Corporation. No. CIV-02-0452-PHX-MHM; 231 F.R.D. 582 (U.S. Dist. 2005). LEXIS 29198. Decided November 17, 2005.

Mark and Dian Workman v. AB Electrolux Corporation et al., Case No. 03-4195-JAR (U.S. Dist. 2005). LEXIS 16306. Decided August 8, 2005.

Theresa M. Zeigler, individually; and Theresa M. Zeigler, as mother and next friend of Madisen Zeigler v. Fisher-Price Inc. No. C01-3089-PAZ (U.S. Dist. 2003). LEXIS 11184; Northern District of Iowa. Decided July 1, 2003 (Unpublished).

Glossary

accelerant. A fuel (usually a flammable liquid) that is used to initiate or increase the intensity or speed of spread of fire.

accident. Unplanned or unintentional event; an event occurring without design or intent.

adiabatic. Conditions of equilibrium of temperature and pressure. Also describing a reaction occurring without loss or gain of heat.

adsorption. Trapping of gaseous materials on the surface of a solid substrate.

aliphatic. Hydrocarbons with their carbon atoms in a straight chain; *normal* hydrocarbon.

alligatoring. Rectangular patterns of char formed on burned wood.

ambient. Surrounding conditions.

ampacity. Current-carrying capacity of electric conductors (expressed as amperes).

ampere. Quantity of electrical charge passing a point in an electrical circuit per unit time (1 coulomb per second).

annealing. Loss of temper in metal caused by heating.

anoxia. Condition relating to an absence of oxygen.

appliance. Equipment, usually nonindustrial, that is installed or connected as a unit to perform one or more functions such as clothes washing, air conditioning, food mixing, or cooking. Normally built in standardized sizes and types.

arc. Flow of current across a gap between two conductors, generally producing high temperatures and luminous gases.

arc mapping. Locating and analyzing the pattern of arc faults in electrical circuits as a means of locating a possible area of origin.

arcing through char. Unintended passage of current through a semiconductive degradation product.

area of confusion. Mixture of fire directional indicators in a wildlands fire.

area of origin. The general locale in which a fire was ignited.

aromatic. Hydrocarbon compound whose structure is based on a benzene ring.

arson. The intentional setting of a fire with intent to damage or defraud.

atom. Smallest unit of an element that still retains its properties.

Auger spectroscopy. A means of identifying elements trapped within a substrate by analyzing the electrons emitted during electron microscopy.

autoignition. Ignition due to sufficient surrounding temperature in the absence of an external source of ignition; nonpiloted ignition.

autoignition temperature. The temperature at which a material will ignite in the absence of any external pilot source of heat; spontaneous ignition temperature.

backdraft. A deflagrative explosion or rapid combustion of gases and smoke from an established fire that has depleted the oxygen content of a structure, most often initiated by introducing oxygen through ventilation or structural failure.

BLEVE. Boiling-liquid expanding-vapor explosion. A mechanical explosion caused by the heating of a liquid in a sealed vessel to a temperature far above its boiling point.

boiling point. The (pressure-dependent) temperature at which a liquid changes to its gas phase and that transition reaches equilibrium.

branch circuit.   The circuit conductors between the outlet(s) and the final overcurrent device protecting that circuit.

brisance.   The amount of shattering effect that can be produced by a high explosive.

Btu.   British thermal unit. A standardized measure of heat, it is the heat energy required to raise the temperature of 1 pound of water by 1 degree Fahrenheit.

calcination.   Loss of water of crystallization caused by heating.

calorie.   The amount of heat necessary to raise the temperature of 1 gram of water by 1 degree Celsius.

char.   Carbonaceous remains of burned organic materials.

chromatography.   Chemical procedure that allows the separation of compounds based on differences in their chemical affinities for two materials in different physical states, such as gas/liquid and liquid/solid.

circuit breaker.   A device designed to open a circuit automatically at a predetermined overcurrent without injury to itself when properly applied within its ratings.

clean burn.   An area of wall or ceiling or other surface where the charred organic residues have been burned away by direct flame contact or other source of high temperature.

combustible liquid.   A liquid having a flash point at or above 38°C (100°F).

combustion.   A self-sustaining oxidation reaction that generates detectable heat and light.

concealed wiring.   Wiring rendered inaccessible by the structure or finish of the building. Wiring in covered raceways is considered concealed.

conduction.   Process of transferring heat through a material or between materials by direct physical contact.

conductor.   Any material capable of permitting the flow of electrons. (1) *Bare.* A conductor having no covering or electrical insulation whatsoever. (2) *Covered.*

A conductor encased within a material whose composition or thickness is not considered insulative. (3) *Insulated.* A conductor encased within a material recognized by the electrical code as insulation.

convection.   Process of transferring heat by movement of a fluid (typically gas). In convective flow, a warm fluid is less dense than the surrounding fluid and rises, inducing a circulation.

corpus delicti.   Literally, the body of the crime. The fundamental facts necessary to prove the commission of a crime.

crazing.   Stress cracks in glass as the result of rapid cooling.

dead load.   The weight of a structure and any equipment and appliances permanently attached.

deep-seated.   Fire that has gained headway and built up sufficient heat in a structure to require great cooling for extinguishment; fire that has burrowed deep into combustible fuels (as opposed to a surface fire); deep charring of structural members.

deflagration.   A very rapid oxidation (typically of fuel gas, vapor, or dust) with the evolution of heat and light and the generation of a low-energy pressure wave that can cause damage. The reaction proceeds between fuel elements at subsonic speeds in the fuel (< 1000 m/s; 3300 ft/s).

detonation.   An extremely rapid reaction that generates very high temperatures and an intense pressure/shock wave that produces violently disruptive effects. It propagates through the material at supersonic speeds (> 1000 m/s; 3300 ft/s).

device.   Any chemical or mechanical contrivance or means used to start a fire or explosion.

diatomic.   Molecules consisting of two atoms of an element.

diffusion flame.   Combustion supported by fuel and oxygen molecules diffusing into one another.

direct attack.   Application of hose streams or other extinguishing agents directly on a fire, rather than attempting extinguishment by generating steam within a structure.

drop-down. The collapse of burning material in a room that induces separate, low-level ignition; fall-down.

duty. Conditions of use in electrical service: (1) *Continuous duty*. Operation at substantially constant load for an indefinitely long time. (2) *Intermittent duty*. Operation for alternating intervals of (a) load/no load, (b) load/rest, or (c) load/no load/rest. (3) *Periodic duty*. Intermittent operation in which the load conditions are regularly recurrent. (4) *Short-time duty*. Operation at substantially constant load for a short and definitely specified time. (5) *Varying duty*. Operation at loads and for intervals of time that are both subject to wide variation.

endothermic. Absorbing heat during a chemical reaction.

entrainment. The mixing of two or more fluids (gases) as a result of laminar flow or movement.

eutectic. An alloy of two materials having special physical or chemical properties, typically having the lowest melting point of any combination of the two.

exothermic. Generating or giving off heat during a chemical reaction.

explosion. The sudden conversion of potential energy (chemical or mechanical) into kinetic energy with the extremely rapid production or release of heat, gases, or mechanical pressure.

explosion-proof apparatus. Apparatus enclosed in a case that is capable of withstanding an explosion of a specified gas or vapor within it or preventing the ignition of a specific gas or vapor surrounding it by sparks, flashes, or explosion of fuels within and that operates at such an external temperature that a surrounding flammable atmosphere will not be ignited by it.

explosive. Any material that can undergo a sudden conversion of physical form to a gas with a release of energy.

explosive limits (flammability limits). The lower and upper concentrations of an air/gas or air/vapor mixture in which

combustion or deflagration will be supported.

exposure. Property that may be endangered by radiant heat from a fire in another structure or an outside fire. Generally, property within 40 feet is considered an exposure risk, but larger fires can endanger property much farther away.

fire. Rapid oxidation with the evolution of heat and light; self-sustaining (glowing or flaming) combustion.

fire behavior. The manner in which a fuel ignites, flame develops, and fire spreads. Unusual fire behavior may reveal the presence of added fuel or accelerants.

fire load. The total amount of fuel that might be involved in a fire, as measured by the amount of heat that would evolve from its combustion (expressed as units of heat).

fire plume. The buoyant convective column of hot gases generated by a fire; may include both flames and nonflaming products.

fire point. See flame point.

fire-resistive. A structure or assembly of materials built to provide a predetermined degree of fire resistance as defined in building or fire prevention codes (calling for 1-, 2-, or 4-hour fire resistance).

firestorm. Overwhelming progression of fire through structures or wildlands caused by a combination of convective and radiative processes.

fire wall. A solid wall of masonry or other noncombustible material capable of preventing passage of fire for a prescribed time (usually extends through the roof with parapets).

flame. A luminous cloud or stream of burning gases, with temperatures high enough that some of the energy produced is released as light in the range visible to the human eye; combustion occurring in a gaseous state.

flame plume. The buoyant column of incandescently hot gases from a fire visible to the unaided eye.

flameover. The flaming ignition of the hot gas layer in a developing compartment fire; rollover.

flame resistant. Material or surface that does not maintain or propagate a flame once an outside source of flame has been removed.

flame point (fire point). The minimum temperature at which (under defined test conditions) a flame is sustained by evaporation or pyrolysis of a fuel after ignition.

flame spread. The rate at which flames extend across the surface of a material (usually under specific conditions).

flammable (same as *inflammable*). A combustible material that ignites easily, burns intensely, or has a rapid rate of flame spread.

flammable liquid. A liquid having a flash point below 38°C (100°F).

flashback. The ignition of a gas or vapor from an ignition source back to a fuel source (often seen with flammable liquids).

flashover. In fire growth, the transition between the fire spreading from item to item and the ignition of all exposed fuel surfaces in the room.

flash point. The minimum temperature at which (under defined test conditions) the vapor being produced by a material can be ignited to produce a brief flash (not sustained) of flame.

fragmentation. The fast-moving solid pieces created by an explosion. Primary fragmentation is that of the explosive container itself; secondary fragmentation is that of the target shattered by an explosion.

fuel load. All combustibles in a fire area, whether part of the structure, finish, or furnishings.

fully involved. The entire area of a fire building so involved with heat, smoke, and flame that entry is not possible until some measure of control has been obtained with hose stream attacks.

ghost marks. Stained outlines of floor tiles produced by the dissolution and combustion of tile adhesive.

glowing combustion. The rapid oxidation of a solid fuel directly with atmospheric oxygen creating light and heat in the absence of flames.

ground. A conducting connection, whether intentional or accidental, between electrical circuit, equipment, and the earth, or to some conducting body that serves in place of the earth.

ground fault. An interruption of the normal ground return path of electricity in a structure that leads to unintended current flows.

heat flux. The rate at which heat energy is transferred to a surface per unit time per unit area.

heat release rate. The rate at which heat is generated by a source, usually measured in watts, joules/second, or Btu/second.

heat transfer. Spread of thermal energy by convection, conduction, or radiation.

high explosive. Any material designed to function by, and capable of, detonation.

hydrocarbon. Chemical compound containing only hydrogen and carbon.

hypoxia. Condition relating to low concentrations of oxygen.

ignitable liquid. A designation for liquid fuels, including those classified by NFPA as either a flammable liquid or a combustible liquid.

ignition energy. The quantity of energy that must be transferred into a fuel/oxidizer combination to trigger a self-sustaining combustion.

ignition temperature (same as **autoignition**). The minimum temperature to which (under specific test conditions) a substance must be heated in air to ignite independently of the heating source (i.e., in the absence of any other ignition source; nonpiloted ignition).

incendiary fire. A deliberately set fire.

incipient. Beginning stages of a fire.

indirect application. A method of firefighting by applying water fog into heated atmospheres to absorb heat and smother the fire by generating steam.

inorganic. Containing elements other than carbon, oxygen, and hydrogen.

isochar. Line drawn on a scene diagram that connects points of similar char depth on wood surfaces.

low explosive. Any material designed to function (i.e., generate gases at high pressures) by deflagration.

Molotov cocktail. A breakable container filled with flammable liquid, usually thrown. It may be ignited by a flaming wick or by chemical means.

nonflammable. Material that will not burn under most conditions.

normal hydrocarbons (*n*-hydrocarbons). Hydrocarbons having straight-chain structures with no side branching; aliphatics.

ohm. Unit of measurement for electrical resistance.

olefinic. Hydrocarbons containing double carbon–carbon bonds (denoted C=C); nonsaturated; alkenes.

open wiring. Uninsulated conductors or insulated conductors without grounded metallic sheaths or shields, installed above ground but not inside enclosures or appliances.

organic. Compounds based on carbon.

outlet. A point on the wiring system at which current is taken to supply equipment or appliances by means of receptacles or direct connections.

overcurrent. Current flow above the intended or design current.

overhaul. The firefighting operation of eliminating hidden flames, glowing embers, or sparks that may rekindle the fire, usually accompanied by the removal of structural contents.

oxidation. Combination of an element with oxygen; chemical conversion involving a loss of electrons.

paraffinic. Hydrocarbon compounds involving no double or triple C—C bonds; alkanes; saturated aliphatic hydrocarbons.

piloted ignition temperature. The minimum temperature at which a fuel will sustain a flame when exposed to a pilot source.

point of origin. The specific location at which a fire was ignited.

pyrolysis. The chemical decomposition of substances through the action of heat, in the absence of oxygen; oxidative py-

rolysis occurs when oxygen comes in contact with the decomposing material.

pyromania. Uncontrollable psychological impulse to start fires.

pyrophoric. Capable of oxidizing on exposure to atmospheric oxygen at normal temperatures.

raceway. Any channel for holding wires, cables, or busbars that is designed expressly for, and is used solely for, this purpose.

radiation. Transfer of heat by electromagnetic waves.

receptacle. A contact device installed as the outlet for the connection of an appliance by means of a plug.

rekindle. Reignition of a fire by latent heat, sparks, or embers.

resistance. Opposition to the passage of an electrical current.

rollover. *See* flameover.

salvage. Procedures to reduce incidental losses from smoke, water, and weather following fires, generally by removing or covering contents.

saturated. Hydrocarbons that have no double or triple C—C bonds.

seat of explosion. The area of most intense physical damage caused by high explosive pressures and shock waves in the vicinity of a solid or liquid explosive.

seat of fire. Area where main body of fire is located, as determined by outward movement of heat, flames, and smoke.

self-heating. An exothermic chemical or biological process that can generate enough heat to become an ignition source; spontaneous ignition.

self-ignition. *See* Autoignition.

service. The conductors and equipment for delivering electricity from the supply system to the equipment of the premises served.

service conductors. The supply conductors that extend from the street main or transformer to the equipment of the premises served.

service drop. The overhead service conductors from the pole, transformer, or other aerial support to the service entrance equipment on the structure, including any splices.

service equipment. The necessary hardware that constitutes the main control and means of cutoff of the electrical supply, usually consisting of circuit breakers, fuses, or switches located in a panel box near the point of entrance of supply conductors.

set. Device or contrivance used to ignite an incendiary fire.

short-circuit. Direct contact between a current-carrying conductor and another conductor.

smoldering combustion. Self-sustained oxidation involving the direct combination of a solid fuel with atmospheric oxygen to generate heat in the absence of gaseous flames. *See* glowing combustion.

soot. The carbon-based solid residue created by incomplete combustion of carbon-based fuels.

spall. Crumbling or fracturing of concrete or brick surface as a result of exposure to thermal or mechanical stress.

spark. Superheated, incandescent particle.

spoliation. Destruction, loss, or alteration of material that is or can be evidence at a judicial proceeding.

spontaneous ignition. Chemical or biological process that generates sufficient heat to ignite the reacting materials. *See* self-heating.

stoichiometry. Balance of chemical reactants and products.

suspicious. An outmoded term for characterizing a fire whose cause has not been determined but where there are indications that it was deliberately set.

thermal inertia. A value calculated from the thermal conductivity, density, and heat capacity that relates to the ignitability of that material.

thermal protector. An inherent device against overheating that is responsive to temperature or current and will protect the equipment against overheating owing to overload or failure to start.

thermoplastics. Organic materials that can be melted and resolidified without chemical degradation.

thermosetting. Plastic or resin materials that once solidified will not melt but will chemically degrade as they are heated.

torch. A professional fire setter.

trailers. Long assemblages of combustible materials used to spread a fire throughout a structure.

vapor. The gaseous phase of a material that is a liquid or solid at normal temperatures and pressures.

vapor density. The ratio of the molecular weight of a given volume of gas or vapor to that of an equal volume of air; or the ratio of the molecular weight of the gas or vapor to that of air (MW = 29).

vented. A fire that has extended outside the structure or compartment by destroying the windows or burning an opening in the roof or walls.

ventilation. A technique for opening a burning building to allow the escape of heated gases and smoke to prevent explosive concentrations (smoke explosions or backdrafts) and to allow the advancement of hose lines into the structure.

volatile. A liquid having a low boiling point; one that is readily evaporated into the vapor state.

volt. The basic unit of electromotive force.

voltage, nominal. A value assigned to a circuit or system for the purpose of conveniently designating its voltage class (120/240, 480Y/277, 600, etc.).

watt. Unit of heat release, 1 watt = 1 joule per second; unit of power or work (in electrical circuits, equivalent to voltage multiplied by amperes).

Mathematics Refresher

C.1 FRACTIONAL POWERS

When a number (or value) is followed by a superscript number or fraction, such as X^n, it means that the number (X) is raised to a power (n). If the power is 2, the number is squared (multiplied by itself) ($3^2 = 3 \times 3 = 9$). If the power is 3, the number is cubed (multiplied by itself and then by itself again) ($3^3 = 3 \times 3 \times 3 = 27$).

The number n can be a fraction, such as $\frac{1}{2}$.

$$X^{1/2} \quad \text{or} \quad X^{0.5} = \sqrt{X} \quad \text{(the square root of } X\text{).} \tag{C.1}$$

The number n can be any value: $n = 2/5, 5/2$, and $3/2$ are common calculations in fire dynamics. Such values must be calculated using a "scientific notation" pocket calculator with a y^x or y^n function (where $y = X$, and x [or n] = the power to which y is raised). If n is a negative number, then X^{-n} denotes $1/X^n$ (X^n is put in the denominator of a fraction).

The following are general examples of mathematical equations using powers:

$$Q^x Q^y = Q^{x+y} \tag{C.2}$$

$$\frac{Q^x}{Q^y} = Q^{x-y} \tag{C.3}$$

$$(Q^x)^y = Q^{xy} \tag{C.4}$$

C.2 LOGARITHMS

Logarithms are numerical equivalents calculated by raising 10 to a power, where $\log_{10}$ = the value to which the number 10 must be raised to be equal to the original number.

The

$$\log_{10} \quad \text{of} \quad 100 = 2 \tag{C.5}$$

because

$$10^2 = 100. \tag{C.6}$$

Also,

$$\log_{10} \quad \text{of} \quad 10 = 1 \tag{C.7}$$

because

$$10^1 = 10. \tag{C.8}$$

This value can be found using a pocket calculator, tables, or a slide rule. For instance,

$$\log_{10} 3.5 = 0.54, \tag{C.9}$$

and

$$\log_{10} 2330 = 3.3673. \tag{C.10}$$

The inverse function is sometimes called the *antilog:*

$$\text{antilog}_{10}(2) = 10^2 = 100$$

When the base unit e is used ($e = 2.71$), the value is called the *natural log* (ln). For example,

$$\log_e 5 = \ln 5 = 1.6094, \tag{C.11}$$

and

$$\ln 10 = 2.3025. \tag{C.12}$$

The antilog e^x is the value calculated by raising e to that power:

$$e^5 = 148.413 \tag{C.13}$$

and

$$e^{10} = 22,026.5 \tag{C.14}$$

Figure 6.1 uses a modified "power of 10" notation as shorthand for very large or very small numbers. For example,

$$3.40E - 05 = 3.4 \times 10^{-5} = 0.000034,$$
$$2.88E + 00 = 2.88 \times 10^0 = 2.88,$$
$$5.00E + 02 = 5.0 \times 10^2 = 500.$$

C.3 DIMENSIONAL ANALYSIS

A useful concept for checking calculations is called *dimensional analysis*. If one keeps track of the units used for variables and constants and applies the rules of canceling units when they appear in both the numerator and the denominator of a function, for instance, one can verify that the correct relationship was used.

For example
$$\dot{Q} = \dot{m}'' A\, \Delta H_c, \tag{C.15}$$

where

$\dot{m}'' = $ mass flux (kg/m^{2}s),
$A = $ area of burning surface (m^2), and
$\Delta H_c = $ heat of combustion (kJ/kg).

Multiplication gives

$$\left(\frac{\text{kg}}{\text{m}^2\text{s}}\right)(\text{m}^2)\left(\frac{\text{kJ}}{\text{kg}}\right) = \frac{\text{kJ}}{\text{s}} = \text{kW}, \tag{C.16}$$

which is the correct unit for $\dot{Q}$.

For conduction calculations,

$$\dot{Q} = k\frac{T_2 - T_1}{l}A, \tag{C.17}$$

where

$$k = \frac{W}{m \cdot K},$$

l = length (m),

$T_2 - T_1 = \Delta T(\text{K})$, and

A = area (m^2).

Carrying out the calculation, we obtain

$$\dot{Q} = \frac{\left(\dfrac{W}{m \cdot K}\right)(\text{K})(\text{m}^2)}{m} = W, \tag{C.18}$$

which is the correct unit for $\dot{Q}$.

Websites

Bureau of Alcohol, Tobacco, Firearms, and Explosives
www.atf.treas.gov

California Conference of Arson Investigators
www.arson.org or www.ccaihq.org

CFI Trainer
www.cfitrainer.net

Consumer Product Safety Commission
www.cpsc.gov

Eastern Kentucky University, Dept. of Fire and Engineering Safety Technology
www.fireandsafety.eku.edu

Fire Findings
www.firefindings.com

Fire Protection Association
www.thefpa.co.uk

Forest Products Laboratory
www.fpl.fs.fed.us

Interactive fire investigation training courses
www.cfitrainer.net

InterFire
www.interfire.com

International Association of Arson Investigators
www.firearson.com

Lightning strike data
www.vaisala.com

Material safety data sheets
msds.pdc.cornell.edu/msdssrch.asp

University of Edinburgh, BRE Centre for Fire Safety and Engineering
www.see.ed.ac.uk/fire/index.html

University of Maryland, Department of Fire Protection Engineering
www.enfp.umd.edu

National Association of Fire Investigators
www.nafi.org

National Association of State Fire Marshals
www.firemarshals.org

National Center for Forensic Science at the University of Florida
Technical Working Group for Fire and Explosions (TWGFEX)
www.ncfs.ucf.edu/twgfex/home.html

National Institute of Justice
www.ojp.usdoj.gov/nij
 (This is the NIJ home page.)

www.ojp.usdoj.gov/nij/pubs.htm
 (This is the publications home page. The Scene Guides can be found in this listing.)

National Institute of Standards and Technology
fire.nist.gov

National Fire Protection Association
www.nfpa.org

Occupational Health and Safety Administration
www.osha.gov

Oklahoma State University, School of Fire Protection and Safety Engineering Technology
fpst.okstate.edu

Society of Fire Protection Engineers
www.sfpe.org

Synergy Technologies
www.synergytech.net

U.S. Fire Administration (FEMA)
www.usfa.fema.gov

Weather Information Network (NOAA)
iwin.nws.nozz.gov

Weather information (historic)
National Climatic Data Center, 1-866-742-3322

www.ncdc.noaa.gov
www.weatherunderground.com

Worcester Polytechnic Institute
Fire Protection Engineering & Center for Firesafety Studies
www.wpi.edu

The FPE & CFS Department at Worcester Polytechnic Institute
www.wpi.edu/Academics/Depts/Fire

Public domain software for wildland fire applications
(includes Behave Plus, FARSITE, and other models)
www.fire.org

VENDORS (CHAPTERS 4 AND 10)

Laser Scanning/Total Stations
Leica Geosystems Inc.
Tony.Grissim@hds.leica-geosystems.com or
tony@lgshds.com

3rdTech/DeltaSphere
www.3rdTech.com

Lighting
www.airstar-light.us

Magellan GPS/Mobile Mapper
mobilemapping@magellangps.com

Microscopes
www.zarbeco.com

Photographic stitching (panoramic)
Hewlett-Packard
www.hp.com

www.ptgui.com
www.roxio.com/Photosuite

Portable X-rays
www.goldenengineering.com

Scene Scanning and Mapping
Scene/works (also Spheron VR)
rsmck@csimapping.com

Crime Scene Virtual Tour
www.crime.scene-vr.com
www.iPix.com

Panoscan
www.panoscan.com

X-ray Fluorescence
www.innovxsys.com
nitron@thermofisher.com

Index